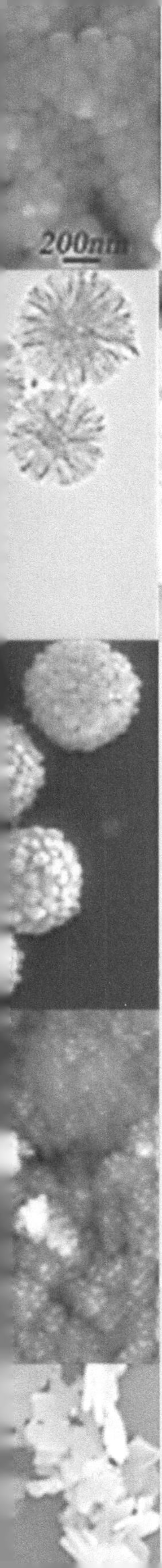

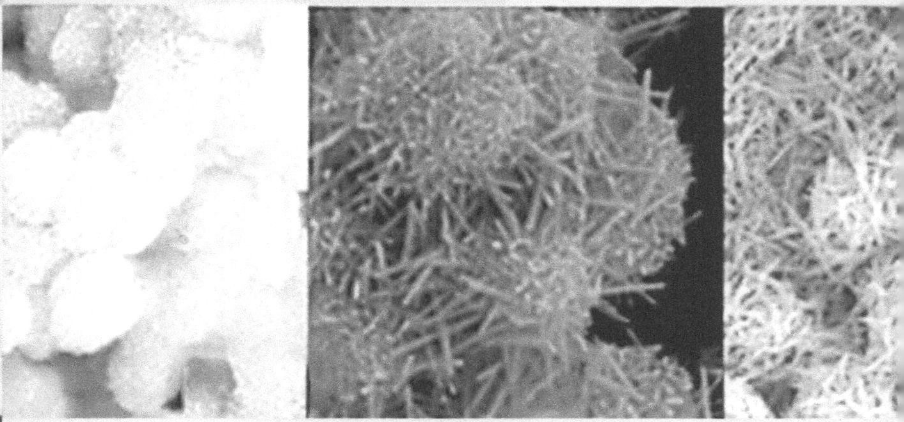

Nanomaterials in Energy and Environmental Applications

Nanomaterials in Energy and Environmental Applications

edited by

Junhui He

Routledge
Taylor & Francis Group
LONDON AND NEW YORK

First published 2013 by Pan Stanford Publishing Pte. Ltd.

Published 2022 by Routledge
2 Park Square, Milton Park, Abingdon, Oxon OX14 4RN
605 Third Avenue, New York, NY 10017

Routledge is an imprint of the Taylor & Francis Group, an informa business

Nanomaterials in Energy and Environmental Applications

Publisher's Note

The publisher has gone to great lengths to ensure the quality of this reprint but points out that some imperfections in the original copies may be apparent.

British Library Cataloguing-in-Publication Data
A catalogue record for this book is available from the British Library.

ISBN 978-9-814-46378-2 (hbk)
ISBN 978-981-4463-79-9 (eBook)

Contents

Preface

Nanoscience and nanotechnology are interdisciplinary fields that bring together physicists, chemists, materials scientists, biochemists, and engineers to meet current and potential future challenges that humankind faces, including searching for renewable energies for sustainable development and new technologies for carbon capture and environmental protection. Among the current subjects in nanoscience and nanotechnology, nanomaterials are developing fast and explosively and attract a huge amount of attention. They have recently shown emerging applications and continue to show promising potentials in technologies such as solar cells, fuel cells, secondary batteries, supercapacitors, air and water purification, and removal of both domestic and outdoor air pollutants. The application of nanomaterials has also drawn attention to their effects on human health. This book invited experts in the fields of nanomaterials, energy, and environmental science and assembled 13 reviews that discuss the design and fabrication of nanostructured materials and their energy and environmental applications.

This is the first book that summarizes the very recent efforts through nanoscience and technology towards meeting the pressing energy and environmental challenges that human beings are facing. It also points out future directions of nanomaterial development and encourages future efforts, especially by the younger generation.

Finally, I would like to take this opportunity to acknowledge all the authors who had spent their precious time in preparing their great contributions to the book. I would also like to thank Dr. Mingqing Yang, who contributed a lot to the communication with the authors and preparation and publication of the book. I am very grateful to Pan Stanford Publishing for providing me an opportunity to publish this book. I hope that the readers will find the contents both useful and enjoyable.

Junhui He

Beijing

June 2016

Chapter 1

Multifunctional Coatings for Solar Energy Applications

Lin Yao[a,b] **and Junhui He**[a]
[a]*Functional Nanomaterials Laboratory and Key Laboratory of Photochemical Conversion and Optoelectronic Materials, Technical Institute of Physics and Chemistry, Chinese Academy of Sciences, Zhongguancundonglu 29, Haidianqu, Beijing 100190, China*
[b]*University of Chinese Academy of Sciences, Beijing 100049, China*
jhhe@mail.ipc.ac.cn

1.1 Introduction

Transparent surface coatings with suitable optical path differences can suppress the reflection of substrates. Such coatings are usually called "antireflective coatings (ARCs)." Many antireflective surfaces exist in nature. Taking some diurnal butterflies as an example, they use arrays of nonclose-packed nipples (about 300 nm) as antireflective structure (ARS) to reduce reflection from their compound eyes [1,2]. The ARS on the cornea can increase light transmission and suppress reflection losses at the interfaces. Besides these excellent optical properties, the ARS arrays also exhibit self-cleaning capability because of the high fraction of air trapped between arrays [3,4].

Nanomaterials in Energy and Environmental Applications
Edited by Junhui He

ISBN 978-981-4463-78-2 (Hardcover), 978-981- 4463-79-9 (eBook)
www.panstanford.com

The antireflection phenomenon provides enormous inspirations for scientists to mimic for many important applications. Antireflection technology has been widely used in some high-precision optical components, solar cells, flat panel displays, and light-emitting diode lighting to increase the transmittance of incident light [5–8]. For solar cells, due to reflection at the air–glass interface of the packaging glass and scattering by accumulated dust on outdoor panels, part of incident energy on solar modules is lost. On the one hand, ARC on the glass can help alleviate reflection in systems; on the other hand, the self-cleaning property can, to some extent, solve the dust accumulation problem [5]. Tseng et al. systematically studied the antireflection and light trapping effects. Their results showed a 76.9% enhancement of short-circuit current density compared with that of bare silicon due to suppression of surface reflection [9].

It is common knowledge that smooth surfaces shine more than rough ones. It contains the very basic idea of antireflection that roughness is necessary to reduce reflection of surfaces. The reflection or optical disturbance will be zero if the medium for light propagation does not change or if the two media have the same refractive index (RI) [10]. Therefore, many materials with micro/nanostructure are perspective to fabricate ARC, including silicon, silica, titania, zirconia, zinc oxide, cobalt oxide, tin oxide, carbon, and poly(ethylene terephthalate) (PET), polystyrene (PS), gallium nitride [11–17]. To date, two kinds of approach are available for fabricating ARCs. One is coating porous or multilayered films on the surface of devices, and the other is fabricating sub-wavelength ARS directly on the substrate [18–21]. The corresponding fabrication routes can be classified into bottom-up and top-down modes. The bottom-up technique usually uses nanoparticles as building block to form ARCs. The top-down approach relies on etching or lithography and so on techniques performed with or without masks. Recently, the two-step method with a combination of top-down and bottom-up approaches also attracts much attention [11].

When contamination or fogging occurs on ARCs, however, their optical properties would dramatically deteriorate. Contaminants accumulate and water molecules condense on the surface, leading to scattering and reflection of light. This problem may be solved by creation of a surface that has special wettability (superhydrophilicity or superhydrophobicity) and photocatalytic property. ARCs with self-cleaning property have developed rapidly in recent years from window glass to various devices [22].

In this review, we aim to provide recent developments in antireflective and self-cleaning surfaces, with particular emphasis on silicon and fused silica materials, as they are both commonly used in many functional devices. The current review is composed of four sections. In the first section, the basic concept and principle of antireflection and self-cleaning are briefly described. In the following section, the fabrication pathways and their mechanisms are discussed. Then, we introduce the latest typical progress in self-cleaning ARCs in recent 5 years. In the last section, some applications of the self-cleaning ARCs surfaces are introduced. A large number of research articles have reported the fabrication of ARCs or superhydrophobic/superhydrophilic surfaces within the past few years. This paper intends to give readers both an integrated tour of antireflective and self-cleaning coatings fabrication technologies and a basic realization of their great application prospects.

1.2 Theoretical Aspects of Antireflection and Self-Cleaning

1.2.1 Principle of Antireflection

1.2.1.1 Basic concept of antireflection

ARCs can effectively enhance the transmission of light. The Fresnel equation offers the basic mathematical model of the reflection of coatings [23]. We learn that RI that quantifies the speed of light in the current medium with respect to that in vacuum is the major parameter for characterizing the transmission of light in the medium.

For single-layer coatings, the Fresnel equation, which offers the basic preliminary mathematical model of reflection and refraction, has the following two assumptions:

(1) The reflected waves have the same intensity and one wave is reflected per interface;

(2) Other optical interactions such as scattering, absorption, and so on are negligible.

As illustrated in Fig. 1.1a, there would be no reflection if there is a destructive interference between light reflected from the coating-substrate and the air-coating interfaces. Therefore, n_c for an ideal homogeneous ARC has to meet the following two conditions:

(1) $n_c = (n_a n_s)^{1/2}$, where n_a and n_s are the refractive indices of the air and the substrate, respectively.

(2) $d = \lambda/4n_c$, d is the thickness of the coating, λ is the wavelength of the incident light.

On the basis of the Fresnel's theory, the antireflection property is also dependent on the *s* and *p*-polarization of the light [23]. The s-polarization and p-polarization have the electric field perpendicular and parallel to the incidence plane, respectively. Sunlight shows appreciable degree of polarization when light reflects at shallow angles due to the fact that light reflecting at shallow angles has the p-polarized light reflecting to the maximum.

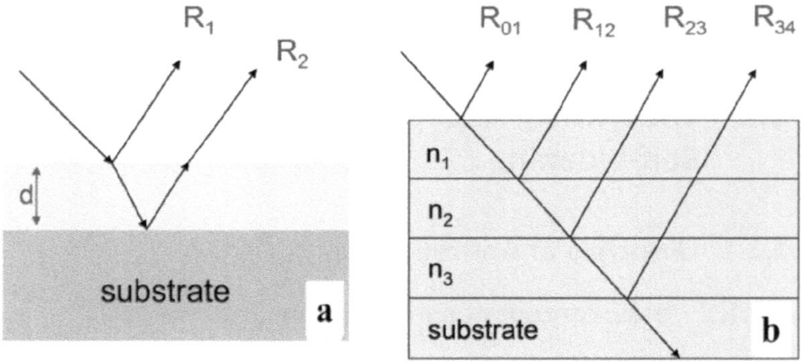

Figure 1.1 Propagation of light rays through (a) a single-layer coating and (b) multi-layer coating on substrate ($n_s > n_n$, where n_s and n_n are the refractive indices of substrate and coatings, respectively) [4]. Copyright 2014 Elsevier.

For multi-layer coatings, the mathematical model has some differences. As shown in Fig. 1.1b, the reflected light from the interface ij between the adjacent layers i and j is [24]:

$$R_{ij} = |R_{mn}| \exp[-2(\delta_i + \delta_j)]$$

in which, $|R_{mn}| = [(n_i - n_j)/(n_i + n_j)]$, $\delta_i = 2\pi n_i d_i \cos\theta_i/\lambda$ (θ_i is the angle of refraction, d_i is the physical thickness of the layer)

For the entire coating:

$$R_{sum} = R_{01} + R_{12} + R_{23} + \ldots + R_{ns}$$

By adjusting the reflective index and the thickness of each layer, a minimum R_{sum} can be obtained, and an ARC can be achieved.

In recent years, gradient-RI ARCs have received great attention. Different profiles for gradient-RI layers, including linear, parabolic, cubic, Gaussian, quintic, exponential, exponential-sine, and Klopfenstein, have been investigated for broadband and omnidirectional ARCs [25–27]. Typical expressions for the continuous gradient-RI with linear, cubic, and quintic profiles are presented below [28]:

Linear index profile: $n = n_o + (n_s - n_o)t,\ 0 \leq t \leq 1$

Cubic index profile: $n = n_o + (n_s - n_o)(3t^2 - 2t^3)$

Quintic index profile: $n = n_o + (n_s - n_o)t(10t^3 - 15t^4 + 6t^5)$

where n_o and n_s are the RIs of the incident and substrate media, respectively, and t is the thickness of graded interface region. Assembling a multilayer heterostructure that has a suited gradient-RI profile is beneficial to obtained AR properties within a wide range of wavelength.

1.2.1.2 Strategies to achieve antireflection

From the above discussion, we know that the antireflection property is obtained when n of a single-layer coating is $(n_{air}n_s)^{1/2}$. As for a glass substrate (n_s =1.5), n of ARCs materials should be 1.22. Unfortunately, natural materials having such a low RI are rare and expensive to obtain. Some insect compound eyes (such as moth eyes, fly eyes) present excellent optical and other properties (high sensitivity, antireflection, and anti-fogging) because of their unique multi-scale structures (Fig. 1.2) [29–31], which inspired continuous efforts of scientists to introduce nanopores into coatings. They imitated compound eyes to fabricate a periodic array of silicon nanotips with a sub-wavelength structure that can suppress the reflection from the ultraviolet, through the visible part of the spectrum, to the terahertz region [32].

Subsequently, the introduction of nanopores becomes an effective method to obtain appropriate RI, and by controlling the quantity and volume of introduced nanopores, the RI of coatings can be well tailored. There are also three requirements, which have to be considered for fabricating an ideal ARC. First, the antireflection property must be maintained over a broad spectrum of wavelength of the incident light in the ultraviolet, visible, and infrared regions. Second, as some coatings need to be oriented to the sun throughout the day, omni-directional antireflection is highly desired. Third, for practical applications, the coatings should be mechanically robust.

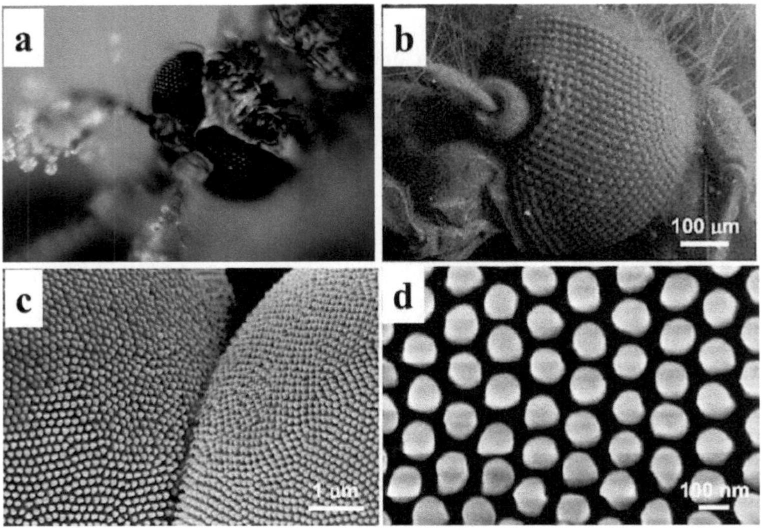

Figure 1.2 (a) A photograph of antireflective and antifogging mosquito eyes. (b) A SEM image of a single mosquito eye. (c) Two neighboring ommatidia. (d) Hexagonally nonclose-packed nanonipples covering an ommatidial surface [31]. Copyright 2007 Wiley-VCH.

It is important to distinguish between ARC and antiglare coating (AGC). Glare is a type of light distortion commonly encountered in our daily life. Discomfort glare reduces the contrast of visibility of the object, while intense glare can lead to temporary vision impairment [33]. For instance, when mineral glass is used for glazing of paintings, the optical appearance of the painting is disturbed by unwanted reflections, which are enhanced by bright daylight or artificial illumination [34]. Therefore, different from ARCs, AGCs are applied to decrease the glare caused by reflecting light coherently from external light source in the environments. Adjustable gloss, low haze, and good clarity are considered as important criteria for AGCs performance [35–37].

1.2.2 Principle of Self-Cleaning

1.2.2.1 Special wettability induced self-cleaning

Wettability is an important property of a solid surface, and contact angle (θ) has been commonly used to characterize the surface wettability.

For a perfectly smooth and chemically homogeneous solid surface, the contact angle of liquid is given by Young's equation [38]:

$$\cos\theta = (\gamma_{SV} - \gamma_{SL}) / \gamma_{LV}$$

where γ_{SV}, γ_{SL}, and γ_{LV} are the interfacial tensions of the solid–vapor, solid–liquid, and liquid–vapor interfaces, respectively (Fig. 1.3a).

For a realistic solid surface that is rough and chemically heterogeneous, the early theoretical works about its contact angle were done by Wenzel and Cassie-Baxter.

I. Wenzel's equation [39]

In the model developed by Wenzel, liquid completely penetrates into the roughness grooves (Fig. 1.3b). The Wenzel's equation is:

$$cos\theta_w = rcos\theta$$

where θ_w is the contact angle on a rough surface, θ is the Young's contact angle on a similar smooth surface, and r is the surface roughness factor, defined as the ratio of the actual surface area to the projected surface area (for a perfectly smooth surface: $r = 1$, and for a rough surface: $r > 1$).

The above equation points out that when θ is less than 90°, the wetting could be enhanced by r; when θ is more than 90°, the wetting is lessened by r.

II. Cassie-Baxter equation [40]

The Cassie-Baxter state considers that under some roughness conditions, air bubbles may be trapped when θ is greater than 90° (Fig. 1.3c). In this case, the liquid-surface interface actually consists of two phases, that is, the liquid–solid interface and the liquid–vapor interface, respectively. Therefore, the apparent contact angle is the sum of all the contributions of the different phases:

$$\cos\theta_c = f_1 \cos\theta_1 + f_2 \cos\theta_2$$

where θ_c is the apparent contact angle, f_1 and f_2 are the surface fractions of phase 1 and phase 2, respectively; θ_1 and θ_2 are the contact angles on phase 1 and phase 2, respectively.

If one of the surfaces is the air-liquid interface and f is defined as the fraction of the solid surface that is wetted by the liquid, the air fraction can be expressed as $(1-f)$. With $\theta = 180°$ for air, the resulting apparent contact angle can be calculated by the following equation:

$$\cos\theta_c = f\cos\theta + (1-f)\cos 180° = f\cos\theta + f - 1$$

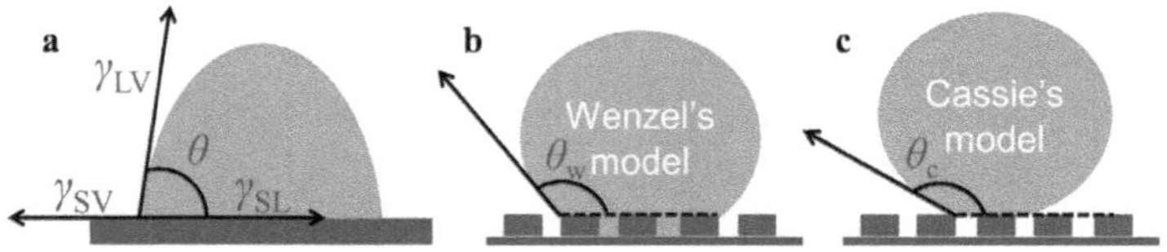

Figure 1.3 (a) A liquid droplet on a smooth surface. (b) A liquid droplet in the Wenzel state. (c) A liquid droplet in the Cassie state [4]. Copyright 2014 Elsevier.

In the above equations, the contact angles are all static. However, owing to chemical heterogeneity, surface roughness, and surface reorganization, different contact angles can coexist along the contact line upon contact with probe liquid [41]. As shown in Fig. 1.4a, the contact angle of the droplet at the front is greater than that at the back, and they correspond to the advancing and receding contact angle, respectively. The difference between the advancing and receding contact angles is defined as the contact angle hysteresis, which is a measurement of how well a drop of liquid sticks to the solid surface (Fig. 1.4b).

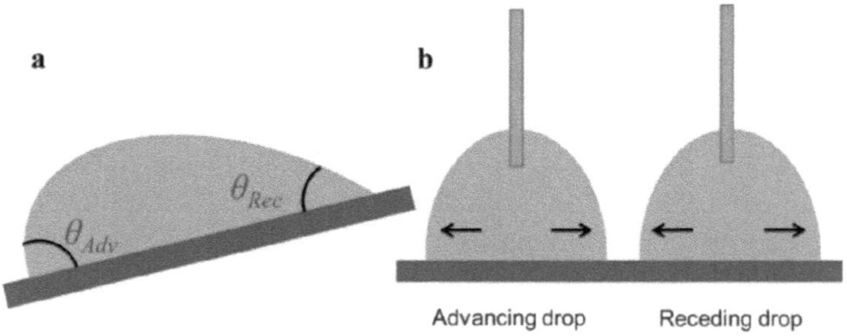

Figure 1.4 (a) Schematic illustration of contact angle hysteresis on a tilted surface, (b) advancing and receding contact angles on a flat surface via increasing and decreasing the volume of droplet, respectively [4]. Copyright 2014 Elsevier.

There are two kinds of surface with special wettability: superhydrophobic surface (water contact angle WCA > 150° and low contact angle hysteresis) and superhydrophilic surface (water contact angle <5° within 0.5 s or less). A self-cleaning superhydrophobic surface should always have a low contact angle hysteresis. The leaf of the lotus plant is a representative model for superhydrophobic and self-cleaning surface. Due to the papillose epidermal cells covered with wax crystals (Fig. 1.5a), the leaf surface is very rough and water-repelling. Therefore, water droplets on the surface are in the Cassie-Baxter state, making them highly spherical and be able to roll off easily. Usually, contaminants are larger than the cellular structure of the leaves, leaving the particles resting on the tips of the surface structure. When a water droplet rolls over the contaminants, dirt particles are adsorbed to the water droplet and moved away from the surface (Fig. 1.5b). On such intrinsic hierarchically structured surfaces, both the contact area and adhesion to the surface are dramatically reduced.

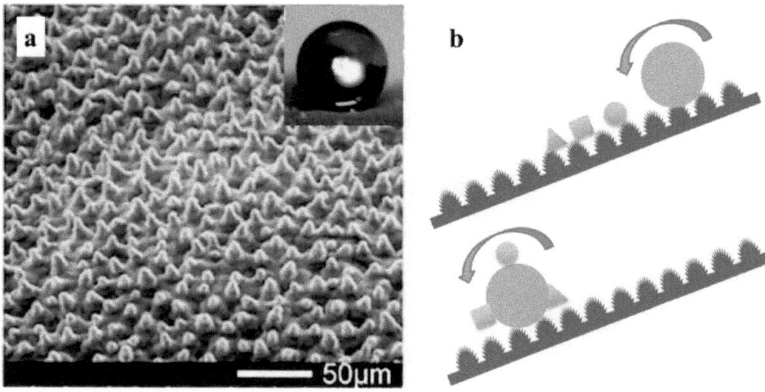

Figure 1.5 (a) SEM image of lotus leaf showing epidermal cells, and (b) diagram showing the self-cleaning process on a rough surface [42]. Copyright 2005 American Chemical Society.

Moreover, special wettability also endows the surfaces with antifogging function. Fogging occurs when water vapor condenses on the surface owing to temporary changes in humidity, temperature, or convection [43]. If the surface has a rather low contact angle, it could form a continuous or near-continuous film, scattering is suppressed, and optical transmission is enhanced. On the contrary, if the surface has a quite high contact angle and low contact angle hysteresis, the droplets may slide from the surface rapidly. Fog may thus be difficult to cover the surface. Superhydrophobic or superhydrophilic coatings are in line with the above requirements and have shown good prospects for practical applications [22].

1.2.2.2 Photocatalysis-induced self-cleaning

Since Frank and Bard first discovered that TiO_2 can decompose cyanide in water in 1977 [44], TiO_2 has been an ideal photocatalyst in decomposing organic contamination for environmental applications. TiO_2 has a strong oxidation power under UV illumination, and various organics accumulating on TiO_2 thin films could be oxidized into CO_2, H_2O, NO_3^-, or other simple basic products [45–47]. Moreover, under UV illumination, the surface of TiO_2 becomes superhydrophilic with a water contact angle close to 0°, making the surface easily cleaned by water. Due to these two characteristics, TiO_2-coated substrates can achieve outstanding self-cleaning property when they are exposed outdoor, as shown in Fig. 1.6.

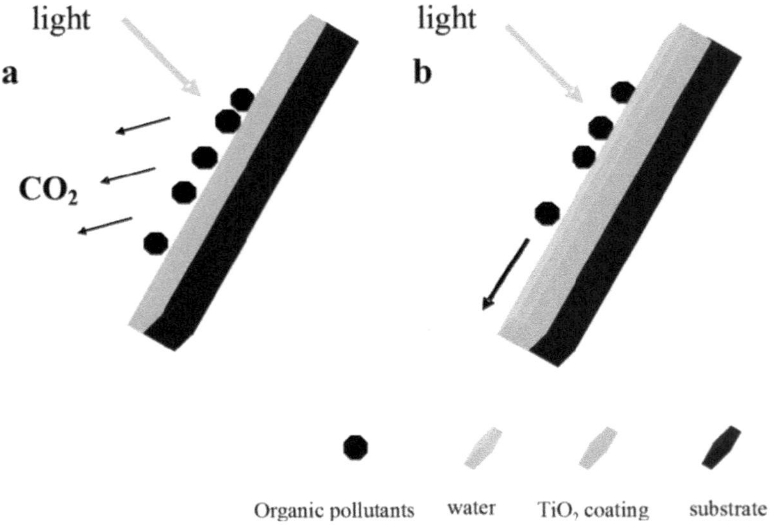

Figure 1.6 Schematic diagram of the decontamination process occurring on TiO_2-coated substrates. (a) Photocatalyic degradation of organics. (b) Removal of contaminants by water sheet due to the superhydrophilicity of TiO_2 surface [4]. copyright 2014 Elsevier.

Mechanism of photocatalysis

The redox potential for TiO_2 photo-generated holes is +2.53 V, which can produce hydroxyl radicals (·OH). The redox potential for conduction band electrons is –0.52 V, which is enough to reduce dioxygen to superoxide, or to hydrogen peroxide [48]. The basic processes for an electron-hole pair reducing or oxidizing an adsorbate and forming a singly oxidized electron donor and singly reduced electron acceptor can be summarized as Fig. 1.7 [49].

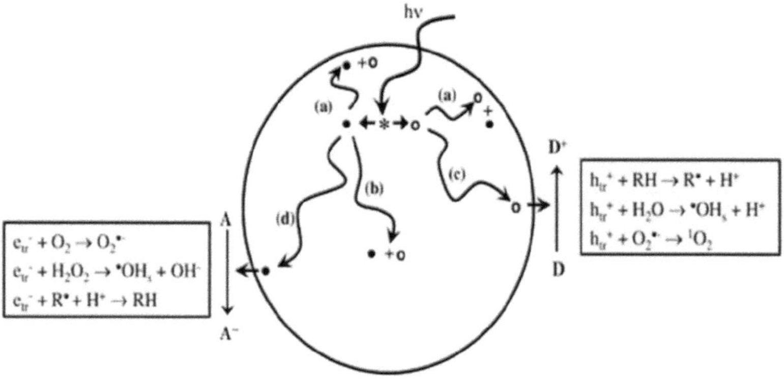

Figure 1.7 The processes occurring on bare TiO_2 particle after UV excitation [49]. Copyright 2008 Elsevier B.V.

Previous results have shown that many liquid and solid organic compounds can be photocatalytically decomposed into nontoxic gas-phase products. The decomposition of organic compounds (alkanes, alcohols, ketones, carboxylic acids, and so on) can be summarized as Scheme 1.1 with alcohols as an example [50,51]. As reported by Schwitzgebel et al. [52], long-chain organic compounds (e.g., C_{16} and C_{17} acids) and even polymers can also be decomposed by chains breaking.

$$RCH_2COOH + h^+ (\cdot OH) \rightarrow RCH_2\cdot + CO_2 + H^+ (H_2O)$$

$$RCH_2\cdot + O_2 \rightarrow RCH_2OO\cdot$$

$$RCH_2OO\cdot + \cdot OOH \rightarrow RCH_2OOOOH$$

$$RCH_2OOOOH \rightarrow RCHO + O_2 + H_2O$$

Scheme 1.1 Processes proposed for decomposition of organic compounds, for example, alcohols, on the surface of TiO_2 [50].

Mechanism of photo-induced superhydrophilicity

TiO_2 surface becomes superhydrophilic via UV light irradiation; however, the reduced WCA gradually revert to origin WCA in the dark. Fujishima's group [48,49,53,54] proposed a mechanism for photoinduced wettability conversion of TiO_2 surface, which has been widely accepted by scientists. The mechanism was proposed on the basis of the reconstruction of surface hydroxyl groups under UV irradiation in air. The density of surface hydroxyl groups is strongly correlated with the reciprocal of WCA. As illustrated in Figs. 1.7 and 1.8, molecular oxygen captures the photoexcited electrons, while the holes diffuse to the TiO_2 surface, being trapped at lattice oxygen atoms. Afterward, the hole weakens the binding energy between the Ti atom and the lattice oxygen, and the bond is interrupted by one adsorbed water molecule, forming new hydroxyl groups. In the dark circumstance, the hydroxyl groups gradually desorb from the surface in the form of H_2O_2 or $H_2O + O_2$. Compared with the OH groups bound to oxygen vacancies, the OH groups produced by UV irradiation are thermodynamically less stable because of the different desorption temperature [53]. As a result, the reconstruction between the stable OH groups and the thermodynamically metastable OH groups changes the surface energy of the TiO_2 and switches the surface between superhydrophilicity and hydrophobicity.

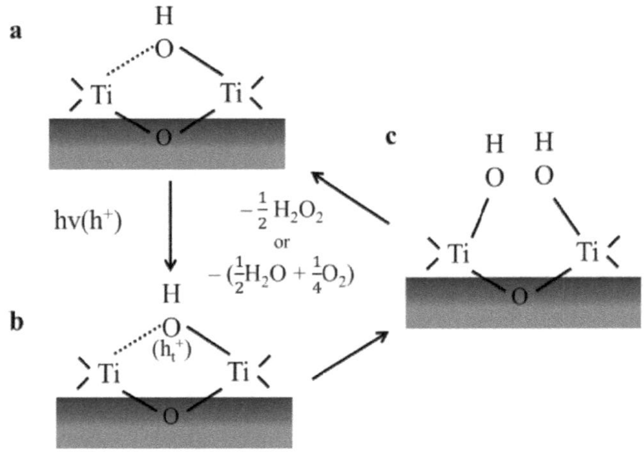

Figure 1.8 Surface structural models for the TiO_2 surface. (a) (Before UV irradiation) the OH group is bound to oxygen vacancy, (b) (at the transition state) the photogenerated hole is trapped at the lattice oxygen, and (c) (after UV irradiation) new OH groups are formed [53]. Copyright 2003 American Chemical Society.

1.3 Fabrication Strategies and Methods

For nearly two centuries, researchers have been seeking novel methods to increase light transmission in optical systems as well as to eliminate unwanted reflection. Meanwhile, self-cleaning surfaces in nature have also attracted more and more attention of scientists due to their unique properties. The technology of fabricating antireflective surfaces with highly efficient and durable self-cleaning properties has an extensive range of applications from glasses, displays, and solar panels to window optical lenses. Different chemical and physical methods for fabricating ARCs and how to form rough surfaces with special surface energy are discussed in the following section.

1.3.1 Fabrication of Antireflective Surfaces

ARCs are a basic way to increase light transmission. Varied methods have been used to generate ARCs, which can be classified into both bottom-up and top-down approaches. Accurate control over the morphology of ARCs is crucial in order to achieve ideal antireflective surfaces. The techniques combining both bottom-up and top-down approaches may pave an effective way toward making the dream of application in industry come true.

1.3.1.1 Materials perspective for antireflectivity

(1) Silicon based

Fabrication of antireflective layers on silicon surfaces has attracted much attention since the 1980s to effectively improve the efficiency of silicon solar cells, both in first-generation photovoltaic cells based on bulk materials and second-generation structures based on thin films [55–60]. Silicon has been explored in almost all types of ARCs, starting from porous silicon on solar cells to Moth's eye inspired biomimetic structures and silicon nanotips [23]. For instance, Hong's group [61] developed a self-masking reactive ion etching process for fabricating the large-area (15 × 15 cm^2) antireflective silicon surface by creating a high density of nano-pyramids on the surface. Large-area poly-crystalline silicon surfaces with low reflectance (<2%) and high uniformity could be reproducibly achieved, which is potential for improving light absorption and energy conversion efficiency of solar cells.

(2) Silicon oxide based

Silicon oxide based materials, such as glass (n = 1.52–1.80), quartz (n = 1.54), and fused silica (n = 1.459), are also outstanding candidates for ARCs [10]. A large number of works have been devoted to their study and application [23,62]. Almost all of the methods discussed later in the current review can be used for the preparation of silica ARCs. Recently, Li et al. [63] prepared sol-gel silica coatings for high-power laser applications, which was optimized by a two-step structure modification through treatment of the sol with poly(ethylene glycol), followed by treatment of the coating with ammonia. The transmittance of coated glass at 1064 nm increased from 91.9% to 99.9%. Furthermore, the win-win situation between antireflection and durability is an important issue for study and application. Scientists made various efforts to enhance mechanical performance or made self-healing coatings to increase durability, which will be introduced in detail in the following chapter.

(3) TiO_2 based

TiO_2 is the most widely used white pigment because of its brightness, high RI, and excellent stability. Due to these favorable characteristics, TiO_2 semiconductor has been intensively investigated for applications in photocatalysis, solar cells, sensors, and biomedical devices [64–66]. Meanwhile, TiO_2 has been widely used as ARCs due to its matching RI (anatase 2.49, rutile 2.903) and low deposition cost in semiconductor detectors and photovoltaics industry [67,68]. Different approaches have been attempted to reduce the RI of TiO_2 for various substrates. Among these methods, Xi et al. reported that the RI of a TiO_2 nanorod layer on aluminum nitride substrates can be controlled in a way that reduces the RI from 2.7 to 1.3 by changing the vapor-deposition incident angle using the oblique-angle deposition method [69]. There are also some reports on the fabrication of ZnO nanostructures as ARCs [70, 71]. However, ZnO is unstable in either acidic or basic solutions, limiting their applications to some extent. Here, ZnO will not to be classified and introduced in detail because there are many similarities in the preparation of TiO_2 and ZnO ARCs.

(4) Polymer based

Polymers are large molecules with multiple structural forms. On the basis of their micro-phase separation, self-assembly, or self-organizing properties, the molecular level structure (chain conformation), secondary structure (aggregation structure), and tertiary structure (self-assembly or self-organization) could be modulated to achieve the polymer film nanostructure [72,73]. There are many excellent characteristics for polymers, such as the ability of controllable morphology and porosity ratio, adherence to the substrate, flexible, and suitable for large-scale preparation. Thus, similar to inorganic materials, polymer materials also have a great practical value and wide applications as ARCs. Taking poly (methyl methacrylate) (PMMA) as an example, it is a very popular material for optical applications with an RI of 1.49. It has low optical absorption and transmits up to 92% in the visible light range due to ca. 4.0% reflection loss at the interface between PMMA and air. Broadband ARS on PMMA have been fabricated by plasma treatment [74–76]. He et al. [77] developed a facile dip-coating approach to fabricate visible/near-IR ARCs on glass substrates from PMMA and hydrophobic hexamethyldisilazane (HMDS)-modified hollow silica nanoparticles. The optical and wetting properties can be controlled by adjusting the PMMA concentration in the dip-coating suspension.

(5) Carbon based

Amorphous carbon and hydrogenated amorphous carbon are materials that can be used in several applications due to their excellent physical and chemical properties, such as high mechanical hardness, high optical transparency, high thermal conductivity, low friction coefficient, and chemical inertness to some corrosive agents [78,79]. Their applications include protective tribological coatings, radiation protection, electron field emitters, and biomedicine [80]. Oliveira, Jr. et al. reported the use of different forms of amorphous carbon films (polymeric-like carbon, diamond-like carbon, and tetrahedral amorphous carbon) as ARCs on crystalline silicon solar cells [81]. Da Silva et al. extended the investigation to double-layer ARCs on polished and texturized silicon surfaces [82]. The well-known "moth's eye" nanostructure has also been fabricated on the flat diamond surface as well. Reflectance of flat diamond surface is around 18% at 10 μm. When the surface is replaced with the "moth's eye" structure, the reflectance is reportedly reduced to 7% at the same wavelength [83]. In addition, carbon nanotubes ARCs have been reported as well. Long (greater than 300 μm) and low-density (0.01–0.02 g/cm^3) vertically aligned carbon nanotubes arrays were produced by Yang et al. using water-assisted chemical vapor deposition (CVD) [84]. The total reflectance at 633 nm was found to be an impressive 0.045%, which pointed to a near-perfect optically absorbing material.

(6) Group III–V based

Group III–V semiconductors, which have high absorption coefficient and electron mobility, are important materials for optoelectronic device applications including vertical-emitting lasers, image sensors, photodetectors, and solar cells [85–87]. For instance, GaAs-based solar cells have attracted considerable attention due to their superior efficiency and high radiation hardness [88,89]. However, owing to the high RI of GaAs ($n_{GaAs} > 3.6$ for visible wavelengths), more than 30% of incident light is reflected back from the substrate surface, which limits the efficiency of GaAs solar cells. An ARS, which suppresses unwanted surface reflection losses and thereby increases the transmission of sunlight into the solar cells, is inevitably necessary. Droplet epitaxy (DE) is useful for the growth of various nanostructures, due to the perfect separation of group III and group V [90,91]. Both the growth of homo-epitaxial nanostructures, such as GaAs quantum dots (QDs) on GaAs substrates, and the formation of ultralarge-sized QDs applicable to antireflective nanostructures are possible. Lee et al. presented the growth of large GaAs islands on $Al_{0.3}Ga_{0.7}As$/GaAs substrate by the DE method. After GaAs island growth, the reflectance of s-polarization at 70° by in situ measurement was reduced down to approximately 2–20% in the wavelength range of 350–900 nm [92]. Moreover, GaP nanorods [93], GaN nanopillars [94], random GaS based sub-wavelength structures [95], and sub-wavelength gratings for GaSb [96] were fabricated with increased transmittance for various applications.

1.3.1.2 Bottom-up approach

1.3.1.2.1 *Layer-by-layer assembly*

In 1966, Iler [97] demonstrated a novel method for creating multilayers of inorganic colloidal particles, which is now known as layer-by-layer (LbL) assembly. It is one of the first studies to disclose the successful assembly of nanoparticle thin films. Unfortunately, Iler's work did not attract much attention at the time. After 25 years, Decher and co-workers [98] using LbL processing fabricated multilayer thin films from oppositely charged polyelectrolytes. Their work has led to a great many studies about films for various potential applications based on LbL methods. Now, LbL has been proven a simple and effective way to fabricate various coatings with tailored chemical compositions and structures. Since then, much more attention has been paid to the LbL assembly technique, whose building blocks could be linear polyelectrolytes [99], nanoparticles [100] or nanowires with low RI [101], co-polymer micelles [102], proteins [103], and so forth. LbL assembly can be driven by multiple weak interactions, including electrostatic interactions, hydrogen-bonds, halogen-bonds, coordination bonds, charge-transfer interactions, biospecific interactions, guest–host interactions, cation–dipole interactions, and the combined interaction of the above forces, and so on [104]. An outline of the LbL assembly technique conducted via electrostatic interaction is illustrated in Fig. 1.9a [105].

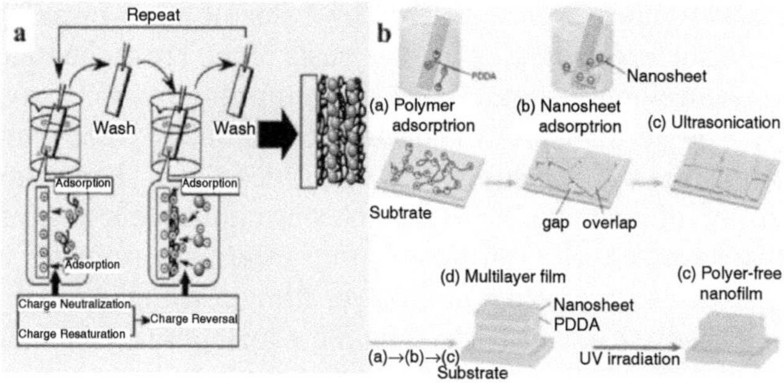

Figure 1.9 (a) Outline of LbL assembly through electrostatic interaction [105]. Copyright 2007 Royal Society of Chemistry. (b) A modified sequential LbL adsorption process combined with ultrasonication treatment. Ultrasonication treatment removed overlapped patches to produce a monolayer film in which nanosheets were densely tiled having a nearly negligible overlap area between them [106]. Copyright 2012 Wiely-VCH.

The LbL assembly technique has several advantages in the preparation of ARCs. First, as mentioned above, varied materials (building blocks) can be used to fabricate ARCs and the LbL processing is substrate-independent. Second, it is possible to deposit ARCs of large areas on non-flat surfaces by the LbL assembly technique. Third, it is easy for the LbL assembly technique to control over the film thickness, and therefore, to tailor the wavelength of maximum transmittance simply by changing the number of film deposition cycles [107,108].

The groundbreaking work of applying the LbL assembly technique to the fabrication of ARCs was carried out by Rubner and coworkers [99]. They described an aqueous-based LbL process controlled at the molecular level that may coat surfaces of almost any material. High-efficiency antireflection coatings were created from phase-separated polyelectrolyte multilayer films that underwent reversible pH-induced swelling transition. After that, many LbL-assembled high-performance ARCs have been fabricated. He's group has done a large number of works to explore simple, economical but efficient ways for preparation of ARCs by LbL technique. They designed and synthesized silica nanoparticles of varied structures, such as solid silica nanoparticles [109], monodisperse raspberry-like PS@SiO_2 composite nanoparticles [110], nonspherical mesoporous silica nanoparticles [111], hollow mesoporous silica nanospheres [100], hierarchically mesoporous silica nanoparticles [112], and so forth, and used these multifarious silica nanoparticles as building block for LbL assembly. Coatings with ideal antireflectivity and special wettability were achieved by tailoring their porosity, thickness, and surface roughness on the basis of the Fresnel equation, the quarter-wavelength design, and the requirements for special wettability. The performance of coatings could be further optimized by combining the LbL assembly with other techniques. For example, the adsorption of oversized nanoparticles or nanosheets always results in a disorganized film. Ultrasonication treatment removes overlapping patches to produce a monolayer film, in which nanosheets are densely tiled with a nearly negligible overlapping area [106]. The application of this technique facilitates the improvement of coating homogeneity and other performance.

1.3.1.2.2 *Dip-coating*

Dip-coating has advantages such as simplicity, controllability, reliability, and reproducibility. The process is generally described as that the substrate is dipped in and withdrawn from the precursor solution at a controlled rate (feed velocity). The steps in the process can be summarized as immersion, start-up, deposition, evaporation, and drainage [113,114], as shown in Fig. 1.10. Dip-coating is frequently selected for the preparation of nano-structured coatings because it is simple and fast, especially for double-functional or multi-functional films.

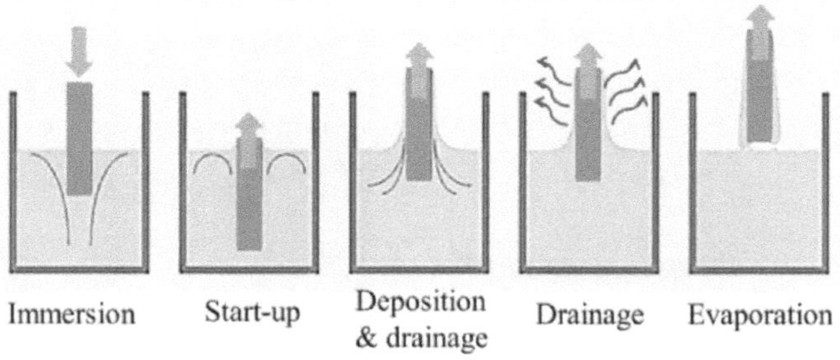

Figure 1.10 Steps involved in dip-coating.

Several literature reports [14,115–118] are dedicated to the utilization of dip-coating technique as a time-saving and low-cost method to fabricate ARCs. Helsch et al. [118] fabricated quarter-wave ARCs on silica glass by a sol–gel dip coating process to minimize reflective losses of the receiver cover materials for solar tower power plants. The coated glass was proved having excellent transmittance (98.6%) and thermal resistance. It has been demonstrated [117] that dip-coating derived ARCs exhibit some excellent properties, such as homogeneity, stoichiometry control, purity, and ability to coat large and complex area substrates, compared with coatings prepared by other techniques. However, there are also some inadequacies. For instance, the thickness and uniformity of fabricated coatings are sensitive to substrates and flow conditions such as liquid bath and gas over-head [114]. Meanwhile, it is still a challenge for scientists to construct roughness-controlled films by dip-coating due to the flowing of sol-gel liquid during the preparation process.

1.3.1.2.3 *Spin-coating*

Spin-coating is widely used in microelectronics industry, where it is used to deposit uniform coatings of organic materials or to uniformly distribute particulate matter on a flat surface [119]. In short, in the procedure, an excess amount of precursor solution is placed on the substrate, which is then rotated at high speed in order to spread the fluid by centrifugal force. The film keeps thinning until disjoining pressure effects cause the film to reach an equilibrium thickness or until it turns solid-like due to a dramatic rise in viscosity resulting from solvent evaporation. The thin film finally forms due to solvent evaporation [121]. The key process and typical operation are illustrated in Fig. 1.11 [120].

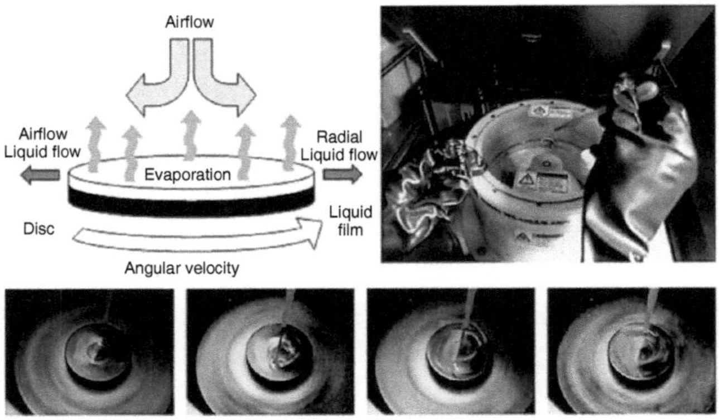

Figure 1.11 Schematic illustration of spin-coating (top left) along with a photograph of a typical spin-coating operation in a glovebox environment (top right) and high-speed images showing application of solution to a rotating substrate and film formation. The images were recorded at 300 images s^{-1} (below). The timing of the images (from left to right) after impact of the first drop is: t = 17, 100, 137, and 180 ms [120]. Copyright 2009 Elsevier.

He's group reported spin-coating fabrication of highly ARCs on Fresnel lenses using solid silica nanoparticles and mesoporous silica nanoparticles as building block [122,123]. The maximum transmittance of the coated Fresnel lenses reached as high as 99.8%, while that of the bare Fresnel lens was only 94.3%. If spin-coating is combined with LbL assembly, the time for inducing significant adsorption of weakly attached polyelectrolytes could be decreased. Dipping-based LbL assembly of multilayered films usually lasts for hours, with each dipping step requiring between 10 and 30 min for the deposition. As an alternative, spin-assisted LbL assembly allows rapid processing of multilayered films. During the spinning, the whole deposition process of a single polyelectrolyte layer usually takes only 10^{-15} s [124]. The significant difference between the dipping and the spinning method is caused by different adsorption mechanisms. In the spinning process, fast elimination of water significantly increases the molar concentration of the polyelectrolyte solution during the short deposition time [125].

1.3.1.2.4 *Spray-coating*

Spray-coating is quite a facile method, which is not specific to a particular substrate and can be easily applied to a large surface area [126]. As spray-coating does not typically require other complicated and costly processes, it is commercially feasible for a wide range of applications. The general process of spray-coating is shown in Fig. 1.12a. This method is simple, fast, material efficient, easy to use, carried out in non-vacuum system, and permits to obtain films with required properties for large area applications [127]. Among diversified spraying techniques, thermal spraying has become an excellent solution to the processing limitations of polymer/ceramic nanocomposites. Coating materials available for thermal spraying include metals, ceramics, alloys, plastics, and composites. Usually, they are fed in powder or wire form, heated to a molten or semimolten state, and accelerated toward substrates in the form of micrometer/nanometer-scale particles. Thermal spray generally is a process in which a material is heated, accelerated, and propelled by a high-temperature jet through a confining nozzle toward a surface [128,129]. Coating quality is determined by its porosity, oxide content, hardness, bond strength, and surface roughness. He's group [130] used spray-coating method to construct coatings on slide glasses with synthesized silica nanoparticles of 20 nm in size. When the number of spray-coating cycles was 3, the transmittance of coating was greatly improved in the visible range. Recently, Zhang's group [131] developed a simple technique to fabricate superhydrophobic coatings by spraying microparticle/polymer suspensions to produce a robust polymer composite coating. Both the surface microstructures and corresponding superhydrophobicity depend on calcium stearate formed on the coating surfaces. The method for preparing a regenerative superhydrophobic coating has the advantage of simplicity in fabrication, easy availability of the materials, and applicability to a large area surface. Moreover, multilayer coatings with multiple components can be fabricated by spray-coating method as shown in

Fig. 1.12b. By continuous spraying of polyanion and polycation solutions onto a charged surface, a uniform film that grows consecutively with time can be obtained [132].

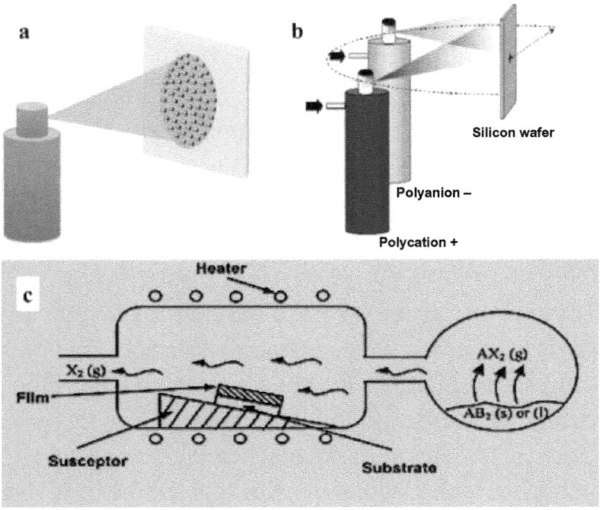

Figure 1.12 (a) Schematic diagram of spray process, (b) continuous simultaneous film formation of multilayer coating by spraying [132]. Copyright 2005 American Chemical Society. (c) Schematic diagram of CVD [135]. Copyright 2003 Elsevier.

Combination of spray-coating and the LbL method for fabricating thin films has attracted much attention in recent years. Aqueous solutions of suitable materials are sequentially sprayed onto surfaces and the self-limiting adsorption processes that are similar to the LbL assembly take place. Therefore, the spray-LbL technique can be used to create multilayer assemblies using less time. The first paper describing spray-LbL assembly was published by Schleno's group [133]. Since then, extensive literatures have been published to report this alternative processing approach and now it is possible to create complex multi-stack optical coatings. Because conventional manual spraying systems are difficult to control a few experimental parameters such as the spraying time, solution flow rate, and volume, Krogman and his co-workers [134] developed an automated process to improve uniformity and versatility of the spray-LbL deposition process.

1.3.1.2.5 *Chemical vapor deposition*

CVD is a widely used technology for the preparation of inorganic thin films. It can produce single layer, multilayer, composite, nanostructured, and functionally graded coating materials with well controlled dimension and unique structure at relatively low processing temperatures [135]. CVD essentially involves the process of dissociation of molecules of gaseous reactants, which subsequently react chemically to form various structures when activated by heat, light, or plasma discharge. Mostly solid-state, stable products are formed as a result of the chemical reactions [136]. Basically, the process of CVD at a surface to generate a thin and solid film can be described as Fig. 1.12c [135]. In addition to conventional CVD method, that is, so-called thermal-activated CVD (TACVD) [137–139], many variants of CVD method are used to fabricate ARCs, such as plasma-enhanced CVD (PECVD) [140], photo-assisted CVD (PACVD) [141], hot-wire CVD (HWCVD) [142,143], and so on.

PECVD has been one of the most industrially preferred technologies for ARCs. It is a glow discharge CVD, which uses electron energy (plasma) as the activation method to enable deposition to occur at a low temperature and at a reasonable rate. By this technique, the precursor molecules can be decomposed by pulsed microwave-induced plasma and finally form antireflective layer (TiO_2/SiO_2) and/or anti-scratch layer (SiO_2) [144]. Remache and co-workers [145] designed a porous silicon/PECVD SiOx ARCs with nice optical properties. They first prepared mesoporous silicon layers on the front surface of the n^+ emitter of n^+–p mono-crystalline silicon junction. The single-layer ARCs achieved around 9% of effective reflectivity on junction n^+–p solar cells at wavelength of 400–1000 nm. Then, they deposited silicon oxide layers by PECVD on mesoporous silicon layers, leading to the effective reflectivity reduction to 3.8%. The multilayer coating was predicted to give an enhancement of the photogenerated current exceeding 44%.

1.3.1.2.6 *Phase-separation*

Polymer ARCs also have considerable practical value and wide-ranging applications because of their ability to adhere to flexible substrates. The preparation of suitable polymer porous structure with tunable maximum transmittance by a simple approach is still, however, a big challenge [146]. In 1999, a method based on the phase separation of a macromolecular liquid to generate nanoporous polymer films was proposed by Steiner's group [147]. Their technique was based on the demixing of a binary polymer blend during spin-coating [148]. PS and PMMA were first dissolved in tetrahydrofuran and spin-cast onto both sides of a microscope slide. During spin-coating, phase separation set in, and after evaporation of the solvent, lateral phase morphology was obtained. Then, the glass slide was exposed to cyclohexane, which selectively dissolves PS. Finally, porous films with high optical transmission were obtained. A major advantage of the approach is its versatility. By using appropriate combinations of polymers and solvents, various monolayer and multilayers could be created, such as copolymer PS-b-poly (dimethylsiloxane) (PS-b-PDMS) surface [149], porous PS/poly (vinylpyrrolidone) (PS/PVP) films, and so on [15].

In addition, Ibn-Elhaj and Schadt [150] presented an alternative approach to ARCs. The process is based on an extension of photo-aligning and photo-patterning technology for liquid-crystal displays and liquid-crystal polymer films with smooth surfaces to optical polymer films with controlled surface topologies. Monomer phase-separation induced nano and micro-corrugated topologies on the polymer surfaces. The products were low-cost ARCs for optical components of virtually any size, shape, and material. However, the wear-resistance is still a major limitation to most polymer ARCs. As far as we know, choosing materials with improved mechanical characteristics and appropriate post-treatment are expected to solve this problem.

1.3.1.3 Top-down fabrication

1.3.1.3.1 *Etching*

Etching technique can be described as a selective dissolution or ablation of a surface or substrate. Basically, etching can be divided into two categories: wet etching and dry etching.

Wet etching can be described as a process using a chemical solution for the dissolution with advantages such as precise selective attack and fast etching rate. Usually, the solution is water-based because water is corrosive to glass to some extent. Although the dissolution mechanism remains unclear, there are some reactions that are widely accepted to occur during dissolution: (1) hydration and hydrolysis; (2) ion-exchange; and (3) network reconstruction. In most glass structures, the voids between oxygen and hydration are not large enough to allow water molecules to penetrate into the surface, so the hydrolysis reaction may be accompanied by a network dissolution by releasing water soluble species such as Si $(OH)_4$ into solution and leaving large voids for further reaction, as shown in Scheme 2. By dehydration, the silanol groups (Si–OH) can condense into a Si–O–Si network, as shown in the reverse reaction of Scheme 1.2, which leads to a reconstruction of the glass surface and forms a porous material [151,152]. He's group [153] reported a method that uses water to create a hierarchical superhydrophobic surface. However, the process temperature is relatively high

$$\text{Si–O–Si} + H_2O \leftrightarrow \text{Si–OH} + \text{OH–Si}$$

$$\text{Si–O–Si(OH)}_3 + OH^- \rightarrow \text{Si–O–Si(OH)}_4^- \rightarrow \text{Si–O}^- + \text{Si(OH)}_4$$

$$\text{Si–OR} + H_3O^+ \rightarrow \text{Si–OH} + R^+ + H_2O$$

$$\text{Si–OR} + H_2O \rightarrow \text{Si–OH} + R^+ + OH^-$$

Scheme 1.2 Chemical reactions in the etching mechanism [152].

(>160°C). Recently, Qi et al. [154] demonstrated a simple method for creating hierarchical pyramidal structures on whole silicon wafer using alkali etching at low temperature (65°C). The fabrication procedure for creating hierarchical structures is shown in Fig. 1.13a. There were three main steps: (1) fabrication of pyramidal structures on silicon surface with KOH etching; (2) electroless deposition of a thin discontinuous layer of Ag nanoparticles on the pyramidal structures in $HF/AgNO_3$ solution; and (3) generation of hierarchical structures by Ag-assisted etching and removal of Ag nanoparticles in nitric acid. Their research provided a cost-effective and facile approach for producing superhydrophobic and antireflective silicon surfaces with hierarchical structure (Fig. 1.13b,c).

Dry etching is currently used in semiconductor industry due to its unique ability over wet etching to carry out anisotropic and directional etching or to create high-aspect ratio structures (e.g., deep via holes or capacitor trenches). Typically, bombardment of ions (plasma of reactive gas such as fluorocarbons, oxygen, chlorine, boron trichloride) is used to dislodge portions of the material from the exposed surface with masked patterns. And other benefits of dry etching are reduced chemical hazard treatment and contamination, and ease of process automation and tool clustering [156].

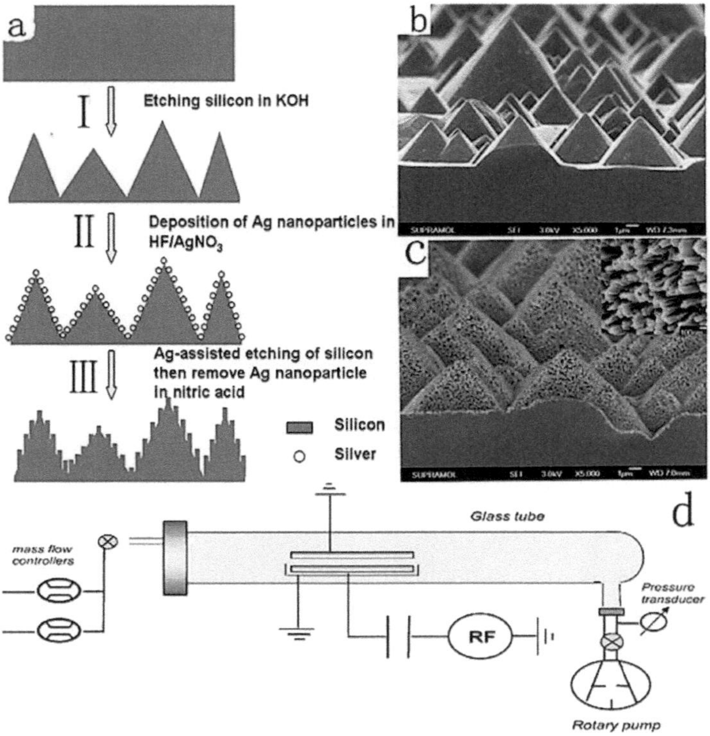

Figure 1.13 (a) Fabrication procedure for creating hierarchical structures on silicon surface. (b) Cross-sectional SEM photograph of silicon pyramids created with KOH etching. (c) Cross-sectional SEM photograph of hierarchical structures generated with Ag-assisted etching. Inset: magnified SEM image [154] Copyright 2009 American Chemical Society. (d) Scheme of the reactor for plasma texturing [155]. Copyright 2011 Wiley-VCH.

Versatility of plasma processing is demonstrated for the preparation of nanotextured surfaces with superior wettability properties. Leem [157] reported the fabrication of broadband and wide-angle antireflective subwavelength structures on silicon substrate by inductively coupled plasma etching using Au nanopatterns as etching masks. Palumbo et al. [155] reported an effective plasma processing for preparing both superhydrophobic and superhydrophilic surfaces from polycarbonate (PC). The polymer nanostructure size can be controlled by tuning the experimental plasma parameters during oxygen etching. The scheme of the reactor for plasma texturing is shown in Fig. 1.13d.

1.3.1.3.2 *Lithography*

The semiconductor industry has been pushing high-precision nanoscale lithography to a rapid development. The previous fabrication strategy relied on costly e-beam lithography [158] and photolithography [159,160]. Recently, unconventional nanofabrication techniques such as colloidal lithography [161,162], nanoimprint lithography [163], soft lithography [7] based on molding, embossing, and printing were explored. The fundamentals of some lithography processes are shown in Fig. 1.14 [164,165]. These techniques have emerged due to high throughput and cost-effective process capabilities, and have been applied into the fabrication of ARCs, microchips, and micro-electromechanical devices. These techniques always resort to expensive tools such as those used in deep-UV projection lithography and electron-beam lithography.

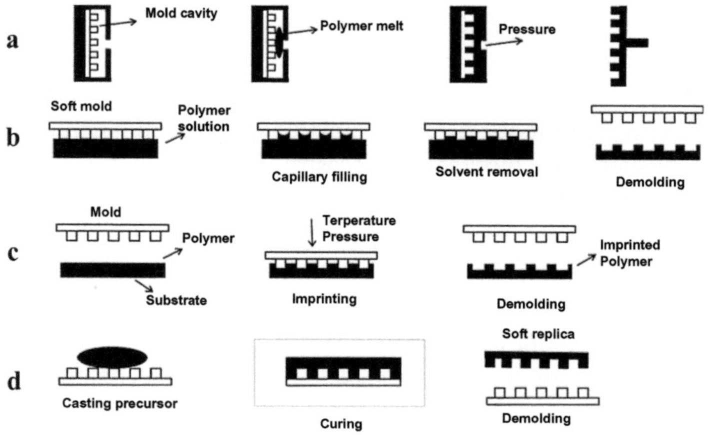

Figure 1.14 Different alternative lithographic processes. (a) Injection molding, (b) solvent-assisted molding, (c) hot embossing, and (d) soft lithography [164,165]. Copyright 2008 American Chemical Society.

Currently, photolithography has been one of the most widely used top-down techniques in the fabrication of antireflective and self-cleaning coatings [162,166,167] due to its several advantages, such as full wafer processing and short exposure time. However, drawbacks still exist such as the limited resolution and the limited possibilities of working with pre-existing topography or curved substrates, which still need further study.

1.3.2 Fabrication of Self-Cleaning Surfaces

Techniques to produce self-cleaning surfaces can be generally classified into two categories: making a rough surface; and modifying a rough surface with a material of low surface energy (for superhydrophobic surfaces) or high surface energy (for superhydrophilic surfaces) [168].

1.3.2.1 Fabrication of superhydrophilic surfaces

Superhydrophilic surfaces can induce "complete spreading" of water, which would have a broad range of application, including self-cleaning, antifogging, water harvesting, and biomedical field [169,170]. On the one hand, the roughness of a surface plays a key role in controlling its wettability. On the other hand, superhydrophilicity can be achieved through photo-induced surface reactions by some semiconductors such as TiO_2, ZnO, WO_3, and V_2O_5 [171–174].

Nano-scale porous structures and some microstructures are known to exhibit superhydrophilicity. However, film porosity also plays an important role in obtaining superhydrophilicity. Wetting can be controlled by changing porosity and/or roughness. There are three types of rough surface: regular (designed) surface, irregular (random) surface, and hierarchically rough surface [175]. Films of nanoparticles are often deposited on substrates by LbL [109,176], spin-coating [177], sputtering [178], hydrothermal synthesis [179], and so on. Microstructures can grow from polymerization [180], lithographic [167], and electrochemical techniques [181]. Photoinduced surface reactions are important for achieving superwetting properties, and a large number of methods were used to prepare superhydrophilic coatings with light-sensitive materials, such as drop casting [173,182], peroxotitanate-complex deposition and liquid-phase deposition [183,184], physical vapor deposition

(PVD) [185], atomic layer deposition [186–188], and so on. To avoid duplication, we just describe a few novel preparation methods in this section.

1.3.2.1.1 *Liquid phase deposition*

Liquid phase deposition (LPD) is a process based on the controlled hydrolysis and has been developed as a low-temperature route to obtain several oxide films, including TiO_2 [189]. LPD processes are environmentally friendly because they use low-cost and low energy-consuming equipment and water as a solvent [190]. Deposition rate and crystal orientation can be controlled by an appropriate choice of experimental parameters such as pH, temperature, surface characteristics, and so on. Organic polymer-core/TiO_2-shell composite particles and TiO_2 hollow spheres could be fabricated at low deposition temperature [191,192]. As the TiO_2 are successfully patterned at the micro and nanoscale, it is very attractive in superhydrophilic applications.

In recent years, some research groups successfully prepared porous TiO_2 films with rough surface structure by combining LPD method with LbL self-assembly process [176,193], and also extended it to films of other metal oxides such as FeOOH [194]. The combination of different technologies enhances its appeal for industrial applications due to its scalability and effectiveness in coating surfaces with complex morphologies.

1.3.2.1.2 *Physical vapor deposition*

Deposition of coatings by PVD techniques, in which a material is vaporized and then condensed onto a substrate to form a thin coating, has been widely used in many industrial sectors such as integrated circuits. Common PVD methods include thermal deposition, ion plating, ion implantation, sputtering, laser vaporization, and laser surface alloying [64]. TiO_2 thin films with a thickness ranging from 300 to 500 nm were prepared by PVD in an electron evaporation system [185]. To control the porosity and the surface roughness, evaporation was carried out at different glancing angles with respect to the evaporation source. The top views of TiO_2 thin films prepared by PVD and the effect of UV irradiation on the wetting angle on the films are shown in Fig. 1.15. The control of deposition conditions by PVD has given thin films in which not only the crystallographic structure but also the microstructure and porosity are controlled. Illumination, surface roughness, surface contamination, type of crystal planes at the surface, and so on play varied roles in the UV-induced wetting behavior of the films. Superhydrophilic TiO_2 thin films can also be fabricated by frequency magnetron sputtering deposition [195,196]. The temporal evolution of the water contact angle was found to depend on the total gas pressure that influences the crystalline structure of the films.

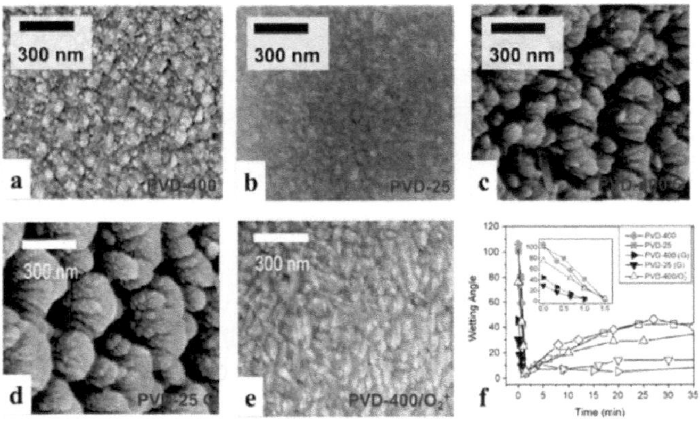

Figure 1.15 (a–e) Top views of TiO_2 thin films prepared by PVD. (f) Dependence of wetting angle on time. In (f), the filled symbols refer to the decay in wetting angle during illumination and the empty symbols refer to the recovery of wetting angle when the illuminated thin films were kept in the dark. The inset in (f) shows an enlarged time scale evolution of wetting angle for the PVD thin films under illumination [185]. Copyright 2009 Elsevier.

1.3.2.1.3 *Hydrothermal treatment*

Hosono et al. reported a fabrication method of TiO_2 nanosheet films by heat treatment [197]. The resulting anatase TiO_2 nanosheet films showed superhydrophilic properties without the need for any UV irradiation treatment. An aqueous solution of urea and the titanium sheet were placed in a Teflon-lined autoclave, which was heated at 100°C for 4 days. After the hydrothermal treatment, the sheet was then transformed to TiO_2 by heating at 400°C or 600°C for 30 min in air. Figure 1.16a shows the presence of perpendicular TiO_2 nanosheets on the titanium sheet, and the perpendicular nanosheet thickness is around 2.5 nm. The high concentration of defects or dangling bonds is a result of the specific morphology and is responsible for the superhydrophilicity without prior UV irradiation (Fig. 1.16c). Hydrothermal treatment also applies to silica glasses. He's group [153] fabricated superhydrophilic surface by simple hydrothermal treatment of slide glass at 180°C for 24 h. After reaction, a predominant three-dimensionally interconnected porous structure (the upper left corner in Fig. 1.16d) was observed together with minor network structures (the lower right corner in Fig. 1.16d), and the coating thickness was about 4 μm (Fig. 1.16e). The rough surface exhibited a WCA of ca. 0° in less than 40 ms.

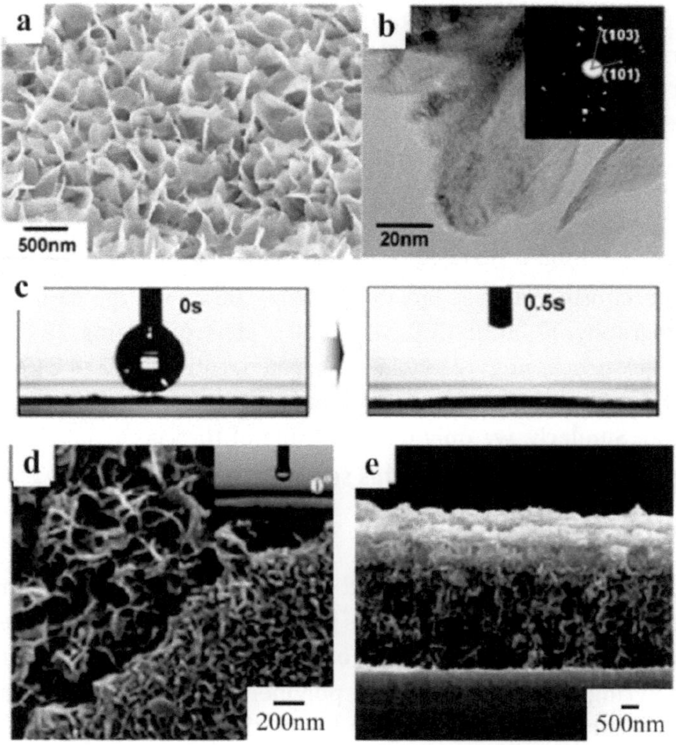

Figure 1.16 SEM (a) and TEM (b) images of perpendicular TiO_2 nanosheet film fabricated via hydrothermal synthesis and heating in air at 400°C for 30 min. (c) Measurement of the contact angle of water on perpendicular TiO_2 nanosheet film [197]. Copyright 2007 American Chemical Society. (d) SEM image of coating prepared at 180°C for 24 h. (e) Cross-sectional image of the coating in (d) [153]. Copyright 2007 American Chemical Society.

1.3.2.2 Fabrication of superhydrophobic surfaces

Superhydrophobic surfaces have a water contact angle greater than 150° and low contact angle hysteresis (lower than 5°). Because of their unique water-repellent property, these surfaces have attracted significant attention over the last two decades, which can essentially influence a lot of cutting-edge topics, including surface structures and fluid motivation, in engineering and biotech researches [198–200]. Learning from nature, we know that the low surface energy of $-CF_3$ groups, silyl compound, or fluorocarbons is not the only condition to obtain superhydrophobicity [201,202]. Regardless of the material used and structure (particles, rod arrays, or pores) formed on the surface, suitable roughness in combination with low surface energy is required. Until now, procedures of roughening the surface followed by hydrophobization or transforming low-surface energy materials into rough surfaces have been commonly used to produce superhydrophobic surfaces [203]. As Tsujii et al. [204] first both theoretically and experimentally studied biomimetic superhydrophobic surfaces in 1996, various smart methods for attaining rough surfaces that exhibit superhydrophobicity have been reported, such as LbL deposition [205,206], dip-coating [207], lithography [7,208], CVD [209,210], electrospinning [211–213], sol–gel processing [214,215], plasma treatment [216,217], one-pot reaction [218,219], electrochemical deposition [220,221], and so on. Similarly, we only discuss some of the classic fabrication of superhydrophobic films in this section.

1.3.2.2.1 *Electrospinning*

Electrospinning is a simple and versatile technique for continuous production of micro/nanofibers from polymers [222,223]. The electrospinning process is carried out by a high electric field between the syringe containing viscous polymer solution and the conducting substrate. To form uniform fibers, the molecular weight of the polymer and the concentration of the solution should be carefully considered and adjusted. By electrospinning a mixture of polymer and hydrophobic material or introducing secondary structures, uniform superhydrophobic fibers can be obtained [222,224,225]. Figure 1.17 shows a coaxial electrospinning operation, in which a coaxial nozzle consists of a central tube surrounded by a concentric

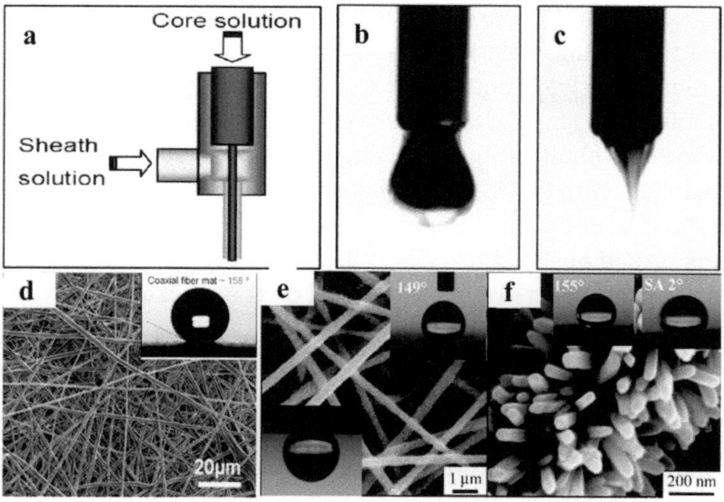

Figure 1.17 Coaxial electrospinning operation: (a) diagram of the coaxial nozzle; (b) core-sheath droplet without bias; and (c) Taylor cone and coaxial jet formation at 12.5 kV; (d) a coaxially electrospun PCL/Teflon AF membrane and contact angle measurement (2 μL water droplet) [226]; Copyright 2009 American Chemical Society. (e) SEM image of fibers prepared using a precursor solution aged for a t_{aging} of 24 h and a $t_{electrospinning}$ of 90 min; insets are contact angles on the membrane; (f) SEM image of titania nanomembrane after hydrothermal treatment for 120 min. Insets are corresponding contact angle and sliding angle (t_{aging} = 24 h and membrane pore size = 0.786 μm) [212]. Copyright 2009 American Chemical Society.

annular tube [226]. Steckl's group produced core-sheath structured nano/microfibers that combine different properties from individual core and sheath materials by coaxial electrospinning. The first Teflon electrospun fibers were reported using coaxial electrospinning with Teflon AF sheath and poly (ε-caprolactone) core materials. Superhydrophobic and oleophobic membranes were successfully produced by these core/sheath fibers. In addition, the combination of nanoparticles and electrospinning process is effective to increase surface area and porosity for superhydrophobic surfaces. In 2009, the He's group [212] reported the preparation of large area superhydrophobic titania membranes by electrospinning, followed by calcination and surface modification with 1H, 1H, 2H, 2H-perfluorooctyltriethoxysilane (POTS). Varied surface structures were fabricated by combination of electrospinning and hydrothermal treatment. It is important to well control precursor aging time (t_{aging}) and electrospinning time ($t_{electrospinning}$). Very interesting, when the titania membrane was obtained by using a $t_{electrospinning}$ of 90 min followed by treatment with POTS, water droplets on the surface appeared spherical and adsorbed firmly onto the substrate even when the substrate was turned upside down (Fig. 1.17e). However, if the titania membrane was subjected to hydrothermal treatment in a $Ti(OBu)_4$ acidic solution for 120 min, the water contact angle became 155° and the sliding angle remarkably changed to 2°. It truly meets the conditions of superhydrophobic self-cleaning. As shown in Fig. 1.17f, nanoparticles grew along the perpendicular direction of fibers and finally became rod-like crystals after 120 min of hydrothermal treatment. The formation of the hierarchical micronanostructure resulted in a thick air layer and kept the water droplet in contact with the fibers.

1.3.2.2.2 *Templating*

Template-based method is an imprint-related way to prepare superhydrophobic surfaces [227]. Templating involves the use of a master with desired features, replication of the features, and subsequent lifting off the replica or dissolution of the templates [200]. Many materials can be used as a template such as natural lotus leaves [228,229], candle soot [230], insect wings [231], reptile skins [231], master prepared by lithographic processes [232], or commercial inorganic membranes [233].

Sheng et al. [234] have demonstrated fabrication of nanostructured superhydrophobic polymeric surfaces by a simple (one-step) reproductive method of anodic aluminum oxide (AAO) template extrusion. By tuning the diameter of the AAO template and the pressure to extrude, high-density polyethylene nanofiber surfaces with different nanometer roughness were prepared. The water contact angle test showed that the surface was superhydrophobic and sliding angles were very important for the dynamic wettability of superhydrophobic surfaces. Park's group [235] also reported a synthetic route for ordering a set of Au nanoparticles on the vertically aligned conducting polymer for superhydrophobic surfaces with low water flow friction. By mimicking the lotus leaves, they combined two scale roughness composed of nanorods and nanoparticles, which is very favorable for the preparation of the superhydrophobic structure.

1.3.2.2.3 *Hydrophobization*

After the surface coating has been roughened, most of them still do not exhibit superhydrophobicity. It is necessary to lower the surface energy of the roughened surfaces. The commonly used reactive molecules for low surface energy modification can be classified mainly into four categories: fluorinated molecules, alkyl molecules, nonfluorinated polymer, and silicon/silane compounds [203].

Due to the extremely low surface free energy and simple reaction of the silane group with the hydroxyl group, fluoroalkylsilanes are the most commonly used chemicals [236,237]. However, most of fluoroalkylsilanes are expensive and potentially harmful to human health and the environment. Therefore, Wang's group used stearic acid to modify TiO_2 or SiO_2 particles on roughened cotton textiles to make them superhydrophobic [238–240]. Polymers with low surface energy can also be used for hydrophobization. Ramaratnam [241] developed a permanent nanocoating to create ultrahydrophobic fibers/textiles with excellent water repellency and self-cleaning ability, which contained 29 wt% styrene and 1.4 wt% reactive maleic anhydride groups on a PET fabric covered with epoxidized silica nanoparticles. The coating exhibited excellent water repellency and self-cleaning ability.

1.3.3 Other Substrates for Antireflective and Self Cleaning Coatings

1.3.3.1 Polymer

Polymer substrates have several advantages over conventional glass or silicon substrates, such as light weight, design freedom, easy processing, mechanical flexibility, and low cost [242,243]. The availability of AR or self-cleaning coatings to increase the transmittance and durability of polymer substrates plays an important role in replacing glass optics in many fields. AR and self-cleaning coatings have been well studied in glass optics for many years, while the coating of polymer surfaces is a fast-growing field at present. Applications include plastic eyeglass lenses, optical lenses for sensors inside cars, photographic equipment of various electronic devices, and covers to protect screens and displays [244]. Substrates can be divided into two categories: rigid substrates and flexible substrates. Fundamental researches were mostly conducted on rigid solid substrates such as silicon wafers, glass, and metal surfaces. Recently, flexible substrates received a great attention for potential industrial applications [245,246]. Polymers, such as PMMA, PC, PET, polycycloolefin, polysulfone, polyethersulfone, and polyamide, are common transparent and flexible substrates [244]. Another example of flexible substrates is fibrous substrates, most of which are woven or nonwoven textiles with natural or synthetic microfibers.

Yun et al. reported the direct coating of a novel AR nanoarray structure of silica nanoparticles on highly flexible polymer substrates by a vacuum coating method using plasma-enhanced CVD [247]. Globular-shaped silica nanoparticles were found to be self-arranged in a periodic pattern on subwavelength scales without the use of artificial assemblies that typically require complicated nanolithography or solution-based nanoparticle fabrication approaches. The transmittance of the substrate increased from 89.4% to 95.8% at a wavelength of 500 nm by coating both sides of the PET substrate with the AR layer, which reduced the specular reflection to 2.6%. The AR characteristics of the coating on the PET substrate exhibited stable behavior even after repeated exposure to extreme bending geometries (Fig. 1.18a,b). Fateh et al. developed a new strategy for the preparation of self-cleaning coatings by depositing highly crystalline, mesoporous TiO_2 films on a PC substrate using a dip-coating process [248]. The self-cleaning process of PC/SiO_2/TiO_2 is shown in Fig. 1.18c. As described by the authors, it is the first report focusing on the synthesis of mesoporous TiO_2 films on PC substrates. The measured photonic efficiency of ζ = 0.078% for these meso-TiO_2 films on PC is among the highest ζ-values reported up to now even when using sodalime glass substrates. The newly prepared photocatalysts films was found about three times higher in the efficiency of the photocatalytic reactions than films prepared from UV-100 or Pilkington Glass Activ. As shown in Fig. 1.18d, the mesoporous films prepared are transparent and homogeneous.

The formation of AR or self-cleaning coatings on polymer substrates is also connected with some problems. First, when polymer substrates come in contact with plasma or with some chemicals, undesired reactions appear sometimes. Second, environmental factors such as UV irradiation and varying humidity is the most a significant threat to the long-term stability of coated polymer substrates, which would result in slow changes of bulk or interface properties of polymers. Third, the permissible post-treatment temperature for polymer is limited, such as PET, which has a glass transition temperature of 80°C. Annealing, heat curing, and the high temperature modification, which are common in the glass film industry, cannot be used for polymer. Therefore, not all the coating technologies, which are well established for glass, can be used for polymer. Further researches are very important and urgent.

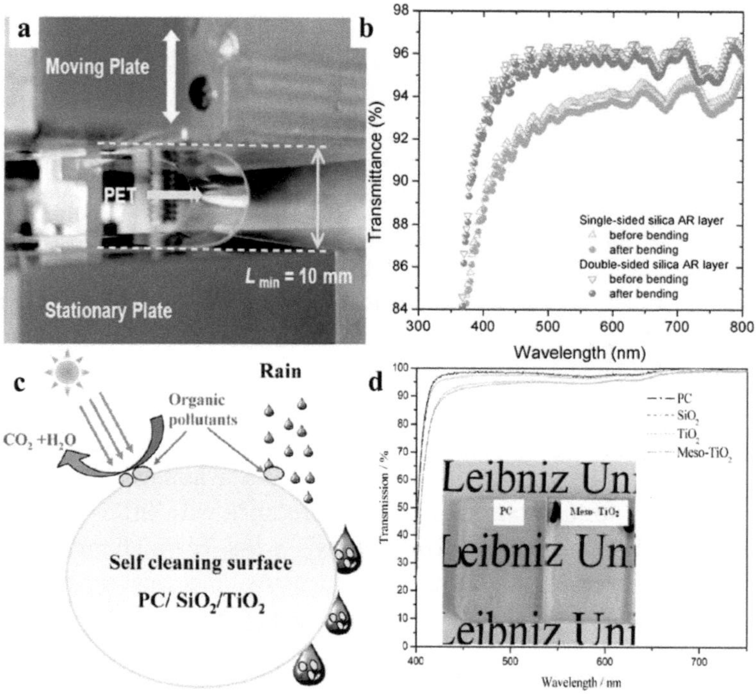

Figure 1.18 (a) Photograph of bending PET thin film with self-cleaning coating; (b) Specular transmission spectra of single-sided and double-sided silica nanoparticle AR layers, consisting of 90 nm thick silica nano-particles for 3 min treatment time, before and after 3000 cycles of curvature-imposed tensile strain [247], Copyright 2012, Royal Society of Chemistry (c) Scheme of the self-cleaning principle of $PC/SiO_2/TiO_2$ coating; (d) Optical transmission spectra of an uncoated PC substrate and a PC substrate precoated with SiO_2, sol-gel TiO_2, and meso-TiO_2. Inset: the digital photograph of a PC substrate (left) and a meso-TiO_2/SiO_2/coating on PC substrate (right) [248] Copyright 2011 American Chemical Society.

To meet the need for ARCs on various flexible transparent polymer substrates in large volume or area applications such as flat panel displays, automotive displays, portable communication systems, or complex optical devices, varied methods have been developed. These methods range from classical ways in the glass industry to some novel methods specifically developed for flexible substrates. Here, we introduce roll-to-roll (R2R) method as an example. When the substrate is a long and flexible sheet, it could be wound on a roll. During printing, the substrate is unwound from the roll and passed through the printing machine, and once through the process, the material is rewound on a roll. The print process could be replaced by heating, drying, UV-curing, and so on. [120] Several researches were focused on the R2R processing of antireflective or self-cleaning coatings on flexible substrates [249–253]. For example, Choi et al. recently fabricated a nanoporous honeycomb polymer film with high throughput and antireflection performance by R2R processable, thermal nanoimprint lithography [249]. Figure 1.19a–c clearly illustrate the whole fabrication process. Contrary to raised structure such as compound eyes of some nocturnal (moths and some butterflies) insects, honeycomb-like structure is defined as negative ARS (N-ARS) (Fig. 1.19d). The film with the N-ARS has a mean transmittance of 95.5%, which is higher than blank PMMA substrate (92%). In addition, the N-ARS in a honeycomb pattern was found to maintain high AR performance with little structural deformation under strong compressive stress. The high mechanical stability endows the film with attractive practical application potential.

1.3.3.2 Metal foil

The important metal engineering materials, such as titanium, aluminum, copper, and alloys, attract more and more attention from researchers around the world for the applications in aerospace, aircraft, national defense, air-conditioner, and so on [66,254–259]. Metal materials, especially the alloy, possess many unique physicochemical properties, such as high-specific strength, excellent heat and electrical conductivities, and low-specific weight. Therefore, fabricating antireflective and self-cleaning coatings on metal materials is of great significance.

Cai et al. reported a facial hydrothermal solution growth of novel mesocrystalline rutile TiO_2 nanorod arrays on Ti foil substrates,

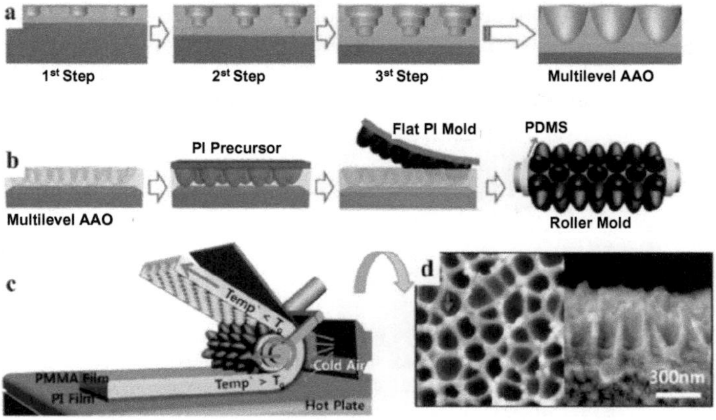

Figure 1.19 (a) Multistep anodization and etching process for the fabrication of multilayer AAO with continuously tapered pores; (b) Schematic process to prepare a PI roller mold. The pores in multilayer AAO were filled with a PI precursor. After thermosetting, a flat PI mold film was detached from the AAO. The PI mold film was wound around a stainless cylinder, which had been covered with a thin PDMS film; (c) Schematic diagram showing the R2R processable, thermal nanoimprint lithography used to prepare the N-ARS on a PMMA surface. (d) Top and side SEM images of the pores in the PMMA AR film. The red arrows in (c) and (d) correspond to the slightly elevated spots at the intersections of neighboring pores [249]. Copyright 2012 Royal Society of Chemistry.

which exhibited efficient decrease of reflection toward wavelengths ranging from visible to the near infrared region at a wide range of incident angles [66]. The 2D arrays of rutile TiO_2 nanorods with a unique tips-on-rod structure led to an optimized graded RI and hence prominent broadband and quasi-omnidirectional AR properties, as shown in Fig. 1.20a. From Fig. 1.20b, we can see the significant decrease of reflection of Ti foils. Moreover, many efforts have been made to fabricate superhydrophobic surfaces on metal surfaces. Wu et al. [259] demonstrated an easy and industry compatible approach to fabricate robust surfaces with super-repellency toward a broad range of liquids, such as hexadecane, silicone oil, and so on. Figure 1.20c(I) shows that the fractal morphology in 3-D was formed with numerous facets that formed a large number of cavities deep in the substrate surface. After anodization etching, very well aligned alumina nanowires were formed on all facets (Fig. 1.20c(II), c(III), c(IV)). Figure 1.20d shows that 1 g of solid paraffin on a super-repellent surface formed globules upon melting above 70°C (below).

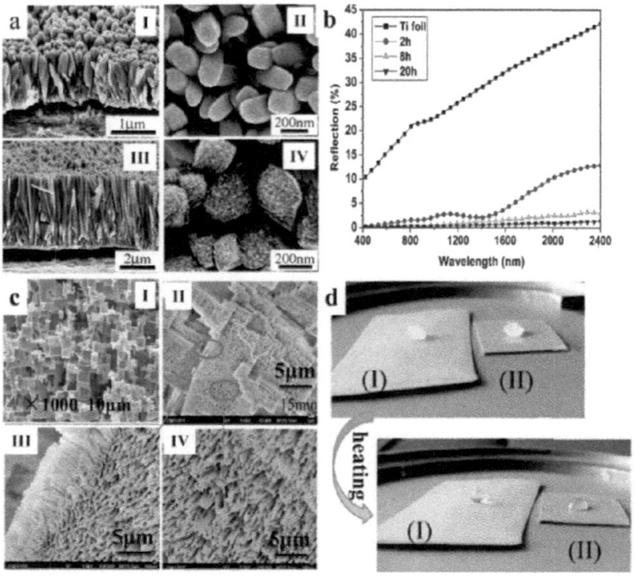

Figure 1.20 (a) Cross-sectional view (a-I and a-III) and top view (a-II and a-IV) SEM images of rutile TiO_2 nanorod arrays grown on Ti foils at 150°C for different times: (a-I and a-II) 2 h and (a-III and a-IV) 8 h. (b) Specular reflectance spectra of rutile TiO_2 nanorod arrays grown on Ti foils for different times at an incident angle of 6° [66], Copyright 2012 Royal Society of Chemistry. (c-I) Multi-facet aluminum by first step anodization, (c-II~c-IV) High field anodic oxidation for 15 min, resulting in nanowire forests on a multi-facet mattress; (d) photographs of low melting point solids (solid paraffin) on a heated super-repellent surface (I) and a control alumina surface (II) [259] Copyright 2009 Royal Society of Chemistry.

1.4 Progress in Antireflective Self-Cleaning Coatings and Other Multifunctional Coatings

1.4.1 Progress in Antireflective Self-cleaning Coatings

In recent decades, surfaces with either antireflective or self-cleaning property have been studied intensively. Scientists have made significant progress and a few excellent reviews have been published to give an integrated introduction [4,23,62,244,260,261]. Coatings with practical significance that combine both antireflective and self-cleaning properties still need, however, vigorous investigation and a consolidated summary. In this section, we introduce the latest typical progress of antireflective self-cleaning coatings in recent 5 years.

Both silicon and glass are the most important materials for solar cells and other optical devices. Unfortunately, due to the high refractive indices, more than 30% of incident light is reflected back from the surface of crystalline silicon and nearly 10% of light is lost from the air–glass interface [262]. In addition to effective improvement of the transmittance and wettability of substrates, low-cost and simplification of the fabrication method are also very important for industrial applications. He's group did lots of work on antireflective self-cleaning coatings [19,100,111,263–268]. The building blocks used for the LbL assembly are from solid SiO_2 nanoparticles of different sizes (Fig. 1.21a) [19], nonspherical mesoporous silica nanoparticles (Fig. 1.21b) [111], and hierarchically mesoporous silica nanoparticles (Fig. 1.21c) [263] to hollow mesoporous silica nanospheres (Fig. 1.21d) [100]. Take the last one as an example, hollow hierarchically mesostructured silica nanoparticles were prepared by a one-step synthesis using dodecanethiol and cetyltrimethylammonium bromide as the dual template, and trimethylbenzene as the swelling agent. Through LbL dip-coating on glass substrates, superhydrophilic and visible/near-IR antireflection coating with appropriate surface roughness was fabricated on a glass substrate (Fig. 1.22a,b). The silica particulate coatings increased the maximum transmittance of slide glass from 90% to 96%, whereas they reduced its minimum reflection from 8% to 2% at the optimized wavelength region that could be adjusted from visible to near-IR with a growing number of deposition cycles (Fig. 1.22c). The inset in Fig. 1.22d shows the shape of a sheet-like water of around 1.2° and the calcined coatings demonstrated excellent superhydrophilicity. In addition, the He's group [267] also took silica nanoparticles of ca. 20 nm in size as a building block for fabricating hierarchically porous silica coatings on PMMA substrates via LbL assembly followed by oxygen plasma treatment. After the fabrication procedure shown in Fig. 1.22e, the porous silica coatings were highly transparent and superhydrophilic. Compared with glass substrates, PMMA has higher transmittance and flexibility. With self-cleaning ARCs decorated on PMMA substrates, they may be used in some special fields wherein glass substrates are not suitable, including lens caps and the walls of green house.

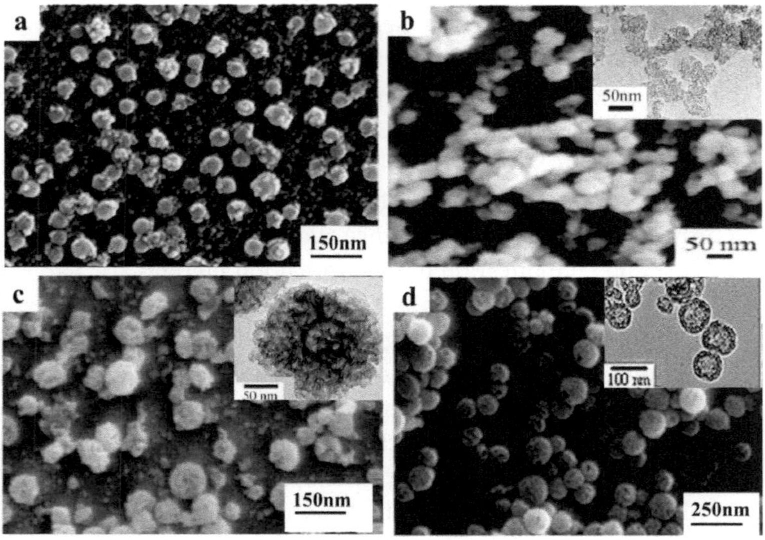

Figure 1.21 SEM images of (a) LbL deposited $(PDDA/S\text{-}150)_1/(PDDA/S\text{-}30)_1$ coating on slide glass [19], Copyright 2009 American Chemical Society. (b) Nonspherical mesoporous silica nanoparticles coating as prepared by deposition of one cycle of PDDA and nonspherical mesoporous silica nanoparticles (PDDA/MPSNP) [111], Copyright 2009 American Chemical Society. (c) Hierarchically mesoporous silica nanoparticles coating on slide glass by alternate LbL deposition of PDDA/MSNs of one cycle [263], Copyright 2009 American Chemical Society. (d) Hollow mesoporous silica nanospheres particulate coating prepared by deposition of one cycle of PDDA/HMSN followed by calcination [100], Copyright 2011 Wiley-VCH.

Recently, Jiang's group developed several simple colloidal templating approaches for fabricating ARCs on both glass and silicon substrates [269–272]. All these approaches are based on a simple and scalable spin-coating technique to produce colloidal crystals with nonclose-packed structures. In 2008, they developed a simple and scalable templating technique for making ARCs on silicon and glass substrates, and investigated the antireflective and nonwetting properties of templated pillar arrays with a high aspect ratio (up to ca. 10) [273]. The outline of the templating procedure for patterning subwavelength pillar arrays on silicon substrate is shown in Fig. 1.23a. They first used spin-coating technique to generate nonclose-packed colloidal layers on silicon wafers. After coating, the wafer and glass exhibited much lower reflection, which were below 2.5% and 0.5%, respectively. Figures 1.23b,c show water drop profiles on fluorosilane-modified silicon and glass pillar arrays, respectively. Both coatings were superhydrophobic and the WCA was ca. 172° for the former and ca. 160° for the latter. They believed that this technique is promising for developing self-cleaning ARCs for a large variety of technological applications and for industrial manufacture.

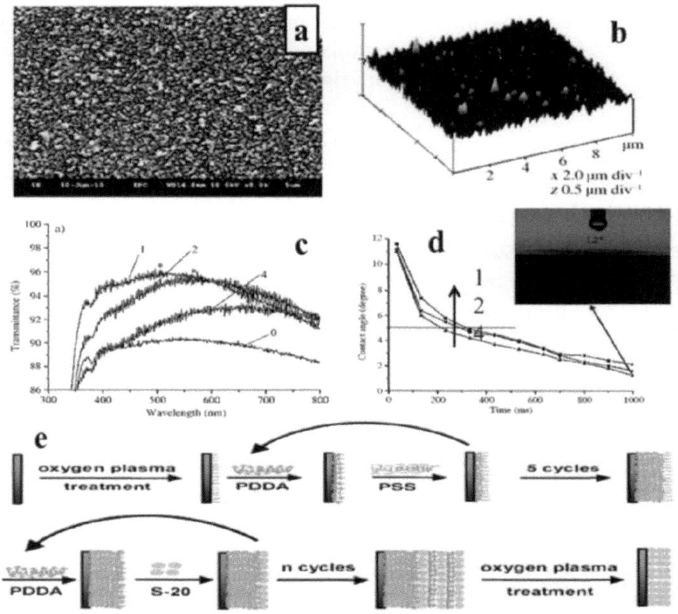

Figure 1.22 (a) SEM image and (b) AFM image of hollow mesoporous silica nanospheres particulate coating prepared by deposition of one cycle of PDDA/HMSN; (c) UV/Vis transmission spectra of control slide glass and slide glasses coated with one to four cycles of PDDA/HMSN [100], Copyright 2011 Wiley-VCH. (d) Time-dependent changes in instant contact angle as a function of the number of deposition cycles. (e) The fabrication procedure of superhydrophilic and antireflective silica coatings on PMMA substrates by layer-by-layer assembly and subsequent oxygen plasma treatment [267], Copyright 2012 Elsevier.

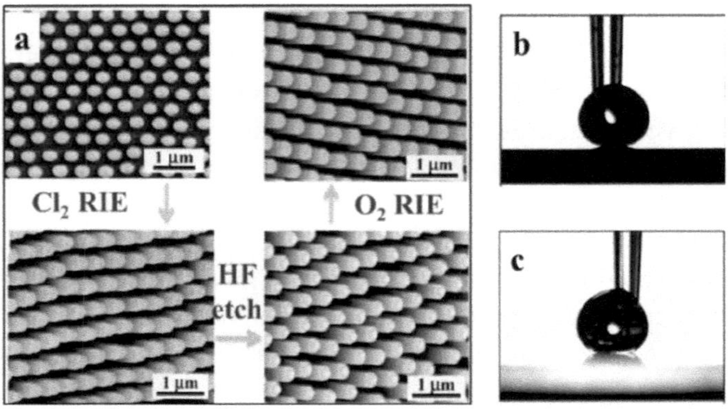

Figure 1.23 (a) Outline of the templating procedure for fabricating antireflective silicon pillar arrays by using nonclose-packed colloidal monolayer as a template. Superhydrophobic coatings achieved on both templated silicon (b) and sol–gel glass pillar arrays (c) [273]. Copyright 2011 Wiley-VCH.

TiO_2-based coatings can be easily applied to transparent substrates such as plastics to provide a self-cleaning function. However, the coatings always increase the surface reflection of the transparent substrate because of the large RI of TiO_2, and the reflection of TiO_2-coated substrates can be as high as 20% [274]. Fujishima et al. [21] fabricated PET films with a moth-eye like surface, which were coated with TiO_2 particles. The films exhibited a high transmittance of 76–95% in the range of 400–800 nm and good superhydrophilicity after irradiation with UV light. They believed that such coated films will be important to improving the performance of solar cells and flat panel displays.

Recently, antireflective and self-cleaning coatings on conductive substrates [such as indium-tin-oxide (ITO) glasses] have gradually attracted the attention of scientists. White organic light-emitting devices (OLEDs), which have many desirable characteristics such as ultrathin, light weight, and good flexibility, have been considered as a new generation of solid-state lighting sources. For the conventional bottom-emitting OLEDs fabricated on a flat glass substrate, however, only a small fraction (<20%) of the light generated in the device can escape and 80% is trapped in the glass substrate (glass mode) and the high index organic layer (waveguide modes). Wu et al. [275] came up with two reference bottom-emitting OLEDs without and with antireflective film, as shown in Figs. 1.21a,b. To improve light out coupling efficiency of conventional OLEDs, they fabricated an antireflective nanopillar arrays film onto the opposite side of the ITO-coated glass substrate by low-cost imprint lithography. Owing to the nanopillar structures and the inherent low-surface energy of PDMS, the film exhibited a high WCA (151°), as shown in Fig. 1.24d. After putting into dirty water, a lot of dust resided on the glass surface, while the nanopillar arrays film still kept clean. The transmittance and SEM characterization are shown in Figs. 1.24c,d,f. The simple, cost-effective, and reproducible method may find great application in illumination and display.

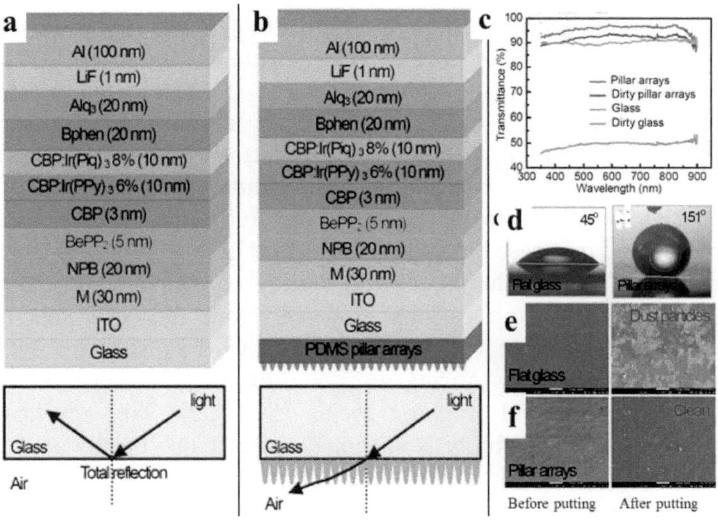

Figure 1.24 (a) External quantum efficiency enhancement of white OLEDs (b) with antireflective film. (c) The optical transmittance before and after being immersed into dirty water. (d) The wetting property of flat glass and pillar arrays. (e) and (f) Comparison of the flat glass and pillar arrays characterized by the SEM images [275]. Copyright 2011 Optical Society of America.

As we know, the enhanced visible light is beneficial to improving the photoelectric conversion efficiency of solar cells. Similarly, antireflection near-infrared optics also has many important applications such as for night vision, sensors, dichroic mirrors, medical detection, body scans, and hot-spot detection [276,277]. Camargo et al. [278] prepared a two-layer film that was composed of multi-scale silica nano-particles and a poly-tetrafluoroethylene (PTFE)-like layer. An additional silica aerogel layer, combined with the silica nanoparticle multi-scale clusters and PTFE-like layer, improves light transmittance further to 99.5% between 1600 and 2100 nm and a WCA of 158±2°.

Recently, Park et al. [279] reported the fabrication of multifunctional surfaces composed of tapered conical nanotextures by interference lithography. They designed an ingenious procedure to precisely, geometrically control the conical shape and slenderness of the features as well as periodicity at the nanoscale. As a result, the desired aspect ratios and periodicities with few defects and large pattern area were achieved. The fabricated nanostructures demonstrated enhanced polarization- independent optical transmission (>98%) over a broad range of bandwidth and incident angles, and superhydrophilicity or robust superhydrophobicity after CVD treatment. However, in our opinion, though the fabricated structure shows good performance, the preparation conditions are too complex and demanding for practical applications. It is meaningful and necessary to further simplify the preparation process.

The performance of antireflective self-cleaning coatings has made good progress in last few years, and is expected to be applied to many fields, including optics, energy conversion and storage, and biomedical applications. However, very few products have been launched using such surfaces due to their susceptibility to mechanical contacts during normal uses, such as finger contact, abrasive wear, and washing cycles [198,280]. Poor mechanical reliability and durability is still one of the major bottlenecks that prevent them from implying in commercial and industrial domains. Some efforts are apparently needed to overcome these drawbacks.

1.4.2 Mechanical Integrity (Durability and Adhesion) of Coatings

Nowadays, the limited mechanical durability has been regarded as a major obstacle to real applications of antireflective self-cleaning coatings [281]. Mechanical wear on the surfaces could destroy the micro/nano-scale roughness features. Meanwhile, mechanical abrasion could also remove the surface layer on substrates, causing a decline in their properties [282]. Thus, enough attention should be paid to seeking a way to mitigate the mechanical damage problem. Fortunately, many methods have been developed to tackle the mechanical damage problem recently.

Many nanoparticle deposition techniques (e.g., dip-coating, spin-coating) add cross-linking agents and monomers into the coating formulation, which need curing steps that eventually lead to extensive interconnectivity and mechanical robustness of the coatings [283]. Whereas for nanoparticle assembly techniques (e.g., LbL, Langmuir Blodgett), the assembly is often driven by Vander Waals forces, electrostatic attraction, hydrogen bonding, and so forth. Such secondary interactions are relatively weak over macroscopic length scales and therefore do not provide sufficient mechanical durability to the entire coating [284]. Auxiliary components (e.g., cross-linking agents) may not be compatible with the assembly technique, and other methods have been attempted, such as infiltration of polymerizable species into preassembled structures [285], CVD [286], and atomic layer deposition (ALD) [287].

Conventional calcination is a post-assembly curing step. Wu et al. [169] used high temperature (550°C) calcination to enhance mechanical durability of thin films. They added calcination in various stages of the assembly process to eliminate all of the polymers from the porous structure and enhance the mechanical robustness. He's group [266] reported a method that combined hydrothermal treatment (at 124°C for 1 h) and annealing treatment (at 700°C for 200 s) to enhance the mechanical durability of LbL ARCs. The significant improvements in mechanical durability were proved by pencil hardness test. After the two post-treatments, 5H pencil could not damage the surface structures on the coated glasses. However, the coating could only resist 4H pencil scratch when only hydrothermal treatment was applied. The specific reason for durability improvement by hydrothermal treatment may be that the steady decline of cations in glass under hydrothermal conditions speeds up the dissolution of adjacent particles, which can enhance the strength of adhesion and abrasion resistance of the coating [288]. Denget al. fabricated superhydrophobic surfaces on the basis of porous core–shell raspberry-like silica particles (Fig. 1.25a) [289]. In order to enhance the mechanical property, the neighboring particles were chemically bound to each other and to the surface via silica bridges by hydrolysis and condensation of TEOS, which were catalyzed by ammonia in a CVD process. The coating exhibited excellent superhydrophobicity even prepared on the solar cell (Fig. 1.25b). The robust surface can withstand the peeling of tape pressed with approximately 10 kPa to the substrate, although almost all particles were removed by peeling the tape off the surface without CVD treatment (Fig. 1.25c). Moreover, sand abrasion test was performed to quantify the mechanical stability of the surface (Fig. 1.25d). After 100–300 μm sized sand grains impacted on the superhydrophobic surface, the shells remained intact for impact heights, *h*, up to 30 cm.

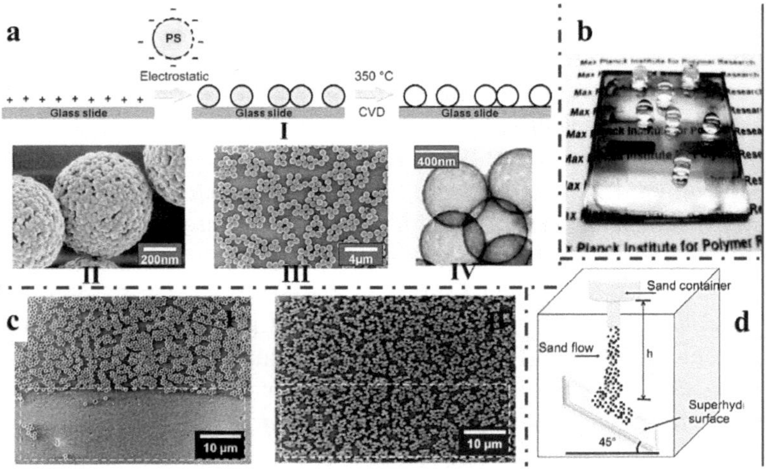

Figure 1.25 (a-I) Sketch of the procedure taken to prepare transparent superhydrophobic surfaces. (a-II) SEM image of PS particles coated with nanometer-sized silica spheres. (a-III) After dip-coating, the glass substrate is partially covered with raspberry-like particles. (a-IV) TEM image taken after the PS core has been removed by calcination. Despite the high porosity (white patches) of the silica shell (dark rim), it remains intact. (b) Water droplets deposited on the solar cell. (c) SEM images of superhydrophobic surfaces that were partially exposed to double-sided tape (white boxes indicate the exposed areas). (c-I) If the particles stick to the surface by van der Waals interaction only, they can easily be removed. (c-II) Particles cannot be removed by double-sided tape after binding them chemically to the surface by silica bridges. (d) Sketch of the setup used to determine the stability of the surface against sand impact [289]. Copyright 2011 Wiley-VCH.

Miguez et al. [286] used CVD to adsorb and hydrolyze $SiCl_4$ monolayers on stacks of silica microspheres, and discovered that the adsorption/hydrolysis cycles were sufficient to strengthen the nanoparticle architecture. However, it also has some disadvantages. For example, some surface functionalities (e.g., catalytic) may be annulled by encapsulation within inert SiO_2 shells, and the CVD processing is complicated due to the requirement of a perfectly anhydrous environment and the extreme reactivity of $SiCl_4$. Dafinone et al. [287] explored how ALD method reinforces nanoparticle thin films on inorganic and organic substrates and affects wetting and optical properties of all-nanoparticle coatings on both inorganic and organic substrates. ALD is a self-limiting gas-phase nanofabrication method for growing atomic scale thin films of metal oxides and metals [290]. They demonstrated that ALD is a versatile method for imparting mechanical durability without drastically changing the structure and functional properties of the coatings. The abrasion test demonstrated that the mechanical durability of TiO_2/SiO_2 nanoparticle LbL films on glass and polycarbonate substrates can be drastically improved by ALD treatment.

As we all know, antireflective or self-cleaning structures in nature such as moth's eyes, plant leaves, and insect wings can sustain their superior performance over their whole lifetime. Instead of relying on extremely tough surfaces, their ability to withstand damage is typically based on continuous repairing of the surfaces as a result of biological growth processes [280], for example, lotus leaves maintain their non-wettability by regenerating the epicuticular wax layer after it is damaged [291]. Therefore, mimicking the self-healing functionality in artificial surfaces has become another interesting prospect. Meanwhile, scientists also endowed prepared coatings with many innovative features such as antimicrobial, anti-icing, and oleophobic to make them more suitable for outdoor applications [230,231,292,293] We call these coatings as smart coatings in this review and will give a brief introduction below.

1.4.3 Smart Coatings

Recently, the fabrication of multifunctional ARCs has attracted much attention. The assembly of multiple nanomaterials into a thin film enables the generation of multifunctional structures with synergistic properties.

1.4.3.1 Self-healing coatings

The function of ARCs or self-cleaning coatings may deteriorate because of adsorption of humidity [111] or contaminants, and could be recovered by exposure to sunlight [294]. ARCs or self-cleaning coatings may also be damaged or even scratched away by sand in the wind or decomposed under sunlight when exposed to an outdoor environment. These situations would lead to the decline of transmittance and change of wettability. Generally, coatings would have to be redeposited and recovered perpetually, which is expensive and inconvenient to accomplish. Therefore, fabricating coatings endowed with self-healing ability is believed to provide an efficient and economical way to maintain optical and other surface physical/chemical properties of coatings for practical applications. Natural plants like lotus leaf can maintain their superhydrophobicity very well by regenerating the epicuticular wax layer after they are damaged in the environment [291]. Although several types of self-healing coatings have been fabricated [131,282,295,296], obtaining excellent self-healing function without changing the surface properties is still confronted with tremendous challenges.

Recently, Sun et al. [282] reported a method to fabricate self-healing superhydrophobic coatings by preserving healing agents consisting of reacted fluoroalkylsilane in porous coatings. Their strategy is shown in Fig. 1.26a. The key step was the fabrication of porous polymer coatings that were rigidly flexible with micro/nano-scaled hierarchical structures. The coatings became superhydrophobic after CVD of a fluoroalkylsilane. These superhydrophobic coatings preserved a large number of reacted fluoroalkylsilane moieties as healing agent. They proposed a self-healing process as follows. The superhydrophobic coating became superhydrophilic and absorbed water in a comparatively humid environment after the decomposition of surface-reacted POTS layer by O_2-plasma treatment. Through rearrangement of polyelectrolyte chains, the covalently attached hydrophobic POTS molecules underneath the damaged surface migrated to the outer surface to minimize the free energy of the interface between the coating and the surrounding air. Simultaneously, the oxygen-containing hydrophilic groups became buried inside the coating. The migrated fluoroalkyl chains, like the wax secreted by plants, healed the impaired superhydrophobicity of the coating.

Men and co-workers [131] demonstrated a simple technique for fabricating superhydrophobic coatings with regenerability via spraying (Fig. 1.26b). After immersed in potassium stearate solution, the as-prepared poly(furfuryl alcohol)/calcium hydroxide (PFA/

$Ca(OH)_2$) composite coatings became superhydrophobic. When the surfaces were destroyed, they could be restored completely after being modified in potassium stearate solution again. Because the spraying is a fairly facile and commercially available method, the process has potential in the preparation of a large area hydrophobic surface.

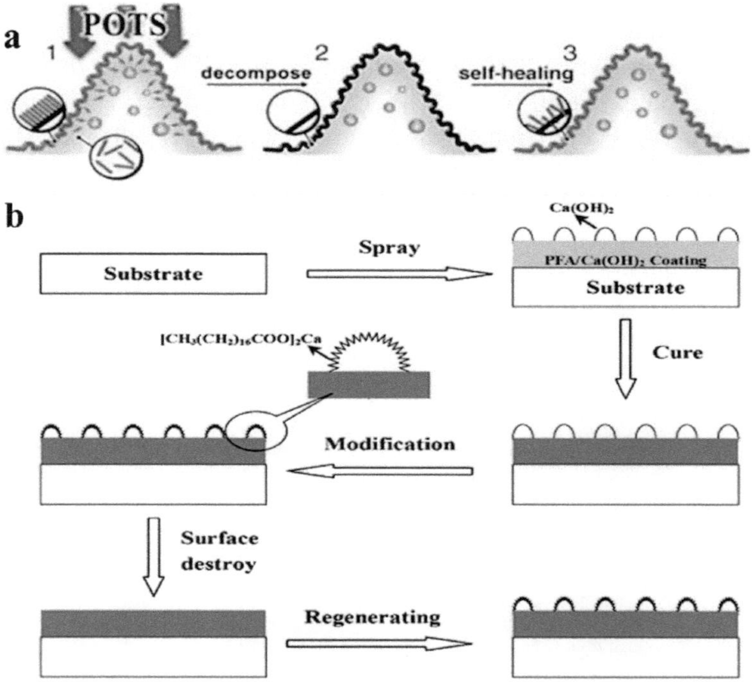

Figure 1.26 (a) Working mechanism of self-healing superhydrophobic coatings: (1) the porous polymer coating with microscaled and nanoscaled hierarchical structures can preserve an abundance of healing agent units of reacted fluoroalkylsilane; (2) the top fluoroalkylsilane layer is decomposed and the coating loses its superhydrophobicity; (3) the preserved healing agents can migrate to the coating surface and heal the superhydrophobicity [282]. Copyright 2010 Wiley-VCH. (b) Schematic illustration of the fabrication process and regeneration of the superhydrophobic coating [131]. Copyright 2011 Royal Society of Chemistry.

Very recently, Won and co-workers [297] reported a strategy to create a slippery liquid-infused porous surface (SLIPS), which has liquid and ice-repellency, pressure stability, enhanced optical transparency, and self-healing properties. They used nano/microstructured substrates to lock in place the infused lubricating oil, which was inspired by Nepenthespitcher plants. The film could rapidly restore the liquid-repellent function following damage of the porous material by abrasion or impact. Surface energy driven capillary action due to the mobile nature of liquid led to the physical voids refilled spontaneously. They observed that the self-recovery time for a 50-mm fluid displacement of the FC-70 lubricating layer on an epoxy-resin based SLIPS is 150 ms (Fig. 1.27a). And more impressively, it can repeatedly restore their liquid-repellent function after large-area physical damage (Fig. 1.27b).

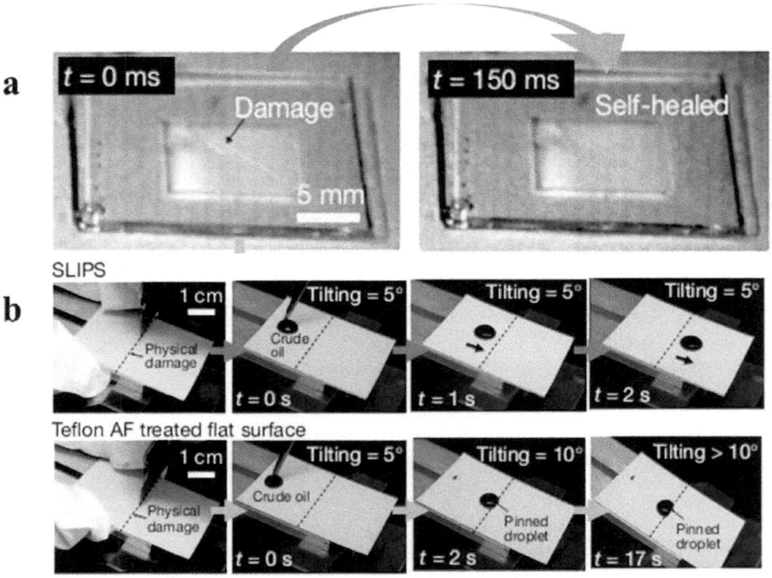

Figure 1.27 Self-healing of SLIPS. (a) Time-lapse images showing the capability of a SLIPS to self-heal from physical damage, 50 mm wide on a timescale of the order of 100 ms. (b) Time-lapse images showing the restoration of liquid repellency of a SLIPS after physical damage, as compared with a typical hydrophobic flat surface (coated with DuPont Teflon AF amorphous fluoropolymers) on which oil remains pinned at the damage site [297]. Copyright 2011 Nature.

1.4.3.2 Antimicrobial coatings

In recent years, antimicrobial coatings have attracted increasing attention for preventing the spread of infection and bacteria. Varied materials were used for this purpose, such as silver, copper, and metal oxides of photocatalytic effects [298]. These substances combined with ARCs make the coatings multifunctional and intelligent to adapt to extended practical applications.

Different antimicrobial additives on surfaces have different reaction mechanisms, as shown in Fig. 1.28. Silver ions, which can interact with thiol groups in proteins and lead to the inactivation of respiratory enzymes and the production of reactive oxygen species (ROS), can prevent DNA replication and affect the structure and permeability of the cell membrane [298,299]. The antimicrobial activity of TiO_2 is also related to the production of ROS, which formed under the UV irradiation via oxidative and reductive pathways, respectively [301]. Photocatalytic treatment gradually increases the cell permeability and subsequently allows free efflux of constituents in the cells. Free TiO_2 particles may also access damaged cells and subsequently directly attack intracellular components, leading to the alteration of protein structure [48]. However, microbe death also occurs in the dark, indicating that other unknown mechanisms may be involved [302]. In the case of ZnO-containing surfaces, except for the production of ROS under illumination with light whose energy is higher than the material bandgap, release of Zn^{+2} ions may also contribute to the antimicrobial activity [300].

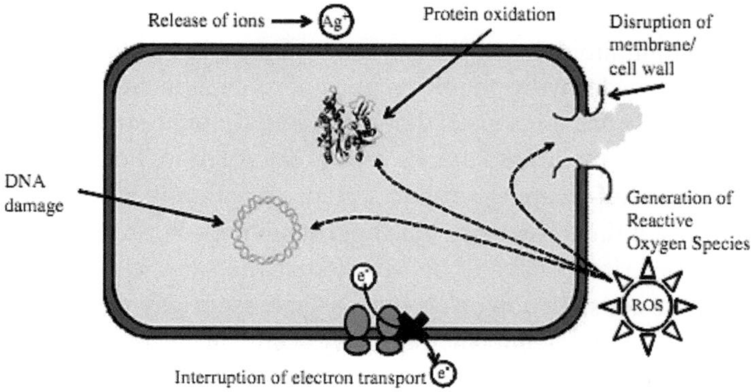

Figure 1.28 Varied mechanisms of antimicrobial activities exerted by nanomaterials [300]. Copyright 2008 Elsevier.

Recently, Treme's group [303] demonstrated a method of fabricating homogeneous, self-cleaning, antimicrobial, transparent, and durable SnO_2 films on glass. Polyamine (spermine) was first immobilized on glass surfaces, and the modified glass was subsequently incubated with hexafluorostannate to form the SnO_2 film. The spermine-mediated SnO_2 coating can form superoxide radicals ($O_2^{\bullet -}$) upon exposure to sunlight. Due to the intrinsic photocatalytic activity, the SnO_2-coated surfaces can degrade organic contaminants such as Rhoda-mine B and act as efficient photobiocides toward bacterial pathogens such as *Escherichia coli*. According to the authors, it is the first time that SnO_2 is reported to have photobiocidal properties. The photocatalytic properties of SnO_2 bring the antimicrobial and self-cleaning functions to the surface, which is helpful to maintain the cleanliness of the substrate and to improve the lifetime of the material. Although the coatings are endowed with good antimicrobial properties, it is quite essential to maintain high transparency, desired wettability, and good environmental stability. In these aspects, choosing appropriate substrate and designing optimized reaction route deserve considerable attention.

1.4.3.3 Superamphiphobic coatings

Superhydrophobic surfaces, which are inspired by some natural phenomena such as lotus leaves and water striders' legs, have extensive applications in self-cleaning, anti-corrosion, improving blood compatibility, and so on. [304–306]. Simultaneously, superoleophobicity is attracting growing attention owing to its great potential in many practical applications. Nevertheless, there are only a few studies on superamphiphobicity, although many papers have been published on either superhydrophobicity or superoleophobicity. It is an increasingly broad consensus that superamphiphobic surfaces are desirable for many areas such as green construction, liquid-repellent materials, self-cleaning products, and microfluidic systems [307]. It is more difficult to construct superoleophobic surfaces, for which two basic criteria have to be satisfied at the same time [308,309]. One is the construction of a rough surface, especially with the structures including "re-entrant geometry", "convex curvature," or "surface overhanging," and so on to prevent air bubbles from escaping from the rough surface [308–311]. The other is the construction of a low energy surface with a surface tension lower than a quarter of that of oils, which is often achieved by using a fluorine-based compound [312,313]. In this section, recent progress of superamphiphobic surfaces fabrication by varied methods is introduced.

Among various multifunctional surfaces, electrically conductive superhydrophobic coatings, which are mainly prepared by electrochemical polymerization methods with fluorinated monomers, are of particular interest owing to their potential applications as electromagnetic interference shielding materials [223,224]. Recently, Wang et al. [314] reported a one-step method to prepare electrically conductive superamphiphobic coatings on fabrics through a vapor-phase polymerization of pyrrole in the presence of a fluorinated alkyl silane (FAS). They applied a $FeCl_3$/FAS–ethanol solution onto the fabric substrate using a dip-coating method, and then put the coated fabric into a pyrrole saturated nitrogen atmosphere to carry out the polymerization reaction. After the treatment, the fabric turned black, indicating the formation of polypyrrole (PPy). The contact angle (CA) measurement indicated that the PPy–FAS coated polyester fabric had a WCA of $165° \pm 2°$ and a hexadecane CA of $154° \pm 2°$. By combining the PPy–FAS treatment with a screen printing technique, the patterns can be used to form a circuit for driving electronic devices. Fabric is a common substrate for fabrication of superamphiphobic surfaces. In addition, Pereira and others [315] demonstrated preparation of the functional textiles by a one-pot co-condensation methodology between tetraethylorthosilicate and tridecafluorooctyltriethoxysilane (F13), with hexadecyltrimethylammonium chloride (CTAC) as the template and triethanolamine as the base. The increase in the organosilane proportion during the co-condensation process led to a decrease of surface silanol groups, and an enrichment of the surface with fluorocarbon moieties, which endowed the textile with superamphiphobicity. Another instance of superamphiphobic surfaces is a coated superamphiphobic paper that was made by Xiong et al [316]. They designed a diblock copolymer poly[3-(triisopropyloxysilyl)propyl methacrylate]-block-poly[2- (perfluorooctyl)-ethyl methacrylate] (P1). P1 was then used to coat silica particles, which were then deposited onto microscope coverslips or printing paper. The paper turned to be superamphiphobic and have substantial resistance when etching by aqueous NaOH solution.

Although some of the preparation methods have made impressive progresses, a general, inexpensive, and operationally convenient method is still a challenge, which would be very advantageous for large-scale industrial production. Jiang's group [317] reported an electrodeposition system for the fabrication of superhydrophobic surfaces on a series of substrates such as copper, titanium, iron, zinc, aluminum, and stannum. Furthermore, this method is easily expanded to fabricate superamphiphobic surfaces when the electrolyte is changed to a low surface energy material. For instance, if the electrolyte is changed to an ethanol solution of nonadecafluorodecanoic acid and the applied volts dc are adjusted, superamphiphobic surfaces can be fabricated on copper substrate with the CAs for water and salad oil being 161.3° and 142.8°, respectively.

The transparency of superamphiphobic surfaces is another research challenge, which is crucial for many outdoor applications, such as solar cells, aircraft canopies and engineering shields, and so on. High surface roughness and good light transmission need an effective balance, because light scattering will be increased when visible light passes through materials with high roughness, directly resulting in reduced transparency. Accordingly, the fabrication of transparent superamphiphobic surfaces needs a very smart control over the surface roughness. He et al. [318] reported an effective method to fabricate a transparent superamphiphobic coating with superhydrophobicity and near-superoleophobicity. The starting materials are PDMS and SiO_2 and the coating was sintered to degrade the PDMS into SiO_2 before it was further oxidized into silanol (Si–OH). Then, the coating was treated with 1H,1H,2H,2H-Perfluorooctyltrichlorosilane (PFTS). The as-treated coating showed transparency, with a water CA of 152.7 ± 2.1° and a diiodomethane CA of 140.7 ± 3.2°. However the transmittance between 400 and 600 nm had a relatively larger reduction, which still needs further studies. Poly (methyl methacrylate) (PMMA) is another significant transparent substrate. In the study by Gogolides et al. [319], an ordered, hierarchical (triple-scale), and superamphiphobic surface on PMMA was fabricated using PS microparticle colloidal lithography, followed by fluorocarbon plasma deposition. After a two-step etching process, an excellent re-entrant topography and good water/oil repellence were obtained, as shown in Figs. 1.29a,b. Lately, Deng and others [230] demonstrated a very creative, simple, and interesting work to make robust, transparent, superamphiphobic coatings. First, they held the glass above a flame of paraffin candle to deposit a soot layer (Fig. 1.29c). The soot, which formed a loose, fractal-like network consisting of carbon particles with a typical diameter of 30–40 nm (Fig. 1.29d), was used as a template to coat with a silica shell making use of CVD of tetraethoxysilane (TES) catalyzed by ammonia. Then, they calcinated the hybrid carbon/silica network at 600°C in air to combust the carbon core (Fig. 1.29e) but keep coating's roughness and network texture. Finally, the hydrophilic silica shells were coated with semi-fluorinated silane by CVD to reduce the surface energy. The as-treated coating showed a water CA of 165° ± 1° and CA of 154° for tetradecane, 162° for diiodomethane (Fig. 1.29f). The surface could keep its superamphiphobicity after annealing at 400°C, showing that it could be applied to a variety of heat-resistant surfaces, such as aluminum, copper, or stainless steel. The superamphiphobic coating showed thermal stability and mechanical robustness by annealing test, water drop impact, and sand abrasion tests. Nevertheless, the transmittance of the coated glass (3 μm thick coating) was reduced by about 10% compared with pristine glass at a wavelength of ca. 500 nm. There are still big challenges to prepare antireflective superamphiphobic coatings, for which almost no relevant results have been reported to date. Very recently, He's group [320] fabricated highly antireflective superamphiphobic coatings on glass substrates, which consisted of three building blocks of different structures (solid silica nanoparticle, hollow silica nanosphere, and mesoporous silica nanosheet). The maximum light transmittance of the coated glass was as high as 96.1% at 530 nm and the CA was 171° for water, 157° for diiodomethane, and 156° for ethylene glycol, respectively. It was the first time of reporting the design and fabrication of coatings with both broadband antireflective and superamphiphobic properties via layer-by-layer assembly on glass. The current results may be promising in a variety of practical applications, such as low-resistance oil pipelines and environmental monitors for off-shore drilling platforms. However, for practical applications, efforts are still needed to improve the mechanical property and durability of coatings.

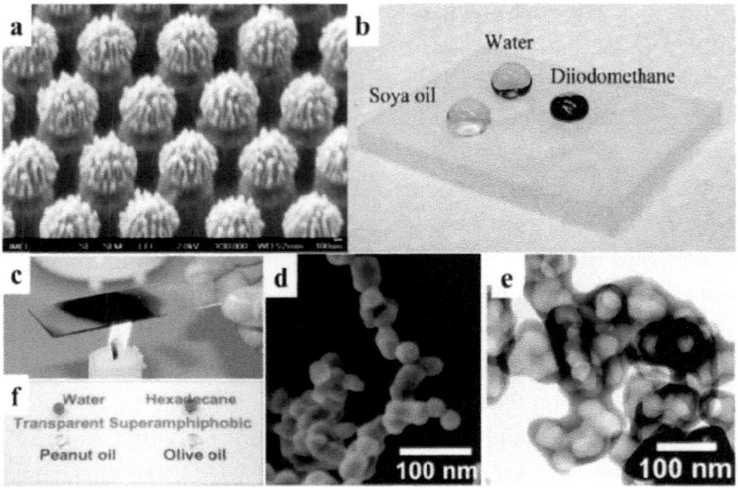

Figure 1.29 (a) Hierarchical (triple-scale) roughness micropillars produced by the combination of colloidal lithography of 1 μm PS particles followed by plasma etching, and dual-scale nanoroughness on top of the pillars formed during plasma etching. (b) Fluorocarbon-coated, oxygen plasma treated superamphiphobic surface, liquid drop volume: 10 μl [319] Copyright 2011, American Chemical Society. (c) Photograph depicting sample preparation. A glass slide is held in the flame of a candle until a few μm thick soot layer is deposited. (d) High-resolution SEM image showing a single particle chain made up of almost spherical carbon beads of 40 ± 10 nm diameter. (e) High resolution transmission electron microscopy image of a cluster after calcination, revealing the silica coating with holes that were previously filled with carbon particles. The silica shell is 20 ± 5 nm thick. (f) Photograph of dyed water (γ_{lv}= 72.1 mN/m; blue), peanut oil (γ_{lv} = 34.5 mN/m; white), olive oil (γ_{lv} = 32.0 mN/m; yellow), and dyed hexadecane drop (γ_{lv} = 27.5 mN/m; red) deposited on a superamphiphobic glass slide. The coated slide was placed on a labeled paper [230]. Copyright 2012 American Association for the Advancement of Science.

1.5 Applications

In the last years, multi-functional coating surfaces have attracted a lot of interest because the additional functions can improve the performance of optical devices. Antireflective and self-cleaning coatings have received a tremendous attention. Antireflective self-cleaning coatings can be employed in a variety of applications, not only optical and electro-optical systems in telecommunications, glass lenses, eyeglasses, medicine, military products, but also lasers, mirrors, solar cells, IR diodes, multipurpose broad and narrow band-pass filters, architectural and automotive glasses, and varied kinds of display [321]. For instance, multi-functional coatings with antireflective, self-cleaning, and antibacterial properties can effectively improve the efficiency of solar cells outdoor, which have a wide range of applications in street lighting, electric vehicle charging stations, solar-powered mobile phone, housing supply system, space shuttle and satellite, and so on, as shown in Fig. 1.30.

Figure 1.30 Potential applications of multifunctional coatings for solar cell.

Solar cells are solid-state electrical devices that convert the energy of sunlight directly into electricity by the photovoltaic effect. Crystalline silicon is the most important material for solar cells. However, a common problem is the high RI of doped silicon and more than 30% of incident light is reflected back from the surface of crystalline silicon [271]. The reflection loss at surface or interfaces is a disadvantage, which negatively affects the energy conversion efficiency of solar cells. For these reasons, scientists devised a number of strategies for reducing the reflection of sunlight on the surface of solar cells, such as appending coatings with intermediate RI [322], embedding metallic nanoparticles to induce surface plasmonic scattering [323], micro-texturing [324], and absorption nanostructures with photonic crystals [325]. Bernhard discovered the moth cornea, which has uniformly arrayed conical protuberances in 1967. This structure was deemed having potential to prevent surface reflection for the whole solar radiation spectrum [326]. After that, various films of inorganic, polymeric, and inorganic/polymer hybrid materials have been fabricated by using diverse techniques. Some of them are assembled on solar cells to carry out detailed performance tests. Simultaneously, several compounds have been used as ARCs on crystalline silicon solar cells such as ITO, SnO_2, SiO_2, MgF_2, ZnS, and ZnO [81]. The coatings can raise the efficiency of silicon-based PV devices by reducing the light energy lost due to reflection.

Nowadays, quarter wavelength silicon nitride (SiNx) films deposited by PECVD are the industrial standard for ARCs on crystalline silicon substrates [273]. This method offers high uniformity, high precision, and high rates of deposition, but on the other side, it is a high cost chemical process that needs vacuum chambers. Other methods such as dip-coating, spray-coating, LbL, and so on offer alternative approaches to low cost and high efficiency fabrication of ARCs. The controllability of thickness, RI, and uniformity in these methods is still, however, a challenge and needs more deep research and development.

Glass is one of the most important optical substrates. Although the reflective loss is ca. 4% from each air–glass interface, which is not as serious as that of silicon, it could still degrade the performance of optical devices especially when multiple components are involved [99,273]. The use of glass with ARCs as cover for solar thermal collectors has been proved increasing the efficiency of the system. For example, calculations have shown that the yearly energy output in a solar heating plant, wherein the temperature of the solar collector fluid is 100°C, can be increased by about 20% by using ARCs on the solar glass cover. Generally, the material used as ARCs on glass is silicon dioxide [327]. However, a drawback limiting the use of ARCs is the adhesion of grime, which reduces transmittance constantly and significantly. For staying clean, outdoor packaging glasses usually mimic the lotus effect wherein dust and water cannot stick to the nano-structured surface. In other words, the ARCs are often endowed with superhydrophilic or superhydrophobic properties to prevent the accumulation of dust on modules that can reduce net efficiency by 32–40% over the first 8-month period and up to 71.8% over a period of 15 years [5,328, 329]. In addition, TiO_2 coatings are also used on solar cells for their photoinduced superhydrophilicity and photocatalytic property, which are very suitable in the humid environments. The self-cleaning properties originate from a combination of the photocatalytic oxidative decomposition of organic contaminants and superhydrophilicity, which causes water droplets to spread on the TiO_2 surface, aiding the cleaning process. Whereas, it was also observed that most of photocatalytic effect was lost within 5.5 years of outdoor exposure [330]. For dry environments, the glass itself should have nano-structured surface so that the dust does not settle on the surface, and water is allowed to spread. Therefore, the combination of both self-cleaning and antireflective properties, which has a strong durability, is very desirable for applications in photovoltaic devices.

There are many other materials that can be used to prepare ARCs. In Chen's articles, the structure of solar selective surfaces is classified into seven different categories, that is, intrinsic selective materials, semiconductor-metal tandems, multilayer absorbers, textured surface absorbers, transition metal oxides, carbon nanoparticles embedded in ZnO and NiO, and metal–dielectric composites [331]. Herein, ZnO single crystal nanorods or nanowires, which are cheap, stable, environmentally-friendly, and nontoxic, have captured the attention of many researchers and show good prospects for optoelectronic applications because of their large exciton energy of 60 meV [332].

1.6 Conclusion and Outlook

The recent developments in typical fabrication routes and applications of the ARCs with self-cleaning property have been described in this review. These surfaces have attracted significant attention within the scientific community over the last few decades because of their unique transparency-enhancing, self-cleaning properties, and their potential for practical applications.

Although varied strategies have been developed to fabricate these functional surfaces, most of them are limited to laboratory research and not suitable for industrial-scale production. Several challenges still remain for practical applications. First, the ARCs are commonly susceptible to mechanical contact during normal use, such as finger contact, abrasive wear, washing cycles, and sand impact [333]. Mechanical wear on the surfaces could destroy the microscopic roughness features, or even remove the coatings from the substrates, finally causing a decline or complete loss of their optical property and wettability. Second, most of the reported preparation techniques involve multistep procedures or harsh conditions, or required special reagents and equipment. Many of those methods are expensive and only applicable to small flat surfaces or substrates of specific materials [203,334]. The limited durability and the high costs of coated surfaces are the two main barriers to the real applications of ARCs.

Most of the current works are still focused on increasing light transmission and wettability. In the near future, the following research directions should be given more consideration. First, by combining theoretical simulation with modern analytical technologies and tools, further exploration and explanation should be conducted to reveal the mechanism of fabrication and the relationship among composition, structure, and property of surfaces [335]. Second, fabricating ARC surfaces of high mechanical strength and abrasion resistance by cost-effective techniques is the key point in future works. It is significantly meaningful for large-area industrial production. Third, environmental issues should be taken into account when preparing ARC surfaces. Not only fly ash but also water slurry, oil may fall on surfaces, and become contaminated species [309]. Organic contaminants may also be adsorbed on the surfaces and change their optical and wetting properties. Therefore, multifunction surfaces with either enhanced durability or self-healing features are preferable for daily use. Fourth, interdisciplinary cooperation is necessary for researchers in the area of science and engineering. The increasing collaboration would be useful for identifying additional challenges, understanding fabrication mechanism, and improving durability and abrasion resistance to adapt to practical applications. Fifth, novel materials and coating techniques should be researched, which would expand the range of application.

In conclusion, further fundamental investigations are still necessary to develop a simple, time, and cost-effective approach to achieve durable antireflective surfaces possessing sufficient self-cleaning and mechanical stability in future work. It will be beneficial for practical applications in large areas.

Acknowledgments

This work was supported by the National High Technology Research and Development Program ("863" Program) of China (Grant No. 2011AA050525), the Knowledge Innovation Program of the Chinese Academy of Sciences (CAS) (Grant Nos. KGCX2-YW-370 and KGCX2-EW-304-2), and the "Hundred Talents Program" of CAS.

References

1. Stavenga D, Foletti S, Palasantzas G, Arikawa K. Light on the moth-eye corneal nipple array of butterflies. *Proc R Soc B*. 2006;273:661–667.
2. Parker AR, Townley HE. Biomimetics of photonic nanostructures. *Nat Nanotechnol*. 2007;2:347–353.
3. Sun T, Feng L, Gao X, Jiang L. Bioinspired surfaces with special wettability. *Accounts Chem Res*. 2005;38:644–652.
4. Yao L, He J. Recent progress in antireflection and self-cleaning technology – from surface engineering to functional surfaces. *Prog Mater Sci*. 2014;61:94–143.
5. Verma L, Sakhuja M, Son J, Danner A, Yang H, Zeng H, et al. Self-cleaning and antireflective packaging glass for solar modules. *Renew Energ*. 2011;36:2489–2493.
6. Wan DH, Chen HL, Tseng TC, Fang CY, Lai YS, Yeh FY. Antireflective nanoparticle arrays enhance the efficiency of silicon solar cells. *Adv Funct Mater*. 2010;20:3064–3075.
7. Zhang JH, Yang B. Patterning colloidal crystals and nanostructure arrays by soft lithography. *Adv Funct Mater*. 2010;20:3411–3424.
8. Olifierczuk M, Zielinski J. Significance of reflection reduction in a TN display for colour visualisation. *Opto-electron Rev*. 2002;10:35–38.
9. Tseng PC, Tsai MA, Yu PC, Kuo HC. Antireflection and light trapping of subwavelength surface structures formed by colloidal lithography on thin film solar cells. *Prog Photovoltaics*. 2011;20:135–142.

10. Chattopadhyay S, Huang Y, Jen Y, Ganguly A, Chen K, Chen L. Antireflecting and photonic nanostructures. *Mat Sci Eng R: Rep.* 2010;69:1–35.
11. Xiong J, Das SN, Shin B, Kar JP, Choi JH, Myoung JM. Biomimetic hierarchical ZnO structure with superhydrophobic and antireflective properties. *J Colloid Interf Sci.* 2010;350:344–347.
12. Wang YJ, Hu FR, Kanamori Y, Wu T, Hane K. Large area, freestanding GaN nanocolumn membrane with bottom subwavelength nanostructure. *Opt Express.* 2010;18:5504–5511.
13. Yang B, Zhang XM, Zhang JH, Ren ZY, Li X, Zhang X, et al. Morphology and wettability control of silicon cone arrays using colloidal lithography. *Langmuir.* 2009;25:7375–7382.
14. Liu ZY, Zhang XT, Murakami T, Fujishima A. Sol-gel SiO2/TiO2 bilayer films with self-cleaning and antireflection properties. *Sol Energ Mat Sol C.* 2008;92:1434–1438.
15. Kuo CY, Chen YY, Lu SY. A facile route to create surface porous polymer films via phase separation for antireflection applications. *Acs Appl Mater Inter.* 2008;1:72–75.
16. Zhu W, Feng X, Feng L, Jiang L. UV-Manipulated wettability between superhydrophobicity and superhydrophilicity on a transparent and conductive SnO2 nanorod film. *Chem Commun.* 2006:2753–2755.
17. Li L, Li Y, Gao S, Koshizaki N. Ordered Co3O4 hierarchical nanorod arrays: tunable superhydrophilicity without UV irradiation and transition to superhydrophobicity. *J Mater Chem.* 2009;19:8366–8371.
18. Lee YJ, Ruby DS, Peters DW, McKenzie BB, Hsu JWP. ZnO nanostructures as efficient antireflection layers in solar cells. *Nano Lett.* 2008;8:1501–1505.
19. Liu X, He J. Superhydrophilic and antireflective properties of silica nanoparticle coatings fabricated via layer-by-layer assembly and postcalcination. *J Phys Chem C.* 2009;113:148–152.
20. Lu N, Wang YD, Xu HB, Shi G, Xu MJ, Lin XW, et al. Biomimetic corrugated silicon nanocone arrays for self-cleaning antireflection coatings. *Nano Res.* 2010;3:520–527.
21. Nakata KN, K., Sakai M, Ochiai T, Murakami T, Takagi K, Fujishima A. Antireflection and self-cleaning properties of a moth-eye-Like surface coated with TiO2 particles. *Langmuir.* 2011;27:3275–3278.
22. Howarter JA, Youngblood JP. Self-cleaning and next generation anti-fog surfaces and coatings. *Macromol Rapid Comm.* 2008;29:455–466.

23. Raut HK, Ganesh VA, Nair AS, Ramakrishna S. Anti-reflective coatings: a critical, in-depth review. *Energy Environ Sci.* 2011;4:3779–3804.

24. Macleod HA. Thin-film optical filters, Third Edition: Taylor & Francis; 2001.

25. Gupta RN. Reflection of sound waves from transition layers. *J Acoust Soc Am.* 1966;39:255–260.

26. Sheldon B, Haggerty J, Emslie A. Exact computation of the reflectance of a surface layer of arbitrary refractive-index profile and an approximate solution of the inverse problem. *JOSA.* 1982;72:1049–1055.

27. Yeh P, Sari S. Optical properties of stratified media with exponentially graded refractive index. *Appl Optics.* 1983;22:4142–4145.

28. Southwell WH. Gradient-index antireflection coatings. *Opt Lett.* 1983;8:584–586.

29. Miller WH, Bernard GD, Allen JL. The optics of insect compound eyes. *Science.* 1968;162:760–767.

30. Huang JY, Wang XD, Wang ZL. Bio-inspired fabrication of antireflection nanostructures by replicating fly eyes. *Nanotechnology.* 2008;19.

31. Gao X, Yan X, Yao X, Xu L, Zhang K, Zhang J, et al. The dry-style antifogging properties of mosquito compound eyes and artificial analogues prepared by soft lithography. *Adv Mater.* 2007;19:2213–2217.

32. Huang YF, Chattopadhyay S, Jen YJ, Peng CY, Liu TA, Hsu YK, et al. Improved broadband and quasi-omnidirectional anti-reflection properties with biomimetic silicon nanostructures. *Nat Nanotechnol.* 2007;2:770–774.

33. Chang CC, Chen CM, Hwang FH, Chen CC, Cheng LP. Preparation of polymer/silica composite antiglare coatings on poly (ethylene terphathalate)(PET) substrates. *J Coat Technol Res.* 2012;9:561–568.

34. Schelle C, Mennig M, Krug H, Jonschker G, Schmidt H. One step antiglare sol-gel coating for screens by sol-gel techniques. *J Non-Cryst Solids.* 1997;218:163–168.

35. Aegerter M, Al-Dahoudi N, Solieman A, Kavak H, Oliveira P. Transparent conducting coatings made by chemical nanotechnology processes. *Mol Cryst Liq Cryst.* 2004;417:105–114.

36. Al-Dahoudi N, Bisht H, Göbbert C, Krajewski T, Aegerter MA. Transparent conducting, anti-static and anti-static–anti-glare coatings on plastic substrates. *Thin Solid Films.* 2001;392:299–304.

37. Tsubota K, Miyake M, Matsumoto Y, Shintani M. Visual protective sheet can increase blink rate while playing a hand-held video game. *Am J Ophthalmol.* 2002;133:704–705.

38. Nakajima A, Fujishima A, Hashimoto K, Watanabe T. Preparation of transparent superhydrophobic boehmite and silica films by sublimation of aluminum acetylacetonate. *Adv Mater.* 1999;11:1365–1368.

39. Wenzel RN. Resisitance of solid Surfaces to wetting by water. *Ind Eng Chem.* 1936;28:988–994.

40. Cassie ABD, Baxter S. Wettability of porous surfaces. *Transact Faraday Soc.* 1944;40:546.

41. Johnston E, Bullock S, Uilk J, Gatenholm P, Wynne KJ. Networks from alpha, omega-dihydroxypoly(dimethylsiloxane) and (tridecafluoro-1,1,2,2-tetrahydrooctyl)triethoxysilane: Surface microstructures and surface characterization. *Macromolecules.* 1999;32:8173–8182.

42. Sun MH, Luo CX, Xu LP, Ji H, Qi OY, Yu DP, et al. Artificial lotus leaf by nanocasting. *Langmuir.* 2005;21:8978–8981.

43. Liu X, He J. Progress in antifogging technology-from surface engineering to functional surfaces. *Prog Chem.* 2010;22:270–276.

44. Frank SN, Bard AJ. Heterogeneous photocatalytic oxidation of cyanide ion in aqueous-solutions at TiO2 powder. *J Am Chem Soc.* 1977;99:303–304.

45. Ohko Y, Hashimoto K, Fujishima A. Kinetics of photocatalytic reactions under extremely low-intensity UV illumination on titanium dioxide thin films. *J Phys Chem A.* 1997;101:8057–8062.

46. Sopyan I, Murasawa S, Hashimoto K, Fujishima A. Highly efficient TiO2 film photocatalyst degradation of gaseous acetaldehyde. *Chem Lett.* 1994:723–726.

47. Lee J, Park H, Choi W. Selective photocatalytic oxidation of NH3 to N2 on plantation TiO2 in water. *Environ Sci Technol.* 2002;36:5462–5468.

48. Fujishima A, Zhang XT. Titanium dioxide photocatalysis: present situation and future approaches. *C R Chim.* 2006;9:750–760.

49. Fujishima A, Zhang X, Tryk DA. TiO2 photocatalysis and related surface phenomena. *Surf Sci Rep.* 2008;63:515–582.

50. Minabe T, Tryk DA, Sawunyama P, Kikuchi Y, Hashimoto K, Fujishima A. TiO2-mediated photodegradation of liquid and solid organic compounds. *J Photoch Photobio A.* 2000;137:53–62.

51. Heller A. Chemistry and applications of photocatalytic oxidation of thin organic films. *Acc Chem Res.* 1995;28:503–508.

52. Schwitzgebel J, Ekerdt JG, Sunada F, Lindquist SE, Heller A. Increasing the efficiency of the photocatalytic oxidation of organic films on aqueous solutions by reactively coating the TiO2 photocatalyst with a chlorinated silicone. *J Phys Chem B.* 1997;101:2621–2624.

53. Sakai N, Fujishima A, Watanabe T, Hashimoto K. Quantitative evaluation of the photoinduced hydrophilic conversion properties of TiO2 thin film surfaces by the reciprocal of contact angle. *J Phys Chem B.* 2003;107:1028–1035.

54. Hashimoto K, Irie H, Fujishima A. TiO_2 photocatalysis: a historical overview and future prospects. *Jpn J Appl Phys Part 1 - Regul Pap Brief Commun Rev Pap.* 2005;44:8269–8285.

55. Geng X, Qi Z, Li M, Duan BK, Zhao L, Bohn PW. Fabrication of antireflective layers on silicon using metal-assisted chemical etching with in situ deposition of silver nanoparticle catalysts. *Sol Energ Mat Sol C.* 2012;103:98–107.

56. Campbell P, Green MA. Light trapping properties of pyramidally textured surfaces. *J Appl Phys.* 1987;62:243–249.

57. Hantschel T, Vandervorst W. Anisotropic etching of inverted pyramids in the sub-100 nm region. *Microelectron Eng.* 1997;35:405–407.

58. Vazsonyi E, De Clercq K, Einhaus R, Van Kerschaver E, Said K, Poortmans J, et al. Improved anisotropic etching process for industrial texturing of silicon solar cells. *Sol Energ Mat Sol C.* 1999;57:179–188.

59. Peng K, Zhang M, Lu A, Wong N-B, Zhang R, Lee S-T. Ordered silicon nanowire arrays via nanosphere lithography and metal-induced etching. *Appl Phys Lett.* 2007;90:163123.

60. Kim K, Dhungel S, Jung S, Mangalaraj D, Yi J. Texturing of large area multi-crystalline silicon wafers through different chemical approaches for solar cell fabrication. *Sol Energ Mat Sol C.* 2008;92:960–968.

61. Lin H-H, Chen W-H, Wang C-J, Hong FC-N. Fabrication of antireflective nanostructures for crystalline silicon solar cells by reactive ion etching. *Thin Solid Films.* 2012;529:138–142.

62. Wang X, Shen J. A review of contamination-resistant antireflective sol–gel coatings. *J Sol-Gel Sci Techn.* 2011:1–7.

63. Li X, Gross M, Oreb B, Shen J. Increased laser-damage resistance of sol–gel silica coating by structure modification. *J Phys Chem C.* 2012;116:18367–18371.

64. Chen X, Mao SS. Titanium dioxide nanomaterials: synthesis, properties, modifications, and applications. *Chem Rev.* 2007;38:2891–2959.

65. Roy P, Berger S, Schmuki P. TiO2 nanotubes: synthesis and applications. *Angew Chem Int Edit.* 2011;50:2904–2939.

66. Cai JG, Ye JF, Chen SY, Zhao XW, Zhang DY, Chen S, et al. Self-cleaning, broadband and quasi-omnidirectional antireflective structures based

on mesocrystalline rutile TiO_2 nanorod arrays. *Energy Environ Sci.* 2012;5:7575–7581.

67. Jayasinghe RC, Perera AGU, Zhu H, Zhao Y. Optical properties of nanostructured Ti02 thin films and their application as antireflection coatings on infrared detectors. *Opt Lett.* 2012;37:4302–4304.
68. Thomson AF, McIntosh KR. Light-enhanced surface passivation of Ti02-coated silicon. *Prog Photovoltaics.* 2012;20:343–349.
69. Xi J-Q, Schubert MF, Kim JK, Schubert EF, Chen M, Lin S-Y, et al. Optical thin-film materials with low refractive index for broadband elimination of Fresnel reflection. *Nat Photonics.* 2007;1:176–179.
70. Tian Y, Hu C, Xiong Y, Wan B, Xia C, He X, et al. ZnO pyramidal arrays: novel functionality in antireflection. *J Phys Chem C.* 2010;114:10265–10269.
71. Chao YC, Chen CY, Lin CA, He JH. Light scattering by nanostructured anti-reflection coatings. *Energy Environ Sci.* 2011;4:3436–3441.
72. Hawker CJ, Russell TP. Block copolymer lithography: merging "bottom-up" with "top-down" processes. *Mrs Bull.* 2005;30:952–966.
73. Li X, Yu X, Han Y. Polymer thin films for antireflection coatings. *J Mater Chem C.* 2013;1:2266–2285.
74. Leitel R, Kaless A, Schulz U, Kaiser N. Broadband antireflective structures on PMMA by plasma treatment. *Plasma Process Polym.* 2007;4:S878–S881.
75. Schulz U, Munzert P, Kaiser N. Plasma surface modification of PMMA for optical applications. *J Adhes Sci Technol.* 2010;24:1283–1289.
76. Hsu JC, Wang PW, Lin YH, Chen HL, Chen YY, Yao YD, et al. Anti-reflective effect of transparent polymer by plasma treatment with end-hall ion source and optical coating. *Opt Rev.* 2010;17:553–556.
77. Xu L, Gao L, He J. Fabrication of visible/near-IR antireflective and superhydrophobic coatings from hydrophobically modified hollow silica nanoparticles and poly(methyl methacrylate). *Rsc Adv.* 2012;2:12764–12769.
78. Robertson J. Diamond-like amorphous carbon. *Mater Sci Eng R: Rep.* 2002;37:129–281.
79. Theye M-L, Paret V. Spatial organization of the sp2-hybridized carbon atoms and electronic density of states of hydrogenated amorphous carbon films. *Carbon.* 2002;40:1153–1166.
80. Zhu H, Wei J, Wang K, Wu D. Applications of carbon materials in photovoltaic solar cells. *Sol Energ Mat Sol C.* 2009;93:1461–1470.

81. Oliveira Jr M, Silva D, Côrtes A, Namani M, Marques F. Diamond like carbon used as antireflective coating on crystalline silicon solar cells. *Diam Relat Mater.* 2009;18:1028–1030.

82. da Silva D, Côrtes A, Oliveira M, Motta E, Viana G, Mei P, et al. Application of amorphous carbon based materials as antireflective coatings on crystalline silicon solar cells. *J Appl Phys.* 2011;110:043510–8.

83. DeNatale J, Hood P, Flintoff J, Harker A. Fabrication and characterization of diamond moth eye antireflective surfaces on Ge. *J Appl Phys.* 1992;71:1388–1393.

84. Kennedy SR, Brett MJ. Porous broadband antireflection coating by glancing angle deposition. *Appl Optics.* 2003;42:4573–4579.

85. Yoon J, Jo S, Chun IS, Jung I, Kim H-S, Meitl M, et al. GaAs photovoltaics and optoelectronics using releasable multilayer epitaxial assemblies. *Nature.* 2010;465:329–333.

86. Huang MC, Zhou Y, Chang-Hasnain CJ. A surface-emitting laser incorporating a high-index-contrast subwavelength grating. *Nat Photonics.* 2007;1:119–122.

87. Na BH, Ju GW, Choi HJ, Cho YC, Park YH, Lee YT. Large aperture asymmetric Fabry Perot modulator based on asymmetric tandem quantum well for low voltage operation. *Opt Express.* 2012:6003–6009.

88. Green MA, Emery K, Hishikawa Y, Warta W, Dunlop ED. Solar cell efficiency tables (version 39). *Progr Photovoltaics: Res Appl.* 2012;20:12–20.

89. Kim B-J, Kim J. Fabrication of GaAs subwavelength structure (SWS) for solar cell applications. *Opt Express.* 2011;19:A326–A30.

90. Yamagiwa M, Mano T, Kuroda T, Tateno T, Sakoda K, Kido G, et al. Self-assembly of laterally aligned GaAs quantum dot pairs. *Appl Phys Lett.* 2006;89:113115–3.

91. Mano T, Kuroda T, Sanguinetti S, Ochiai T, Tateno T, Kim J, et al. Self-assembly of concentric quantum double rings. *Nano Lett.* 2005;5:425–428.

92. Lee EH, Song JD, Yoon JJ, Bae MH, Han IK, Choi WJ, et al. Formation of self-assembled large droplet-epitaxial GaAs islands for the application to reduced reflection. *J Appl Phys.* 2013;113:154308.

93. Yamaji H, Terada Y, Yoshida H, Miyake H, Hiramatsu K. Improved optical properties using self-organized GaN nanotip structure. *Phys Status Solidi C.* 2003:2566–2569.

94. Huang Y-F, Chattopadhyay S, Jen Y-J, Peng C-Y, Liu T-A, Hsu Y-K, et al. Improved broadband and quasi-omnidirectional anti-reflection properties with biomimetic silicon nanostructures. *Nat Nanotechnol.* 2007;2:770–774.
95. Leem J, Yu J, Song Y, Lee Y. Antireflective characteristics of disordered GaAs subwavelength structures by thermally dewetted Au nanoparticles. *Sol Energ Mat Sol C.* 2011;95:669–676.
96. Kanamori Y, Kobayashi K-i, Yugami H, Hane K. Subwavelength antireflection gratings for GaSb in visible and near-infrared wavelengths. *Jpn J Appl Phys.* 2003;42:4020.
97. Iler RK. Multilayers of colloidal particles. *J Colloid Interf Sci.* 1966;21:569–594.
98. Decher G, Hong JD. Buildup of unltrathin multilayer films by a self-assembly process. 1. Consecutive adsorption of anionic and cationic bipolar amphiphiles on charged surfaces. *Makromol Chem-M Symp.* 1991;46:321–327.
99. Hiller J, Mendelsohn JD, Rubner MF. Reversibly erasable nanoporous anti-reflection coatings from polyelectrolyte multilayers. *Nat Mater.* 2002;1:59–63.
100. Du X, He J. Facile fabrication of hollow mesoporous silica nanospheres for superhydrophilic and visible/near-IR antireflection coatings. *Chem-Eur J.* 2011;17:8165–8174.
101. Podsiadlo P, Sui L, Elkasabi Y, Burgardt P, Lee J, Miryala A, et al. Layer-by-layer assembled films of cellulose nanowires with antireflective properties. *Langmuir.* 2007;23:7901–7906.
102. Cho JH, Hong JK, Char K, Caruso F. Nanoporous block copolymer micelle/micelle multilayer films with dual optical properties. *J Am Chem Soc.* 2006;128:9935–9942.
103. Ariga K, Lvov Y, Kunitake T. Assembling alternate dye-polyion molecular films by electrostatic layer-by-layer adsorption. *J Am Chem Soc.* 1997;119:2224–2231.
104. Li Y, Wang X, Sun J. Layer-by-layer assembly for rapid fabrication of thick polymeric films. *Chem Soc Rev.* 2012;41:5998–6009.
105. Ariga K, Hill JP, Ji QM. Layer-by-layer assembly as a versatile bottom-up nanofabrication technique for exploratory research and realistic application. *Phys Chem Chem Phys.* 2007;9:2319–2340.
106. Osada M, Sasaki T. Two-dimensional dielectric nanosheets: novel nanoelectronics from nanocrystal building blocks. *Adv Mater.* 2012;24:210–228.

107. Sun JQ, Zhang LB, Li Y, Shen JC. Mechanically stable antireflection and antifogging coatings fabricated by the layer-by-layer deposition process and postcalcination. *Langmuir.* 2008;24:10851–10857.

108. Ariga K, Ji Q, Hill JP, Bando Y, Aono M. Forming nanomaterials as layered functional structures toward materials nanoarchitectonics. *NPG Asia Mater.* 2012;4:e17.

109. Liu X, He J. Hierarchically structured superhydrophilic coatings fabricated by self-assembling raspberry-like silica nanospheres. *J Colloid Interf Sci.* 2007;314:341–345.

110. Du X, Liu X, Chen H, He J. Facile fabrication of raspberry-like composite nanoparticles and their application as building blocks for constructing superhydrophilic coatings. *J Phys Chem C.* 2009;113:9063–9070.

111. Li X, Du X, He J. Self-cleaning antireflective coatings assembled from peculiar mesoporous silica nanoparticles. *Langmuir.* 2010;26:13528–13534.

112. Du X, He J. Hierarchically mesoporous silica nanoparticles: extraction, amino-functionalization, and their multipurpose potentials. *Langmuir.* 2011;27:2972–2979.

113. Brinker CJ, Hurd AJ. Fundamental of sol-gel dip-coating. *J Phys III.* 1991;4:1231–1242.

114. Scriven L. Physics and applications of dip coating and spin coating. Cambridge Univ Press; 1988. p. 717–729.

115. Faustini M, Nicole L, Boissière C, Innocenzi P, Sanchez C, Grosso D. Hydrophobic, antireflective, self-cleaning, and antifogging sol–gel coatings: an example of multifunctional nanostructured materials for photovoltaic cells. *Chem Mater.* 2010.

116. Kesmez Ö, Erdem Çamurlu H, Burunkaya E, Arpaç E. Sol–gel preparation and characterization of anti-reflective and self-cleaning SiO2–TiO2 double-layer nanometric films. *Sol Energ Mat Sol C.* 2009;93:1833–1839.

117. Bautista M, Morales A. Silica antireflective films on glass produced by the sol–gel method. *Sol Energ Mat Sol C.* 2003;80:217–225.

118. Helsch G, Mos A, Deubener J, Holand M. Thermal resistance of nanoporous antireflective coatings on silica glass for solar tower receivers. *Sol Energ Mat Sol C.* 2010;94:2191–2196.

119. Lawrence C. The mechanics of spin coating of polymer films. *Phys Fluids.* 1988;31:2786.

120. Krebs FC. Fabrication and processing of polymer solar cells: a review of printing and coating techniques. *Sol Energ Mat Sol C.* 2009;93:394–412.

121. Hall DB, Underhill P, Torkelson JM. Spin coating of thin and ultrathin polymer films. *Polym Eng Sci.* 1998;38:2039–2045.

122. Zhou G, He J. Antireflective coatings on fresnel lenses by spin-coating of solid silica nanoparticles. *J Nanosci Nanotechnol.* 2013;13:5534–5541.

123. Zhou G, He J, Xu L. Antifogging antireflective coatings on Fresnel lenses by integrating solid and mesoporous silica nanoparticles. *Micropor Mesopor Mat.* 2013;176:41–47.

124. Patel PA, Dobrynin AV, Mather PT. Combined effect of spin speed and ionic strength on polyelectrolyte spin assembly. *Langmuir.* 2007;23:12589–12597.

125. Cho J, Char K, Hong JD, Lee KB. Fabrication of highly ordered multilayer films using a spin self-assembly method. *Adv Mater.* 2001;13:1076–1078.

126. Steele A, Bayer I, Loth E. Inherently superoleophobic nanocomposite coatings by spray atomization. *Nano Lett.* 2009;9:501–505.

127. Babu BJ, Maldonado A, Velumani S, Asomoza R. Electrical and optical properties of ultrasonically sprayed Al-doped zinc oxide thin films. *Mater Sci Eng B-Adv.* 2010;174:31–37.

128. Fender T. Thermal spray high performance polymer coatings. *Mater Technol.* 1996;11:16–20.

129. Schmidt T, Gärtner F, Assadi H, Kreye H. Development of a generalized parameter window for cold spray deposition. *Acta Mater.* 2006;54:729–742.

130. Gao L, He P, Li X, He J. Construction of nanostructured silica surfaces by spray-coating and their optical and wetting properties. *Imaging Sci Photochem.* 2012;30:260–268.

131. Men XHMXH, Zhang ZZ, Yang J, Zhu XT, Wang K, Jiang W. Spray-coated superhydrophobic coatings with regenerability. *New J Chem.* 2011;35:881–886.

132. Porcel CH, Izquierdo A, Ball V, Decher G, Voegel JC, Schaaf P. Ultrathin coatings and (poly(glutamic acid)/polyallylamine) films deposited by continuous and simultaneous spraying. *Langmuir.* 2005;21:800–802.

133. Schlenoff JB, Dubas ST, Farhat T. Sprayed polyelectrolyte multilayers. *Langmuir.* 2000;16:9968–9969.

134. Krogman KC, Zacharia NS, Schroeder S, Hammond PT. Automated process for improved uniformity and versatility of layer-by-layer deposition. *Langmuir.* 2007;23:3137–3141.

135. Choy KL. Chemical vapour deposition of coatings. *Prog Mater Sci.* 2003;48:57–170.

136. Biswas A, Bayer IS, Biris AS, Wang T, Dervishi E, Faupel F. Advances in top–down and bottom–up surface nanofabrication: techniques, applications & future prospects. *Adv Colloid Interfac.* 2012;170:2–27.

137. Randich E, Pettit R. Solar selective properties and high temperature stability of CVD ZrB2. *Sol Energ Mater.* 1981;5:425–435.

138. Yu C, Sandhu G, Doan T. Antireflection coatings for advanced semiconductor device metallization using laser reflow and chemical mechanical planarization. *Appl Phys Lett.* 1991;59:1978–1980.

139. Mollart T, Lewis K. Transition metal oxide anti-reflection coatings for airborne diamond optics. *Diam Relat Mater.* 2001;10:536–541.

140. Yang SM, Hsieh YC, Jeng CA. Optimal design of antireflection coating and experimental verification by plasma enhanced chemical vapor deposition in small displays. *J Vac Sci Technol A.* 2009;27:336–341.

141. Mukhopadhyay S, Jana T, Ray S. Development of low temperature silicon oxide thin films by photo-CVD for surface passivation. *J Vac Sci Technol A.* 2005;23:417–422.

142. Van der Werf C, Goldbach H, Löffler J, Scarfo A, Kylner A, Stannowski B, et al. Silicon nitride at high deposition rate by Hot Wire Chemical Vapor Deposition as passivating and antireflection layer on multicrystalline silicon solar cells. *Thin Solid Films.* 2006;501:51–54.

143. Verlaan V, Van der Werf CHM, Flouweling ZS, Romijn G, Weeber AW, Dekkers HFW, et al. Multi-crystalline si solar cells with very fast deposited (108 nm/min) passivating hot wire CVD silicon nitride as antireflection coating. *Prog Photovoltaics.* 2007;15:563–573.

144. Kuhr M, Bauer S, Rothhaar U, Wolff D. Coatings on plastics with the PICVD technology. *Thin Solid Films.* 2003;442:107–116.

145. Remache L, Fourmond E, Mahdjoub A, Dupuis J, Lemiti M. Design of porous silicon/PECVD SiOx antireflection coatings for silicon solar cells. *Mater Sci Eng B-Adv.* 2011;176:45–48.

146. Li X, Han YC. Tunable wavelength antireflective film by non-solvent-induced phase separation of amphiphilic block copolymer micelle solution. *J Mater Chem.* 2011;21:18024–18033.

147. Walheim S, Schäffer E, Mlynek J, Steiner U. Nanophase-separated polymer films as high-performance antireflection coatings. *Science.* 1999;283:520–522.

148. Böltau M, Walheim S, Mlynek J, Krausch G, Steiner U. Surface-induced structure formation of polymer blends on patterned substrates. *Nature.* 1998;391:877–879.

149. Zhao N, Xie Q, Weng L, Wang S, Zhang X, Xu J. Superhydrophobic surface from vapor-induced phase separation of copolymer micellar solution. *Macromolecules*. 2005;38:8996–8999.

150. Ibn-Elhaj M, Schadt M. Optical polymer thin films with isotropic and anisotropic nano-corrugated surface topologies. *Nature*. 2001;410:796–799.

151. Bunker BC. Molecular mechanisms for corrosion of silica and silicate-glasses. *J Non-Cryst Solids*. 1994;179:300–308.

152. Myoung JM, Xiong J, Das SN, Kar JP, Choi JH. A multifunctional nanoporous layer created on glass through a simple alkali corrosion process. *J Mater Chem*. 2010;20:10246–10252.

153. Liu X, He J. One-step hydrothermal creation of hierarchical microstructures toward superhydrophilic and superhydrophobic surfaces. *Langmuir*. 2009;25:11822–11826.

154. Qi D, Lu N, Xu H, Yang B, Huang C, Xu M, et al. Simple approach to wafer-scale self-cleaning antireflective silicon surfaces. *Langmuir*. 2009;25:7769–7772.

155. Palumbo F, Di Mundo R, Cappelluti D, d'Agostino R. Superhydrophobic and superhydrophilic polycarbonate by tailoring chemistry and nano-texture with plasma processing. *Plasma Process Polym*. 2011;8:118–126.

156. Pearton S. High ion density dry etching of compound semiconductors. *Mater Sci Eng B*. 1996;40:101–118.

157. Leem JWLJW, Yu JS. Broadband and wide-angle antireflection subwavelength structures of Si by inductively coupled plasma etching using dewetted nanopatterns of Au thin films as masks. *Thin Solid Films*. 2011;519:3792–3797.

158. Gamo K. Nanofabrication by FIB. *Microelectron Eng*. 1996;32:159–171.

159. Ito R, Okazaki S. Pushing the limits of lithography. *Nature*. 2000;406:1027–1031.

160. Wallraff G, Hinsberg W. Lithographic imaging techniques for the formation of nanoscopic features. *Chem Rev*. 1999;99:1801–1821.

161. Yang B, Li YF, Zhang JH, Zhu SJ, Dong HP, Jia F, et al. Bioinspired silica surfaces with near-infrared improved transmittance and superhydrophobicity by colloidal lithography. *Langmuir*. 2010;26:9842–9847.

162. Park SJ, Lee SW, Lee KJ, Lee JH, Kim KD, Jeong JH, et al. An antireflective nanostructure array fabricated by nanosilver colloidal lithography on a silicon substrate. *Nanoscale Res Lett*. 2010;5:1570–1577.

163. Tommila J, Polojarvi V, Aho A, Tukiainen A, Viheriala J, Salmi J, et al. Nanostructured broadband antireflection coatings on AlInP fabricated by nanoimprint lithography. *Sol Energ Mat Sol C.* 2010;94:1845–1848.

164. del Campo A, Arzt E. Fabrication approaches for generating complex micro-and nanopatterns on polymeric surfaces. *Chem Rev.* 2008;108:911–945.

165. Acikgoz C, Hempenius MA, Huskens J, Vancso GJ. Polymers in conventional and alternative lithography for the fabrication of nanostructures. *Eur Polym J.* 2011;47:2033–2052.

166. Irie S, Shigematsu M, Sakamoto R, Mizusawa K, Nakajima Y, Itani T. A study of an organic bottom antireflective coating for 157-nm lithography. *J Photopolym Sci Tec.* 2003;16:565–572.

167. Lai Y, Lin C, Wang H, Huang J, Zhuang H, Sun L. Superhydrophilic-superhydrophobic micropattern on TiO2 nanotube films by photocatalytic lithography. *Electrochem Commun.* 2008;10:387–391.

168. Sharma P, Hanumantha Rao K. Adhesion of Paenibacillus polymyxa on chalcopyrite and pyrite: surface thermodynamics and extended DLVO theory. *Coll Surf B: Biointerfaces.* 2003;29:21–38.

169. Wu Z, Lee D, Rubner MF, Cohen RE. Structural color in porous, superhydrophilic, and self-cleaning SiO_2/TiO2 bragg stacks. *Small.* 2007;3:1445–1451.

170. Lai Y, Tang Y, Gong J, Gong D, Chi L, Lin C, et al. Transparent superhydrophobic/superhydrophilic TiO2-based coatings for self-cleaning and anti-fogging. *J Mater Chem.* 2012;22:7420–7426.

171. Ganesh VA, Nair AS, Raut HK, Walsh TM, Ramakrishna S. Photocatalytic superhydrophilic TiO2 coating on glass by electrospinning. *RSC Adv.* 2012;2:2067–2072.

172. Lai Y, Lin Z, Huang J, Sun L, Chen Z, Lin C. Controllable construction of ZnO/TiO2 patterning nanostructures by superhydrophilic/superhydrophobic templates. *New J Chem.* 2010;34:44–51.

173. Lim HS, Kwak D, Lee DY, Lee SG, Cho K. UV-driven reversible switching of a roselike vanadium oxide film between superhydrophobicity and superhydrophilicity. *J Am Chem Soc.* 2007;129:4128–4129.

174. Miyauchi M, Shibuya M, Zhao ZG, Liu Z. Surface wetting behavior of a WO3 electrode under light-irradiated or potential-controlled conditions. *J Phys ChemC.* 2009;113:10642–10646.

175. Blossey R. Self-cleaning surfaces-virtual realities. *Nat Mater.* 2003;2:301–306.

176. McDonald BT, Cui T. Superhydrophilic surface modification of copper surfaces by layer-by-layer self-assembly and liquid phase deposition of TiO2 thin film. *J Colloid Interf Sci.* 2011;354:1–6.

177. Denison KR, Boxall C. Photoinduced "stick-slip" on superhydrophilic semiconductor surfaces. *Langmuir.* 2007;23:4358–4366.

178. Liu Y, Qian L, Guo C, Jia X, Wang J, Tang W. Natural superhydrophilic TiO2/SiO_2 composite thin films deposited by radio frequency magnetron sputtering. *J Alloy Compd.* 2009;479:532–535.

179. Tang D, Cheng K, Weng W, Song C, Du P, Shen G, et al. TiO2 nanorod films grown on Si wafers by a nanodot-assisted hydrothermal growth. *Thin Solid Films.* 2011;519:7644–7649.

180. Kobayashi M, Terayama Y, Yamaguchi H, Terada M, Murakami D, Ishihara K, et al. Wettability and antifouling behavior on the surfaces of superhydrophilic polymer brushes. *Langmuir.* 2012;28:7212–7222.

181. Zang J, Li CM, Bao SJ, Cui X, Bao Q, Sun CQ. Template-free electrochemical synthesis of superhydrophilic polypyrrole nanofiber network. *Macromolecules.* 2008;41:7053–7057.

182. Xie XN, Lim SX, Wang Y, Gao X, Lee KK, Sow CH, et al. A nanosegregant approach to superwettable and water-attracting surfaces. *Macromol Chem Physic.* 2010;211:2187–2192.

183. Gao Y, Masuda Y, Koumoto K. Light-excited superhydrophilicity of amorphous TiO2 thin films deposited in an aqueous peroxotitanate solution. *Langmuir.* 2004;20:3188–3194.

184. Pizem H, Sukenik CN, Sampathkumaran U, McIlwain AK, De Guire MR. Effects of substrate surface functionality on solution-deposited titania films. *Chem Mater.* 2002;14:2476–2485.

185. Rico V, Romero P, Hueso J, Espinós J, González-Elipe A. Wetting angles and photocatalytic activities of illuminated TiO2 thin films. *Catal Today.* 2009;143:347–354.

186. Lee CS, Kim J, Gu GH, Jo DH, Park CG, Choi W, et al. Photocatalytic activities of TiO2 thin films prepared on galvanized iron substrate by plasma-enhanced atomic layer deposition. *Thin Solid Films.* 2010;518:4757–4761.

187. Lee CS, Kim J, Son J, Choi W, Kim H. Photocatalytic functional coatings of TiO2 thin films on polymer substrate by plasma enhanced atomic layer deposition. *Appl Catal B: Environ.* 2009;91:628–633.

188. Pore V, Kivelä T, Ritala M, Leskelä M. Atomic layer deposition of photocatalytic TiO2 thin films from TiF4 and H2O. *Dalton Trans.* 2008:6467–6474.

189. Tang Y, Liu Y, Sampathkumaran U, Hu MZ, Wang R, De Guire MR. Particle growth and particle–surface interactions during low-temperature deposition of ceramic thin films. *Solid State Ionics.* 2002;151:69–78.

190. Gutiérrez-Tauste D, Domènech X, Hernández-Fenollosa MA, Ayllón JA. Alternative fluoride scavengers to produce TiO2 films by the liquid phase deposition (LPD) technique. *J Mater Chem.* 2006;16:2249–2255.

191. Aoi Y, Kambayashi H, Kamijo E, Deki S. Synthesis of mesoscopic hollow spheres and inner surface functionalized hollow spheres of titanium dioxide by the liquid phase deposition method. *J Mater Res.* 2003;18:2832–2836.

192. Strohm H, Sgraja M, Bertling J, Löbmann P. Preparation of TiO2-polymer hybrid microcapsules. *J Mater Sci.* 2003;38:1605–1609.

193. Tsuge Y, Kim J, Sone Y, Kuwaki O, Shiratori S. Fabrication of transparent TiO2 film with high adhesion by using self-assembly methods: application to super-hydrophilic film. *Thin Solid Films.* 2008;516:2463–2468.

194. Deki S, Aoi Y, Okibe J, Yanagimoto H, Kajinami A, Mizuhata M. Preparation and characterization of iron oxyhydroxide and ironoxide thin films by liquid-phase deposition. *J Mater Chem.* 1997;7:1769–1772.

195. Sirghi L, Aoki T, Hatanaka Y. Hydrophilicity of TiO2 thin films obtained by radio frequency magnetron sputtering deposition. *Thin Solid Films.* 2002;422:55–61.

196. Yamagishi M, Kuriki S, Song P, Shigesato Y. Thin film TiO2 photocatalyst deposited by reactive magnetron sputtering. *Thin Solid Films.* 2003;442:227–231.

197. Hosono E, Matsuda H, Honma I, Ichihara M, Zhou H. Synthesis of a perpendicular TiO2 nanosheet film with the superhydrophilic property without UV irradiation. *Langmuir.* 2007;23:7447–7450.

198. Zhang X, Shi F, Niu J, Jiang Y, Wang Z. Superhydrophobic surfaces: from structural control to functional application. *J Mater Chem.* 2008;18:621–633.

199. Bhushan B, Jung YC, Koch K. Self-cleaning efficiency of artificial superhydrophobic surfaces. *Langmuir.* 2009;25:3240–3248.

200. Li XM, Reinhoudt D, Crego-Calama M. What do we need for a superhydrophobic surface? A review on the recent progress in the preparation of superhydrophobic surfaces. *Chem Soc Rev.* 2007;36:1350–1368.

201. Guo Z, Liu W, Su BL. Superhydrophobic surfaces: from natural to biomimetic to functional. *J Colloid Interf Sci.* 2011;353:335–355.

202. Neinhuis C, Barthlott W. Characterization and distribution of water-repellent, self-cleaning plant surfaces. *Ann Bot.* 1997;79:667–677.

203. Xue CH, Jia ST, Zhang J, Ma JZ. Large-area fabrication of superhydrophobic surfaces for practical applications: an overview. *Sci Technol Adv Mat.* 2010;11.

204. Onda T, Shibuichi S, Satoh N, Tsujii K. Super-water-repellent fractal surfaces. *Langmuir.* 1996;12:2125–2127.

205. Yoon M, Kim Y, Cho J. Multifunctional colloids with optical, magnetic and superhydrophobic properties derived from nucleophilic substitution-induced layer-by-layer assembly in organic media. *ACS Nano.* 2011;5:5417–5426.

206. Li Y, Liu F, Sun J. A facile layer-by-layer deposition process for the fabrication of highly transparent superhydrophobic coatings. *Chem Commun.* 2009:2730–2732.

207. Xu L, He J. Fabrication of highly transparent superhydrophobic coatings from hollow silica nanoparticles. *Langmuir.* 2012;28:7512–7518.

208. Feng J, Tuominen MT, Rothstein JP. Hierarchical superhydrophobic surfaces fabricated by dual-scale electron-beam-lithography with well-ordered secondary nanostructures. *Adv Funct Mater.* 2011;21:3715–3722.

209. Ishizaki T, Hieda J, Saito N, Takai O. Corrosion resistance and chemical stability of super-hydrophobic film deposited on magnesium alloy AZ31 by microwave plasma-enhanced chemical vapor deposition. *Electrochim Acta.* 2010;55:7094–7101.

210. Lau KKS, Bico J, Teo KBK, Chhowalla M, Amaratunga GAJ, Milne WI, et al. Superhydrophobic carbon nanotube forests. *Nano Lett.* 2003;3:1701–1705.

211. Miyauchi Y, Ding B, Shiratori S. Fabrication of a silver-ragwort-leaf-like super-hydrophobic micro/nanoporous fibrous mat surface by electrospinning. *Nanotechnology.* 2006;17:5151–5156.

212. Tang H, Wang H, He J. Superhydrophobic titania membranes of different adhesive forces fabricated by electrospinning. *J Phys Chem C.* 2009;113:14220–14224.

213. Acatay K, Simsek E, Ow-Yang C, Menceloglu YZ. Tunable, superhydrophobically stable polymeric surfaces by electrospinning. *Angew Chem Int Edit.* 2004;43:5210–5213.

214. Shang H, Wang Y, Limmer S, Chou T, Takahashi K, Cao G. Optically transparent superhydrophobic silica-based films. *Thin Solid Films.* 2005;472:37–43.

215. Shirtcliffe NJ, McHale G, Newton MI, Perry CC, Roach P. Porous materials show superhydrophobic to superhydrophilic switching. *Chem Commun.* 2005:3135–3137.

216. Cortese B, Morgan H. Controlling the wettability of hierarchically structured thermoplastics. *Langmuir.* 2012;28:896–904.

217. Fernández-Blázquez JP, Fell D, Bonaccurso E, Campo A. Superhydrophilic and superhydrophobic nanostructured surfaces via plasma treatment. *J Colloid Interf Sci.* 2011;357:234–238.

218. Nicolas M. Acids and alkali resistant sticky superhydrophobic surfaces by one-pot electropolymerization of perfluoroalkyl alkyl pyrrole. *J Colloid Interf Sci.* 2010;343:608–614.

219. Yang X, Zhuang J, Li X, Chen D, Ouyang G, Mao Z, et al. Hierarchically nanostructured rutile arrays: acid vapor oxidation growth and tunable morphologies. *ACS Nano.* 2009;3:1212–1218.

220. Hsieh CT, Yang SY, Lin JY. Electrochemical deposition and superhydrophobic behavior of ZnO nanorod arrays. *Thin Solid Films.* 2010;518:4884–4889.

221. Darmanin T, Taffin de Givenchy E, Amigoni S, Guittard F. Hydrocarbon versus fluorocarbon in the electrodeposition of superhydrophobic polymer films. *Langmuir.* 2010;26:17596–17602.

222. Ma M, Hill RM, Lowery JL, Fridrikh SV, Rutledge GC. Electrospun poly (styrene-block-dimethylsiloxane) block copolymer fibers exhibiting superhydrophobicity. *Langmuir.* 2005;21:5549–5554.

223. Reneker DH, Yarin AL. Electrospinning jets and polymer nanofibers. *Polymer.* 2008;49:2387–2425.

224. Jiang L, Zhao Y, Zhai J. A lotus-leaf-like superhydrophobic surface: a porous microsphere/nanofiber composite film prepared by electrohydrodynamics. *Angew Chem Int Edit.* 2004;116:4438–4441.

225. Agarwal S, Horst S, Bognitzki M. Electrospinning of fluorinated polymers: formation of superhydrophobic surfaces. *Macromol Mater Eng.* 2006;291:592–601.

226. Han D, Steckl AJ. Superhydrophobic and oleophobic fibers by coaxial electrospinning. *Langmuir.* 2009;25:9454–9462.

227. Shi F, Liu Z, Wu GL, Zhang M, Chen H, Wang Z, et al. Surface imprinting in layer-by-layer nanostructured films. *Adv Funct Mater.* 2007;17:1821–1827.

228. Jeffree C. The cuticle, epicuticular waxes and trichomes of plants, with reference to their structure, functions and evolution. Insects and the plant surface Edward Arnold: London. 1986:23–64.

229. Jung YC, Bhushan B. Mechanically durable carbon nanotube-composite hierarchical structures with superhydrophobicity, self-cleaning, and low-drag. *ACS Nano*. 2009;3:4155–4163.

230. Deng X, Mammen L, Butt HJ, Vollmer D. Candle soot as a template for a transparent robust superamphiphobic coating. *Science*. 2012;335:67–70.

231. Lee Y, Yoo Y, Kim J, Widhiarini S, Park B, Park HC, et al. Mimicking a superhydrophobic insect wing by argon and oxygen ion beam treatment on polytetrafluoroethylene film. *J Bionic Eng*. 2009;6:365–370.

232. Zhang X, Zhang J, Ren Z, Li X, Zhu D, Wang T, et al. Morphology and wettability control of silicon cone arrays using colloidal lithography. *Langmuir*. 2009;25:7375–7382.

233. Cho WK, Choi IS. Fabrication of hairy polymeric films inspired by geckos: wetting and high adhesion properties. *Adv Funct Mater*. 2008;18:1089–1096.

234. Sheng X, Zhang J. Superhydrophobic behaviors of polymeric surfaces with aligned nanofibers. *Langmuir*. 2009;25:6916–6922.

235. Bok HM, Shin TY, Park S. Designer binary nanostructures toward water slipping superhydrophobic surfaces. *Chem Mater*. 2008;20:2247–2251.

236. Hoefnagels H, Wu D, G. de With a, Ming W. Biomimetic superhydrophobic and highly oleophobic cotton textiles. *Langmuir*. 2007;23:13158–13163.

237. Qian B, Shen Z. Fabrication of superhydrophobic surfaces by dislocation-selective chemical etching on aluminum, copper, and zinc substrates. *Langmuir*. 2005;21:9007–9009.

238. Xue CH, Jia ST, Zhang J, Tian LQ, Chen HZ, Wang M. Preparation of superhydrophobic surfaces on cotton textiles. *Sci Technol Adv Mat*. 2008;9:035008.

239. Xue CH, Jia ST, Chen HZ, Wang M. Superhydrophobic cotton fabrics prepared by sol–gel coating of TiO2 and surface hydrophobization. *Sci Technol Adv Mat*. 2008;9:035001.

240. Xue CH, Jia ST, Zhang J, Tian LQ. Superhydrophobic surfaces on cotton textiles by complex coating of silica nanoparticles and hydrophobization. *Thin Solid Films*. 2009;517:4593–4598.

241. Ramaratnam K, Tsyalkovsky V, Klep V, Luzinov I. Ultrahydrophobic textile surface via decorating fibers with monolayer of reactive nanoparticles and non-fluorinated polymer. *Chem Commun.* 2007:4510–4512.

242. Leterrier Y. Durability of nanosized oxygen-barrier coatings on polymers. *Prog Mater Sci.* 2003;48:1–55.

243. Lewis J. Material challenge for flexible organic devices. *Mater Today.* 2006;9:38–45.

244. Schulz U. Review of modern techniques to generate antireflective properties on thermoplastic polymers. *Appl Optics.* 2006;45:1608–1618.

245. Ko YH, Kim MS, Yu JS. Controllable electrochemical synthesis of ZnO nanorod arrays on flexible ITO/PET substrate and their structural and optical properties. *Appl Surf Sci.* 2012;259:99–104.

246. Jiang H, Zhao W, Li C, Wang Y. Polymer nanoparticle-based porous antireflective coating on flexible plastic substrate. *Polymer.* 2011;52:778–785.

247. Yun J, Bae T-S, Kwon J-D, Lee S, Lee G-H. Antireflective silica nanoparticle array directly deposited on flexible polymer substrates by chemical vapor deposition. *Nanoscale.* 2012;4:7221–7230.

248. Fateh R, Ismail AA, Dillert R, Bahnemann DW. Highly active crystalline mesoporous TiO2 films coated onto polycarbonate substrates for self-cleaning applications. *J Phys Chem C.* 2011;115:10405–10411.

249. Choi K, Park SH, Song YM, Cho C, Lee HS. Robustly nano-tailored honeycomb structure for high-throughput antireflection polymer films. *J Mater Chem.* 2012;22:17037–17043.

250. Kleinhempel R, Wahl A, Thielsch R. Large area AR coating on plastic substrate using roll to roll methods. *Surf Coat Tech.* 2011;205:S502–S505.

251. Ting C-J, Chen C-F, Chou C. Subwavelength structures for broadband antireflection application. *Opt Commun.* 2009;282:434–438.

252. Lopez JEO, Jacoby WA. Microfibrous mesh coated with titanium dioxide: a self-sterilizing, self-cleaning filter. *J Air Waste Manage.* 2002;52:1206–1213.

253. Freschauf LR, McLane J, Sharma H, Khine M. Shrink-induced superhydrophobic and antibacterial surfaces in consumer plastics. *PLoS One.* 2012;7:e40987.

254. Song J, Xu W, Lu Y. One-step electrochemical machining of superhydrophobic surfaces on aluminum substrates. *J Mater Sci.* 2012;47:162–168.

255. Grignard B, Vaillant A, de Coninck J, Piens M, Jonas AM, Detrembleur C, et al. Electrospinning of a functional perfluorinated block copolymer as a powerful route for imparting superhydrophobicity and corrosion resistance to aluminum substrates. *Langmuir.* 2010;27:335–342.

256. Burmeister F, Kohn C, Kuebler R, Kleer G, Bläsi B, Gombert A. Applications for TiAlN-and TiO2-coatings with nanoscale surface topographies. *Surf Coat Tech.* 2005;200:1555–1559.

257. Chen X, Kong L, Dong D, Yang G, Yu L, Chen J, et al. Fabrication of functionalized copper compound hierarchical structure with bionic superhydrophobic properties. *J Phys Chem C.* 2009;113:5396–5401.

258. Wang H, Dai D, Wu X. Fabrication of superhydrophobic surfaces on aluminum. *Appl Surf Sci.* 2008;254:5599–5601.

259. Wu W, Wang X, Wang D, Chen M, Zhou F, Liu W, et al. Alumina nanowire forests via unconventional anodization and super-repellency plus low adhesion to diverse liquids. *Chem Commun.* 2009:1043–1045.

260. Genzer J, Efimenko K. Recent developments in superhydrophobic surfaces and their relevance to marine fouling: a review. *Biofouling.* 2006;22:339–360.

261. Ganesh VA, Raut HK, Nair AS, Ramakrishna S. A review on self-cleaning coatings. *J Mater Chem.* 2011;21:16304–16322.

262. Handbook of Photovoltaic Science and Engineering EAL, S. Hegedus, John Wiley & Sons, Chichester 2003.

263. Du X, Li X, He J. Facile fabrication of hierarchically structured silica coatings from hierarchically mesoporous silica nanoparticles and their excellent superhydrophilicity and superhydrophobicity. *Acs Appl Mater Inter.* 2010;2:2365–2372.

264. Li X, He J. In situ assembly of raspberry-and mulberry-like silica nanospheres toward antireflective and antifogging coatings. *Acs Appl Mater Inter.* 2012;4:2204–2211.

265. Xu L, He J. Anti-fogging and-reflection coatings fabricated by integrating solid and mesoporous silica nanoparticles without any post-treatments. *Acs Appl Mater Inter.* 2012;4:3293–3299.

266. Xu L, Li X, He J. Mechanically durable superhydrophilic antireflective nanoparticles coatings via layer-by-layer assembly followed by post-treatments. *Acta Chim Sinica.* 2011;69:2648–2652.

267. Geng Z, He J, Xu L. Fabrication of superhydrophilic and antireflective silica coatings on poly (methyl methacrylate) substrates. *Mater Res Bull.* 2012;45:1562–1567.

268. Yao L, He J. Multifunctional surfaces with outstanding mechanical stability on glass substrates by simple H2SiF6-based vapor etching. *Langmuir.* 2013;29:3089–3096.

269. Linn NC, Sun CH, Jiang P, Jiang B. Self-assembled biomimetic antireflection coatings. *Appl Phys Lett.* 2007;91:101108.

270. Sun CH, Gonzalez A, Linn NC, Jiang P, Jiang B. Templated biomimetic multifunctional coatings. *Appl Phys Lett.* 2008;92:051107.

271. Sun CH, Jiang P, Jiang B. Broadband moth-eye antireflection coatings on silicon. *Appl Phys Lett.* 2008;92:061112.

272. Sun CH, Min WL, Linn NC, Jiang P, Jiang B. Templated fabrication of large area subwavelength antireflection gratings on silicon. *Appl Phys Lett.* 2007;91:231105.

273. Min WL, Jiang B, Jiang P. Bioinspired self-cleaning antireflection coatings. *Adv Mater.* 2008;20:3914–3918.

274. Zhang XT, Sato O, Taguchi M, Einaga Y, Murakami T, Fujishima A. Self-cleaning particle coating with antireflection properties. *Chem Mater.* 2005;17:696–700.

275. Wu D, Zhao YB, Wu SZ, Liu YF, Zhang H, Zhao S, et al. Simultaneous efficiency enhancement and self-cleaning effect of white organic light-emitting devices by flexible antireflective films. *Opt Lett.* 2011;36:2635–2637.

276. Park MS, Kim JK. Broad-band antireflection coating at near-infrared wavelengths by a breath figure. *Langmuir.* 2005;21:11404–11408.

277. Zhang L, Li Y, Sun J, Shen J. Layer-by-layer fabrication of broad-band superhydrophobic antireflection coatings in near-infrared region. *J Colloid Interf Sci.* 2008;319:302–308.

278. Camargo KC, Michels AF, Rodembusch FS, Horowitz F. Multi-scale structured, superhydrophobic and wide-angle, antireflective coating in the near-infrared region. *Chem Commun.* 2012;48:4992–4994.

279. Park KC, Choi HJ, Chang CH, Cohen RE, McKinley GH, Barbastathis G. Nanotextured silica surfaces with robust superhydrophobicity and omnidirectional broadband supertransmissivity. *ACS Nano.* 2012;6:3789–3799.

280. Verho T, Bower C, Andrew P, Franssila S, Ikkala O, Ras RHA. Mechanically durable superhydrophobic surfaces. *Adv Mater.* 2011;23:673–678.

281. Roach P, Shirtcliffe NJ, Newton MI. Progess in superhydrophobic surface development. *Soft Matter.* 2008;4:224–240.

282. Sun JG, Li Y, Li L. Bioinspired self-healing superhydrophobic coatings. *Angew Chem Int Edit.* 2010;49:6129–6133.

283. Fan HY, Wright A, Gabaldon J, Rodriguez A, Brinker CJ, Jiang YB. Three-dimensionally ordered gold nanocrystal/silica superlattice thin films synthesized via sol-gel self-assembly. *Adv Funct Mater.* 2006;16:891–895.

284. Whitesides GM, Kriebel JK, Mayers BT. Self-assembly and nanostructured materials. Nanoscale Assembly: Springer; 2005. p. 217–239.

285. Rouse JH, MacNeill BA, Ferguson GS. Sol-gel processing of ordered multilayers to produce composite films of controlled thickness. *Chem Mater.* 2000;12:2502–2507.

286. Miguez H, Tetreault N, Hatton B, Yang SM, Perovic D, Ozin GA. Mechanical stability enhancement by pore size and connectivity control in colloidal crystals by layer-by-layer growth of oxide. *Chem Commun.* 2002:2736–2737.

287. Lee D, Dafinone MI, Feng G, Brugarolas T, Tettey KE. Mechanical reinforcement of nanoparticle thin films using atomic layer deposition. *ACS Nano.* 2011;5:5078–5087.

288. Gemici Z, Shimomura H, Cohen RE, Rubner MF. Hydrothermal treatment of nanoparticle thin films for enhanced mechanical durability. *Langmuir.* 2008;24:2168–2177.

289. Deng X, Mammen L, Zhao Y, Lellig P, Müllen K, Li C, et al. Transparent, thermally stable and mechanically robust superhydrophobic surfaces made from porous silica capsules. *Adv Mater.* 2011;23:2962–2965.

290. George SM. Atomic layer deposition: an overview. *Chem Rev.* 2010;110:111–131.

291. Neinhuis C, Koch K, Barthlott W. Movement and regeneration of epicuticular waxes through plant cuticles. *Planta.* 2001;213:427–434.

292. Choi BG, Park HS. Superhydrophobic graphene/nafion nanohybrid films with hierarchical roughness. *J Phys Chem C.* 2012;116:3207–3211.

293. Cao L, Jones AK, Sikka VK, Wu J, Gao D. Anti-icing superhydrophobic coatings. *Langmuir.* 2009;25:12444–12448.

294. Xu L, He J. A novel precursor-derived one-step growth approach to fabrication of highly antireflective, mechanically robust and self-healing nanoporous silica thin films. *J Mater Chem C.* 2013;1:4655–4662.

295. White SR, Sottos NR, Geubelle PH, Moore JS, Kessler MR, Sriram SR, et al. Autonomic healing of polymer composites. *Nature.* 2001;409:794–797.

296. Shchukin DG, Zheludkevich M, Yasakau K, Lamaka S, Ferreira MGS, Mohwald H. Layer-by-layer assembled nanocontainers for self-healing corrosion protection. *Adv Mater.* 2006;18:1672–1678.

297. Wong TS, Kang SH, Tang SKY, Smythe EJ, Hatton BD, Grinthal A, et al. Bioinspired self-repairing slippery surfaces with pressure-stable omniphobicity. *Nature.* 2011;477:443–447.

298. Page K, Wilson M, Parkin IP. Antimicrobial surfaces and their potential in reducing the role of the inanimate environment in the incidence of hospital-acquired infections. *J Mater Chem.* 2009;19:3819–3831.

299. Feng QL, Wu J, Chen GQ, Cui FZ, Kim TN, Kim JO. A mechanistic study of the antibacterial effect of silver ions on Escherichia coli and Staphylococcus aureus. *J Biomed Mater Res.* 2000;52:662–668.

300. Li QL, Mahendra S, Lyon DY, Brunet L, Liga MV, Li D, et al. Antimicrobial nanomaterials for water disinfection and microbial control: potential applications and implications. *Water Res.* 2008;42:4591–4602.

301. Kikuchi Y, Sunada K, Iyoda T, Hashimoto K, Fujishima A. Photocatalytic bactericidal effect of TiO2 thin films: dynamic view of the active oxygen species responsible for the effect. *J Photoch Photobio A.* 1997;106:51–56.

302. Adams LK, Lyon DY, Alvarez PJJ. Comparative eco-toxicity of nanoscale TiO2, SiO_2, and ZnO water suspensions. *Water Res.* 2006;40:3527–3532.

303. André R, Natalio F, Tahir MN, Berger R, Tremel W. Self-cleaning antimicrobial surfaces by bio-enabled growth of SnO2 coatings on glass. *Nanoscale.* 2013;5:3447–3456.

304. Xie Q, Xu J, Feng L, Jiang L, Tang W, Luo X, et al. Facile creation of a super-amphiphobic coating surface with bionic microstructure. *Adv Mater.* 2004;16:302–305.

305. Sun T, Tan H, Han D, Fu Q, Jiang L. No platelet can adhere-largely improved blood compatibility on nanostructured superhydrophobic surfaces. *Small.* 2005;1:959–963.

306. Gao X, Jiang L. Biophysics: water-repellent legs of water striders. *Nature.* 2004;432:36-.

307. Wang X, Hu H, Ye Q, Gao T, Zhou F, Xue Q. Superamphiphobic coatings with coralline-like structure enabled by one-step spray of polyurethane/carbon nanotube composites. *J Mater Chem.* 2012;22:9624–9631.

308. Steele A, Bayer I, Loth E. Inherently superoleophobic nanocomposite coatings by spray atomization. *Nano Lett.* 2008;9:501–505.

309. Tuteja A, Choi W, Ma M, Mabry JM, Mazzella SA, Rutledge GC, et al. Designing superoleophobic surfaces. *Science.* 2007;318:1618–1622.

310. Rajendra Kumar RT, Mogensen KB, Bøggild P. Simple approach to superamphiphobic overhanging silicon nanostructures. *J Phys Chem C.* 2010;114:2936–2940.

311. Cao L, Price TP, Weiss M, Gao D. Super water-and oil-repellent surfaces on intrinsically hydrophilic and oleophilic porous silicon films. *Langmuir.* 2008;24:1640–1643.

312. Yan H, Kurogi K, Mayama H, Tsujii K. Environmentally stable super water-repellent poly (alkylpyrrole) films. *Angew Chem Int Edit.* 2005;44:3453–3456.

313. Darmanin T, Guittard F. Molecular design of conductive polymers to modulate superoleophobic properties. *J Am Chem Soc.* 2009;131:7928–7933.

314. Wang HX, Xue YH, Lin T. One-step vapour-phase formation of patternable, electrically conductive, superamphiphobic coatings on fibrous materials. *Soft Matter.* 2011;7:8158–8161.

315. Pereira C, Alves C, Monteiro A, Magen C, Pereira AM, Ibarra A, et al. Designing novel hybrid materials by one-pot co-condensation: from hydrophobic mesoporous silica nanoparticles to superamphiphobic cotton textiles. *Acs Appl Mater Inter.* 2011;3:2289–2299.

316. Xiong D, Liu G, Hong L, Duncan EJS. Superamphiphobic diblock copolymer coatings. *Chem Mater.* 2011;23:4357–4366.

317. Xi J, Feng L, Jiang L. A general approach for fabrication of superhydrophobic and superamphiphobic surfaces. *Appl Phys Lett.* 2008;92:053102–053103.

318. He Z, Ma M, Lan X, Chen F, Wang K, Deng H, et al. Fabrication of a transparent superamphiphobic coating with improved stability. *Soft Matter.* 2011;7:6435–6443.

319. Ellinas K, Tserepi A, Gogolides E. From superamphiphobic to amphiphilic polymeric surfaces with ordered hierarchical roughness fabricated with colloidal lithography and plasma nanotexturing. *Langmuir.* 2011;27:3960–3969.

320. Geng Z, He J, Xu L, Yao L. Rational design and elaborate construction of surface nano-structures toward highly antireflective superamphiphobic coatings. *J Mater Chem A.* 2013;1:8721–8724.

321. Duyar O, Durusoy HZ. Design and preparation of antireflection and reflection optical coatings. *Turk J Phys.* 2004;28:139–144.

322. Doshi P, Jellison GE, Rohatgi A. Characterization and optimization of absorbing plasma-enhanced chemical vapor deposited antireflection coatings for silicon photovoltaics. *Appl Optics*. 1997;36:7826–7837.

323. Shin BK, Lee TI, Xiong J, Hwang C, Noh G, Cho JH, et al. Bottom-up grown ZnO nanorods for an antireflective moth-eye structure on CuInGaSe2 solar cells. *Sol Energ Mat Sol C*. 2011;95:2650–2654.

324. Müller J, Rech B, Springer J, Vanecek M. TCO and light trapping in silicon thin film solar cells. *Sol Energy*. 2004;77:917–930.

325. El Daif O, Drouard E, Gomard G, Kaminski A, Fave A, Lemiti M, et al. Absorbing one-dimensional planar photonic crystal for amorphous silicon solar cell. *Opt Express*. 2010;18:A293–A299.

326. Bernhard C, Kolmodin G, Meyerson B. On the prenatal development of function and structure in the somesthetic cortex of the sheep. *Prog Brain Res*. 1967;26:60–77.

327. Furbo S, Jivan Shah L. Thermal advantages for solar heating systems with a glass cover with antireflection surfaces. *Sol Energy*. 2003;74:513–523.

328. Salim A, Huraib F, Eugenio N. PV power-study of system options and optimization. 1988:688–692.

329. Ward JC, Lof GOG. Long-term (18 years) performance of a residential solar heating system. *Sol Energy*. 1976;18:301–308.

330. Motohashi K, Inukai T. Self-cleaning performance evaluation of commercial photocatalyst coating materials through 5 years outdoor exposure. RILEM Int Symp on Photocatalysis, Environment and Construction Materials; 2007; Italy. p. 307–313.

331. Chen J, Ye H, Aé L, Tang Y, Kieven D, Rissom T, et al. Tapered aluminum-doped vertical zinc oxide nanorod arrays as light coupling layer for solar energy applications. *Sol Energ Mat Sol C*. 2011;95:1437–1440.

332. Lee C, Lee T, Lyu S, Zhang Y, Ruh H, Lee H. Field emission from well-aligned zinc oxide nanowires grown at low temperature. *Appl Phys Lett*. 2002;81:3648.

333. Zhu X, Zhang Z, Men X, Yang J, Wang K, Xu X, et al. Robust superhydrophobic surfaces with mechanical durability and easy repairability. *J Mater Chem*. 2011;21:15793–15797.

334. Levkin PA, Svec F, Fréchet JMJ. Porous polymer coatings: a versatile approach to superhydrophobic surfaces. *Adv Funct Mater*. 2009;19:1993–1998.

335. Liu K, Yao X, Jiang L. Recent developments in bio-inspired special wettability. *Chem Soc Rev*. 2010;39:3240.

Chapter 2

Functionalization of Polyelectrolyte Multilayers via Counterions

Xin Zhang and Zhaohui Su
State Key Laboratory of Polymer Physics and Chemistry,
Changchun Institute of Applied Chemistry,
Chinese Academy of Sciences,
Changchun 130022, China
zhsu@ciac.ac.cn

2.1 Introduction

Functional thin films have attracted tremendous interest over the years because of their application in a multitude of fields ranging from coatings to optoelectronics. Various strategies have been developed for construction of functional thin films, such as Langmuir-Blodgett technique[1,2] and the self-assembly monolayer method.[3] Compared with these traditional methods, layer-by-layer (LbL) deposition technique is a facile and versatile method for construction of thin films with controlled architecture and compositions that has become extremely popular in recent two decades.[4] The technique

Nanomaterials in Energy and Environmental Applications
Edited by Junhui He

ISBN 978-981-4463-78-2 (Hardcover), 978-981- 4463-79-9 (eBook)
www.panstanford.com

was first reported by Iler in 1966, who fabricated multilayer films by alternating deposition of positively and negatively charged inorganic colloidal nanoparticles.[5] However, Iler's work went largely unnoticed, and in the early 1990s, the LbL technique was rediscovered by Decher and Hong using polyelectrolytes and other charged species.[6,7] The principle of the LbL technique is illustrated schematically in Fig. 2.1. A negatively charged substrate is immersed in a polycation solution, upon which a polycation layer is deposited onto the substrate, and the surface now is positively charged. The substrate is rinsed with water to remove the residual solution and immersed in a polyanion solution to deposite a layer of polyanion, restoring negative surface charges, and after a second water rinse, the substrate is ready for the next deposition cycle. The process is repeated for a desired number of cycles and a multilayer film is fabricated.[8] The technique is simple and straightforward, and requires no sophisticated apparatus or tedious operating procedures. It can be applied to various substrates ranging from metals and oxides to polymers including polytetrafluoroethylene (PTFE), of different sizes and shapes, including cells and viruses.[9–11] In addition to polyelectrolytes, a great variety of organic and inorganic building blocks can be assembled, such as polyelectrolyte complexes,[12] proteins,[13,14] DNA,[15,16] dye molecules,[17] silica colloids,[18] metal nanoparticles,[19] metal oxides,[20] clays,[21] and various carbon materials.[22] Live cells have been assembled into multilayers.[23] Various types of complexes such as electrostatic complexes, hydrogen bonding complexes, block copolymer micelle encapsulation, and π-π interaction complexes have been utilized, and these largely enhance the concept of LbL assembly.[24] In addition, multilayer films can be constructed on the basis of driving forces other than electrostatic interactions, including hydrogen bonding, molecular recongnition, charge-transfer interactions, step-by-step reactions, sol-gel processes and noncovalent modification.[25]

Traditionally, the multilayer films are built by dipped solution adsorption (dip-LbL), as described above, which takes relatively long time, and then researchers developed spray-assisted LbL, spin-LbL and spin-spray LbL techniques, which allow much faster construction of multilayers and are easier to adapt for industrial production.[26–29] These different techniques have been compared and reviewed by Hammond[30] and Schaaf.[31]

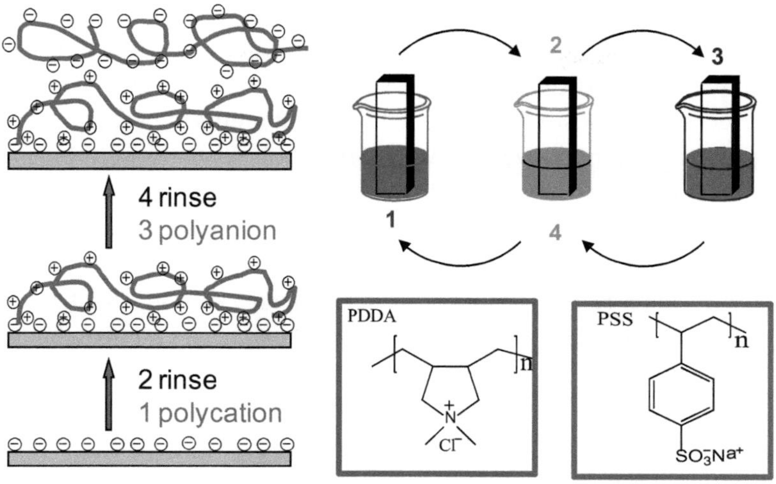

Figure 2.1 Schematic illustration of the LbL deposition process.

The extremely simple and versatile LbL technique has been empolyed extensively to build functional thin films. Various charged species that are readily available or designed and synthesized have been assembled directly to impart desired functionalities to the multilayer films. For example, gelatin was directly assembled into multilayer films, which were used as coatings to mediate cytocompatibility;[32] thin films used as biosensors and bioreactors were fabricated using different kinds of proteins, enzymes and DNAs;[33] polyelectrolyte was modified with phosphorylcholine moities to create multilayers resistant to protein adsorption even under mechanical stretching;[34] polyelectrolyte-metal ion complex was prepared and then assembled into a multilayer film, in which the metal ions subsequently can be reduced to yield catalytic metal nanoparticles;[35] polyelectrolytes of opposite charges were mixed in aqueous solution to form soluble complexes, which were used as building blocks for rapid fabrication of thick LbL films used as self-healing hydrophobic coatings or transparent and scratch-resistant coatings.[12] Alternatively, functionalities can be introduced to PEMs prepared from common polyelectrolytes via post-treatment. For instance, multilayers assembled from poly(allylamine hydrochloride) (PAH) and poly(acrylic acid) (PAA), both weak polyelectrolytes, can be induced by appropriate acid treatment to form pores and honeycomb-like structure, which upon deposition of SiO_2 nanoparticles and a fluorosilane top layer can impart superhydrophobicity.[36] By controlling the pH in the assembly process or in post-treatment, significant amounts of carboxylic acid groups were preserved in PAA/PAH multilayers and used to bind metal ions, which via subsequent reduction or other reactions formed nanoparticles.[37] Recently, we discovered another avenue to functional multilayers provided by the counterions existing in the multilayers. These counterions can be used as vehicles to introduce different functional species into the multilayer films via ion exchange. Because counterions are present universally in all multilayers assembled via electrostatic interactions, this is a general approach to fabrication of functional multilayer films.

2.2 Counterions in Polyelectrolyte Multilayers

In general, salt of different types and concentrations is usually maintained in polyelectrolyte solutions used for PEM fabrication. In these solutions, polyelectrolyte charges are compensated by the oppositely charged salt ions as counterions to some degree. The salts in these solutions play an important role in the multilayer formation process.[38] They can affect the adsorption kinetics,[39] layer thickness,[40,41] structure of the multilayer,[42] mechanical stiffness,[43] swelling behavior,[44] and permeability of the film.[45] For example, Barrett et al. reported faster adsorption kinetics of polyelectrolyte with larger counterions in multilayer buildup process.[39] Salomaki et al. studied the effects of anions on the growth of PEMs and found that the chaotropic ions can better screen the polyelelctrolyte charges, resulting in a loopy conformation of the polyelectrolyte chains in the deposition process and formation of thicker films, whereas the cosmotropic ions do not screen the polyelectrolyte charges well and lead to more planar films.[41] When divalent anions are used in the PEM fabrication, they can promote intramolecular and intermolecular bridging in the polyelectrolyte, and influence profoundly the film structure, leading to formation of void defects in the PEM and porous multialyer films.[42] The effect of counterion concentration and identity on the multilayer film thickness was studied systematically by Schlenoff and coworkers in 1999.[40] The counterions can also affect the elastic properties and swelling behavior of the multilayer films, and the film stiffness was found to increase with charge screening produced by the counterions.[43,44] Furthermore, it is noted that counteranions generally exhibit greater effect on the properties of the multilayer films than countercations.[41,46]

In a PEM, a charge on a polyelectrolyte can be balanced either by an oppositely charged unit on another polyelectrolyte or by a small counterion from polyelectrolyte solutions, which is termed intrinsic charge compensation and extrinsic charge compensation, respectively (Fig. 2.2).[47,48] From Fig. 2.1, it is clear that there are excess polyelectrolyte units (excess charges) of one type with associated counterions, that is, extrinsic compensation, present in a PEM, at least at the surface. However, whether counterions are present or not in a PEM remained controversial for quite some time. An early study by X-ray photoelectron spectroscopy (XPS) showed a 1:1 stoichiometry between the polycation and the polyanion, and detected no counterion in poly(butanyl viologen) dibromide/poly(styrene sulfonate) (PBV/PSS) multilayer films.[49] Yet in a different study by Losche and coworkers, counterions were found to be present in the PSS/PAH system via X-ray reflectivity.[50] Schlenoff and coworkers then further investigated counterion contents in PEMs assembled from poly(diallyldimethylammonium chloride) (PDDA) and PSS using a radioanalytical method, and concluded that charges on polyions balanced each other quantitatively, and excess charges and counterions were present only at the surface of the PEM.[47] Another study by Riegler and Essler, on the contrary, showed that the polyelectrolyte charge stoichiometry was not 1:1 in PAH/PSS multilayers, and the films contained some small counterions for charge compensation, the content of which depending on the charge density of the polyelelctrolyte.[51] So far, there have been extensive reports confirming the presence of counterions in multilayer films, including a series of papers by our group to be described below in this chapter, and the counterion content in a PEM has been shown to be strongly affected by the washing and drying during each deposition step.[52]

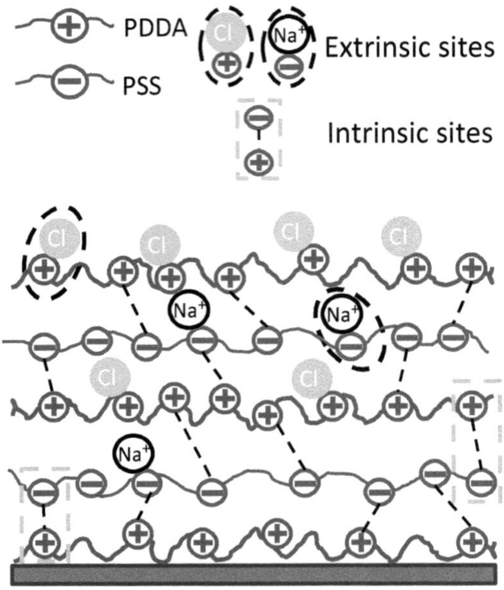

Figure 2.2 Intrinsic and extrinsic charge compensation.

With the presence of counterions in PEMs confirmed, the next and more complicated question is the distribution of counterions in the film, or more specifically, whether the counterions are present or not in the bulk of a PEM. Barrett and coworkers used neutron reflectometry to explore ion distribution in PAA/PAH PEMs, and found that in films assembled with salt, counterions localize near the substrate, and the counterion distribution is greatly influenced by the capping layer.[53] Schollmeyer and coworkers employed standing-wave X-ray fluorescence to map the ion profile in PAH/PSS PEMs using bromide ions as probe entities, and showed that the sign of the charge of the outmost layer as well as the washing procedure decisively impact the ion concentration and distribution in the PEM; the PEMs capped with PSS were found to contain fewer counteranions than PAH terminated ones.[54] Schlenoff and coworkers showed that in multilayers assembled from PDDA and PSS in the presence of NaCl of 1 M or lower concentrations, wherein the PEM growth is linear, the counterions reside only at the surface of the PEM and not in the bulk.[47] Zan and coworkers revealed that for PSS-terminated PDDA/PSS PEMs, at low salt concentrations, the countercations are confined at the surface, similar to Schlenoff's finding, whereas in the PEMs assembled at high salt concentration, wherein the PEM thickness grows exponentially with number of layers, there exist a significant amount of excess PSS units and associated countercations in the bulk of the PEM.[55]

We then systematically studied the counterion content and distribution in PDDA/PSS multilayers as functions of salt concertration, number of layers deposited, and the identity of the capping layer, wherein quartz crystal microbalance (QCM), QCM with dissipation (QCM-D), UV-vis spectroscopy, and XPS were used to determine the dispositions of polyelectrolytes and counterions across the PEM thickness.[56] PEMs are essentially charge neutral, so a mismatch in the number of ionizable units between the two polyelectrolytes mandates an excess of one of their counterions. Such mismatches are immediately manifested in the QCM data. The incremental mass depositions during LbL assembly were quantified by QCM, which uncovered as functions of deposition number (N) excess polyelectrolyte charge (Σ) and excess polyelectrolyte charge density (σ), defined by the following two equations:

$$\Sigma = \sum_{iodd}^{N} \frac{\Delta m^i}{M_{PDDA}} - \sum_{ieven}^{N} \frac{\Delta m^i}{M_{PSS}}$$

$$\sigma = \frac{\Sigma}{h} = \frac{\Sigma\rho}{m_f}$$

where M_{PDDA} and M_{PSS} are the respective PDDA and PSS repeat unit molecular weights, and h, m_{f}, and ρ are PEM thickness, areal mass, and density, respectively. Σ and σ were then used to study the counterions distribution. Counterion dispositions depend strongly on salt concentration, and trends of the two PEM charge parameters have been established for three salt concentration regimes: zero to near zero salt ([NaCl] $\leq$ 0.1 M), low salt (0.1 M $\leq$ [NaCl] $\leq$ 0.75 M), and high salt ([NaCl] $\geq$ 1.5 M). The first two are associated with linear LbL growth while the latter is associated with exponential LbL growth. At zero or near zero salt concentration, flexible polyelectrolytes such as PDDA and PSS adopt extended chain conformations due to the repulsive electrostatic interactions among their chain segments; the LbL assembly of extended polyelectrolytes leads to thin PEMs, and the polyelectrolytes deposited in relatively flat conformations, and for an incoming polyelectrolyte, this provides effective ion pairing with the oppositely charged polyelectrolytes already at/near the surface. As a result, excess polyelectrolyte charge is localized to the PEM surface and the accumulation of counterions in the PEM bulk is negligible. At low salt, although Σ oscillates with *N* due to capping layer alternation, a more obvious trend is the large growth of Σ with *N*; however, σ oscillations with *N* superimpose on a slow mean decay that asymptotes to a finite, positive average value, which indicates the presence of a constant (i.e., *N*-independent) density of counterions in the PEM bulk (Fig. 2.3). Note that both excess charge and excess charge density are positive for all *N*, indicating that

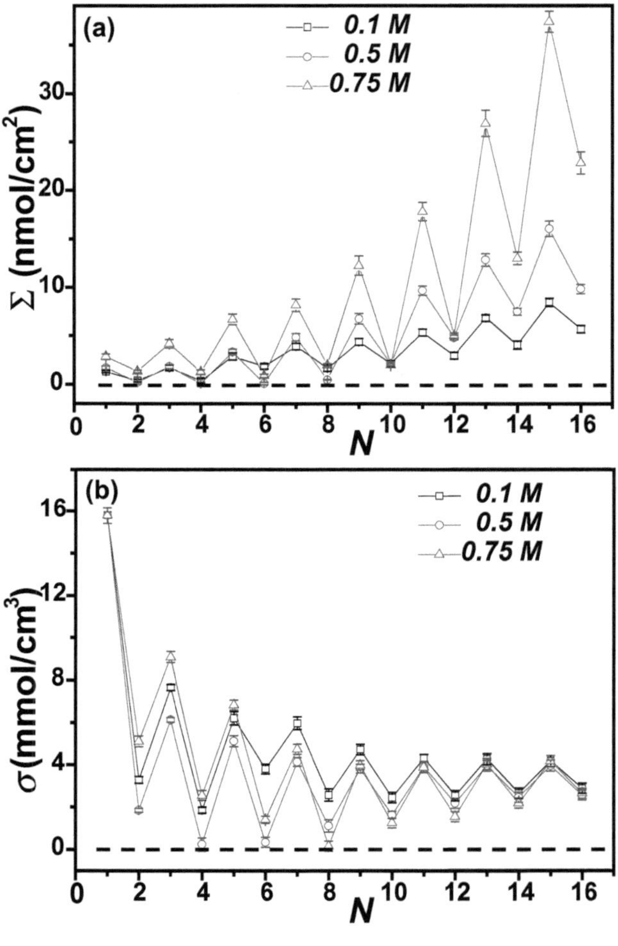

Figure 2.3 (a) Excess polyelectrolyte charge per area Σ and (b) excess polyelectrolyte charge density σ plotted as functions of deposition number *N* for PEMs fabricated at three low NaCl concentrations. Reprinted with permission from Ref. [56] Copyright 2012 American Chemical Society.

the identity of the capping polyelectrolyte does not affect the bulk accumulation of excess PDDA charges (with counteranions) across the bulk. At high salt regime, however, at high enough *N* (~6–12), an excess accumulation of PSS in the PEM bulk is indicated, coinciding with the presence of countercations (Fig. 2.4), due to deposited PSS permeating the PEM bulk conveying an excess of countercations.[57] At all salt concentrations, the PEM surface charge alternates according to the capping polyelectrolyte's identity, yet accumulations of small ions in the PEM bulk occur in the presence of salt that can be ascribed to property asymmetries between the two deposited polyelectrolytes. The counterion dispositions in PDDA/PSS PEMs under different conditions are schematically illustrated in Fig. 2.5.[56]

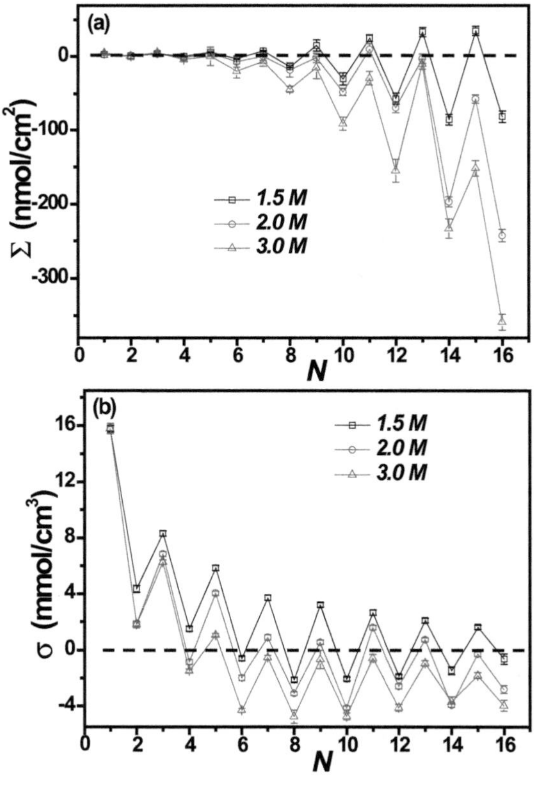

Figure 2.4 (a) Excess polyelectrolyte charge per area Σ and (b) excess polyelectrolyte charge density σ plotted as functions of deposition number *N* for PEMs fabricated at three high NaCl concentrations. Reprinted with permission from Ref. [56] Copyright 2012 American Chemical Society.

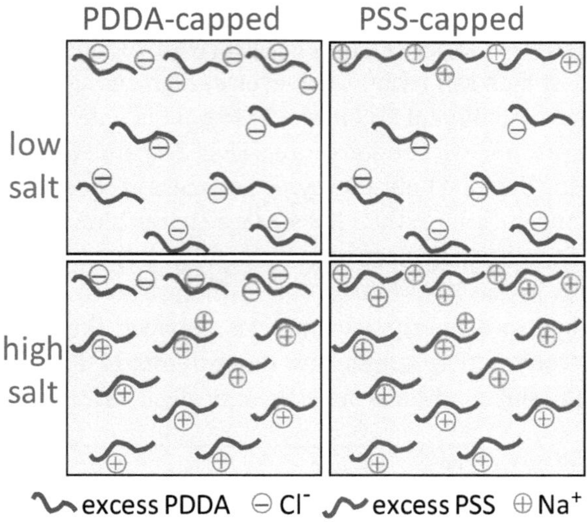

Figure 2.5 Schematic illustration of counterion distribution under different conditions. Reprinted with permission from Ref. [56] Copyright 2012 American Chemical Society.

Recently, Schlenoff and coworkers reported similar overcompensation asymmetry in PDDA/PSS PEMs.[48] They used radioactive counterions to track the contents of the counterions and the ratio of positive to negative polymer repeat units within PDDA/PSS multilayers that grow linearly, and found that although surface charge overcompensation occurs upon PDDA addition, PSS merely compensates the PDDA. As a result, positive sites begin to accumulate in the PEM, and the density of positive charges grows with the number of layers, consistent with our results at low salt. After about 12–14 layers are deposited, the PEM contains counteranions whether the capping layer is PDDA or PSS.

2.3 Surface Wettability Modulation

As discussed earlier, counterions are present in multilayer films constructed by strong or weak polyelelctrolytes. The existence of counterions can significantly affect the properties of PEMs, especially surface properties. Wettability is one of the most important surface properties because of their extensive implication in various fields such as self-cleaning surfaces, antifogging coatings, antiadhesive coatings, microfluidics, and optical lenses. Here by going through some examples, we show that PEMs can be utilized as versatile coatings with interesting surface properties due to counterions in the PEMs.

2.3.1 Surfaces with Tunable Wettability

Design and fabrication of smart surfaces with controllable wetting properties has attracted significant interest in recent years because of their application in many fields, such as coatings, adhesion, microfluidics, and marine anti-fouling. The wetting properties of these surfaces can be manipulated by external stimuli, such as pH, electrochemical potential, irradiation, solvent, or ionic strength. Recently, a potentially more versatile route has been demonstrated, wherein the counterions in self-assembled monolayers[58] or grafted polyelectrolyte brushes[59] are utilized to control the surface wettability. However, the self-assembled monolayers and polyelectrolyte brush grafting require specific substrate chemistry for the approaches to work. The LbL techinque on the contrary, is simple, versatile, and inexpensive, and is generally applicable to a variety of substrates. Thus PEMs, carrying counterions as discussed in the previous section that are readily available for ion exchange, provide a more general and facile approach to wettability modulation.[60,61]

PDDA and PSS, typical polyelectrolytes used in LbL studies, were employed to build the multilayers at 0.25 M NaCl concentration. On a flat Si wafer substrate, the $(PDDA/PSS)_3PDDA$ PEM as assembled, with Cl^- as the counterion for the capping PDDA, exhibited a water contact angle of nearly 19°, indicating a hydrophilic surface. The PEM was then immersed in an aqueous solution of a sodium salt with a different anion (0.10 M) for several minutes, removed and rinsed thoroughly, and dried with a stream of nitrogen. XPS analysis of the surface verified that the counteranion in the PEM was exchanged for the anion in the solution. The water contact angle for the PEM carrying the new counteranion was then measured. As seen in Fig. 2.6a, the PEM surface exhibits different water contact angles depending on the identity of the counteranion. For example, the contact angle for the surface treated with perfluorooctanoate (PFO) is 114°, indicative of a hydrophobic surface. Thus, the ultrathin PEM coating renders tunable wettability to the substrate surface. The wettability of the PEM further depends on the salt concentration in the polyelectrolyte solutions used for the assembly; the exchange of the counterion is rapid and completely reversible (Fig. 2.6 b,c).[61]

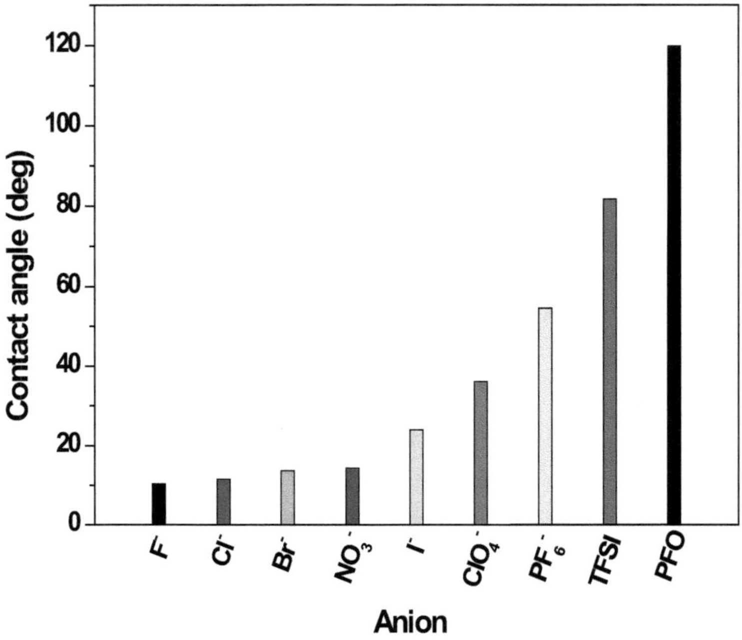

Figure 2.6 (a) Tuning of the PEM wetting characteristics via the counteranion.

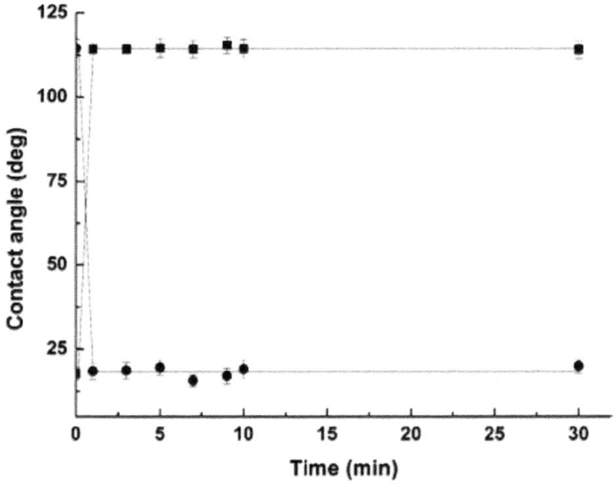

Figure 2.6 (b) Kinetics for Cl^- in the PEM exchanging for PFO (■) and PFO in the PEM exchanging for Cl^- (•).

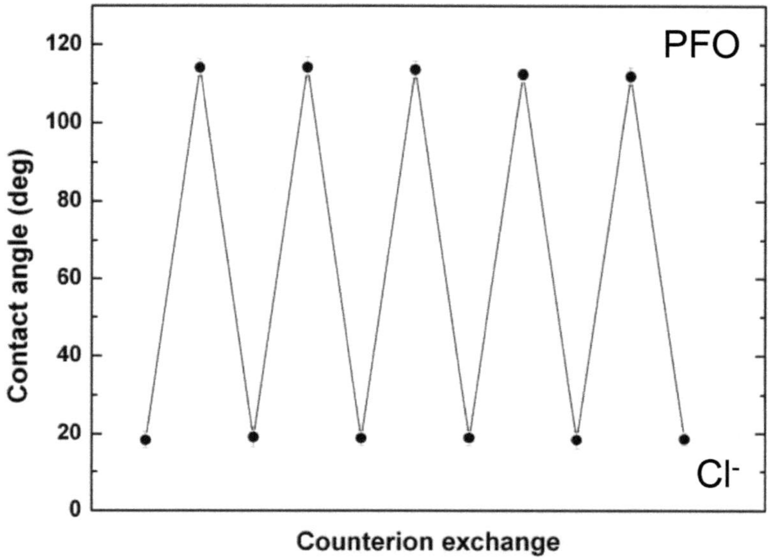

Figure 2.6 (c) Water contact angle as a function of the counterion identity, showing the reversible behavior of the PEM over repeated cycles of Cl– and PFO counterion exchange.

The process is depicted in Fig. 2.7. The use of PEM for wettability modulation is extremely simple, involving no chemical modification of the substrate or the polyelectrolytes used, and is expected to be applicable to any substrate conferred by the LbL technique.

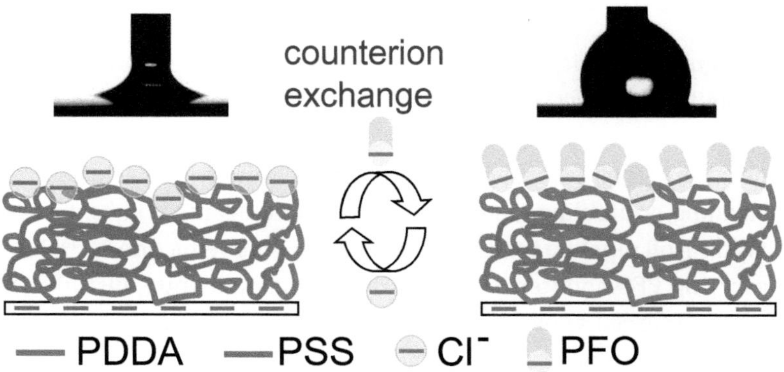

Figure 2.7 Schematic illustration of the counterion exchange in the PEM.

2.3.2 Superhydrophobic and Superhydrophilic Surfaces

Wettability is not only determined by surface chemical composition but also affected by surface topography. Surface roughness is a necessary factor for achieving superhydrophobicity. On a silicon wafer, a rough gold structure was created using the galvanic exchange reaction. Field-emission scanning electron microscopy (FESEM) revealed nano-scaled papillae on micro-scaled gold clusters present at the surface, exhibiting roughness at two different length scales, which as observed in lotus leaf is known to amplify the wetting properties of the surface.[62] The PEM assembled on the rough substrate was about 31 nm in thickness, thin enough to largely preserve the surface topography of the rough substrate. In this case, the deposited PEM with Cl^- counterion exhibited superhydrophilicity with a water contact angle of less than 5°, much lower than that for the flat surface as a result of the 3D capillary effect. After the Cl^- counterions in the PEM were exchanged for PFO anions, the surface became superhydrophobic, showing a water contact angle of 164° and a small sliding angle of less than 5°. The ion exchange process was completely reversible. This example demonstrate that combination of PEM with micro/nanostructured substrate provides a facile approach to surfaces with dynamically and reversibly tunable wettability between superhydrophilicity and superhydrophobicity (Fig. 2.8).[63]

2.3.3 Dual Superoleophobic Surfaces

Design and fabrication of superoleophobic surfaces, surfaces that exhibit oil contact angle greater than 150° and low contact angle hysteresis, has aroused worldwide interest in the past several years because of their extensive application in fields such as marine oil spill cleaning, antifouling coating, industrial metal cleaning, oil/water separation, oil pipeline treatment, and fluidic drag reduction.[64–66] PEMs carrying PFO counterion coated on a rough substrate, just like the ones discussed above, can exhibit superoleophobicity in air. On the contrary, in some practical applications, such as coatings for aquatic devices operating in seawater, the upper part of the device is exposed in air and the lower part submerges in seawater at the same time, and oil repellency both in air and underwater is highly desirable. It

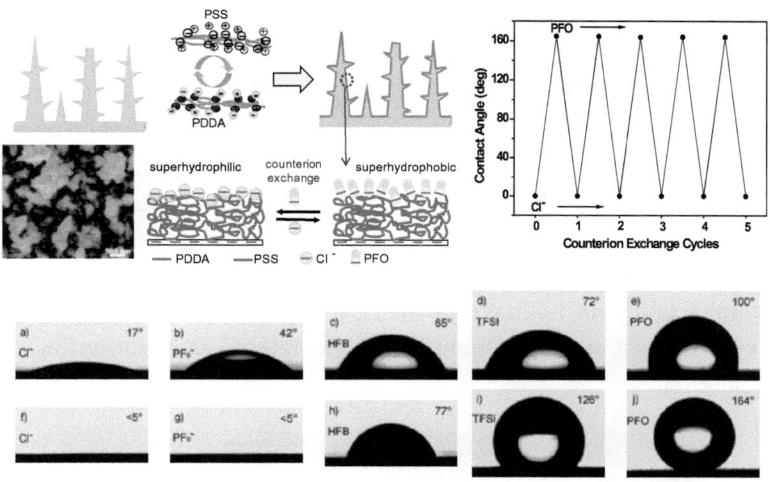

Figure 2.8 Reversible wettability modulation from superhydrophilicity to superhydrophobicity via counterion exchange in PEM deposited on a rough substrate. The SEM image shows the surface morphology of the substrate. The optical images show the wetting behavior of water on the PEM with different counterions on flat (upper) and rough (lower) substrates. Reprinted with permission from Ref. [63] Copyright 2010 American Chemical Society.

has been well established that roughness with dual length scales is necessary for superoleophobicity both in air and underwater; the surface energy requirements for the two cases, however, contradict one another. Superoleophobicity in air requires low surface energy materials such as fluoro compounds, whereas hydrophilic materials are needed for superoleophobicity underwater, where water wets and is trapped in the hierarchically rough structure to repel the oil, just like air does in the former. As a result, a surface that is superoleophobic in air loses its oil repellency when going into water, and vice versa. Taking advantage of the PEM ion exchange chemistry, we are able to achieve superoleophobicity both in air and underwater. A PDDA/PSS PEM coating of nearly 10 nm in thickness was deposited onto an aluminum plate with micro/nano hierarchical surface structure, and the counterion was exchanged for PFO. In air, this surface was superhydrophobic (water contact angle 155°) and superoleophobic, with a 1,2-dichloroethane contact angle of 151°. This plate when submerged in seawater showed a 1,2-dichloroethane contact angle of 164° underwater (Fig. 2.9). Similar dual superoleophobicity was observed when other oils, such as n-hexadecane and rapeseed oil, were tested. That is, the PEM coating is superoleophobic both in air and in seawater.[67] This is because when the coating is submerged in seawater, the PFO counteranion in the PEM is exchanged spontaneously by the rich Cl^- and SO_4^{2-} anions in the seawater, making the surface superhydrophilic and hence superoleophobic underwater. Superoleophobicity in air for the PEM coating can be easily restored by treating it with a PFO solution to reinstall the PFO counterion into the PEM.

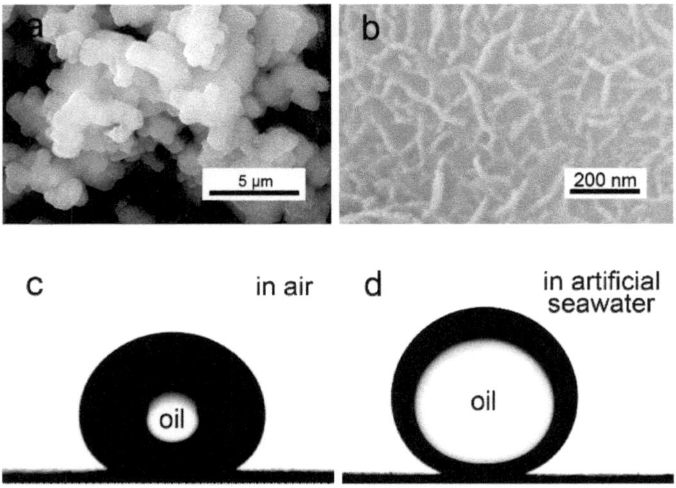

Figure 2.9 (a, b) SEM images of a PEM coating on a rough substrate at different scales. (c, d) Optical image of a 1,2-dichloroethane droplet on the surface (c) in air with a contact angle of 151°, and (d) in artificial seawater with a contact angle of 164°. Reprinted with permission from Ref. [67] Copyright 2013 American Chemical Society.

2.3.4 Rewritable Wettability Gradient

Wetting properties of a surface depend on the chemical composition of the surface, as indicated by the Cassie–Baxter equation. For a PEM surface as described above, by continuously varying the content of a particular counterion along the surface, gradient wettability can be created. This can be conveniently accomplished by controlling the ion-exchange kinetics to manipulate the extent of the ion exchange reaction.[63] Figure 2.10 displays the water contact angle of the PDDA/PSS PEM surface as functions of time the surface is in contact with PFO solutions of different concentrations, showing the dependence of ion-exchange kinetics on the concentration of the PFO solution. When the PFO concentration is 0.025 mmol/l, the ion-exchange kinetics is sufficiently slow, and the contact angle increases gradually with time over a period of more than 40 s, a time scale suitable and convenient for fabricating the wettability gradient. A rough substrate coated with the PEM with Cl^- as the counterion, as discussed above in Section 2.3.2, was slowly dipped into the PFO solution of 0.025 mmol/l concentration at a constant rate, so positions along the latitude of the substrate were gradually exposed to the PFO solution and thus corresponded to a continuously changing reaction time. The substrate was then quickly withdrawn from the solution and rinsed with water. As seen in Fig. 2.10, the water contact angle varies continuously along the surface from higher than 150° at the bottom of

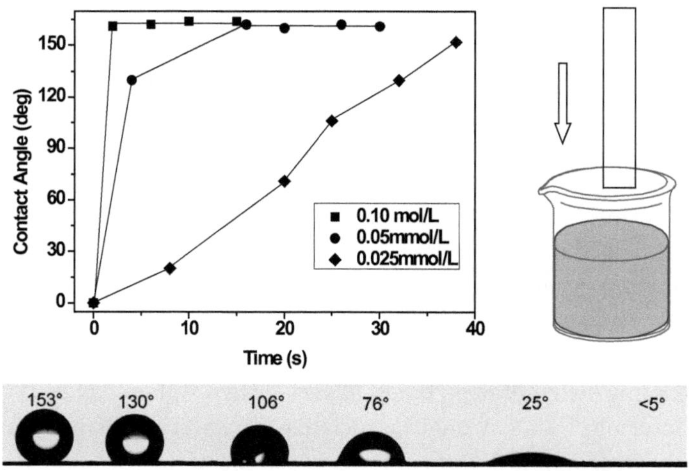

Figure 2.10 Counterion exchange kinetics at different PFO concentrations (left), a schematic illustration of the gradient formation by variation of the counterion exchange time (right), and an optical image of water droplets sitting on different locations of the surface, indicating the gradient wettability along the surface (bottom). Reprinted with permission from Ref. [63] Copyright 2010 American Chemical Society.

the slide, which was immersed into the PFO solution first and stayed in contact with the solution for the longest time, to lower than 5° at the top of the slide, which was barely exposed to the solution. Thus, we have fabricated a wettability gradient from superhydrophilicity to superhydrophobicity in less than 1 min via counterion exchange of the PEM. Because the counterion exchange involved is completely reversible, as demonstrated above, the gradient wettability can be erased easily by dipping the substrate into a NaCl solution to reinstall the Cl^- counterions; then, a same or different gradient can be created on the same substrate using the counterion exchange process again.

2.3.5 Sticky Superhydrophobic Surfaces

The term superhydrophobic usually indicates surfaces that display a water contact angle higher than 150°, and these surfaces often show low contact angle hysteresis as well, on which a water droplet can easily roll off, exhibiting anti-adhesion behavior. However, McCarthy and co-workers discovered a different kind of superhydrophobic surface with water contact angle hysteresis as high as 161°,[68] and this kind of "sticky superhydrophobic" surface has generated increasing interest in recent years. A rough substrate was prepared by coating a flat silicon wafer with submicron-scaled silica spheres, which were crosslinked, and then a $(PDDA/PSS)_3PDDA$ PEM was assembled at 1.0 M NaCl concentration on the substrate, the counterion of which was subsequently exchanged for PFO. The advancing and static contact angle was 156° and 152°, respectively, indicating superhydrophobicity of the surface. The receding contact angle, however, was nearly 0°. The extremely high contact angle hysteresis value of greater than 150° was due to hydrophilic defects, randomly distributed on the PEM surface hydrophobized by the PFO counterions[69] that strongly associate with water molecules. Pinning by the hydrophilic defects of the contact line of the receding liquid is enhanced on the microstructured substrate because the hydrophilic defects can cause penetration of the water droplet between the surface asperities on the rough substrate, resulting in extremely high values of the contact angle hysteresis and "sticky superhydrophobicity." In fact, a water droplet remains pinned to the surface even when the substrate is turned upside down (Fig. 2.11).[70]

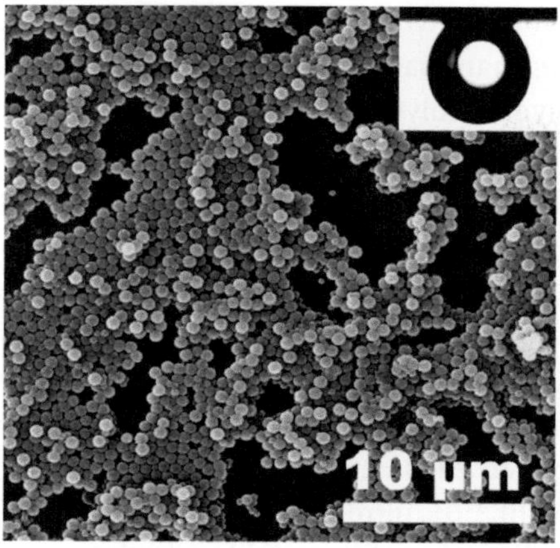

Figure 2.11 SEM micrograph of silica spheres deposited on a Si wafer, creating a rough substrate. The $(PDDA/PSS)_3PDDA$ PEM assembled on the substrate with PFO counterion exhibiting sticky superhydrophobicity. Inset shows a water droplet pinned to the surface turned upside down. Reprinted with permission from Ref. [70] Copyright 2011 American Chemical Society.

2.4 PEM Supported Metal Nanoparticles

Various inorganic nano materials, including metal nanoparticles, semiconductor nanoparticles, and magnetic nanoparticles, have been investigated intensely due to their unique properties and extensive application. Generally, the nanoparticles are synthesized in liquid media, however, for many practical applications, such as catalysis,[71] surface-enhanced Raman scattering (SERS),[72] optoelectronic and sensor devices, the nanoparticles often need to be immobilized on a substrate or embedded in a matrix. When the solution-dispersed nanoparticles are deposited on the support or embedded in the matrix, it is quite challenging to retain their intrinsic structure, size, and morphology. Therefore, in situ synthesis of nanoparticles in polymer matrixes is highly desirable.[73] The porous and supramolecular structure of PEMs affords excellent nanoreactors for fabrication of nanoparticles in situ. For example, Stroeve and co-workers used PEMs composed of strong polyelectrolytes as templates for in situ synthesis of nanoparticles of inorganic materials, such as β-FeOOH,[74] γ-FeOOH,[75] cobalt hydroxide,[76] nickel hydroxide,[77] and PbS.[78]

Rubner and co-workers synthesized different metal and semiconducting nanoparticles in PEMs built from PAA and PAH.[37] As weak polyelectrolytes, the linear charge density and the concentration of free carboxylic acid groups of PAA and free amine groups of PAH can be easily controlled by adjusting the pH of the dipping PAA and PAH solutions. The free carboxylic acid groups are able to bind metal cations (e.g., Ag^{+}) for preparation of Ag nanoparticles,[37,79] whereas the free amine groups in PAH are used to bind anionic precursors such as gold complex ions (e.g., $AuCl_4^{-}$) for synthesis of Au nanoparticles.[80] The metallic ions loaded in the PEMs are exposed to H_2 atmosphere,[79] irradiated with ultraviolet or ambient light,[81,82] immersed in a reductant solution,[83,84] or electrochemically reduced[85] to yield metallic nanoparticles. The key element in this approach is free ionic groups in the PEMs that can bind the metallic ion precusors. Actually, one needs not manipulate the pH in order to find free ionic groups in PEMs. As discussed above in Section 2.2, there always exist excess charged groups and corresponding counterions in a PEM.[86] Taking advantage of these counterions, precursor ions can be introduced via ion exchange and then various metallic nanoparticles can be synthesized in the PEM. The syntheses of Ag,[55,87] Au,[88] Pt[89] monometallic and Au@Pt,[88] Au@Ag[90,91] core-shell, and hollow Ag–Au[92] bimetallic nanoparticles in PDDA/PSS PEMs are given as examples.

2.4.1 Monometallic Nanoparticles

A $(PDDA/PSS)_3$, PEM was assembled on a cleaned silicon wafer from PDDA and PSS solutions with 3.0 M NaCl present. XPS analysis revealed the presence of abundant Na^{+} counterions and absence of Cl in the film, consistent with the results presented in Fig. 2.5 that PDDA/PSS PEMs assembled at high salt concentrations are in excess of PSS with associated countercation, and further, there was no absorption of NaCl salt by the multilayer. The PEM was then immersed in an $AgNO_3$ solution for several minutes to exchange the Na^{+} counterion for Ag^{+}. The ion exchange kinetics was rapid and the time used was sufficient to complete the exchange. In the XPS spectrum, the Na peaks disappeared and Ag peaks emerged, and the N1s peak for NO_3^{-} ion was not observed. Furthermore, the Ag^{+} content was found to be independent of the $AgNO_3$ solution concentration. These results clearly proved that the Ag^+ in the PEM was introduced via counterion exchange, and not by absorption of $AgNO_3$.[55,87]

The Ag^+ counterions in the PEM were then reduced in situ with $NaBH_4$ to produce silver nanoparticles. The surface plasmon resonance peak at nearly 400 nm in the UV-vis spectrum as well as the transmission electron microscopy (TEM) image and selected-area electron diffraction (SAED) pattern confirmed the presence of the silver nanoparticles in the PEM. As the counterion content in the PEM is strongly influenced by the concentration of salt in the polyelectrolyte solutions used to build the PEM as well as the number of layers in the assembly, these factors were utilized to tune the size and content of the Ag nanoparticles synthesized. Nanoparticles of bigger sizes and in greater quantities were produced in the PEMs assembled at higher salt concentrations and with more layers. The PEMs containing Ag nanoparticles thus fabricated may be used as versatile transparent antimicrobial coatings.[93] The antibacterial performance of the composite films can be enhanced by increasing the silver content via repeating the ion-exchange/reduction cycle. In addition, films containing silver in the ionic form exhibit higher initial bactericidal activity, but the silver ions often are consumed quickly, whereas the coatings containing silver nanoparticles show better long-term antimicrobial effects. Therefore, by leaving the Ag^+ exchanged into the PEM in the last cycle in the ionic form, PEMs containing high contents of both ionic silver and silver nanoparticles can be obtained, which should have high initial antimicrobial efficacy as well as long-term performance.

Furthermore, countercations exist only in the top layers and not in the bulk of the PEMs assembled at low salt concentrations, but are present throughout the entire thickness in the PEMs fabricated at high salt concentrations. In other words, $(PDDA/PSS)_n$ blocks assembled at high salt concentrations are capable of binding metal ions in the entire block, whereas the $(PDDA/PSS)_m$ blocks assembled at low salt or salt-free contain no free ionic groups within the block that can bind metal ions, and thus can act as the spacers.[55] Figure 2.12d shows cross-sectional TEM image of a sandwiched structure composed of two $(PDDA/PSS)_n$ blocks (n = 4 and 6, respectively) assembled at 3.0 M NaCl and loaded with silver nanoparticles separated by a $(PDDA/PSS)_{10}$ block assembled at 0.75 M NaCl containing no silver particles. This kind of thin films with precisely controlled heterostructure may find application in optical and photonic devices.

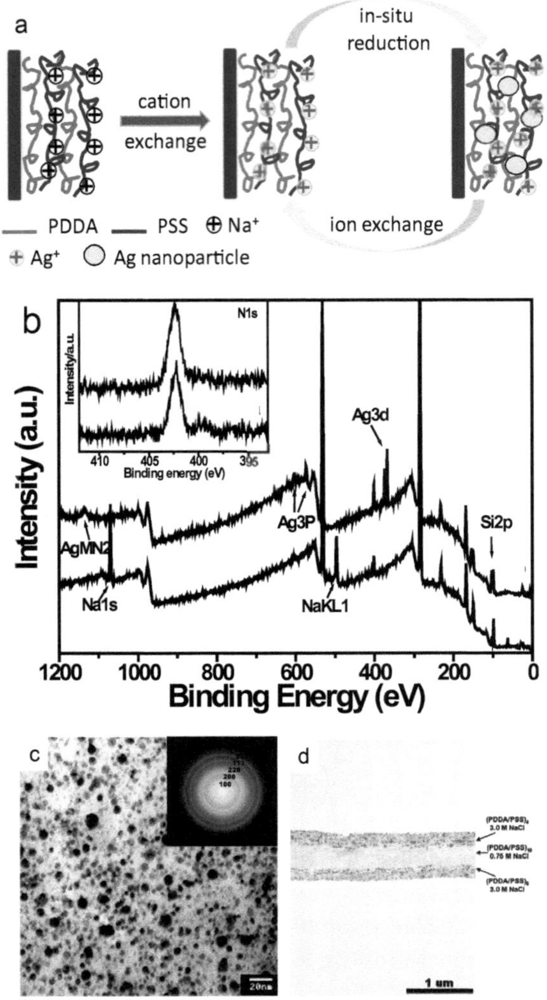

Figure 2.12 (a) Schematic illustration of the countercation exchange and in situ reduction in the PEM. (b) XPS spectra of $(PDDA/PSS)_3$ PEM as assembled (lower) and after treated with $AgNO_3$ solution (upper), revealing the exchange of Na^+ for Ag^+. The inset is the N1s region, wherein the absence of N1s peak at ~405 eV for NO_3^- excludes the absorption of $AgNO_3$ by the multilayer film. (c) TEM image of a $(PDDA/PSS)_5$ film containing silver nanoparticles. The inset is the electron diffraction pattern. (d) TEM image of the cross-section of a sandwiched structure, wherein the blocks containing silver nanoparticles and the spacer block were assembled at 3.0 M and 0.75 M NaCl, respectively. Reprinted with permission from Ref. [55] Copyright 2009 American Chemical Society.

Direct methanol fuel cells (DMFCs) have shown tremendous potential as portable power devices due to their high fuel efficiency and power density, low operating temperature, and low/zero pollutant emission. In DMFCs, the platinum anode catalyst for methanol oxidation is immobilized on a support material, usually carbon black, for it to function properly. The support material has a strong influence on the morphology, size and size distribution, and dispersion of the metal particles, which in turn impact the catalytic activity, stability, and cell performance of DMFCs, and various polymers have been explored as catalyst supports in DMFCs. We attempted to synthesize Pt nanoparticles in PEM thin films and study the catalytic performance of these composite films for methanol oxidation.[89] For $(PSS/PDDA)_5$ PEMs assembled at 1.5 M NaCl concentration terminated with PDDA, the dominent counterion in the PEM is Cl^-. The PEM was immersed in a H_2PtCl_6 solution to exchange the Cl^- counteranion for $PtCl_4^{2-}$ and $PtCl_6^{2-}$ in the solution, which were then reduced with $NaBH_4$ to make Pt nanoparticles in the PEM. The Pt nanoparticles were dispersed uniformly throughout the PEM matrix. The Pt content in the composite film, which is critical to its performance as a catalyst, can be readily controlled by the number of the exchange/reduction cycle. Cyclic voltammetry was employed to investigate electrocatalytic performance of the Pt catalyst for methanol oxidation reaction. The PEM-supported Pt catalysts exhibited forward mass-specific peak current densities about six times greater than Pt black. With high catalytic activity, excellent tolerance to poisoning, and long-term stability, the PEM/Pt composites are promising candidates for anode catalysts in DMFCs (Fig. 2.13).[89]

2.4.2 Core-Shell Nanoparticles

In a PDDA/PSS multilayer assembled at high salt concentrations, in particular at 1.5 M NaCl, as shown in Fig. 2.4, when the outmost layer of the PEM is PDDA, the counterions existing in the film are mainly anions, normally Cl^-, whereas for the PEMs capped with PSS, the predominant counterion present within the PEM is cation, typically Na^+. Therefore, both anionic and cationic precursor ions can be exchanged into the PEM for synthesis of metal nanoparticles, as demonstrated in Section 2.4.1. Furthermore, two different metals can be introduced sequentially to make core-shell nanoparticles, as illustrated in Fig. 2.14. First, a PEM terminated with PDDA is loaded

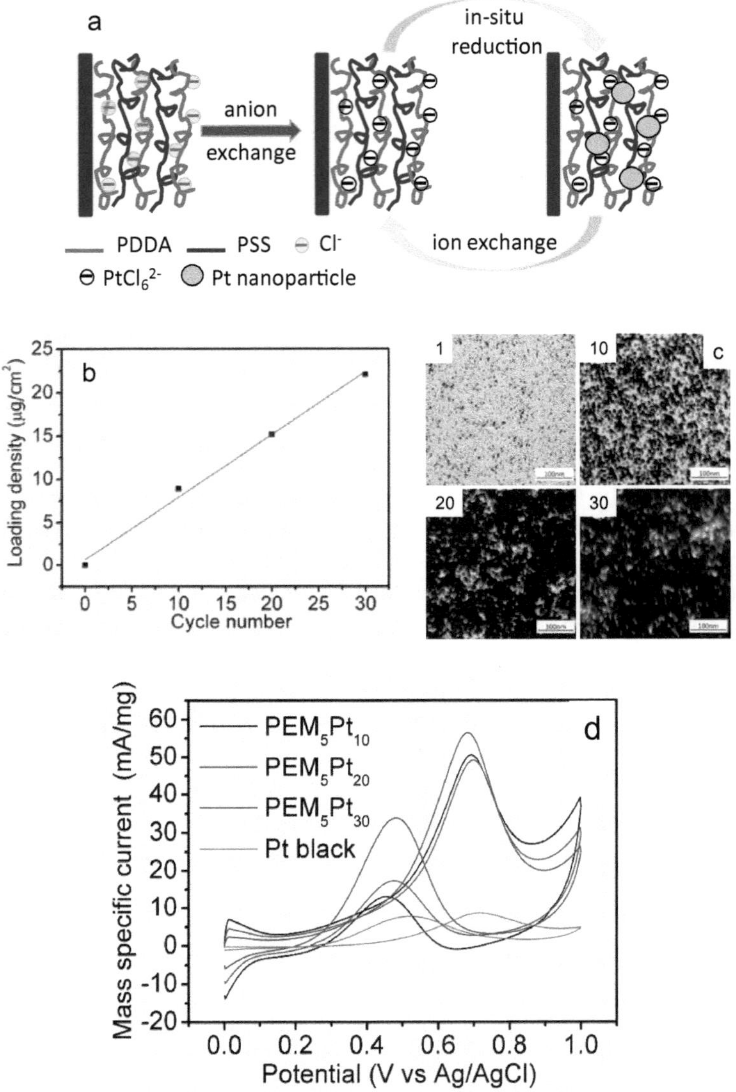

Figure 2.13 (a) Schematic illustration of the counteranion exchange and in situ reduction in the PEM. (b) Pt content in the $(PSS/PDDA)_5$, denoted PEM_5, as a function of the number of exchange/reduction cycles m. (c) TEM images of Pt nanoparticles dispersed in PEM_5Pt_m, with the exchange/reduction cycle m of 1, 10, 20, and 30. (d) Mass-specific cyclic voltammetry curves of methanol oxidation for PEM_5Pt_m (m = 10, 20, 30) and Pt black.[89] Reproduced by permission of The Royal Society of Chemistry.

with $AuCl_4^-$ ions via counteranion exchange, which are reduced to produce Au nanoparticles. And then, a layer of PSS is deposited onto the PEM to switch the capping layer, so that the PEM now is rich in countercations. Then, Ag^+ ions can be introduced into the PEM by countercation exchange, and subsequently reduced with a weak reducing agent to yield Au@Ag core-shell nanoparticles. A strong reducing agent, $NaBH_4$, is used only in the first reduction cycle to produce Au nanoparticles as seeds, and in all subsequent reductions of $AuCl_4^-$ and Ag^+, a weak reducing agent, ascorbic acid, is employed so that the metal tends to deposit on the existing nanoparticles, resulting in core-shell structures (Fig. 2.14a). Both ion-exchange/reduction cycles can be repeated multiple times so that the size of the Au core and thickness of the Ag shell can be controlled independently.[91]

In the field of catalysis, bimetallic nanoparticles are of particular interest because they often show superior catalytic performance than their corresponding monometallic counterparts and bulk materials. The catalytic activity of the PEM-supported Au, Ag, and Au@Ag core-shell nanoparticles was compared for the reduction of 4-nitrophenol by $NaBH_4$. The reaction kinetics was followed by UV-vis absorption spectroscopy. In this catalytic reaction, when $NaBH_4$ is in large excess compared with 4-nitrophenol, the kinetics can be considered pseudo-first order with respect to 4-nitrophenol. Figure 2.14d displays the kinetic plots for the reaction catalyzed by Au, Ag, and Au@Ag nanoparticles embedded in a $(PDDA/PSS)_5$ PEM, each fabricated by two exchange/reduction cycles, respectively, and the Au@Ag core-shell bimetallic nanoparticles clearly exhibit higher catalytic activity than the Au and Ag monometallic nanoparticles.[90] More recently, we also synthesized AuPt alloy nanoparticles in PDDA/PSS PEMs via the ion-exchange/in-situ reduction process.[94] The PEM nanoreactor facilitated the alloy formation by suppressing the Au-Pt phase separation, and the alloy nanoparticles produced exhibited superior catalytic activity to the monometallic ones.

2.4.3 Hollow Nanoparticles

Among the nanoparticles of various shapes and structures, hollow nanoparticles offer some advantages over others, such as increased surface area, lower density, saving of materials, reduction in cost, and tunable optical properties. Hollow particles are often synthesized via galvanic replacement reaction, which is an effective

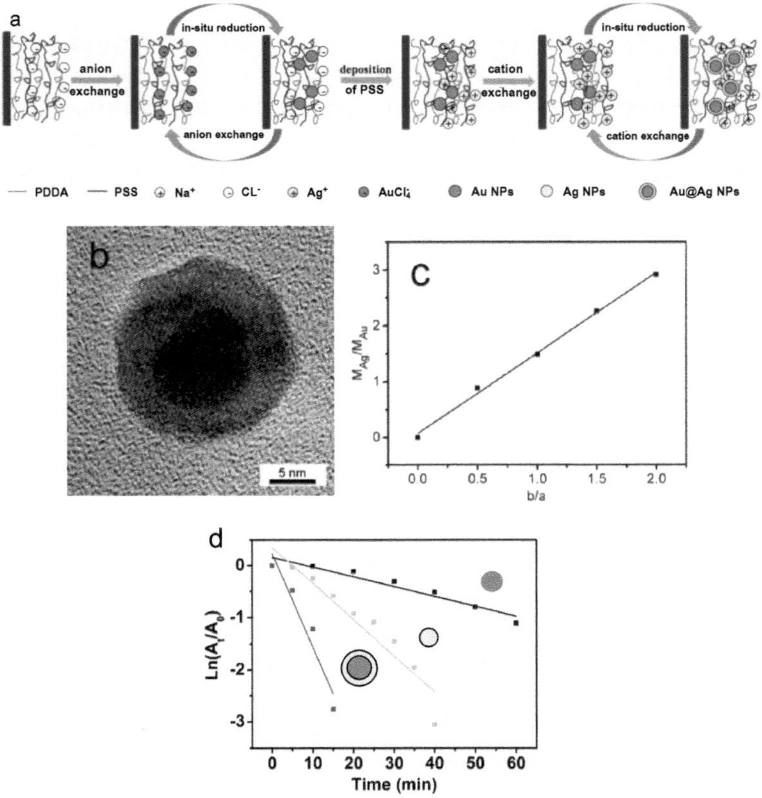

Figure 2.14 (a) Process of fabrication of Au@Ag core–shell nanoparticles in the PEM. (b) High-resolution TEM image of Au-Ag bimetallic particle dispersed in the PEM, confirming the core-shell structure. (c) Au/Ag molar ratio for the nanoparticles in the PEM as a function of the ratio between the Au and Ag exchange/reduction cycles, showing composition control for the nanoparticles. Reprinted with permission from Ref. [91] Copyright 2012 American Chemical Society. (d) Plot of $\ln(A_t/A_0)$, where A_0 and A_t is proportional to the reactant concentration initially and at time t, respectively, as a function of time for the reaction catalyzed by Au, Ag, and Au@Ag nanoparticles in $(PDDA/PSS)_5$.

and simple method for fabrication of metallic nanostructures with scalable cavity, wherein sacrificial template particles are prepared first. Nanoparticles prepared in PEMs in general are small and have a rather broad size distribution. To obtain more uniform template nanoparticles of suitable sizes, we adopted a two-step protocol.

First, Ag seeds were introduced into a $(PDDA/PSS)_3$ film via ion-exchange/reduction, as described in Section 2.4.1. The PEM of only three bilayers had a low counterion content, resulting in only a small number of Ag seeds. Then, additional PDDA and PSS layers were deposited on this seeded PEM, dramatically increasing the counterion content, so that much more Ag metal can be produced in each ion-exchange/reduction cycle and bigger particles can be synthesized in fewer cycles. Also, ascorbic acid was used after the first cycle so that Ag reduced in all subsequent cycles would grow the existing seeds (produced in the first cycle) rather than generate new ones, leading to more uniform particles. The Ag nanoparticles thus synthesized were then used as sacrificial templates to make hollow Ag-Au bimetallic nanoparticles via galvanic replacement reaction (Fig. 2.15). For hollow nanostructures, the sizes of the particle and

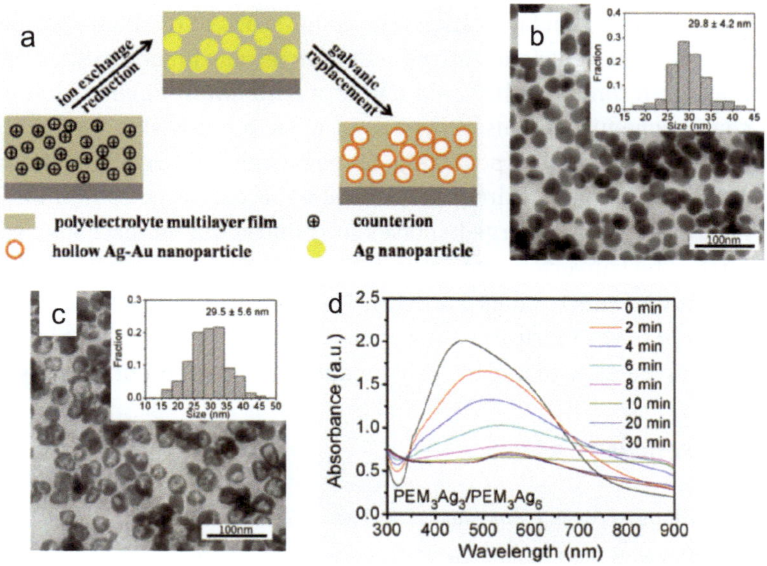

Figure 2.15 (a) Fabrication of Au-Ag bimetallic hollow nanoparticles in the PEM. (b, c) TEM images of (b) Ag template nanoparticles and (c) Au-Ag bimetallic hollow nanoparticles synthesized in $(PDDA/PSS)_6$. Insets are corresponding size distribution. (d) PEM_3Ag_3/PEM_3Ag_6 (reaction time 3 min). Insets are corresponding size distribution. (d) UV-vis absorption spectrum as a function of reaction time for the $(PDDA/PSS)_6$ PEM containing Ag nanoparticles treated with $HAuCl_4$ solution, showing the red-shift in the SPR peak corresponding to the formation of the hollow structure. Reprinted with permission from Ref. [92] Copyright 2013 American Chemical Society.

the cavity are important for their properties. The geometry of the Ag-Au bimetallic nanoparticles can be readily controlled by using the Ag template particles of different sizes and varying the time of reaction between the Ag templates and the $HAuCl_4$ solution. From the UV-vis spectra, it can be seen that the SPR characteristics of the particles in the PEMs are shifted to the near-infrared region.[92] The strategy described here can be utilized to synthesize hollow bimetallic nanoparticles of other elements in multilayer thin films.

2.5 Concluding Remark

PEMs can be conveniently assembled from a large variety of charged species, onto almost any kinds of substrate, with well controlled thickness and composition, making it an extremely versatile technique for thin film fabrication. The counterions in the PEMs provide a convenient vehicle for introduction of various functionalities into these thin films via the rapid and straightforward aqueous ion exchange chemistry. This approach in principle can also be applied to incorporate any (positively or negatively) charged functional species to impart desired functionalities to the polymer thin films. Alternatively, the charged species introduced via ion exchange can further be converted into other species in situ, such as alloy, oxide, and semiconductor nanoparticles. This provides a versatile and facile route to the fabrication of a wide variety of complex thin film nanocomposites with molecularly tunable structure and properties.

References

1. Blodgett, K. B., *J. Am. Chem. Soc.* 1935, 57, 1007.
2. Blodgett, K. B., *Langmuir, I. Phys. Rev.* 1937, 51, 964.
3. Netzer, L., Sagiv, J., *J. Am. Chem. Soc.* 1983, 105, 674.
4. Ariga, K., Hill, J. P., Ji, Q. M., *Phys. Chem. Chem. Phys.* 2007, 9, 2319.
5. Iler, R. K., *J. Colloid Interface Sci.* 1966, 21, 569.
6. Decher, G., Hong, J. D., *Makromol. Chem.-M Symp.* 1991, 46, 321.
7. Decher, G., Hong, J. D., *Ber. Bunsen Phys. Chem.* 1991, 95, 1430.
8. Decher, G., *Science* 1997, 277, 1232.
9. Lee, H., Lee, Y., Statz, A. R., Rho, J., Park, T. G., Messersmith, P. B., *Adv. Mater.* 2008, 20, 1619.

10. Aljabali, A. A. A., Lomonossoff, G. P., Evans, D. J., *Biomacromolecules* 2011, 12, 2723.
11. Nishiguchi, A., Yoshida, H., Matsusaki, M., Akashi, M., *Adv. Mater.* 2011, 23, 3506.
12. Li, Y., Wang, X., Sun, J., *Chem. Soc. Rev.* 2012, 41, 5998.
13. Lvov, Y., Ariga, K., Ichinose, I., Kunitake, T., *J. Am. Chem. Soc.* 1995, 117, 6117.
14. Shutava, T. G., Kommireddy, D. S., Lvov, Y. M., *J. Am. Chem. Soc.* 2006, 128, 9926.
15. Lvov, Y., Decher, G., Sukhorukov, G., *Macromolecules* 1993, 26, 5396.
16. Johnston, A. P. R., Mitomo, H., Read, E. S., Caruso, F., *Langmuir* 2006, 22, 3251.
17. Tokuhisa, H., Hammond, P. T., *Adv. Funct. Mater.* 2003, 13, 831.
18. Li, Y., Liu, F., Sun, J. Q., *Chem. Commun.* 2009, 2730.
19. Cho, J., Caruso, F., *Chem. Mater.* 2005, 17, 4547.
20. Priya, D. N., Modak, J. M., Raichur, A. M., *ACS Appl. Mater. Interfaces* 2009, 1, 2684.
21. Podsiadlo, P., Michel, M., Lee, J., Verploegen, E., Wong Shi Kam, N., Ball, V., Lee, J., Qi, Y., Hart, A. J., Hammond, P. T., Kotov, N. A., *Nano Lett.* 2008, 8, 1762.
22. Michel, M., Taylor, A., Sekol, R., Podsiadlo, P., Ho, P., Kotov, N., Thompson, L., *Adv. Mater.* 2007, 19, 3859.
23. Matsusaki, M., Kadowaki, K., Nakahara, Y., Akashi, M. *Angew. Chem. Int. Ed.* 2007, 46, 4689.
24. Xu, H., Schönhoff, M., Zhang, X., *Small* 2012, 8, 517.
25. Zhang, X., Chen, H., Zhang, H., *Chem. Commun.* 2007, 0, 1395.
26. Schlenoff, J. B., Dubas, S. T., Farhat, T., *Langmuir* 2000, 16, 9968.
27. Chiarelli, P. A., Johal, M. S., Casson, J. L., Roberts, J. B., Robinson, J. M., Wang, H. L., *Adv. Mater.* 2001, 13, 1167.
28. Cho, J., Char, K., Hong, J. D., Lee, K. B., *Adv. Mater.* 2001, 13, 1076.
29. Merrill, M. H., Sun, C. T., *Nanotechnology* 2009, 20, 075606.
30. Hammond, P. T., *AIChE J.* 2011, 57, 2928.
31. Schaaf, P., Voegel, J.-C., Jierry, L., Boulmedais, F., *Adv. Mater.* 2012, 24, 1001.
32. Lin, Y., Wang, L. L., Zhang, P. B., Wang, X., Chen, X. S., Jing, X. B., Su, Z. H., *Acta Biomater.* 2006, 2, 155.

33. Tang, Z., Wang, Y., Podsiadlo, P., Kotov, N. A., *Adv. Mater.* 2006, 18, 3203.
34. Reisch, A., Hemmerle, J., Voegel, J. C., Gonthier, E., Decher, G., Benkirane-Jessel, N., Chassepot, A., Mertz, D., Lavalle, P., Mesini, P., Schaaf, P., *J. Mater. Chem.* 2008, 18, 4242.
35. Dai, J. H., Bruening, M. L., *Nano Lett.*2002, 2, 497.
36. Zhai, L., Cebeci, F. Ç., Cohen, R. E., Rubner, M. F., *Nano Lett.* 2004, 4, 1349.
37. Joly, S., Kane, R., Radzilowski, L., Wang, T., Wu, A., Cohen, R. E., Thomas, E. L., Rubner, M. F., *Langmuir* 2000, 16, 1354.
38. von Klitzing, R., *Phys. Chem. Chem. Phys.* 2006, 8, 5012.
39. Mermut, O., Barrett, C. J., *J. Phys. Chem. B* 2003, 107, 2525.
40. Dubas, S. T., Schlenoff, J. B., *Macromolecules* 1999, 32, 8153.
41. Salomäki, M., Tervasmäki, P., Areva, S., Kankare, J., *Langmuir* 2004, 20, 3679.
42. Dressick, W. J., Wahl, K. J., Bassim, N. D., Stroud, R. M., Petrovykh, D. Y., *Langmuir* 2012, 28, 15831.
43. Salomäki, M., Laiho, T., Kankare, J., *Macromolecules* 2004, 37, 9585.
44. Salomäki, M., Kankare, J., *Macromolecules* 2008, 41, 4423.
45. Ball, V., Voegel, J.-C., Schaaf, P., *Langmuir* 2005, 21, 4129.
46. Wong, J. E., Zastrow, H., Jaeger, W., von Klitzing, R., *Langmuir* 2009, 25, 14061.
47. Schlenoff, J. B., Ly, H., Li, M., *J. Am. Chem. Soc.* 1998, 120, 7626.
48. Laurent, D., Schlenoff, J. B., *Langmuir* 1997, 13, 1552.
49. Schmitt, J., Gruenewald, T., Decher, G., Pershan, P. S., Kjaer, K., Loesche, M., *Macromolecules* 1993, 26, 7058.
50. Riegler, H., Essler, F., *Langmuir* 2002, 18, 6694.
51. Lourenço, J. M. C., Ribeiro, P. A., Botelho do Rego, A. M., Raposo, M., *J. Colloid Interface Sci.* 2007, 313, 26.
52. Ghostine, R. A., Markarian, M. Z., Schlenoff, J. B., *J. Am. Chem. Soc.* 2013, 135, 7636.
53. Tanchak, O. M., Yager, K. G., Fritzsche, H., Harroun, T., Katsaras, J., Barrett, C. J., *J. Chem. Phys.* 2008, 129, 084901.
54. Schollmeyer, H., Guenoun, P., Daillant, J., Novikov, D. V., *J. Phys. Chem. B* 2007, 111, 4036.
55. Zan, X., Su, Z., *Langmuir* 2009, 25, 12355.

56. Zan, X. J., Hoagland, D. A., Wang, T., Su, Z. H., *Macromolecules* 2012, 45, 8805.

57. Zan, X., Peng, B., Hoagland, D. A., Su, Z., *Polym. Chem.* 2011, 2, 2581.

58. Lee, B. S., Chi, Y. S., Lee, J. K., Choi, I. S., Song, C. E., Namgoong, S. K., Lee, S.-G., *J. Am. Chem. Soc.* 2004, 126, 480.

59. Azzaroni, O., Brown, A. A., Huck, W. T. S., *Adv. Mater.* 2007, 19, 151.

60. Wang, L., Lin, Y., Peng, B., Su, Z., *Chem. Commun.* 2008, 5972.

61. Wang, L., Lin, Y., Su, Z., *Soft Matter* 2009, 5, 2072.

62. Gao, L., McCarthy, T. J., *Langmuir* 2009, 25, 14105.

63. Wang, L., Peng, B., Su, Z., *Langmuir* 2010, 26, 12203.

64. Liu, M., Wang, S., Wei, Z., Song, Y., Jiang, L., *Adv. Mater.* 2009, 21, 665.

65. Lin, L., Liu, M., Chen, L., Chen, P., Ma, J., Han, D., Jiang, L., *Adv. Mater.* 2010, 22, 4826.

66. Ganesh, V. A., Dinachali, S. S., Nair, A. S., Ramakrishna, S., *ACS Appl. Mater. Interfaces* 2013, 5, 1527.

67. Zhang, G., Zhang, X., Huang, Y., Su, Z., *ACS Appl. Mater. Interfaces* 2013, 5, 6400.

68. Chen, W., Fadeev, A. Y., Hsieh, M. C., Oner, D., Youngblood, J., McCarthy, T. J., *Langmuir* 1999, 15, 3395.

69. Wang, L. M., Wang, L., Su, Z. H., *Soft Matter* 2011, 7, 4851.

70. Wang, L. M., Wei, J. J., Su, Z. H., *Langmuir* 2011, 27, 15299.

71. Dotzauer, D. M., Dai, J. H., Sun, L., Bruening, M. L., *Nano Lett* 2006, 6, 2268.

72. Abalde-Cela, S., Ho, S., Rodriguez-Gonzalez, B., Correa-Duarte, M. A., Alvarez-Puebla, R. A., Liz-Marzan, L. M., Kotov, N. A., *Angew. Chem. Int. Ed.* 2009, 48, 5326.

73. Shi, X. Y., Shen, M. W., Mohwald, H., *Prog. Polym. Sci.* 2004, 29, 987.

74. Dante, S., Hou, Z., Risbud, S., Stroeve, P., *Langmuir* 1999, 15, 2176.

75. Dutta, A. K., Jarero, G., Zhang, L., Stroeve, P., *Chem. Mater.* 2000, 12, 176.

76. Zhang, L., Dutta, A. K., Jarero, G., Stroeve, P., *Langmuir* 2000, 16, 7095.

77. Fojas, A. M., Murphy, E., Stroeve, P., *Ind. Eng. Chem. Res.* 2002, 41, 2662.

78. Dutta, A. K., Ho, T., Zhang, L., Stroeve, P., *Chem. Mater.* 2000, 12, 1042.

79. Wang, T. C., Rubner, M. F., Cohen, R. E., *Langmuir* 2002, 18, 3370.

80. Chia, K. K., Cohen, R. E., Rubner, M. F., *Chem. Mater.* 2008, 20, 6756.

81. Veletanlic, E., Goh, M. C., *J. Phys. Chem. C* 2009, 113, 18020.
82. Machado, G., Beppu, M. M., Feil, A. F., Figueroa, C. A., Correia, R. R. B., Teixeira, S. R., *J. Phys. Chem. C* 2009, 113, 19005.
83. Logar, M., Jancar, B., Suvorov, D., Kostanjsek, R., *Nanotechnology* 2007, 18, 325601.
84. Gong, X. A., Han, L. L., Yue, Y. A., Gao, J. R., Gao, C. Y., *J. Colloid. Interface Sci.* 2011, 355, 368.
85. Vago, M., Tagliazucchi, M., Williams, F. J., Calvo, E. J., *Chem. Commun.* 2008, 5746.
86. Schlenoff, J. B., Dubas, S. T., *Macromolecules* 2001, 34, 592.
87. Zan, X. J., Su, Z. H., *Thin Solid Films* 2009, 518, 116.
88. Zhang, X., Chu, C., Huang, K., Su, Z., *Chinese J. Appl. Chem.* 2012, 29, 1433.
89. Zhang, X., Zan, X., Su, Z., *J. Mater. Chem.* 2011, 21, 17783.
90. Zhang, X., Su, Z., *Adv. Mater.* 2012, 24, 4574.
91. Zhang, X., Wang, H., Su, Z., *Langmuir* 2012, 28, 15705.
92. Zhang, X., Zhang, G., Zhang, B., Su, Z., *Langmuir* 2013, 29, 6722.
93. Zan, X. J., Su, Z. H., *Thin Solid Films* 2010, 518, 5478.
94. Chu, C. C., Su, Z. H., *Langmuir* 2014, 30, 15345.

Chapter 3

Glass-Based Proton Exchange Membranes for Fuel Cell Applications

Haibin Li, Qiang Xie, and Xiaojing Chen
Institute of Fuel Cell, School of Mechanical Engineering, Shanghai Jiao Tong University, Shanghai 200240, China
haibinli@sjtu.edu.cn

3.1 Introduction

With the development of economy and society, the need for energy is even-increasing. However, the continuous decreasing of conventional fossil fuels as well as the pollutions caused by the conventional energy makes it hard to satisfy our needs for environmental-friendly energy, which forces us to develop other alternative energy. Fuel cells have attracted considerable attention owing to their many merits such as high efficiency, low-pollution, or pollution-free characteristics. Of all different kinds of fuel cells, proton exchange membrane fuel cells (PEMFCs) are regarded as the most promising alternative energy technology and primary power source applicable to vehicular and stationary applications because of their high energy

Nanomaterials in Energy and Environmental Applications
Edited by Junhui He

ISBN 978-981-4463-78-2 (Hardcover), 978-981- 4463-79-9 (eBook)
www.panstanford.com

conversion efficiency, high power density, and innocuous impact on environment [1–4]. Most crucial part in the fuel cells is the proton exchange membrane that separates the anode and cathode in space and mediates the electrochemical reactions by transporting protons [5]. The current state-of-the-art proton electrolyte membranes for polymer electrolyte membrane fuel cell with the operating temperature below 100°C are perfluorinated sulfonic acid ionomers such as Nafion [6, 7]. Nafion has received great attention as a proton conductor owing to its high proton conductivity, excellent thermal, and mechanical stability. Nonetheless, it has several shortcomings. For instance, it is a water-based proton conductor that falls dry quickly at temperature above 80°C due to evaporation of water from the polymer matrix. Also, Nafion membrane in a fuel cell suffers from the stress that is caused by the swelling and deswelling of the membrane in humid atmosphere. Furthermore, the methanol permeability of Nafion membrane is too high to be employed in direct methanol fuel cells (DMFCs).

Nogami et al. have proposed the sol–gel preparation of inorganic glass membranes with a promising proton conductivity of 10 mS/cm at room temperature [8, 9]. Diverse inorganic proton conductive glasses including acid/glass hybrids have been prepared and investigated [10–13]. However, most of the acids are both water and alcohol soluble; therefore, their applications are limited. Inorganic glass membranes exhibit excellent thermal stability at the temperatures above 400°C and no decrease of conductivity is observed up to 120°C [14]. The advantages of high proton conductivity, excellent thermal stability, and low cost make the inorganic glasses very promising to substitute polymer electrolytes in the practical application of fuel cells. However, most of the inorganic glasses, such as P_2O_5–SiO_2 glasses, show deficient flexibility for the membrane electrode assembling as well as the fuel cell assembling. Thus, the feasibility of the application of inorganic glass to fuel cells is challenged because of its brittleness.

In this chapter, on the basis of our previous study, the various performances of the glass-based proton exchange membranes, which are fabricated via the hydrothermally assisted sol-gel approach under different parameters such as the materials of membranes such as organics or inorganics, content of these materials, temperature, relative humidity, and so forth, will be discussed.

3.2 Proton-Conducting Glasses Synthesized via the Hydrothermally Assisted Sol-Gel Approach

In contrast to the conventional hydrothermal treatment temperature higher than 400°C, our group proposed a novel approach of hydrothermal treatment to anneal the phosphosilicate gel at a temperature lower than 150°C. The highlights of the hydrothermal treatment process lie in that it leads to higher proton conductivity and makes possible preparing polymer/glass hybrids via the sol-gel method while averting thermal degradation of the organic component that happens at high temperatures.

3.2.1 High Proton-Conducting Monolithic Phosphosilicate Glass Membranes

3.2.1.1 Preparation of PSMs

Firstly, the phosphosilicate glass membranes were prepared by using $Si(OC_2H_5)_4$(TEOS), deionized water and chloric acid(HCL) in a molar ratio of 1:4:0.004 (TEOS:H_2O:HCl) via the sol-gel method, followed by a hydrothermal process under water vapor atmosphere, where the HCl is the catalyst. The resultant mixture was stirred for 30 min until it had become a transparent solution.

Secondly, the different amounts of H_3PO_4 were added to the solution in the different molar ratios of H_3PO_4 and TEOS, 10:95 and 20:90, designated as 10P and 20P, and then the mixtures were stirred at room temperature for 1 h. The resultant solutions were transferred to Teflon vessels with covers and aged for 3 days at room temperature to form xerogels.

Last, the xerogels were hydrothermally treated in a drying box with the temperature at 150°C and water vapor at a pressure of 1 atm. After these procedures, the phosphosilicate membrances were obtained.

3.2.1.2 Characterization, analysis and discussion of PSMs

3.2.1.2.1 *Morphology and appearance*

Figure 3.1 shows the photograph of a sample with the molar ratio of H_3PO_4 and TEOS 20:90. It is conspicuous that the sample is a transparent, crack-free, and smooth glass membrane. In a typical sol–gel process, the hydrolysis reaction of TEOS is incomplete and various hydrophilic oligomers ($Si(OR)_x(OH)_{4-x}$, x = 0–4) are generated by partial hydrolysis [15]. During the gelation process, the resultant gel then shrinks to form a soft xerogel. In the hydrothermal process used here, however, water vapor is introduced and the xerogel is heated at 150°C, which accelerates the hydrolysis of the previously unhydrolyzed ethoxy groups. Subsequently, the dehydration reaction leads to the formation of a network structure, giving rise to improved mechanical strength of the monolithic membrane. In contrast, traditional gel-derived glass is calcined at high temperature, which leads to the appearance of macrocracks due to tensile stresses arising from anisotropic shrinkage rates within the exterior and interior of the gel body [16]. Hence, we consider the hydrothermal treatment employed here to be a good approach for preparing large-sized and crack-free monolithic membranes. Moreover, by comparing the samples with different molar ratio of H_3PO_4 and TEOS, we can find that the friability of the sample will increase with the increase of content of H_3PO_4.

Figure 3.1 Optical image of a PSM sample [51]. Copyright 2011 Elsevier.

3.2.1.2.2 *Thermal properties*

The thermal properties of the sample 20P are investigated by both thermogravimetry (TG) and differential scanning calorimetry (DSC). The results are shown in Fig. 3.2. A weight loss of about 20% is observed between room temperature and 100°C and a sharp endothermic peak is seen at 64°C, which are mainly due to the evaporation of physisorbed water from within the pores of the glass membrane. In the range 100°C–200°, the weight loss is approximately 2%, and a weak endothermic peak is seen at ca. 170°C. These features can be attributed to dehydration of chemisorbed water molecules involved in hydrogen bonding of Si–OH and P–OH groups in the glass network. The weight loss is negligible in the temperature range above 200°C, indicating good thermal stability of the PSM samples after the hydrothermal treatment.

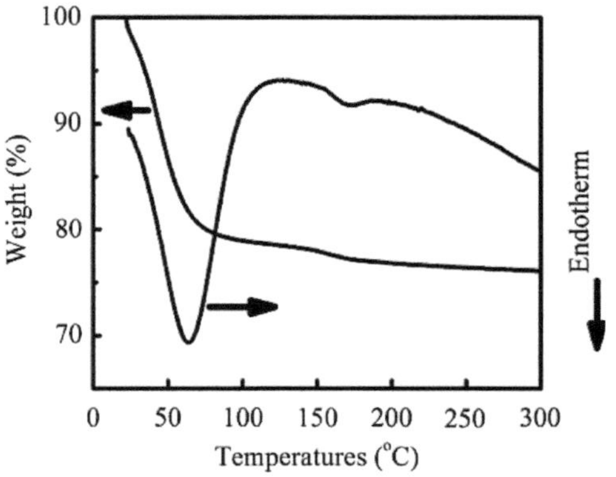

Figure 3.2 TG and DSC curves for sample 20P after hydrothermal treatment at 150°C [51]. Copyright 2011 Elsevier.

3.2.1.2.3 *Structural characterization*

In order to investigate the gelation phenomena involved in the formation of the phosphosilicate network structure as well as the state of the molecular water and hydroxyl groups in the membrane, an FTIR study was performed. Typical IR spectra of sample 20P before and after hydrothermal treatment are illustrated in Fig. 3.3. The peaks at 800 and 1053 cm^{-1} can be attributed to symmetric and asymmetric stretches of the Si–O–Si unit, respectively [17]. An absorption shoulder at 960 cm^{-1} arises from the Si–O stretch of silanol groups, Si–OH [18]. An absorption peak at 1630 cm^{-1} can be assigned to the bending mode of absorbed water molecules, H–O–H [19]. The absorption band at around 2336 cm^{-1} corresponds to ambient CO_2 in the IR measurement. Before hydrothermal treatment, an absorption peak appearing at around 2950 cm^{-1} could be assigned to the C–H stretching vibration of the unhydrolyzed ethoxy groups (OC2H5) of TEOS [18]. After the hydrothermal treatment, this peak had almost disappeared, indicating the further removal of ethoxy groups resulting from complete hydrolysis of the Si–O–C_2H_5 groups during the hydrothermal treatment. A broad absorption band appearing at 3360 cm^{-1} in both spectra can be attributed to the stretching vibrations of Si–O–H and P–O–H involved in hydrogen bonding with absorbed water molecules, H–O–H [20,21]. Moreover, a difference in the peak intensity at 3360 cm^{-1} between before and after the hydrothermal treatment was noted, which indicated that the population of hydroxyl groups in the membrane was increased by the hydrothermal treatment. The shoulder peak appearing at 3600 cm^{-1} is related to Si–OH and P–OH groups not involved in hydrogen bonding [20].

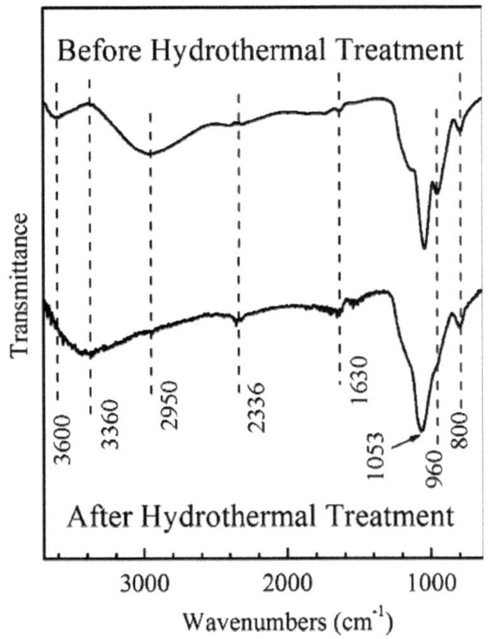

Figure 3.3 Infrared spectra of sample 20P before and after hydrothermal treatment at 150°C [51]. Copyright 2011 Elsevier.

The porosity and accessibility of the phosphosilicate membranes were evaluated by N_2 adsorption–desorption measurements. In Fig. 3.4a, it can be seen that the two membranes obtained after hydrothermal treatment showed similar hysteresis, indicating typical mesoporous character. Interconnection between the pores is indispensable for achieving high proton conductivity. The width of the hysteresis gives an indication of the accessibility of the pores [22]. As shown in Fig. 3.4a, samples 10P and 20P had quite large hysteresises, implying their high degree of interconnection. Figure 3.4b shows the pore size distributions of the samples. Sample 20P showed a peak pore size of 18 nm, while sample 10P exhibited a peak pore size of 9 nm. It is clear that sample 20P had a larger pore size and hence a superior accessibility, which would be beneficial for the migration of water molecular groups adsorbed. The results also indicate that the porous properties of the membranes are dependent on the H_3PO_4 content incorporated into the phosphosilicate network.

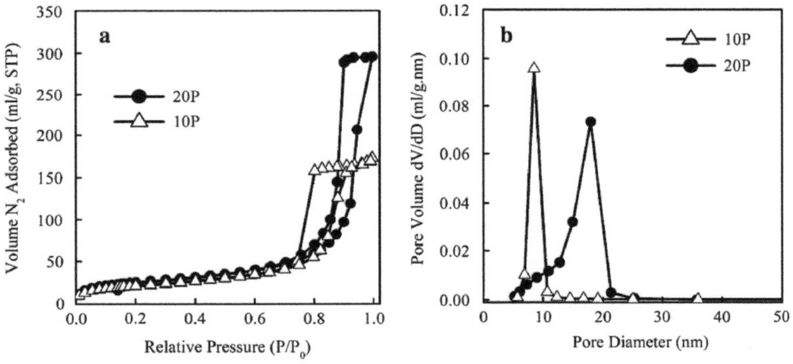

Figure 3.4 (a) N_2 adsorption–desorption isotherms for PSMs obtained by hydrothermal treatment at 150°C. (b) Pore size distributions for PSMs obtained by hydrothermal treatment at 150°C [51]. Copyright 2011 Elsevier.

3.2.1.2.4 *Proton-conductive properties*

The proton-conducting properties of the PSMs were examined by AC impedance spectroscopy. A representative example of a set of complex impedance plots is given in Fig. 3.5a. The membrane resistances of the PSMs were obtained from the intercept on the real axis in the high-frequency range. For sample 10P measured at various temperatures, it can be seen that the membrane resistance decreased with increasing test temperature. Proton conductivity was calculated by introducing the membrane resistance R into the following formula:

$$\sigma = \frac{L}{AR}$$

where L is the distance between the two parallel electrodes (0.5 cm), A is electrode length (0.5 cm) membrane thickness, and R is the membrane resistance determined from the complex impedance plot. The sample impedance was measured over the frequency range from 2.5 MHz to 50 Hz at an AC amplitude of 10 mV.

The temperature dependences of the proton conductivities measured under 70% RH for samples 10P and 20P are illustrated in Fig. 3.5b. It was also found that the conductivities of each of the membranes increased with increasing temperature as a function of reciprocal temperature, and that sample 20P showed higher proton conductivity.

In the Grotthuss mechanism for water-cooperative proton conduction, a proton dissociates from a hydroxyl group (Si–OH or P–OH) to form $H_9O_4^+$ or $H_5O_2^+$ cluster ions with adsorbed molecular water and migrates by hopping between neighboring water cluster sites [23]. Compared with the SiO–H bond, which has purely covalent character, the PO–H group has higher ionic character, as a result of which it has higher cation-exchange capacity and water affinity. Hence, the incorporation of PO–H groups into a membrane helps to promote the dissociation and migration of protons, thereby enhancing proton conduction [9]. A higher proton conductivity of sample 20P than samples 10P can therefore be expected. Figures 3.5b and 3.6 show the apparent effects of H_3PO_4 content on proton conductivity against temperature and relative humidity, respectively. It can be seen that among the two membranes, sample 20P has higher proton conductivity. This could be due to the fact that, besides membrane composition, the pore structure and pore size also play important roles in proton conduction [24,25]. A high degree of rotational freedom of $H_9O_4^+$ or $H_5O_2^+$ cluster ions within the pores is required for fast proton conduction [23]. The cluster ions should be able to migrate with diffusing water through open pores, and the diffusion coefficient of water in porous glasses decreases with decreasing pore size [26]. From the N_2 adsorption–desorption curve, it is evident that the pore size of sample 20P was about 18 nm, larger than that of samples 10P. The large pore size and high degree of pore interconnection mean that the formation of continuous channels for proton migration is more likely in the presence of absorbed water. As a result, sample 20P exhibited higher proton conductivity.

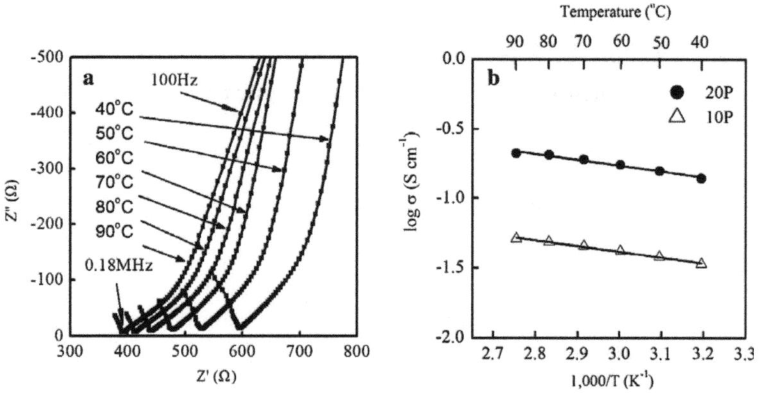

Figure 3.5 (a) Typical impedance responses (Cole–Cole plots) of the 10P PSM measured at various temperatures under 70% RH. (b) Temperature dependences of the proton conductivities of PSMs with different compositions measured under 70% RH [51]. Copyright 2011 Elsevier.

For comparison, the temperature dependence of the conductivity reported for a commercially employed Nafion112 membrane is shown in Fig. 3.7 [27]. It can be seen that the proton conductivities of sample 20P in the temperature range from 30°C to 90°C are higher than those of the Nafion112 membrane. It is worth noting that the surprisingly high proton conductivity of 0.21 S/cm at 90°C under 70% RH is about twice as high as that of the Nafion112 membrane. In addition, the plots in Fig. 3.5b are seen to be exactly linear in the measured temperature range. In the case of sample 20P, the Arrhenius activation energy of the proton conduction was calculated to be about 0.08 eV, which is related to the energy necessary for the dissociation of the proton from the hydroxyl group [30]. It is comparable to values ranging from 0.05 to 0.14 eV reported for other Nafion membranes [27–29]. The proton conductivity of the Nafion112 increased continuously up to about 0.11 S/cm at 80°C under 95% RH. However, a dramatic decrease in the proton conductivity was observed above 80°C because of the inability of the membrane to retain water [27]. For sample 20P, proton conductivity increased with increasing temperature up to 90°C suggesting that our PSMs may have the potential to operate at higher temperatures. There have been numerous reports on the preparation of phosphosilicate membranes, and in many cases, $PO(OEt)_3$ has been used as a phosphorus dopant for the introduction of P–OH groups [9, 31–33]. However, $PO(OEt)_3$ is only slowly hydrolyzed to $PO(OH)_3$, leading to a low concentration of the latter [33]. In our case, phosphoric acid ($O{=}P(OH)_3$) is introduced to provide a phosphorus component with three P–OH per phosphorus, giving rise to a higher population of P–OH than that derived from $PO(OEt)_3$. As indicated by the IR spectra, the hydrothermal treatment served to maintain a high content of free pendant hydroxyl groups in the mesoporous membrane for high proton conduction. On the contrary, the traditional calcinations procedure conducted at high temperatures above 350°C causes extensive dehydration and loss of surface hydroxyl groups [18], resulting in reduced proton conductivity. Thus, it is believed that the unique hydrothermal treatment that we employed and the utilization of H_3PO_4 as a raw material are responsible for the high proton conduction of the obtained PSMs.

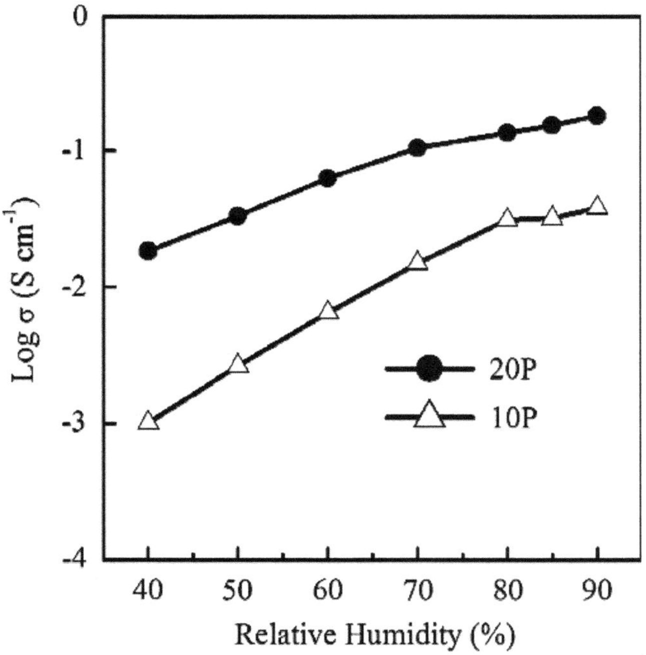

Figure 3.6 Dependences on relative humidity of the proton conductivities of PSMs with different compositions measured at 30°C [51]. Copyright 2011 Elsevier.

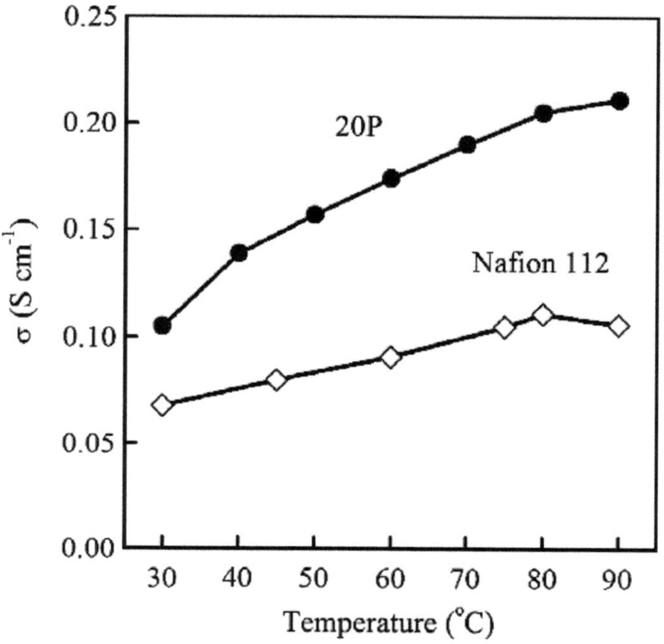

Figure 3.7 Comparison of the proton conductivities of sample 20P under 70% RH and the Nafion112 membrane [51]. Copyright 2011 Elsevier.

3.2.2 Fast Proton-Conducting Glass Membrane Based on Porous Phosphosilicate and Perfluorosulfonic Acid Polymer

3.2.2.1 Preparation of membranes

The glass membranes were prepared via an optimized sol-gel method. A mixture of TEOS, deionized water, and hydrochloric acid was firstly prepared and stirred for a period time. Then, certain amount of Nafion and phosphoric acid solution was slowly added to the mixture above with further stirring. The obtained transparent solution was then transferred into a Teflon vessel and aged at room temperature until gelation happens. Last, the obtained gel is hydrothermally treated. The sample names of the obtained Nafion/phosphosilicate glasses (NPS) membranes and their components were listed in Table 3.1.

Table 3.1 Sample names and the corresponding component [39]. Copyright 2011 Elsevier.

File name of the sample	Si: P in glass (molar ratio)	Feed ratio of TEOS: Nafion®(weight ratio)
NPSi-1	9:1	9:1
NPSi-2	8:2	9:1
NPSi-3	7:3	9:1
NPSi-4	9:1	8:2
NPSi-5	8:2	8:2
NPSi-6	9:1	7:3
NPSi-7	8:2	7:3

3.2.2.2 Characterization, analysis, and discussion of NPS membranes

3.2.2.2.1 *Morphology and structure of NPS membranes*

With the modified sol–gel approach, homogenous and crack-free NPS hybrid glass membranes were obtained (see Fig. 3.8). The shape and size of the membrane can be controlled by changing the shape

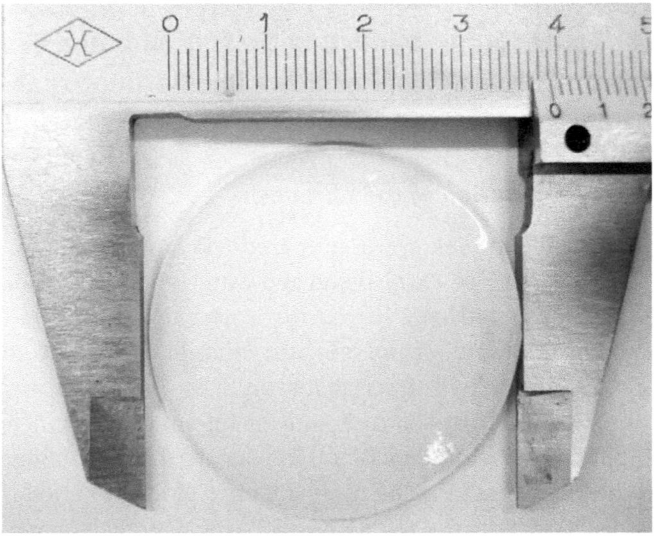

Figure 3.8 Digital photo of NPS hybrid glass membrane [39]. Copyright 2011 Elsevier.

of the vessel in which the membrane is made. The thickness of the membrane can be as thin as 200 μm. The hybrid membrane treated via the conventional method (annealing at high temperatures above 300°C) was easily getting cracked and the color of the membrane turned to be dark brown or black due to the degradation of the polymer. According to the TGA analysis in Fig. 3.9, the membranes are thermally unstable above 270°C. The TGA curves shown in Fig. 3.9 are typical TGA curves of Nafion according to a thermal degradation study of Nafion [34].

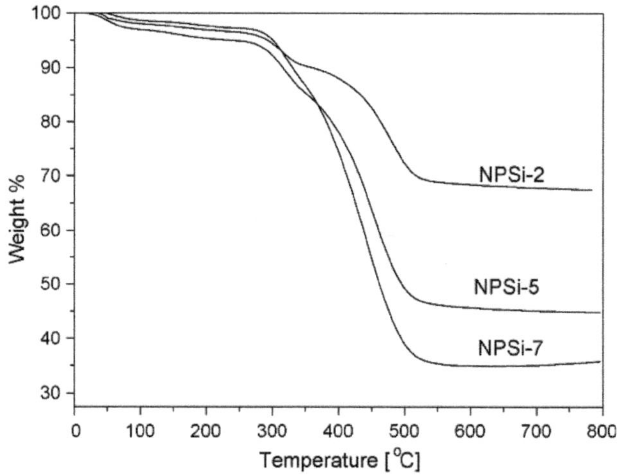

Figure 3.9 TGA curves of the hybrid glass membranes [39]. Copyright 2011 Elsevier.

The morphology of the NPS glass membranes was observed by field emission SEM. Figure 3.10 shows the SEM images of NPSi-1 (a), NPSi-2 (b), NPSi-3 (c), and NPSi-6 (d), respectively. It is apparent that the hybrid membrane consists of nanoparticles (a few nanometers in diameter) with porous structure. The SEM images indicate the homogeneous structure of the membrane and no aggregations of polymer can be observed. The polyelectrolyte is entrapped within and highly dispersed throughout the SiO_2 network. The homogeneous structure of the SiO_2 framework and the highly dispersed polymer filler also give the hybrid glass membrane an enhanced mechanical property that is important for the membrane electrode assembly.

The surface of all the NPS glass membranes has the pore size ranging from several to decades of nanometers. The pore size increases notably as the H_3PO_4 content increases (compared with Fig.

3.10a–c). Figure 3.10d shows the morphology of NPSi-6 that has the same molar ratio of H_3PO_4 to TEOS as NPSi-1. Compared with Fig. 3.10a, NPSi-6 membrane seems less porous than NPSi-1 because NPSi-6 contains more Nafion than NPSi-1 (see Table 3.1) where the polymer plays a role of pore-filler in the hybrids.

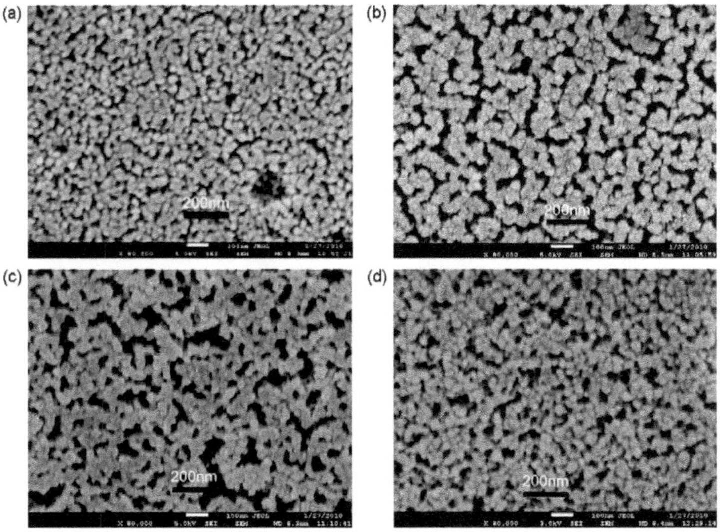

Figure 3.10 SEM images of NPSi-1 (a), NPSi-2 (b), NPSi-3 (c), and NPSi-6 (d), respectively [39]. Copyright 2011 Elsevier.

3.2.2.2.2 *Pore structure of the hybrid glass membranes*

The porous structure of the hybrid glass membranes was evaluated by N_2 adsorption–desorption isotherms and BJH pore size distribution plot. Figure 3.11 shows the pore-size distribution curves determined by BJH equation using the desorption branch of the N_2 gas adsorption–desorption isotherms. In Fig. 3.11a, NPSi-1, NPSi-2, and NPSi-3 have the average diameter of 13.2, 15.4, and 17.5 nm, respectively. Along with the increase of H_3PO_4 content, the average pore diameter of the membrane increases. This result was in accordance with the SEM analysis that further proved that H_3PO_4 concentration has an influence on the pore distribution of the glass membranes. The average pore size of the membrane increases with H_3PO_4 concentration. In Fig. 3.11b, the average diameters of NPSi-1, NPSi-4, and NPSi-6 are 13.2, 18.6, and 31.6 nm, respectively. Along with the increase of Nafion content, the average pore diameter of the membrane increases. The reason is not clear yet. A possible explanation would be that most of the small pores are filled with Nafion and the large pores remain that leads to an increase in average pore diameter. The total pore volume (0.36, 0.22 and 0.02 cm^3/g for NPSi-1, NPSi-4, and NPSi-6, respectively) decreases dramatically with increasing Nafion content because Nafion acts as pore filler in the porous glass matrix as expected. The optimized structure of the hybrid membrane that can be controlled by changing the amount of H_3PO_4 and Nafion is crucial for the high-performance hybrid membrane.

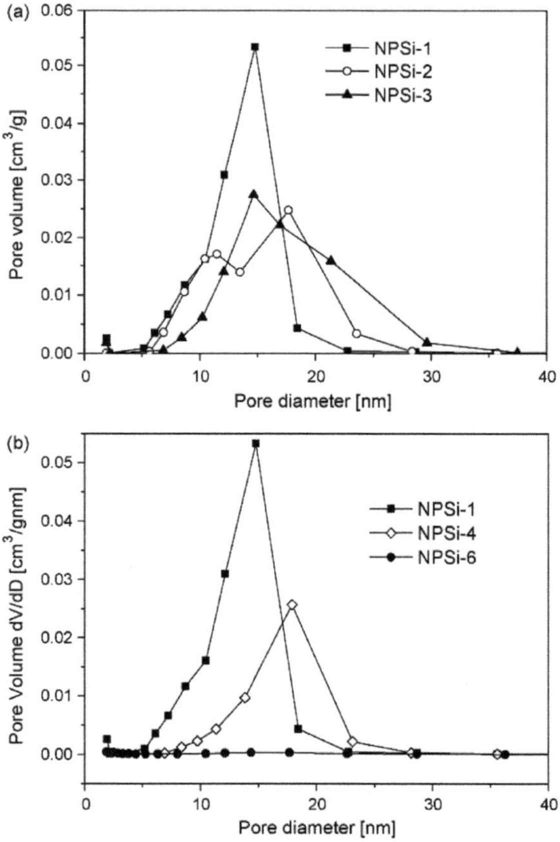

Figure 3.11 The pore-size distribution curves determined by BJH equation using the desorption branch of the N_2 gas adsorption–desorption isotherms [39]. Copyright 2011 Elsevier.

In N_2 adsorption–desorption isotherms, the width of the hysteresis indi cates the accessibility of the pores. Normally, the wider the hysteresis is, the more interconnected the pores are [35]. In bare phosphosilicate glass membrane, the interconnection between pores is indispensable for achieving high proton conductivity. However, the interconnected pore may leads to severe fuel leakage. In the NPS hybrid membrane, Nafion was introduced to reduce the pore volume without notable sacrifice in proton conductivity. In Fig. 3.12a, NPSi-1, NPSi-2, and NPSi-3 show similar hysteresis, which indicates that H_3PO_4 has inconspicuous effect on pore connectedness. In Fig. 3.12b, the pore interconnection of NPSi-4 is similar with that

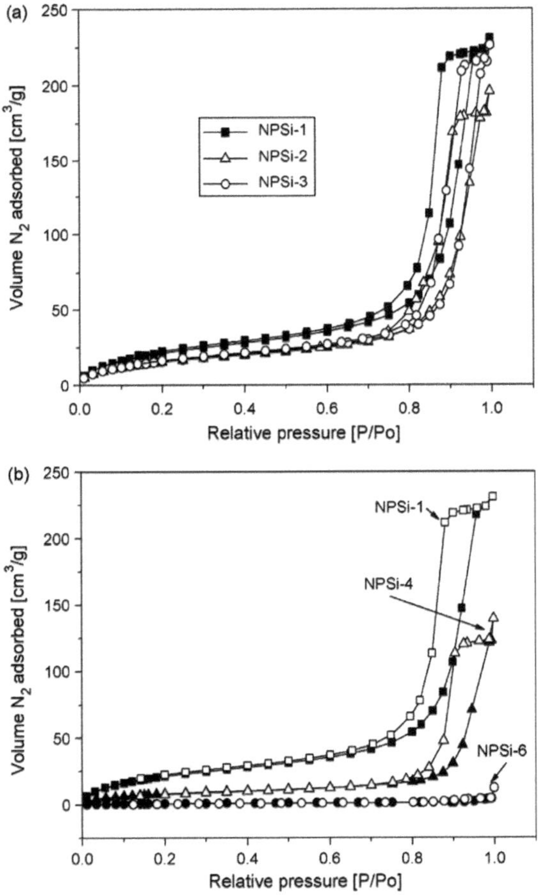

Figure 3.12 N_2 adsorption–desorption isotherms. Solid and hollow marks indicate the adsorption and desorption branches, respectively [39]. Copyright 2011 Elsevier.

of NPSi-1, which demonstrates that the Nafion content in NPSi-4 is not sufficient to block the interconnection of the pores. In NPSi-6, most of the pores are filled up with Nafion, and therefore, no pore interconnection can be observed.

3.2.2.2.3 *Water uptake property of the hybrid glass membrane*

Water uptake property of the hybrid glass membranes was compared in Figs. 3.13 and 3.14. The samples for both water uptake and conductivity measurements were placed simultaneously in the constant temperature and humidity chamber. Figure 3.13 shows the water uptake property of NPSi-1, NPSi-2, and NPSi-3 under various relative humidity measured at 50°C. We can see that the membrane with more H_3PO_4 content is more hygroscopic. The discrepancy is notable at a high relative humidity. In Fig. 3.14, the water uptake property of the glass membranes with different Nafion content was compared. We can see that the water uptake of the membranes decreases with increasing Nafion content. This is because the pore volume of the hybrid membrane decreases with increasing Nafion content and the amount of water that can be adsorbed by the membrane is heavily dependent on the pore volume of the membrane.

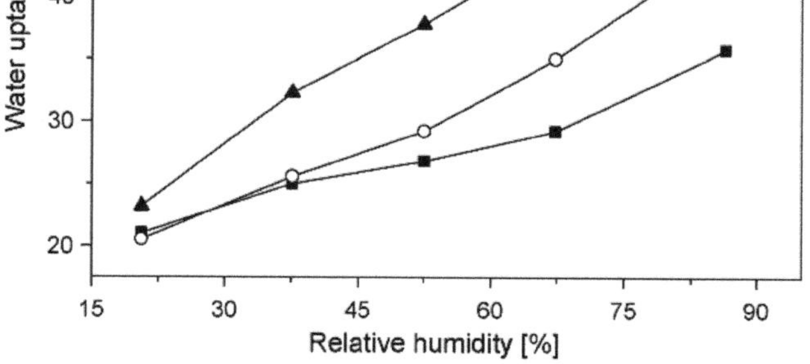

Figure 3.13 Water uptake of NPSi-1, NPSi-2, and NPSi-3 under various relative humidity [39]. Copyright 2011 Elsevier.

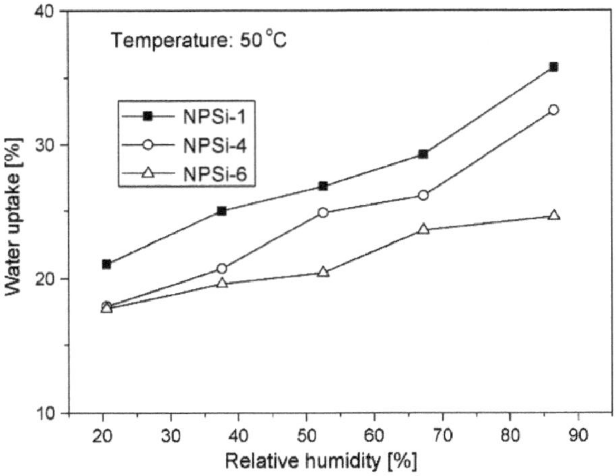

Figure 3.14 Water uptake of NPSi-1, NPSi-4, NPSi-6 under various relative humidity [39]. Copyright 2011 Elsevier.

3.2.2.2.4 *Proton conductivity of the hybrid glass membrane*

Figure 3.15 shows the proton conductivities of the hybrid glass membranes against relative humidity at 50°C. We can see that the increasing water content (see the water uptake data displayed in Fig. 3.13) leads to a rapid increase in conductivity. NPSi-3 shows the highest proton conductivity that can reach 0.1 S/cm with 85% relative humidity, at 50°C. The membranes show high proton conductivity even at 20% relative humidity (>10^{-3} S/cm). From 20 to 85% relative humidity, the conductivity of NPSi-3 increases from 0.01 to 0.1 S/cm. Conductivities of the three membranes increase rapidly with relative humidity, which indicates that water is significantly important for proton transport in these membranes. Figure 3.15 shows the apparent effect of H_3PO_4 content on proton conductivity when plotted against relative humidity. However, the fact how H_3PO_4 content influences the proton conductivity of the membrane cannot be read from the data in Fig. 3.15. In Fig. 3.16, we plotted the conductivity versus water content (water uptake) in the glass membrane on the basis of the data in Figs. 3.13 and 3.15. Interestingly, a "master curve" of the relation between proton conductivity and water content can be obtained for the membranes with the same Nafion content. It is demonstrated that, with the water uptake higher than 20%, proton conduction is a water-governed process that happens in the water salvation. Proton conductivity is exclusively determined by water content in the membranes, where the direct effect of underlying structure and composition of the membrane on proton conduction can be neglected. This phenomenon was also observed in previous reported work [36].

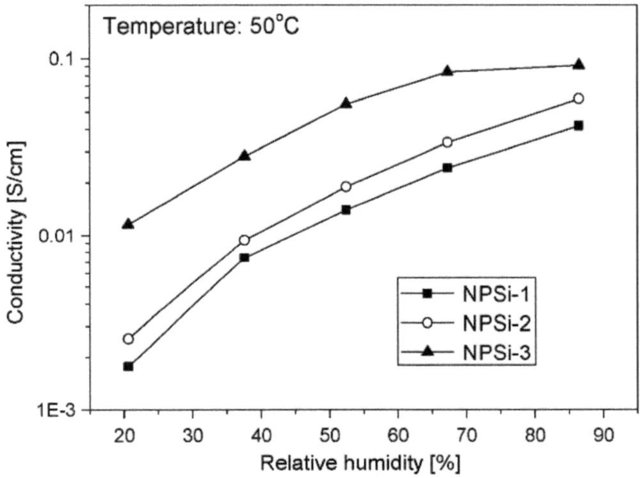

Figure 3.15 Proton conductivities of NPSi-1, NPSi-2, and NPSi-3 against relative humidity at 50°C [39]. Copyright 2011 Elsevier.

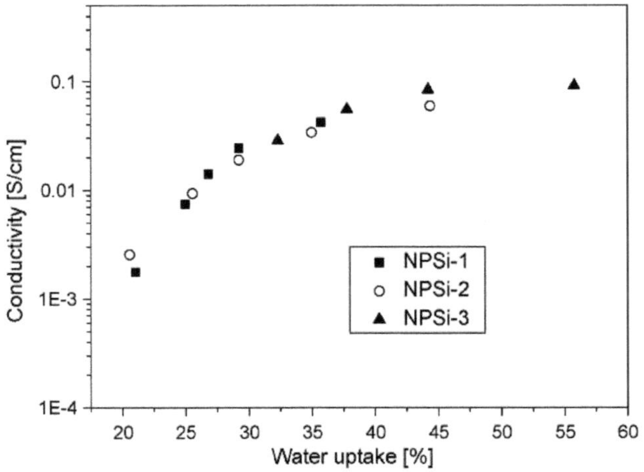

Figure 3.16 Plot of conductivity versus water content in the hybrid glass membranes (at 50°C) [39]. Copyright 2011 Elsevier.

Proton conductivities of NPSi-1, NPSi-4, and NPSi-6 under various relative humidity were measured at 50°C. The result was shown in Fig. 3.17. We can see that NPSi-4 shows higher proton conductivity than NPSi-1, though NPSi-4 is less hygroscopic. As discussed above, water plays an important role in proton conduction in the membrane. Therefore, a lower proton conductivity of NPSi-4 than NPSi-1 should be expected according to the water uptake property shown in Fig. 3.14. However, Nafion itself is a fast proton conductor and could promote proton conduction of the hybrid membranes. The overall effect of Nafion on proton conductivity depends on its content in the hybrid membranes. For example, in NPSi-4 where the feed ratio of Nafion to TEOS is 1:4, Nafion shows positive effect on proton conductivity as compared with NPSi-1. However, in NPSi-6, the water uptake of the membrane is significantly lowered by increasing the Nafion content in the hybrid membrane, and this became the dominant effect on proton conductivity; therefore, NPSi-6 shows the lowest proton conductivity among the three membranes.

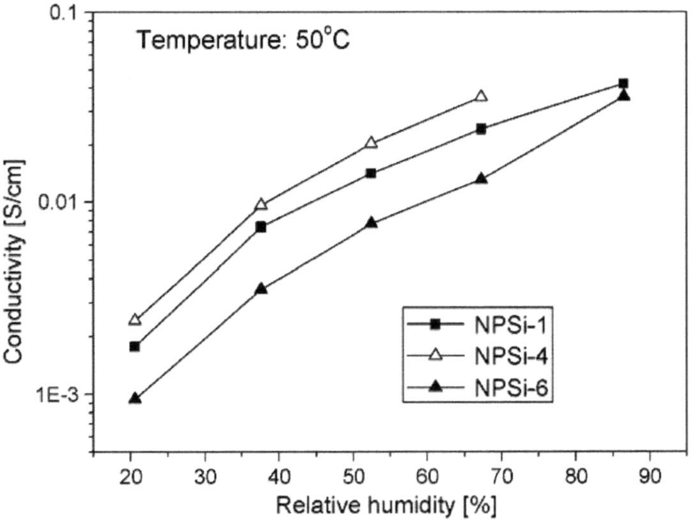

Figure 3.17 Proton conductivities of NPSi-1, NPSi-4, and NPSi-6 against relative humidity at 50°C [39]. Copyright 2011 Elsevier.

Figure 3.18 demonstrates the temperature dependence of proton conductivities of NPSi-1, NPSi-2, and NPSi-3 measured with 70% relative humidity. All the membranes possess promising proton conductivity (above 0.01 S/cm) during the measuring temperature range. Notably, NPSi-3 shows extremely high proton conductivity of about 0.07 S/cm at 30°C, 70% relative humidity, and exceeds 0.1 S/cm at 80°C, 70% relative humidity.

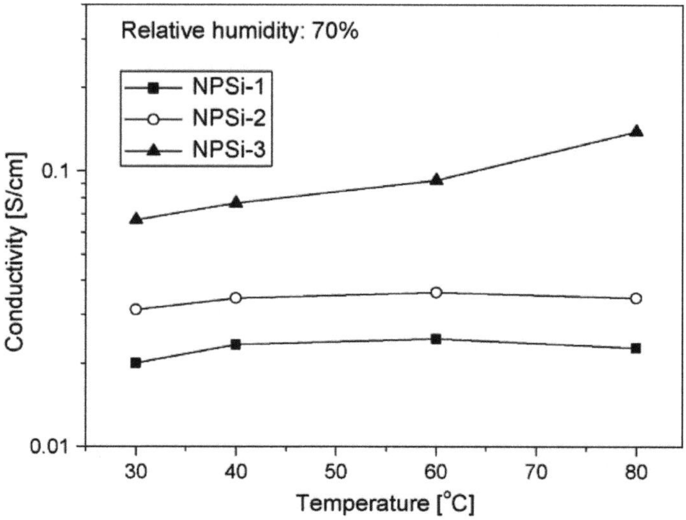

Figure 3.18 Temperature dependence of proton conductivities of NPSi-1, NPSi-2, and NPSi-3 measured with 70% relative humidity [39]. Copyright 2011 Elsevier.

3.2.2.2.5 *H_2/Air fuel cell testing*

The H_2/air fuel cell test was carried out with humidified H_2 and air fed into anode and cathode, respectively, at room temperature. The gas flow rates of H_2 and air were 20 and 100 sccm, respectively. On the basis of the consideration of conductivity and fuel permeation, the sample of NPSi-3 was selected for fuel cell testing.

The open circuit voltage (OCV) of the fuel cell was maintained at 0.94 V after a period of constant fuel supplement and cyclic activation. Polarization curve was then recorded to show the overall performance of the fuel cell, as shown in Fig. 3.19. Power density was simultaneously derived from the I–V curve. A drop of the cell voltage even under an initial small load current shows a great activation potential loss that can be frequently observed at low operating temperatures. The obtained power density was 42.6 mW/cm^2.

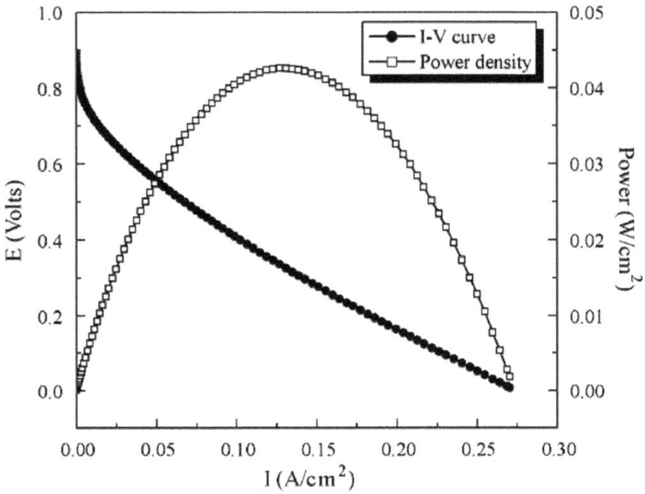

Figure 3.19 Power density derived from I–V curve for NPSi-3 based fuel cell at room temperature [39]. Copyright 2011 Elsevier.

An impedance analysis of the MEA was carried out with a voltage load of 0.4 V and the impedance spectra are shown in Fig. 3.20. The ohmic resistance (R_Ω) of the fuel cell shown in Fig. 3.20 was 1.27 Ωcm^2, which consists of the electronic resistance and the membrane resistance. An estimation of the proton conductivity of the membrane can be made using the following equation:

$$\sigma = \frac{L}{ASR}$$

where σ is the proton conductivity of the membrane, and L and ASR are the thickness and area-specific resistance of the membrane, respectively. The thickness of the used membrane was around 0.5 mm. Suppose ASR of the membrane is 1.27 Ωcm^2 (ignoring the electronic resistance), and the calculated conductivity was around 0.04 S/cm, which is in accordance with that discussed in Section 2.2.2.4.

The indicated total electrode resistance (R_e) of both the cathode and the anode is about 2 Ωcm^2, which is rather high than that of Nafion-based MEA. The electrode resistance can be effectively reduced by optimizing the structure and component of the electrodes.

On the basis of the promising proton conductivity, this membrane could be possibly used as novel proton exchange membranes in fuel cells. Another possible application for this kind membrane is to be used in humidity sensors.

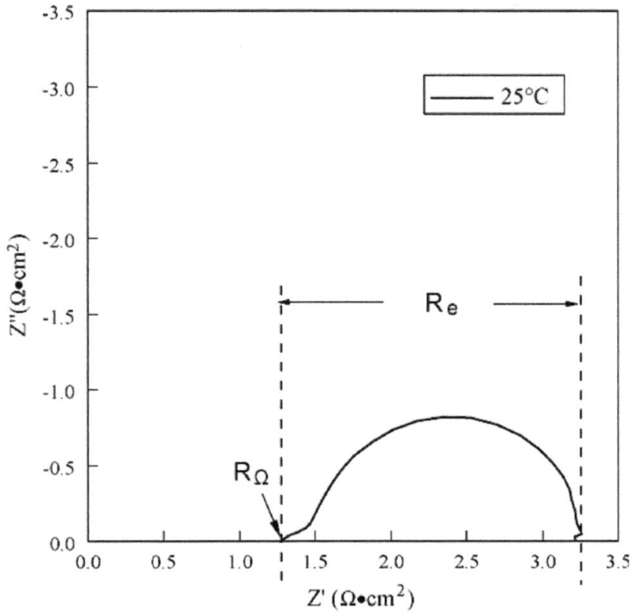

Figure 3.20 Impedance spectra of NPSi-3 based fuel cell at 0.4 V [39]. Copyright 2011 Elsevier.

3.3 Flexible Proton-Conducting Perfluorosulfonate Glass-Based Composite Membranes

It is believed that a key issue is that sol–gel derived glasses are inflexible and fragile, and thus cannot withstand the pressure required to press them into a membrane–electrode assembly (MEA) for favorable interface contact between the electrolyte and electrode. As a result, a high interface resistance arises, which is responsible for an undesirable power loss. The use of these inorganic proton conductors in fuel cells is thereby hampered, and so endowing them with sufficient flexibility is strongly desired.

Herein, we propose a synthetic strategy whereby a glass-based composite membrane is fabricated by mixing a sol–gel derived glass powder and an organic polymer by a simple mechanical ball-milling method. In the glass-based composite membrane, the inorganic glass component affords high proton conduction and dimensional stability; while the organic component provides sufficient mechanical strength and flexibility. In the present work, the NPS glass with 10 wt% Nafion developed by our group [39], showing high proton conductivity of ca. 10^{-1} S/cm, was chosen as the glass component of the composite membrane, and sulfonated poly(ether ether ketone) (SPEEK) was utilized as organic polymeric component due to its adequate proton conductivity, low cost, superior mechanical strength, and high thermal and chemical stabilities [40]. The NPS glass in powder form was incorporated into the SPEEK polymer so that a large-size, flexible, glass-based composite membrane was fabricated.

Such a composite membrane exhibits efficient proton conduction capability. We showed that a single fuel cell equipped with the composite membrane achieved a peak power density of 322 mW/cm^2, which is much higher than the values of 6–165 mW/cm^2 for fuel cells using glass-based electrolytes reported previously [8, 37–38, 41].

3.3.1 Preparation of NPS/SPEEK Membrane

The NPS glass was prepared by a sol–gel method as reported previously [39]. Briefly, a mixture of $Si(OC_2H_5)_4$ (TEOS), deionized water, and hydrochloric acid in a molar ratio of $1:4:4 \times 10^{-3}$ (TEOS:H2O:HCl) was firstly prepared and stirred for 30 min at room temperature. Then, phosphoric acid solution and Nafion® resin in molar ratios of 7:3 (H_3PO_4:TEOS) and 1:9 (Nafion®:TEOS) were slowly added to the above mixture and stirring was continued for 20 min. The obtained transparent solution was then transferred to a Teflon vessel and left to stand at room temperature until gelation occurred. The gel was then hydrothermally treated at 150°C for 24 h under saturated water vapor. After these procedures, an NPS glass monolith was obtained, which was subsequently crushed to powder form using an agate mortar. PEEK was sulfonated using concentrated sulfuric acid according to a procedure described elsewhere [42]. PEEK powder (5 g) was added to 95–97% sulfuric acid (100 ml) under vigorous stirring and the mixture was stirred for 3 h at 50°C. The polymer was then precipitated by pouring the solution into a large excess of iced water. The precipitated polymer was collected by filtration, washed repeatedly with distilled water until the washings were of neutral pH, and dried overnight. The degree of sulfonation (DS) was determined by titration with NaOH followed by back-titration with HCl.

The composite membranes were prepared by solvent casting. The SPEEK polymer was firstly dissolved in dimethylformamide (DMF) to prepare a 10 wt% solution, and then different amounts of the NPS glass powder were added. To indicate the different weight ratios of NPS glass and SPEEK polymer in the samples, such as 7:3, 6:4, and 5:5, they were designated as 7NPS/3SPEEK, 6NPS/4SPEEK, and 5NPS/5SPEEK, respectively. The resultant mixtures were milled using a planetary ball-mill apparatus for 3 days. Thereafter, the obtained suspensions were cast onto glass plates and then dried at 120°C for 6 h to form NPS/SPEEK composite membranes ranging from tens to hundreds of micrometers in thickness. A schematic illustration of the preparation of the NPS/SPEEK composite membranes is shown in Fig. 3.21.

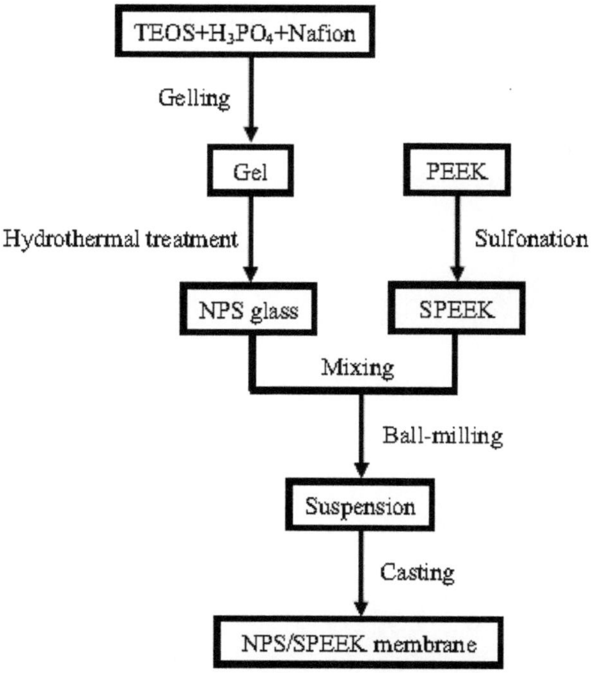

Figure 3.21 Preparative procedure for obtaining the NPS/SPEEK composite membranes [76]. Copyright 2012 Elsevier.

3.3.2 Characterization, Analysis, and Discussion

3.3.2.1 Appearance and chemical structure of the NPS/SPEEK composite membranes

It is known that the proton conductivity of SPEEK strongly depends on the degree of sulfonation (DS) [43]. A high DS with many hydrophilic groups leads to high conductivity, but is associated with a greater susceptibility to swelling and fragility. On the contrary, a low DS gives low conductivity but high mechanical and thermal stability [44]. In this study, the DS of SPEEK was set at 0.6 with the aim of achieving a compromise between the requirements of adequate proton conductivity and reasonable stability.

The NPS/SPEEK composite membranes with various weight ratios of NPS glass powder to SPEEK polymer were prepared. Figure 3.22 depicts a photograph of a 6NPS/4SPEEK composite membrane, showing sufficient flexibility and a large size of more than 150 cm^2. The composite membranes had a yellowish, opaque, and glassy appearance, while a pure SPEEK membrane was observed to be yellow and transparent. In addition, the higher the SPEEK content the NPS/SPEEK composite membranes contain, the more flexible they become.

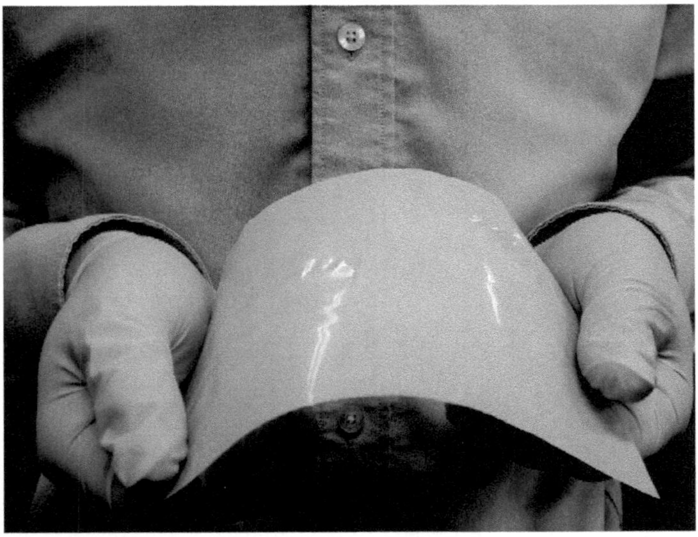

Figure 3.22 Optical image of an NPS/SPEEK composite membrane [76]. right 2012 Elsevier.

FTIR spectra of the NPS glass, the NPS/SPEEK composites, and pure SPEEK were recorded to confirm their chemical structures, and they are shown in Fig. 3.23. For the NPS glass, the peaks at 800 and 1053 cm^{-1} may be assigned as the characteristic peaks of Si–O–Si [45], which provides evidence for the formation of the Si–O–Si skeleton by the condensation reaction. An absorption shoulder at 960 cm^{-1} can be assigned to the Si–O stretch of silanol groups, Si–OH [18]. Two peaks at 1210 and 1150 cm^{-1} are attributed to the C–F stretching modes of the perfluorosulfonic acid polymer (Nafion®) [46]. For pure SPEEK and the SPEEK incorporated into the NPS/SPEEK composites, successful sulfonation was verified by the absorptions at 1224, 1080, and 1020 cm^{-1}, attributable to O=S=O vibrations of the sulfonic acid group [47]. The absorption split into two peaks at 1470 and 1490 cm^{-1} may be attributed to C–C stretches of the newly sulfonated aromatic ring [48], and these are also characteristic peaks confirming the structure of SPEEK. The absorption at 1650 cm^{-1} corresponds to the carbonyl group of SPEEK [42].

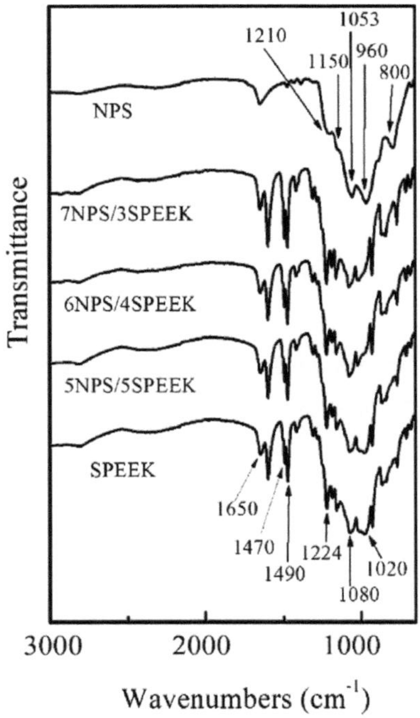

Figure 3.23 Infrared spectra of the NPS/SPEEK composites, the NPS glass, and the SPEEK polymer [76]. Copyright 2012 Elsevier.

Figure 3.24 shows the XRD patterns of the NPS glass and the 6NPS/4SPEEK composite membrane. A typical amorphous reflection peak ranging from 15° to 30° is observed, suggesting the existence of an amorphous matrix in the NPS glass. PEEK is a semi-crystalline polymer, showing sharp crystalline peaks in the 2θ range 20°–30° [49]. Sulfonation strongly decreases its crystallinity [47]. For the 6NPS/4SPEEK composite membrane, a weak reflection peak at 22.5° overlaying the broad amorphous peak is observed, which may be ascribed to the SPEEK component.

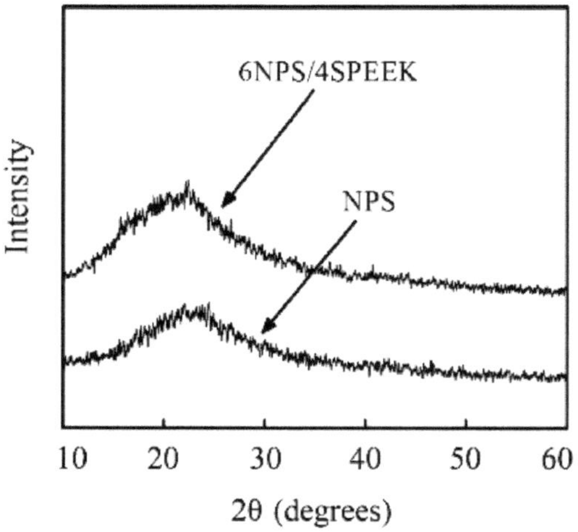

Figure 3.24 XRD pattern of the 6NPS/4SPEEK composite and the NPS glass [76]. Copyright 2012 Elsevier.

3.3.2.2 Morphology and pore structure of the NPS/SPEEK composite membranes

The morphology of the NPS/SPEEK composite membranes was observed by the field-emission SEM. The cross-section SEM images in Figs. 3.25a,c,e reveal that the 5NPS/5SPEEK, 6NPS/4SPEEK, and 7NPS/3SPEEK composite membranes had thicknesses of 47, 175, and 40μm, respectively. Figures 3.25b,d are high-magnification SEM images of 5NPS/5SPEEK and 6NPS/4SPEEK membranes, respectively. Figure 3.25d shows that NPS glass powder particles were entrapped within a continuous matrix, in which no aggregation of the glass particles was observed. As shown schematically in Fig. 3.25f, the NPS/SPEEK composite membrane consisted of a SPEEK polymer matrix and dispersed NPS glass powder particles. Such a continuous matrix of SPEEK polymer containing highly dispersed NPS glass particles could endow the composite membrane with good mechanical flexibility, which is crucial for MEA preparation for fuel cells.

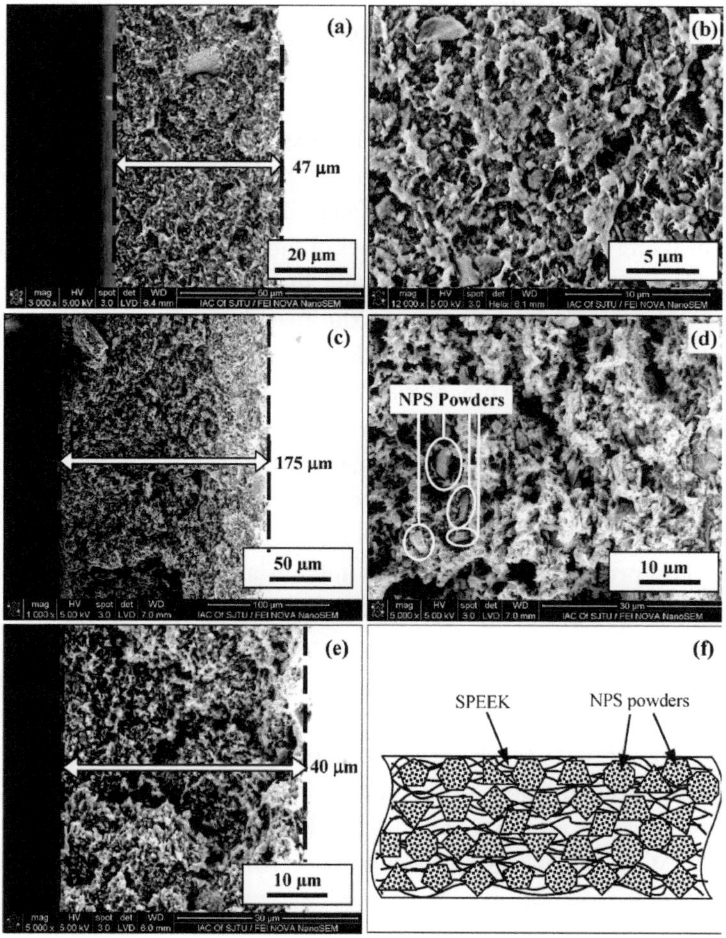

Figure 3.25 SEM images of the 5NPS/5SPEEK (a), the 6NPS/4SPEEK (c), and the 7NPS/3SPEEK (e) composite membranes; (b) and (d) high-magnification SEM images of the 5NPS/5SPEEK and the 6NPS/4SPEEK membranes, respectively; (f) schematic illustration of the NPS/SPEEK composite membranes [76]. Copyright 2012 Elsevier.

The porous structure and accessibility of the NPS glass and the NPS/SPEEK composite membranes were evaluated by N_2 adsorption–desorption measurements. It is known that the width of the hysteresis gives an indication of the accessibility of the pores, and the wider the hysteresis, the more interconnected the pores [50]. As can be seen in Fig. 3.26a, the NPS glass had a large hysteresis, which is a typical mesoporous character due to capillary condensation, indicating a high degree of pore interconnection and corresponding gas accessibility, whereas the 7NPS/3SPEEK and 6NPS/4SPEEK composite membranes possessed an extremely small hysteresis, implying restricted interconnection and gas inaccessibility. The pore-size distributions of the samples were determined by the BJH equation, using the desorption branches of the N_2 adsorption–desorption isotherms. Figure 3.26b shows that the NPS glass had an average pore diameter of 14.6 nm, and the inset therein reveals that the 7NPS/3SPEEK and 6NPS/4SPEEK composites had average pore diameters of 9 and 6.9 nm, respectively. In addition, the NPS glass had a pore volume of 0.343 mL/g, while the two composites had pore volumes of 0.034 and 0.007 mL/g, respectively. Clearly, the NPS glass had higher pore volume and larger pore size, and hence greater gas accessibility than the NPS/SPEEK composites.

For proton-conducting glass membranes, interconnection between the pores is essential for achieving high proton conductivity [51]. However, interconnected pores may lead to severe fuel leakage. N_2 adsorption–desorption analysis has demonstrated that in the NPS/SPEEK composite membranes, the incorporated SPEEK polymer may serve to block pores and to reduce the pore volume and the pore interconnection. In the composite membranes, in which NPS glass particles were encapsulated with SPEEK polymer, as confirmed by SEM analysis, the SPEEK content was evidently sufficient to block the gas accessibility.

3.3.2.3 Proton-conducting properties, water uptake, and swelling ratio

The proton-conducting properties of the samples were examined by AC impedance spectroscopy. The membrane resistance, R, was obtained from the intercept on the real axis in the high frequency

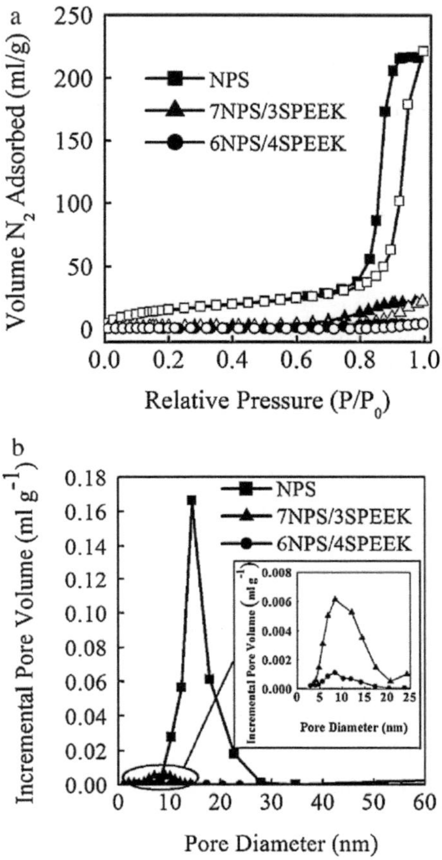

Figure 3.26 (a) N_2 adsorption–desorption isotherms of NPS glass and the NPS/SPEEK composite membranes. Empty and filled symbols indicate the adsorption and desorption branches, respectively. (b) Pore-size distribution curves for NPS glass and the NPS/SPEEK composite membranes [76]. Copyright 2012 Elsevier.

range, and it was imputed into the following formula to calculate proton conductivity.

$$\sigma = \frac{l}{AR}$$

where l is the thickness of the membrane, A is the area of the electrode, and R is the membrane resistance determined from the complex impedance plot.

The obtained proton conductivities were compared with that of the NPS glass reported by our group [39]. Figure 3.27a shows the proton conductivities of the NPS/SPEEK composites, pure SPEEK, and NPS glass under various relative humidity at 50°C. The conductivities of all of the membranes are seen to increase evidently with increasing relative humidity. This tendency indicates that water plays an important role in proton conduction of these membranes. This is due to the fact that increasing water content provides more proton-jumping sites [18]. In this study, proton conductivities of the pure SPEEK made with DMF casting solvent amount to 10^{-5}–10^{-4} S/cm within the relative humidity (RH) in the range from 70% to 90%. The values are much lower than 10^{-3}–10^{-2} S/cm of SPEEK using N,N-dimethylacetamide (DMAC) casting solvent [52], but are comparable to nearly 10^{-5} S/cm of SPEEK (0.65 DS) with identical DMF casting solvent [53]. The large discrepancies in conductivity data may be due to chemical interactions of SPEEK with the various solvents [54]. In comparison with DMAC, DMF as a casting solvent has a strong interaction with SPEEK, which leads to its lesser number of available acid groups, and consequently results in the reducing proton conductivity [42].

Figure 3.27b demonstrates the temperature dependences of the proton conductivities of the NPS/SPEEK composites and of pure SPEEK measured at 90% RH. It can be seen that the conductivities of all of the membranes increased with increasing temperature. Within the temperature range examined, the NPS/SPEEK composite membranes have quite high proton conductivities of above 10^{-3} S/cm. In particular, the 7NPS/3SPEEK membrane, with the highest content of NPS glass, shows the highest proton conductivity of 0.015 S/cm at 80°C and 90% RH.

Figure 3.27a,b show the effects of SPEEK content on proton conductivity against temperature and relative humidity, respectively. In such a composite system, the macroscopic conductivity should depend on the conductivities of the individual components and their contents. As shown in Fig. 3.27a, the NPS glass has high proton conductivities of 10^{-2}–10^{-1} S/cm^{-1} [39], which are much higher than that of the SPEEK polymer used in this study. As a result, the higher the NPS glass content, the higher the proton conductivity. Indeed, as can be seen in Figs. 3.27a,b, the 7NPS/3SPEEK composite with the highest content of NPS glass exhibited the highest proton conductivity. It is noteworthy that the SPEEK polymer itself was shown to have reasonable proton-conducting capability; hence, its incorporation into the composite membrane did not lead to a notable sacrifice in proton conductivity.

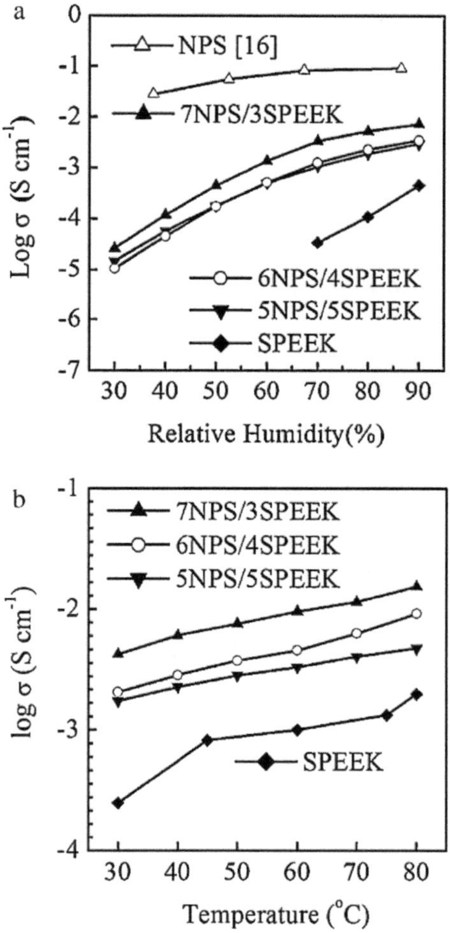

Figure 3.27 (a) Proton conductivities of the NPS/SPEEK composites, the pure SPEEK, and the NPS glass plotted against relative humidity at 50°C. (b) Temperature dependences of the proton conductivities of the NPS/SPEEK composites and the pure SPEEK measured at 90% RH [76]. Copyright 2012 Elsevier.

Water uptake and swelling ratio of the membranes are important parameters in the study of PEMFCs. Proton conductivity depends on water content of the membrane. However, very high water uptake may lead to excessive swelling, resulting in fragility and mechanical instability of membranes. Water uptake and swelling ratio of the pure SPEEK and the 6NPS/4SPEEK membranes were measured. The experimental results show that the pure SPEEK membrane has water uptake of 30% and swelling ratio of 5.1%, while the 6NPS/4SPEEK membrane possesses water uptake of 9.8% and very small swelling ratio of 0.5%. Clearly, the incorporation of the NPS glass powders within the composite membrane plays a very important role for restraining the swelling of membrane body, and the NPS glass powders may serve as nano-scale reinforcements and fixtures.

Good mechanical strength is essential for fuel cell application. It was determined that the 6NPS/4SPEEK membrane has a tensile strength of 18 MPa, and a breaking elongation of 1.4% showing a typical brittle behavior. The tensile strength value is much smaller than 51 MPa of the pure SPEEK membrane, but is comparable to 15.5 MPa of a recasted Nafion membrane [55].

3.3.2.4 H_2/O_2 fuel cell testing

From among the NPS/SPEEK composites, the 6NPS/4SPEEK membrane was selected to prepare a single fuel cell. The 6NPS/4SPEEK membrane used had a thickness of ca. 30μm. For comparison, a fuel cell using the NPS glass was also prepared. Both fuel cells were tested by feeding humidified H_2 and O_2 to the anode and cathode, respectively, at 65°C. The gas flow rates of H_2 and O_2 were 30 sccm and 60 sccm, respectively. A polarization curve was recorded to assess the overall performance of the fuel cell, as shown in Fig. 3.28. Power density was simultaneously derived from the I–V curve. The fuel cell based on the 6NPS/4SPEEK membrane has an open-circuit voltage of 0.93 V. A small drop of the cell voltage under an initial small load current was indicative of a low activation potential loss. The flexible NPS/SPEEK composite membrane could endure sufficient hot-pressing to assemble the MEA, so that good interfacial contacts between the electrolyte membrane and gas-diffusion electrodes (GDEs) with Pt/C catalyst were formed. The peak power density of the fuel cell reached 322 mW/cm^2, which is much higher than the values of 6–163 mW/cm^2 previously reported for fuel cells utilizing glass-based electrolytes [8, 37–39, 41].

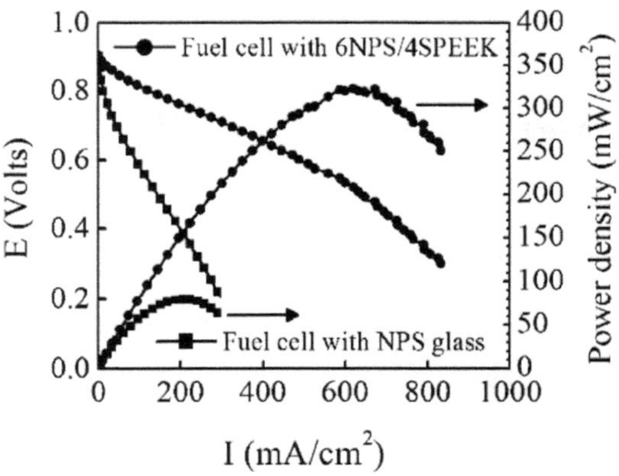

Figure 3.28 I–V and power density curves of both fuel cells operated at 65°C, which are based on the 6NPS/4SPEEK membrane, and the NPS glass, respectively [76]. Copyright 2012 Elsevier.

To the best of our knowledge, this value is the highest output power obtained for a fuel cell with a glass-based electrolyte. In comparison, the fuel cell based on the NPS glass without the hot-pressing procedure, due to the fact that hot-pressing would crush the NPS glass, released a peak power density of 80 mW/cm^2, as shown in Fig. 3.28. This value is much lower than that of the fuel cell with the 6NPS/4SPEEK composite membrane. The large discrepancy in output power may be due to the difference of interface resistance between the electrolyte membrane and electrodes for both fuel cells. The NPS glass has high proton conductivities of 10^{-2}–10^{-1} S/cm^1, giving a low electrolyte resistance. However, the high interface resistance that arises between the NPS glass and electrodes due to poor interface contact without hot-pressing procedure may result in the remarkable reduction in output power.

3.4 Flexible Proton-Conducting Perfluorosulfonate-Free Glass-Based Composite Membranes

In this section, the flexible perfluorosulfonate-free composite PEMs, which are based on NPS (SiO_2–P_2O_5) glass powder and SPEEK via a mechanical ball-milling and synthesized by sol-gel method, will be introduced. Instead, perfluorosulfonic polyer, SPEEK, was incorporated into the composite membrane as an organic polymeric component. The resulting SiO_2–P_2O_5/SPEEK composite membranes have been found to exhibit high proton conductivity, good thermal stability, and sufficient mechanical strength combined with flexibility.

3.4.1 Preparation of SiO_2-P_2O_5/SPEEK Composite Membranes

The SiO_2-P_2O_5 glass was prepared by a modified sol–gel method as reported previously [32]. Briefly, a mixture of TEOS, deionized water, and hydrochloric acid in a molar ratio of 1:4:4 $\times 10^{-3}$ (TEOS:H_2O:HCl) was firstly prepared. Different amounts of H_3PO_4 were then added to the above mixture. The obtained solution was then transferred to a Teflon® container under constant room temperature until gelation occurred. Hydrothermal treatment of the gel was then conducted by annealing it at 140°C for 24 h under saturated water vapor. The obtained SiO_2-P_2O_5 glass was crushed to a powder using an agate mortar. To indicate the different molar ratios of TEOS and H_3PO_4 in the SiO_2-P_2O_5 glasses, such as 7:3 and 6:4, they are designated as Si_7P_3 and Si_6P_4 glasses, respectively.

The sulfonation of PEEK using sulfuric acid was followed by the method described elsewhere [42]. PEEK powder (5 g) was slowly added to the preheated concentrated sulfuric acid (100 ml) under vigorous stirring for 3 h at 50°C. The sulfonated polymer solution was then poured into to a large amount of ice water bath to terminate the reaction and form fiber-type SPEEK polymer. The collected polymer was washed repeatedly with deionized water until neutral pH was achieved and followed an overnight drying process under vacuum below 103 Pa at 80°C. Acid–base titration method using standard NaOH and HCl solutions was chosen to determine the sulfonation degree (DS) of the polymer.

The composite membranes were prepared by solvent casting. Different amounts of SiO_2-P_2O_5 glass powder with different Si/P ratios were added to the SPEEK polymer solution using dimethylacetamide (DMAc) as the solvent. The resultant mixtures were then milled using a planetary ball-mill apparatus for 48 h at speed of 350 r/min.

The obtained slurry was cast onto glass plates using a doctor blade, and then dried at 90°C for 3 h under vacuum to form SiO_2-P_2O_5 / SPEEK composite membranes. The composite membranes could be easily detached from the glass plates after exposure in a humidity chamber for 1 h at 40°C under 90% RH. The samples are designated as *x*Si*a*P*b*/*y*SPEEK, where *x* and *y* represent the weight ratio of the SiO_2-P_2O_5 glass and SPEEK, and *a* and *b* denote the molar ratio of TEOS andH3PO4 in the SiO_2-P_2O_5 glass, respectively. The names and compositions of all of the samples in this study are listed in Table 3.2.

Table 3.2 Sample names and the corresponding components [80]. Copyright 2013 John Wiley and Sons.

Sample name	Feed ratio of Si: P in SiO_2-P_2O_5 glass (molar ratio)	Ratio of SiO_2-P_2O_5 glass: SPEEK in the composite memberane (weight ratio)
Si_7P_3	7:3	1:0
$7Si_7P_3$/3SPEEK	7:3	7:3
$6Si_7P_3$/4SPEEK	7:3	6:4
Si_6P_4	6:4	1:0
$7Si_6P_4$/3SPEEK	6:4	7:3
$6Si_6P_4$/4SPEEK	6:4	6:4

3.4.2 Characterization, Analysis, and Discussion of the Composite Membranes

3.4.2.1 Structure of the SiO_2-P_2O_5/SPEEK composite membranes

In this study, SiO_2-P_2O_5/SPEEK composite membranes with various compositions were prepared by mixing SiO_2-P_2O_5 glass powder and SPEEK polymer by mechanical ball milling. The SiO_2-P_2O_5 glass was prepared by sol–gel synthesis from TEOS and H_3PO_4 with subsequent hydrothermal treatment. The obtained SiO_2-P_2O_5 glass has been confirmed to have proton conductivities of the order of 10^{-2}–10^{-1} S/cm, comparable to those of commercially available Nafion® membranes [32]. For another thing, it is known that a high DS of SPEEK with many hydrophilic groups leads to high conductivity, but is associated with a greater susceptibility to swelling and fragility; conversely, a low DS gives low conductivity, but high mechanical and thermal stability [44]. SPEEK with a DS of 0.65 was used for the preparation of the composite membrane. The ion exchange capacity (IEC) corresponding to DS of 0.65 was determined to be 1.285 mmol/g. The chemical structure of SPEEK polymer is shown schematically in Fig. 3.29.

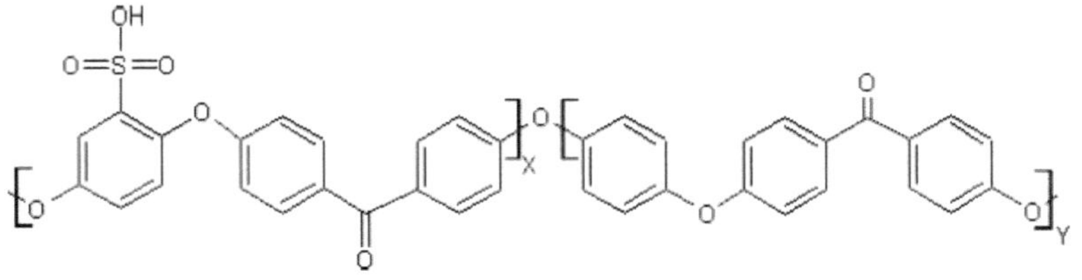

Figure 3.29 Chemical structure of the SPEEK, in which X and Y denote the sulfonated and unsulfonated portions, respectively [80]. Copyright 2013 John Wiley and Sons.

Figure 3.30 shows a photograph of a $7Si_7P_3/3SPEEK$ composite membrane and the corresponding MEA. As can be seen in Fig. 3.30, the composite membrane had a yellowish, opaque appearance, and was smooth and crack-free, indicating that mechanical ball-milling is an efficient approach for preparing composite membranes consisting of a glass powder and a polymer, even when the proportion of the glass component is as high as 70 wt.%. As can be seen in Fig. 3.30, the MEA based on the $7Si_7P_3//3SPEEK$ composite membrane had a square active area of 1 cm × 1 cm.

Figure 3.30 Optical image of the $7Si_7P_3/3SPEEK$ composite membrane and corresponding MEA [80]. Copyright 2013 John Wiley and Sons.

Figure 3.31a,b shows SEM images of the $7Si_7P_3/3SPEEK$ composite membrane, of which Fig. 3.31b is a high-magnification image. It can be seen that SiO_2-P_2O_5 glass powder particles were entrapped within a continuous matrix. The composite membrane thus consisted of glass powder in an SPEEK polymer matrix. It is suggested that the SPEEK network surrounding the SiO_2-P_2O_5 glass particles forms a strong backbone resulting in flexibility of the composite membrane, and it provides adequate proton conduction between the high proton-conducting SiO_2-P_2O_5 glass particles. The porous structure and accessibility of the composite membranes were evaluated by N_2 adsorption–desorption isotherms and BJH pore-size distribution plots. It is well known that the width of the hysteresis gives an indication of the accessibility of the pores, and the wider the hysteresis, the more interconnected the pores [50]. As can be seen in Fig. 3.32a, the SiO_2-P_2O_5 glasses showed a much larger hysteresis than the SiO_2-P_2O_5/SPEEK composite membranes. This is a characteristic of mesoporous materials with capillary condensation, indicating a high degree of pore interconnection and corresponding gas accessibility. It is interesting to note that the Si_6P_4 glass had a higher degree of pore interconnection than the Si_7P_3 glass. The difference is due to the fact that H_3PO_4 also acted as a pore inducer during the gelling process for preparing the glass powder. Figure 3.32b shows that the Si_7P_3 and Si_6P_4 glasses had peak pore diameters of 12.4 and 15.5 nm, respectively. The SiO_2-P_2O_5/ SPEEK composite membranes had pore diameters ranging from 5.9 to 11.5 nm. In addition, the Si_7P_3 and Si_6P_4 glasses had pore volumes of 0.419 and 0.433 ml/g, respectively, while the composite membranes had pore volumes in the range of 0.002–0.129 ml/g. Clearly, the SiO_2-P_2O_5/SPEEK composite membranes had lower pore volume, smaller pore size, less interconnected channels, and hence lower gas accessibility than the SiO_2-P_2O_5 glass powder. This implies that the incorporated SPEEK polymer acts as a pore blocking agent, reducing the overall pore volume and the number of interconnected channels.

Figure 3.31 (a) SEM image and (b) corresponding high-magnification image of the $7Si_7P_3$/3SPEEK composite membrane [80]. Copyright 2013 John Wiley and Sons.

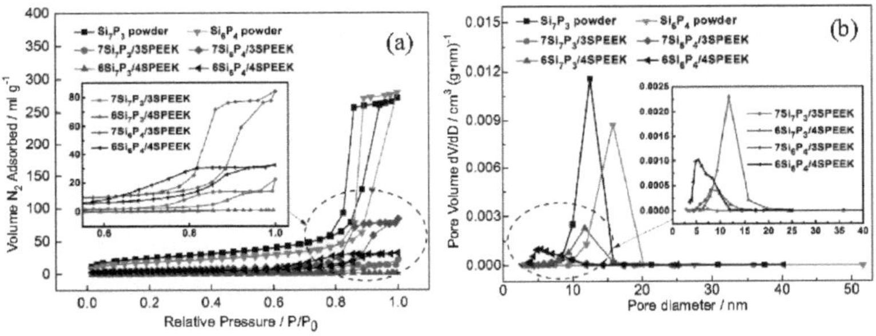

Figure 3.32 (a) N_2 adsorption–desorption isotherms of the SiO_2-P_2O_5 glasses and the SiO_2-P_2O_5/SPEEK composite membranes; (b) pore-size distribution curves for the SiO_2-P_2O_5 glasses and the SiO_2-P_2O_5/SPEEK composite membranes[80]. Copyright 2013 John Wiley and Sons.

FTIR spectra of the SiO_2-P_2O_5 glasses, the SiO_2-P_2O_5/SPEEK composite membranes, and pure SPEEK are shown in Fig. 3.33. For the Si7P3 and Si6P4 glasses, peaks at m = 800 and 1.053 cm^{-1} may be assigned as the characteristic peaks of Si–O–Si [45], which provides evidence for the formation of the Si–O–Si skeleton during the gelation and subsequent hydrothermal treatment. For the pure SPEEK and the SiO_2-P_2O_5/SPEEK composite membranes containing various amounts of SPEEK, sulfonation of the PEEK was verified by absorptions at m = 1.220, 1.072, and 1.012 cm^{-1} attributable to O=S=O vibrations of the sulfonic acid group [47]. The absorption split into two peaks at m = 1.472 and 1.492 cm^{-1} may be attributed to C–C stretches of the newly sulfonated aromatic ring [48], and the absorption at m = 1.645 cm^{-1} corresponds to the carbonyl group of SPEEK [42].

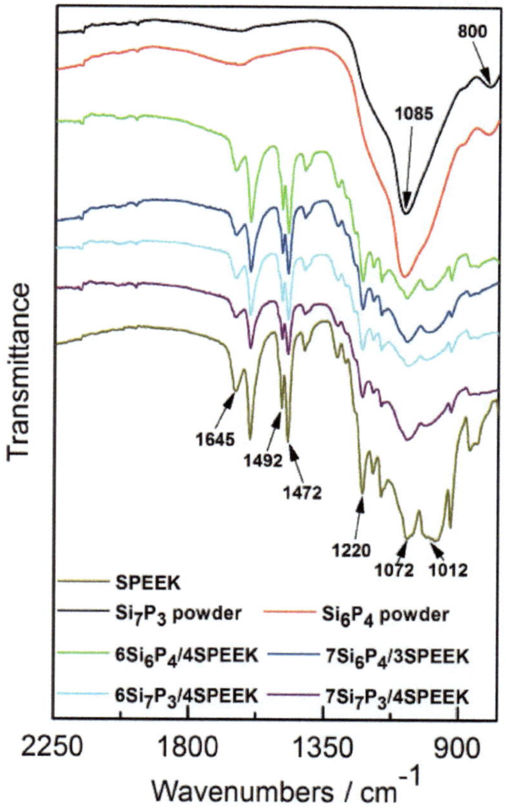

Figure 3.33 Infrared spectra of the SiO_2–P_2O_5 glasses, SiO_2–P_2O_5/SPEEK composite membranes, and the SPEEK polymer [80]. Copyright 2013 John Wiley and Sons.

3.4.2.2 Thermal and mechanical properties of the SiO_2–P_2O_5/SPEEK composite membranes

The thermal stabilities of the SiO_2-P_2O_5/SPEEK composite membranes and SPEEK polymer were determined by thermal gravimetric analysis (TGA) conducted under a nitrogen atmosphere. The weight losses of samples versus temperature are shown in Fig. 3.34. Each of the TGA curves features two main degradation stages, corresponding to degradation of the sulfonic acid groups and thermal oxidation of the polymer main chain. For the $7Si_7P_3$/3SPEEK and

$7Si_6P_4/3SPEEK$ composite membranes, the weight loss of approximately 10% in the range 250°C–400°C may be attributed to dehydration of chemisorbed water molecules involved in hydrogen bonding and the thermal degradation of sulfonic acid groups in SPEEK, and the weight loss in the temperature range above 500°C may be attributed to thermal degradation of the polymer [56]. For SPEEK, the weight loss associated with dehydration of chemisorbed water molecules and thermal degradation of sulfonic acid groups starts at 150°C and reaches 25% at 400°C [56]. Clearly, due to the incorporation of the SiO_2-P_2O_5 glass powder, the degradation of the SiO_2-P_2O_5/SPEEK composite membranes is delayed compared with that of pure SPEEK. The results confirm that incorporation of the inorganic component improves the thermal stability of the composite membranes, allowing it to operate at a higher temperature range [57].

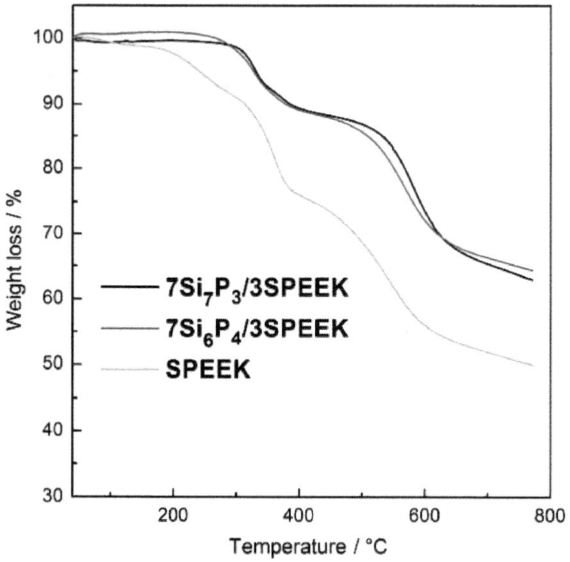

Figure 3.34 TGA curves of the $7Si_7P_3/3SPEEK$ (a) and $7Si_6P_4/3SPEEK$ (b) composite membranes and pure SPEEK (c) [80]. Copyright 2013 John Wiley and Sons.

Stress–strain behavior at room temperature is illustrated in Fig. 3.35. The $7Si_6P_4/3SPEEK$ composite had a tensile strength of 14.6 MPa and a breaking elongation of 0.6%, while the $6Si_6P_4/4SPEEK$ and $6Si_7P_3/4SPEEK$ composites had tensile strengths of 27.6 and 26.6 MPa and breaking elongations of 0.75% and 1.5%, respectively. It is evident that a higher content of SiO_2-P_2O_5 glass powder in the composite membranes leads to reductions in both tensile strength and breaking elongation. It has been reported that, for a recast Nafion® film, a typical tensile strength is around 10 MPa and a yield point can be observed [58]. The tensile strength of 27.6 MPa for the $6Si_6P_4$/4SPEEK composite membrane is higher than that of a recast Nafion® membrane.

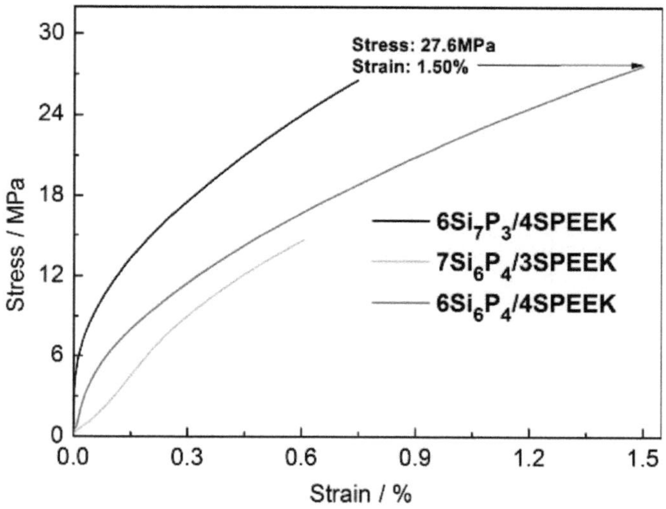

Figure 3.35 Stress–strain curves of $6Si_6P_4$/4SPEEK, $6Si_7P_3$/4SPEEK, and $7Si_6P_4$/3SPEEK composite membranes [80]. Copyright 2013 John Wiley and Sons.

3.4.2.3 Proton-conducting properties, water uptake, and swelling ratios of the SiO_2–P_2O_5/SPEEK composite membranes

Figure 3.36a shows the proton conductivities of the SiO_2-P_2O_5/SPEEK composites and of pure SPEEK under various relative humidity at 60°C. The conductivities of all of the membranes are seen to increase with increasing relative humidity. This can be attributed to the fact that increasing water content, which is absorbed into the membrane from the humid ambience, provides more proton-jumping sites [18]. Among the studied composite membranes with various compositions, $7Si_6P_4$/3SPEEK showed the highest proton conductivities, attaining a maximum value of 1.52×10^{-2} S/cm at 60°C and 90% RH.

Figure 3.36b shows the temperature dependences of the proton conductivities of the SiO_2-P_2O_5/SPEEK composites and of pure SPEEK measured at 80% RH. It can be seen that the conductivities of all of the membranes increased with increasing temperature, as a function of reciprocal temperature, and that the values measured for the SiO_2-P_2O_5/SPEEK composite membranes were in the range 10^{-3}–10^{-2} S/cm.

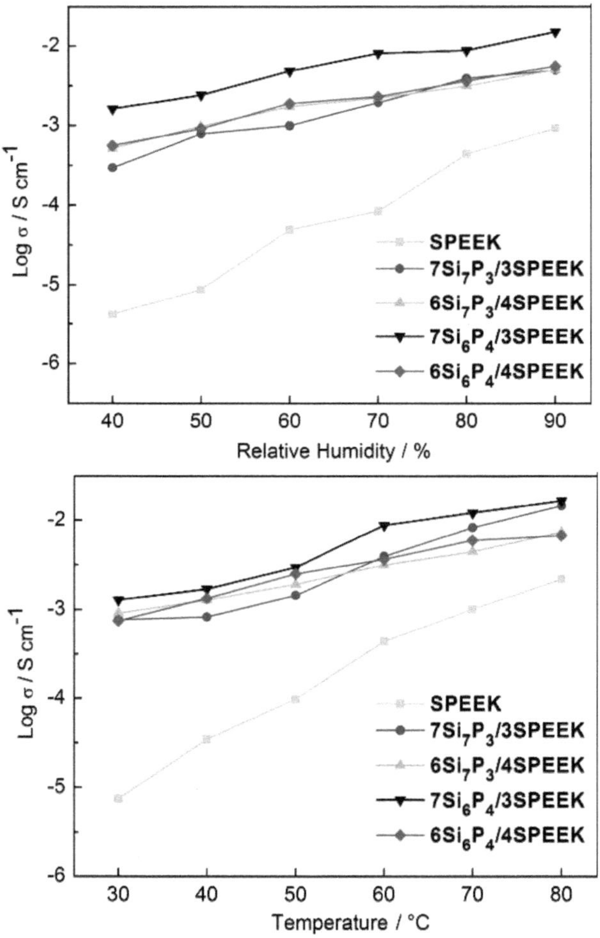

Figure 3.36 (a) Proton conductivities of the SiO_2–P_2O_5/SPEEK composite membranes and pure SPEEK against relative humidity at 60°C; (b) temperature dependence of proton conductivities of the SiO_2–P_2O_5/SPEEK composite membranes and pure SPEEK at 80% RH [80]. Copyright 2013 John Wiley and Sons.

In such a composite system, the macroscopic conductivity should depend on the conductivities of the individual components and their contents. SiO_2-P_2O_5 glass has a high proton conductivity of 10^{-2}–10^{-1} S/cm [32], much higher than that of the SPEEK polymer used in this study. As a result, as can be seen in Fig. 3.36a,b, the proton conductivities of the SiO_2-P_2O_5/SPEEK composite membranes containing a high proportion of SiO_2-P_2O_5 glass were much higher than that of SPEEK, and the $7Si_6P_4/3SPEEK$ composite membrane with the highest content of SiO_2-P_2O_5 glass exhibited the highest proton conductivity. It is noteworthy that the proton conductivity of $7Si_6P_4/3SPEEK$ is higher than that of $7Si_7P_3/3SPEEK$, even though they have equivalent SiO_2-P_2O_5 glass contents in the composite membrane. This implies that the utilization of a SiO_2-P_2O_5 glass with a higher phosphorus content in the composite membrane helps to promote proton conduction. This may be attributed to the fact that in SiO_2-P_2O_5 glasses derived from TEOS and H_3PO_4, the PO–H group has higher ionic character than the SiO–H bond, as a result of which it has higher cation-exchange capacity and water affinity, thereby enhancing proton conduction [32]. It was determined that the pure SPEEK membrane had a water uptake of 32.5% and a swelling ratio of 5.7%, while the $6Si_7P_3/4SPEEK$ membrane had a water uptake of 8.3% and a very small swelling ratio of 0.4%. Clearly, the incorporation of SiO_2-P_2O_5 glass powder into the composite membrane plays a very important role in restraining the swelling of the membrane body.

3.4.2.4 H_2/O_2 single fuel cell testing

For the assembly of a single fuel cell, the $6Si_7P_3/4SPEEK$ membrane used had a thickness of ca. 45 μm. The operating temperature of the single fuel cell was maintained at 68°C. The polarization curve and the impedance spectrum of the fuel cell were recorded and are shown in Fig. 3.37a,b, respectively. As can be seen in Fig. 3.37a, the fuel cell had an open-circuit voltage of 0.96 V and a peak power density of 355.6 mW/cm^2. This peak power density is much higher than the value of 20.8 mW/cm^2 for a fuel cell based on SiO_2–P_2O_5 glass reported previously by our group [60], which may be rationalized as follows. The flexible SiO_2–P_2O_5/SPEEK composite membrane could endure sufficient hot-pressing to assemble the MEA, so that good interfacial contacts between the electrolyte membrane and the gas diffusion electrode (GDE) with Pt/C catalyst were formed, resulting in high output power. On the contrary, for the fuel cell with the SiO_2–P_2O_5 glass, hot-pressing to improve interfacial contacts could not be conducted, as this would have crushed the SiO_2–P_2O_5 glass.

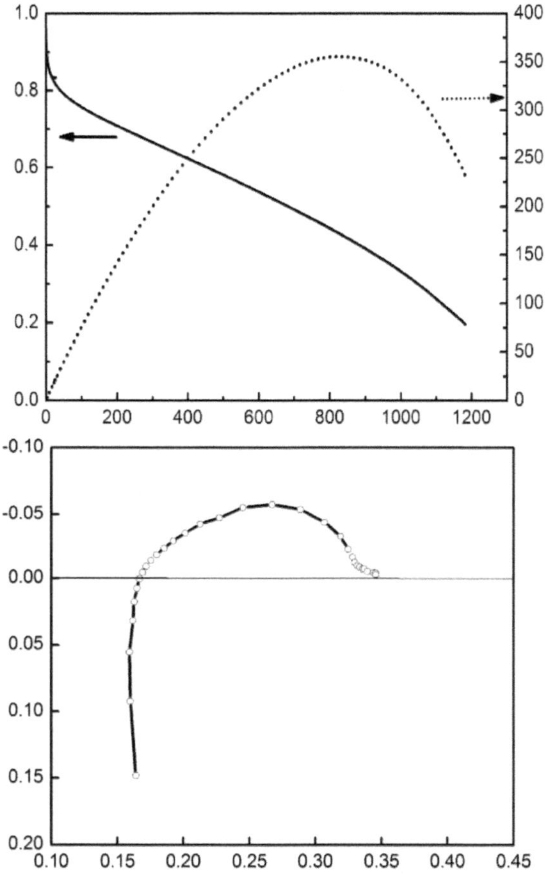

Figure 3.37 (a) *I*–*V* and power density curves of the $6Si_7P_3/4SPEEK$ composite membrane; (b) impedance spectrum of $6Si_7P_3/4SPEEK$-based fuel cell operated at 0.4 V [80]. Copyright 2013 John Wiley and Sons.

Impedance analysis of the single fuel cell was carried out with a voltage load of 0.4 V, and the impedance spectrum is shown in Fig. 3.37b. The intercept in the high-frequency domain on the real axis represents the Ohmic resistance (R_Ω) of the composite membrane used, which is the protic resistance, while the semicircle in the low-frequency domain represents the electrode resistance (R_e) from both the anode and cathode. As can be seen in Fig. 3.37b, the cell had an Ohmic resistance of 0.17 Ωcm^2 and an electrode resistance of 0.18 Ωcm^2. The electrode resistance could be effectively reduced by optimizing the structure and components of the electrode [59]. The previous report has demonstrated the high stability of the proton-conducting glass membrane under fuel cell conditions during the continuous measurement up to 140 h [59].

3.5 Anhydrous Proton-Conducting Glass Membranes Doped with Ionic Liquid for Intermediate-Temperature Fuel Cells

During the past years, much effort has been focused on developing proton exchange membranes for intermediate-temperature PEMFCs that operate at 120°C–200°C [61, 62]. Compared with low-temperature PEMFCs (<100°C), PEMFCs operating above 100°C under anhydrous conditions have many advantages, such as higher tolerance to significant quantities of CO, faster electrode kinetics, no cathode flooding, and simpler system design [63, 64].

Ionic liquids (ILs) are molten salts at ambient temperature and possess unique properties, such as high thermal stability, low vapor pressure, high ionic conductivity, and wide electrochemical windows. ILs are the promising electrolytes for applications in electrochemical devices, such as lithium batteries [65], supercapacitors [66], and solar cells [67]. Protic ILs exhibit high anhydrous proton conductivity and can be employed as novel electrolytes for intermediate-temperature PEMFCs. Watanabe et al. have investigated more than 80 different types of protic ILs [68]. They have found that [dema][TfO] had a wide and stable liquid-temperature range (T_m = −6°C and T_d = 360°C) as well as a high anhydrous ionic conductivity of 5.3 × 10^{-2} S/cm^1 at 150°C and 1.0 × 10^{-2} S/cm at room temperature. Hydrogen oxidation reactions (HORs) and oxygen reduction reactions (ORRs) occur with a small over potential in [dema][TfO], and the measured OCV in the H_2/O_2 fuel cell is 1.03 V[68]. They also studied the sulfonated polyimides/[dema][TfO] hybrid membrane in a fuel cell [69, 70].

Inorganic glasses have the advantages of excellent thermal and chemical stability, high mechanical strength, and low cost. Hybrid proton exchange membranes based on ILs and inorganic glasses are promising for fuel cell applications at temperatures above 100°C. Nogami et al. have prepared glass proton conducting membranes based on 1-butyl-3-methylimidazolium-bis (trifluoromethylsulfonyl) imide [BMI][BTSI][71] and 1-ethyl-3-methylimidazoliumbis (trifluoromethanesulfonyl) imide [EMI][TFSI][72]. The conductivities of the membranes obtained were measured at up to 150°C under anhydrous conditions. The maximum proton conductivity was on the order of 10^{-3} S/cm at 150°C, which was insufficient for application in PEMFCs.

In this section, the [dema][TfO]/SiO_2 hybrid glass membranes fabricated via the sol-gel approach will be introduced and discussed, which includes its anhydrous ionic conductivity, across the temperature range 120°C–220°C and the stability of the ionic conductivity at intermediate temperature.

3.5.1 Preparation of [Dema][TfO]/SiO_2 Hybrid Glass Membranes

3.5.1.1 Preparation of [dema][TfO]ionic liquid

The [dema][TfO] IL was prepared via neutralization of N,N-diethylmethylamine and trifluoromethanesulfonic acid. Around 4.36 g of N,Ndiethylmethylamine and 7.5 g of trifluoromethanesulfonic acid were each diluted with 100 ml of absolute ethanol. The trifluoromethanesulfonic acid solution was then slowly added to the N,N-diethylmethylamine solution under stirring. The reaction mixture was stirred for an additional 6 h at room temperature, then distilled at 80°C to remove the solvent, and dried under reduced pressure at 180°C for 12 h to obtain pure [dema][TfO] protic IL. Figure 3.38 shows a schematic representation of the chemical structure of the [dema][TfO] IL.

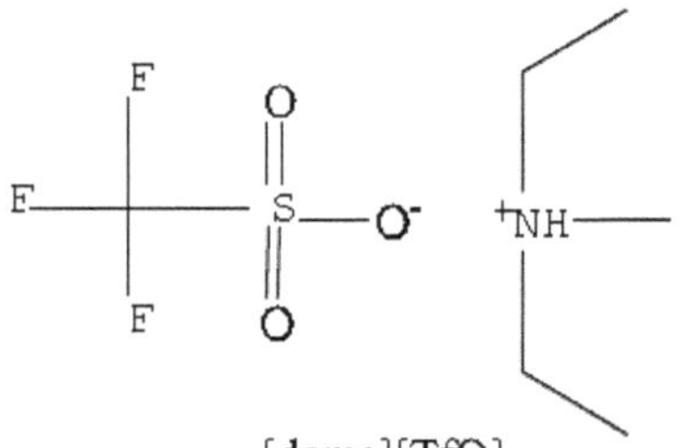

Figure 3.38 Schematic representation of the chemical structure of [dema][TfO] IL [81]. Copyright 2012 Elsevier.

3.5.1.2 Preparation of [dema][TfO]/SiO_2 glass membranes

The glass membranes were prepared via a sol–gel method. A mixture of tetraethylorthosilicate (TEOS), deionized water, and hydrochloric acid (TEOS:H2O:HCl) in a molar ratio of 1:4:4 × 10^{-3} was prepared and stirred for 20 min at room temperature. Then, [dema][TfO] IL was slowly added to the mixture and stirred for another 10 min. The resulting transparent mixture was transferred into a Teflon vessel and aged at 30°C for 3 weeks. Then, the samples were annealed at 120°C in an oven for 12 h. For simplification, the [dema][TfO]/SiO_2 glass membranes are abbreviated as ILSiO_2(x), where x represents the weight portion of the [dema][TfO].

3.5.2 Characterization, Analysis, and Discussion

3.5.2.1 Characterization of [dema][TfO]ionic liquid

The purity of the [dema][TfO] IL obtained was examined by ^{1}H NMR. Figure 3.39 shows the NMR spectrum of the [dema][TfO]IL. The peak at 8.9 ppm is assigned to the NH groups (proton d) resulting from acid–base salt formation. In Fig. 3.39, the peaks at 1.2 ppm, 2.7 ppm, and 3.0 ppm are attributed to protons marked as a, b, and c, respectively.

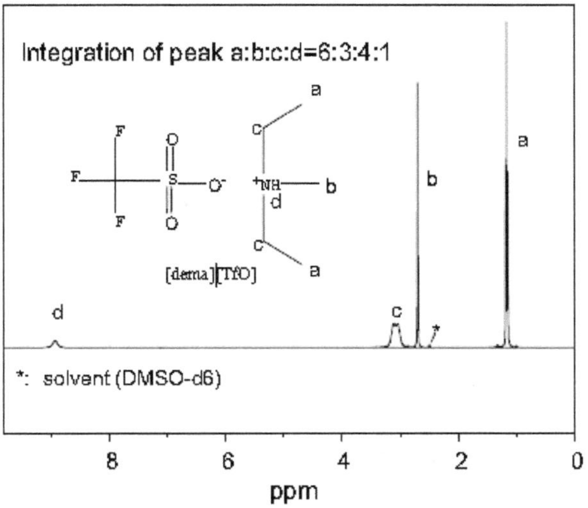

Figure 3.39 ^{1}H NMR spectrum of [dema][TfO] [81]. Copyright 2012 Elsevier.

3.5.2.2 Morphology and structure of the [dema][TfO]/SiO_2 glass membranes

Using the modified sol–gel approach, homogenous and crackfree [dema][TfO]/SiO_2 hybrid glass membranes appeared to be smooth and transparent, as shown in Fig. 3.40, indicating good compatibility between the IL and the SiO_2 matrix. The uniform thickness and homogeneity of the membrane structure are critical for the use of a proton-conducting membrane in fuel cells. Crack-free membranes can be easily prepared via the approach described in Section 5.1.2. However, cracks may appear if gelation occurs at an elevated temperature or in a glass vessel. The shape and size of the membrane can be controlled by the shape of the parent vessel in which the membranes are formed, and the thickness can be adjusted by the amount of sol solution added to the vessel. The membrane shown in Fig. 3.40 was prepared in a round Teflon vessel, and the diameter of the membrane was approximately 5 cm. In general, a uniform thickness for the membrane in the range of 0.2–0.8 mm can be maintained, which is thicker than Nafion® membranes (25–125 μm). Due to the brittle characteristic of the glass membrane, a thickness larger than 0.2 mm might be necessary to obtain an acceptable mechanical strength. However, the increase in membrane thickness may lead to a high resistance for proton conduction across the membrane. Therefore, an optimum thickness for the glass membrane might be 0.3–0.5 mm.

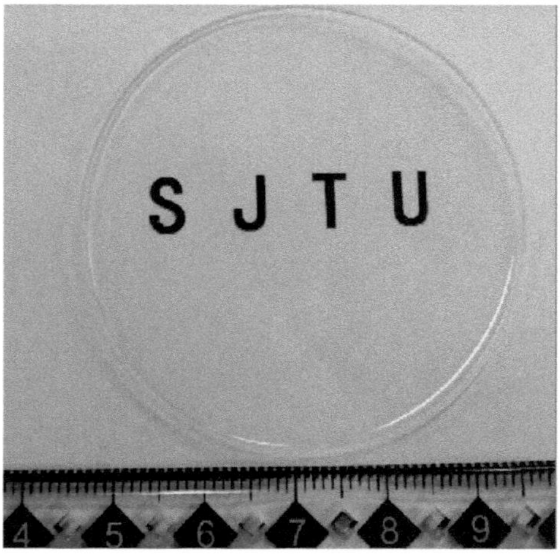

Figure 3.40 Digital photo of the [dema][TfO]/SiO_2 glass membranes [81]. Copyright 2012 Elsevier.

In the hybrid glass, the SiO_2 skeleton forms a matrix with the IL dispersed within the SiO_2 network. Figure 3.41 shows the SEM micrographs of the $ILSiO_2$ (0.56), which demonstrates the homogeneous and dense features of the membrane, with no visible pores on the surface. For application to proton exchange membranes in fuel cells, the membrane has to be compact to suppress fuel crossover though the membrane. Figure 3.41b shows the inorganic skeleton consisting of compact and cross-linked SiO_2 nanoparticles with an average diameter of a few nanometers. The IL was retained in the "gaps" within the SiO_2 particles and formed channels for proton transport. N_2 adsorption–desorption isotherms were measured to investigate the membrane pore structure. A pore volume was not detected from the isotherms, which suggested that the "gaps" are completely filled.

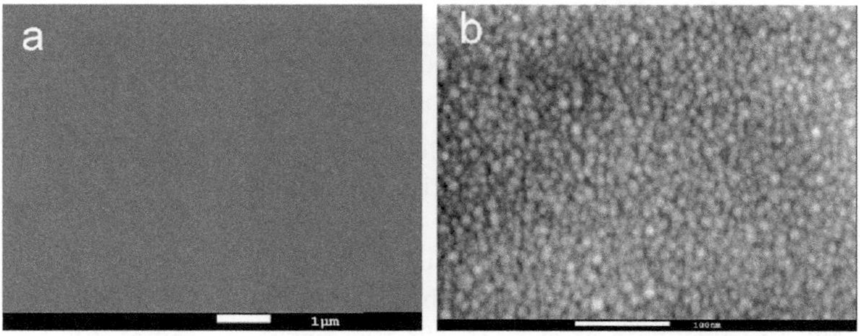

Figure 3.41 SEM micrographs of $ILSiO_2$ (0.56) [81]. Copyright 2012 Elsevier.

3.5.2.3 FTIR analysis of the [dema][TfO]/SiO_2 glass membranes

The IR spectra of pure [dema][TfO], pure SiO_2 and the [dema][TfO]/SiO_2 composites are shown in Fig. 3.42. In the spectrum of pure SiO_2, the peaks at 800 and 1080 cm^{-1} are the characteristic peaks of Si–O–Si [45], which provides evidence for the formation of the Si–O–Si skeleton in the condensation reaction. An absorption peak at 942 cm^{-1} can be assigned to the Si–O stretch in the silanol groups, Si–OH [18]. The band at 1628 cm^{-1} corresponds to the bending frequency of water [19] and has been used to measure water content by determining either its integrated absorbance or peak absorbance. The bands located between 4000 and 3000 cm^{-1}, which consist of several subbands, are related to liquid water clusters. A broad absorption band appearing at 3448 cm^{-1} can be attributed to the stretching vibrations of Si–O–H and absorbed water molecules, H–O–H [21].

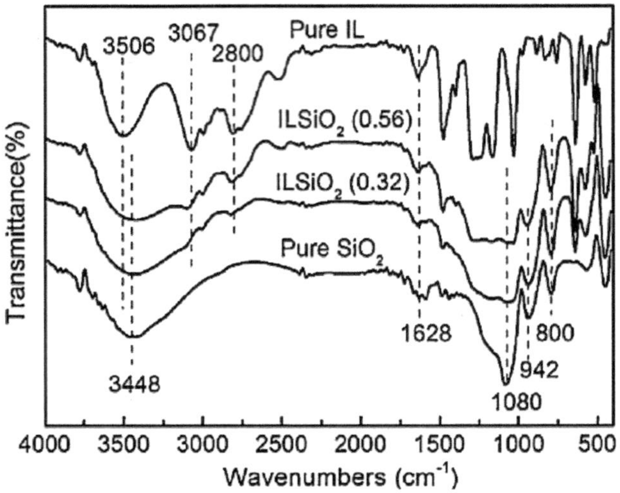

Figure 3.42 FTIR spectra of pure [dema][TfO], SiO_2, and [dema][TfO]/SiO_2 composites [81]. Copyright 2012 Elsevier.

In the spectrum of the pure IL, the water peak appears at 3506 cm^{-1}. A detailed FTIR study of the [dema][TfO] IL has been carried out by Sakakibara et al. [73]. The two peaks located at 3067 and 2800 cm^{-1} can be assigned to N–H stretching vibrations. These two peaks indicate that there are two different vibrational modes for the N–H bond in [dema][TfOH]. It has been proposed that [dema][TfOH] formed a cluster-like structure in the bulk phase, which resulted in the existence of two different N–H stretching modes [73]. The strong peaks at approximately 1150–1300 cm^{-1} in [dema][TfOH] originate from the N–H out-of-plane bending deformation mode coupled with neighboring C–H bending and S–O stretching vibrations in the triflate anion [73]. The strong signals from the N–H stretching and N–H out-of-plane bending deformation mode indicate the dynamic nature of the N–H protons that are important for proton conductivity.

3.5.2.4 Ionic conductivity of the [dema][TfO]/SiO_2 glass membranes

Figure 3.43 shows the Arrhenius plots of the ionic conductivities of the hybrid membranes. The conductivity of the hybrid membrane increases as the amount of [dema][TfO]/SiO_2 in the glass matrix increases. According to Equation: $\sigma(T) = n(T)\mu(T)q$ the conductivity (σ) is proportional to the number of charge carriers (n) per unit volume (i.e., the charge carrier number density), the mobility of ions in an electric field (μ), and the number of charges per carrier (q). As the IL weight portions increased in the matrix, the charge carrier number density increased. Therefore, the ionic conductivity increases accordingly. The conductivity of the glass membranes containing less than 48 wt% [dema][TfO] is on the order of 10^{-3} S/cm from 120°C to 220°C. Lower amounts of [dema][TfO] may result in membranes with conductivity that is too low for them to be used as proton exchange membranes in PEMFCs [e.g., $ILSiO_2$ (0.32)]. The hybrid glasses containing ILs more than 70% exhibited high anhydrous ionic conductivity exceeding 10^{-2} S/cm in the temperature range studied.

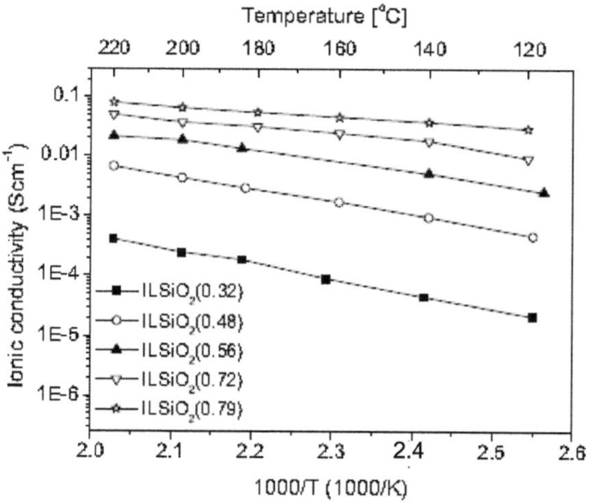

Figure 3.43 Anhydrous ionic conductivities of the [dema][TfO]/SiO_2 glass membrane as a function of temperature [81]. Copyright 2012 Elsevier.

In a fuel cell, water is constantly produced at the cathode, resulting in the fuel cell operating under humidified conditions. Therefore, understanding the effect of humidity on ionic conductivity is important. In Fig. 3.44, the conductivities of $ILSiO_2$ (0.56) at 60°C under atmospheres of various degrees of relative humidity (RH) are summarized. The conductivity of $ILSiO_2$ (0.56) increases by one order of magnitude (from 10^{-3} S/cm to 10^{-2} S/cm) when the relative humidity changes from 0% to 90%, which may be attributed to the absorbed water forming abundant hydrogen bonds with the ILs and facilitating proton hopping via a hydrogen bond breaking and forming process [36]. In addition, water is a good solvent for [dema][TfO], and dilution of [dema][TfO] can reduce the viscosity of the IL leading to improved ion mobility.

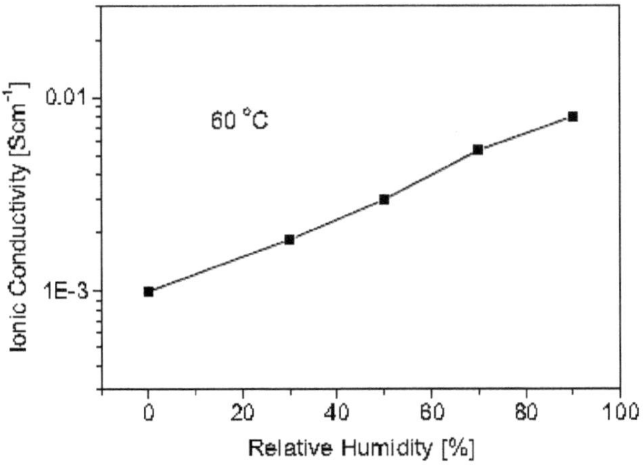

Figure 3.44 Effect of humidity on ionic conductivity of the $ILSiO_2$ (0.56) glass membrane [81]. Copyright 2012 Elsevier.

One of the motivations for preparing [dema][TfO]/SiO_2 glass membrane as an intermediate-temperature proton conductor is that both the SiO_2 monolith and the [dema][TfO] IL exhibit excellent thermal stability. The stability of the ionic conductivity of the [dema][TfO]/SiO_2 glass membrane was investigated, and the results are shown in Fig. 3.45. The samples of [dema][TfO]/SiO_2 glass membrane were kept in an oven continuously at 120°C for 180 h. The ionic conductivity was recorded at specific time intervals. There was no decrease in the ionic conductivity after annealing the samples at 120°C for 180 h. This result is evidence of the thermal stability of the membranes. Until now, the phosphoric acid based membranes have been an important class of membranes for intermediate-temperature fuel cells. However, phosphoric acid undergoes self-condensation at elevated temperatures, which leads to a decrease in conductivity [36, 74]. Therefore, the results shown in Fig. 3.45

indicate an important advantage of the IL-based membranes over the phosphoric acid based membranes.

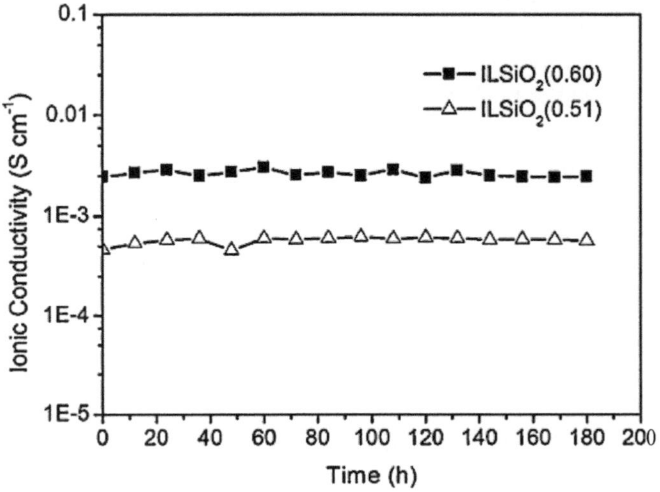

Figure 3.45 Ionic conductivities of the [dema][TfO]/SiO_2 glass membranes as a function of time [81]. Copyright 2012 Elsevier.

3.6 Performance of Fuel Cells (FCs) with Novel Proton-Conducting Glass Membrane or Composite Membrane

In this section, the applications of the two types of proton-conducting composite membranes, which are based on our previous study, to the direct methanol fuel cell (DMFC) and H_2/O_2 fuel cell will be introduced. The content chiefly refers to the performance of the fuel cells equipped with the composite membranes. Meanwhile, the impact of the changes of some parameters of composite membranes on the performance of cells will also be discussed.

3.6.1 Improved Performance of Fuel Cell with Proton-conducting Glass Membrane

It is believed that a key issue in the fabrication of membranes is that sol–gel derived glasses are inflexible and brittle, and thus cannot withstand the pressure required to press them into membrane electrode assemblies (MEAs) for a favorable interface contact between the electrolyte and electrode. As a result, a high interface resistance arises, which is responsible for an undesirable power loss. Therefore, decreasing interface resistance is essential in order to improve the performance of a fuel cell with a glass membrane. In addition, it is well known that increasing the operating temperature and adopting a thinner electrolyte membrane could improve the performance of PEMFCs. It is expected that these methods will be effective for fuel cells with glass membranes.

In this part, a NPS glass membrane based on porous phosphosilicate glass and a perfluorosulfonic acid polymer will be discussed, which is fabricated by using a sol–gel process with a low-temperature hydrothermal treatment (below 150°C). The selection of fabrication process and materials not only endows the process with some highlights, but it also boosts the performance of the membrane, simultaneously. For example, the low-temperature hydrothermal treatment prevents the thermal degradation of the organic component [39]. The incorporated highly proton conductive polyelectrolyte (Nafion®) plays a role in reducing the fuel permeation as a pore filler and in supplying additional proton conductive passes. Also, the NPS glass membrane with 10 wt.% Nafion® exhibits a high proton conductivity of ca. 10^{-1} S/cm [39] Furthermore, a single fuel cell equipped with the NPS membrane achieves a peak power density of 207 mW/cm^2 higher than many other values reported previously. These advantages along with the simple synthesis processes and low manufacturing costs, altogether, make the NPS glass membrane very promising to substitute polymer electrolytes for the practical application in fuel cells.

3.6.1.1 Preparation

3.6.1.1.1 *Preparation of NPS glass membrane*

The NPS glass membrane was prepared by a sol–gel method as reported previously [39]. Briefly, a mixture of $Si(OC_2H_5)_4$ (TEOS), deionized water, and hydrochloric acid in a molar ratio of 1:4:4 × 10^{-3} (TEOS:H_2O:HCl) was initially prepared and stirred for 30 min at room temperature. Then, phosphoric acid solution and Nafion resin in molar ratios of 3:7 (H_3PO_4:TEOS) and 1:9 (Nafion®:TEOS) were slowly added to the above mixture and stirring was continued for 20 min. The obtained transparent solution was then transferred to a Teflon vessel and left to stand at room temperature until gelation occurred. Teflon vessels with xerogels were introduced into a drying box. The gel was then hydrothermally treated at 150°C for 24 h, and a water vapor at a pressure of 1 atm was maintained within the drying box by providing an excess of deionized water. After these procedures, the NPS glass membrane was obtained. The thickness of the membrane can be controlled by changing the solution amount cast into the vessel.

3.6.1.1.2 *Preparation of fuel cell with the NPS glass membrane*

GDEs with a Pt loading of 0.5 mg/cm^2 were prepared by spraying the catalyst ink onto wet-proofed Toray carbon paper. Using a process called Method A, a sandwiched MEA with a circular 0.8 cm diameter active region was made by attaching the GDEs onto the glass membrane, and then, the MEA held between two graphite bipolar plates was placed between two gold plates. As an alternative called Method B, the MEA with the same active region was made by attaching the GDEs onto the glass membrane using Nafion® resin (10 wt.% solution in a mixture of lower aliphatic alcohols and water) as an adhesive, and then, the MEA sandwiched between two graphite bipolar plates was placed between two gold plates. In addition, Method B was improved, in which a thin-film pressure sensor as an additional part was inserted between the gold plate and the insulation plate for assembling the fuel cell. A schematic illustration for assembling the fuel cell is shown in Fig. 3.46.

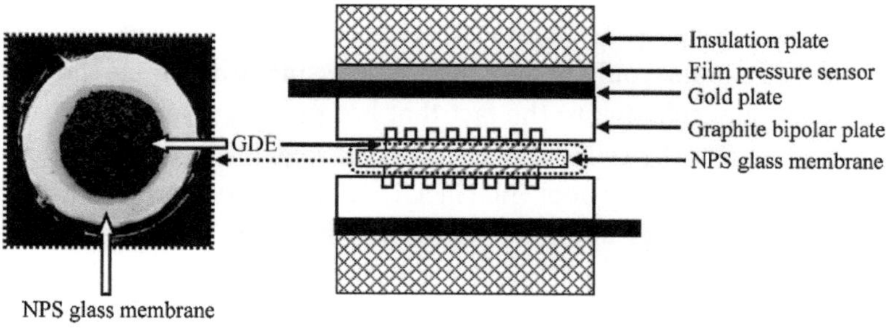

Figure 3.46 Schematic illustration for assembling the fuel cell based on the NPS glass membrane, and optical image of MEA with the NPS glass membrane [59]. Copyright 2012 Elsevier.

3.6.1.2 Characterization, analysis, and discussion

3.6.1.2.1 *Effect of MEA preparation process*

The NPS glass membrane was prepared by a hydrothermal assisted sol–gel method. A photo of the MEA, which are based on the NPS glass membrane, is shown in Fig. 3.46. Polarization and power output curves were recorded to assess the fuel cell performance and their improvement, as shown in Fig. 3.47. The fuel cells operating at room temperature, which are prepared with Method A and Method B, released maximum output powers of 14.4 and 45.4 mW/cm^2, respectively. Obviously, the cell using Method B showed much better performance than that with Method A. The large discrepancy in output power should be due to the difference of internal resistance including ohmic and electrode resistances for both fuel cells. An impedance analysis of the fuel cells was carried out and impedance spectra are shown in Fig. 3.48. The intercept in the high-frequency domain on the real axis represents ohmic resistance (R_Ω), and it is mainly ascribed to the membrane resistance and interfacial resistance between membrane and electrodes, both of which are protonic resistances. The semicircle in the low-frequency domain represents the electrode resistance, which is mainly charge–transfer resistance (R_c) on the anode and cathode, with the main contribution being the cathode ORR process.

The cell prepared using Method A had an ohmic resistance of 2.3 Ω cm^2 and a charge–transfer resistance of 2.7 Ω cm^2, while the cell prepared using Method B gave an ohmic resistance of 1.3 Ω cm^2 and charge–transfer resistance of 2 Ω cm^2. Compared with that of Method A, ohmic resistance and charge–transfer resistance for the fuel cell using Method B were evidently reduced. The NPS glass membrane could not endure hot pressing to assemble the MEA and the fuel cell, due to the fact that hot pressing would crush the NPS glass. For the cell with Method A, the high interfacial resistance developed with very poor interfacial contact between the NPS glass and the electrodes, and thus it gave rise to a large ohmic resistance. In contrast, due to the fact that the Nafion® resin can provide better adhesion and a stronger binding force between the glass membrane and the electrodes, better interfacial contact could be formed, leading to the reduction of interfacial resistance. In addition, it is well known that creating the so-called triplephase boundaries, in

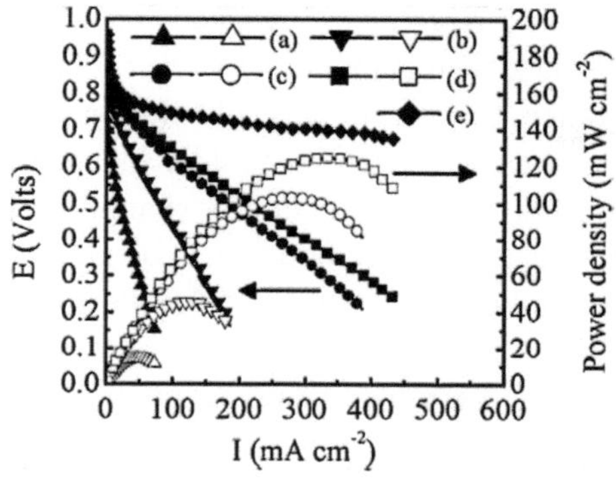

Figure 3.47 I–V and power density curves of fuel cells, which are based on the NPS glass membrane with a thickness of 800 _m. The performance of a fuel cell operating at room temperature, which was prepared by Method A, is shown. The improved performances of fuel cells operating at room temperature by using Nafion® resin as an adhesive (b), operating at 70°C by increasing operation temperature (c), and operating at 70°C by adopting a thin-film pressure sensor (d) are also given, which were prepared by Method B. Curve (e) iR-correction result for the initial measurement of the fulel cell operating at 70°C with the thin-film pressure sensor [59]. Copyright 2012 Elsevier.

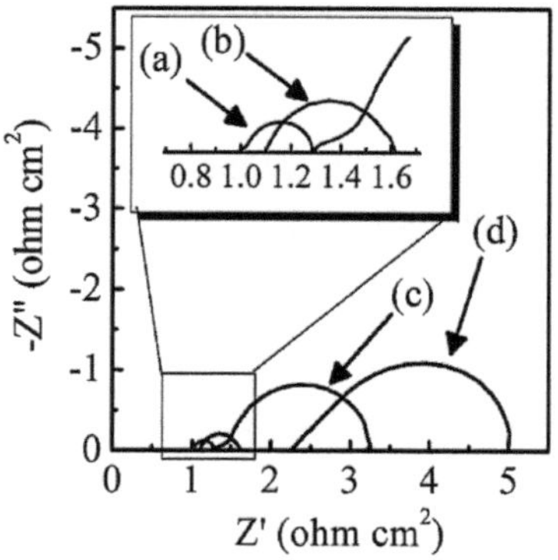

Figure 3.48 Impedance spectra of fuel cells measured at 0.4 V, which are based on the NPS glass membranes with a thickness of 800 μm. The impedance spectrum of a fuel cell operating at room temperature, which was prepared by Method A, was shown. Impedance spectra of fuel cells operating at room temperature by using Nafion® resin as an adhesive (b), operating at 70°C by creasing operation temperature (c), and operating at 70°C by adopting a thin-film pressure sensor (d) are also given, which were prepared by Method B [59]. Copyright 2012 Elsevier.

which the protonic electrolyte, catalytic particles, and reactants are in close contact with each other, is critical in reducing charge-transfer resistance. The introduction of a Nafion® layer between the rigid NPS glass and the electrodes could achieve better triplephase boundaries, leading to the reduction of charge-transfer resistance. Therefore, the utilization of Nafion® resin by Method B leads to the reduction of both ohmic resistance and charge-transfer resistance of the fuel cell, and thus, the increased output power was obtained.

Figure 3.49 shows cross-sectional scanning electron microscopy (SEM) image and corresponding energy dispersive X-ray spectroscopy (EDS) composition profiles of a Nafion®-painted NPS glass membrane. As shown in Fig. 3.49a, the Nafion® layer attached on the NPS glass membrane has a thickness of 800 nm. For another thing, EDS composition profiles in Fig. 3.49b give the distributions of C, O, and F elements within the NPS glass membrane. The sectional distribution of the Nafion® infiltrated into the NPS glass membrane is determined by checking F species. In Fig. 3.49b, the F species gradually decreased from the surface toward the interior, and tend to reach a constant level onto the depth of ca. 4 μm. These results suggest that the Nafion® could infiltrate into the surface layer of the NPS glass membrane at an infiltration depth of ca. 4μm. The Nafion® was introduced to improve interface adhesion between the glass membrane and the electrodes. For another thing, the introduction of the Nafion should be also helpful for promoting proton conduction within the surface layer of the glass membrane and for reducing fuel crossover to some extent.

3.6.1.2.2 *Effect of operating temperature*

Increasing the operation temperature has been considered an effective approach for improving fuel cell performance. As expected, the fuel cell performance shown in Fig. 3.47 increased dramatically from 45.4 to 103 $mWcm^2$ with increasing operation temperature from 25°C to 70°C. The resultant impedance spectra at both temperatures are shown in Fig. 3.48. It is observed that ohmic resistance decreased slightly from 1.3 to 1.1 Ωcm^2, while electrode resistance decreased significantly from 2 to 0.5 Ωcm^2 when the temperature increased from 25°C to 70°C. Ohmic resistance decreased as the temperature

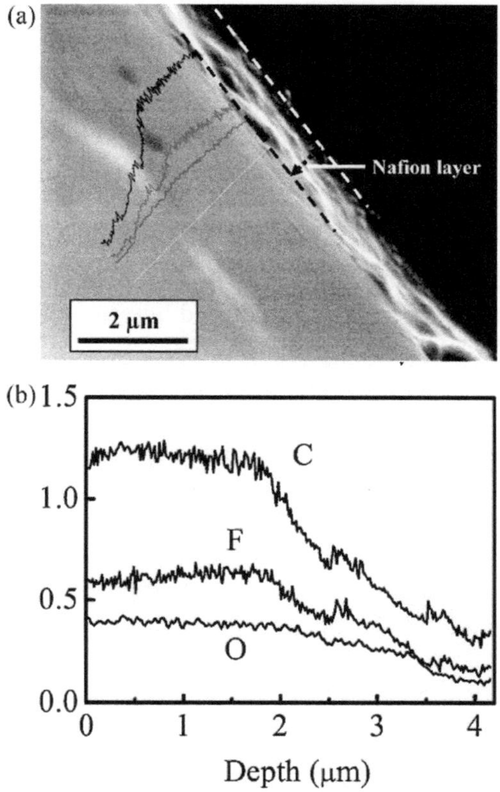

Figure 3.49 Cross-sectional SEM image (a) and corresponding EDS composition profiles (b) of a Nafion®-painted NPS glass membrane [59]. Copyright 2012 Elsevier.

increased, which may have been due to an increase of the membrane conductivity, as reported in the literature [39], in which proton conductivities of ca. 0.07 S/cm at 30°C and ca. 0.1 S/cm at 70°C under 70% relative humidity were given. According to electrochemical reaction kinetics, increasing the temperature could increase the rate of the electrode reaction. Electrode resistance results mainly from charge-transfer resistance and is a direct function of the slow ORR kinetic at the cathode, which becomes faster as the temperature increases [75]. In addition, the increased operation temperature could also improve the mass-transfer process, resulting in improved performance.

3.6.1.2.3 *Effect of assembly pressure*

To reduce interfacial resistance, assembly pressure should be increased as high as possible. For the present study, the NPS glass membrane used could not endure hot pressing to assemble the MEA due to its friability, so that the applied assembly pressure is critical for the reduction of interfacial resistance. In order to accurately control the assembly pressure, and to increase it as high as possible under the prerequisite conditions without crushing the NPS glass, a thin-film pressure sensor was inserted between the gold plate and the insulation plate, as shown in Fig. 3.46.

During the fuel cell assembly, progressively increasing force was applied, and a maximum contact pressure of 4.5 N was used to perform measurements due to the fact that higher pressure tends to crush the glass membrane. As shown in Fig. 3.47, an increased output power of 124.7 mW/cm for the fuel cell with the thin-film pressure sensor was obtained at 70°C, compared with the value of 103 mW/cm at 70°C for that assembled on the basis of operator experience, without the thin-film pressure sensor. Meanwhile, Fig. 3.48 shows that the ohmic resistance and electrode resistance decreased from 1.1 and 0.5 Ωcm^2 to 1 and 0.3 Ωcm^2, respectively, by the adoption of the thin-film pressure sensor.

In order to assess the stability of the NPS glass membrane under fuel-cell conditions, the longevity experiment was carried out. Figure 3.50 gives the stability of output power as a function of time under a cell votage of 0.5 V. The assembling configuration and operating condition of the fuel cell for stability measurement are identical to those of the fuel cell with the thin-film pressure sensor in Fig. 3.47. After an initial drop from ca. 90 to 85 mW/cm^2, during the continuous measurement up to 140 h, the fuel cell showed remarkable stability. The result demonstrates the high stability of the NPS glass membrane under fuel cell conditions. The reason for the drop in output power in the first few hours is still not clear. Thus, further study is definitely required.

3.6.1.2.4 *Effect of the NPS glass thickness*

The iR-correction is given from the following formula: $iR_{correction} = E - iR_{\Omega}$, where E is the measured cell voltage, i is the current density, and R_{Ω} is the ohmic resistance determined from impedance spectrum.

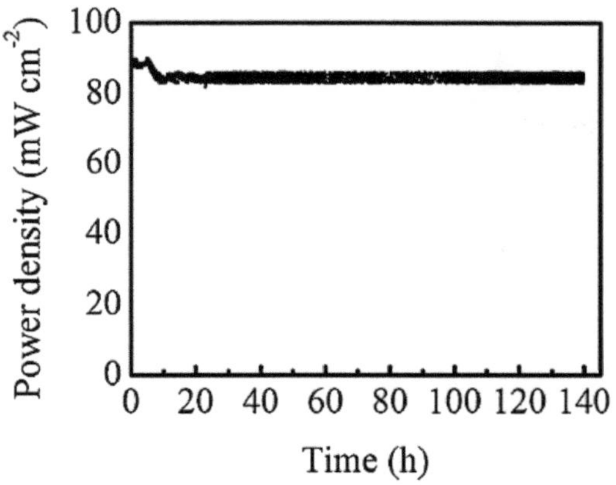

Figure 3.50 Output power as a function of time under a cell votage of 0.5 V for 140 h [59]. Copyright 2012 Elsevier.

The iR-correction result for the initial measurement of the fuel cell operating at 70°C with the thin-film pressure sensor is shown in Fig. 3.47 as curve (e). In comparison to conventional PEMFCs using the polymer electrolyte membranes, whose thicknesses range from the teens to 180 μm, performance limitations were believed to be related to membrane resistance of the NPS glass membrane, which is expected to be fairly large due to the utilization of an 800 μm thick glass membrane. Decreasing the NPS glass membrane thickness could, therefore, be expected to increase the performance. Thus, we constructed a fuel cell by using 500 μm thick NPS glass. As shown in Fig. 3.51, the fuel cell operating at 70°C gave an open-circuit voltage of 0.96 V and released a peak power density of 207 mW/cm^2, which is evidently greater than that with an 800 μm thick glass membrane. In addition, this value is much higher than the values of peak power density previously reported for fuel cells utilizing glass electrolytes. Figure 3.52 shows that the ohmic resistance decreased to 0.6 Ωcm^2 by using the thinner 500 μm thick NPS glass membrane to substitute the 800 μm thick membrane.

3.6.2 Performance of a Direct Methanol Fuel Cell Using Flexible Proton-Conducting Glass-Based Composite Membrane

In this section, a single direct methanol fuel cell (DMFC) using methanol as fuel and based on the NPS glass and SPEEK polymer

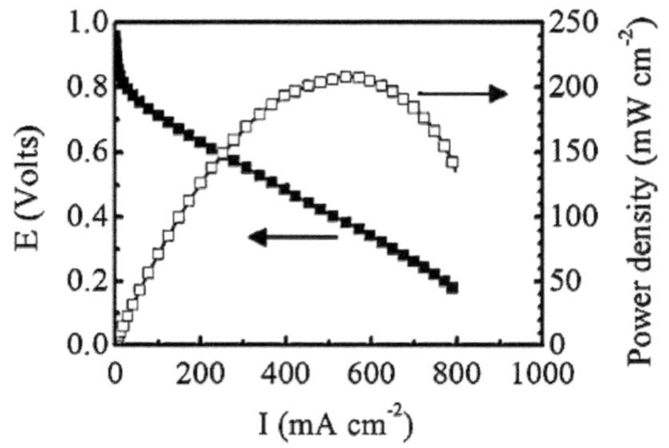

Figure 3.51 I–V and power density curves of fuel cell with a 500 μm thick NPS glass membrane oprating at 70°C, which was prepared by Method B [59]. Copyright 2012 Elsevier.

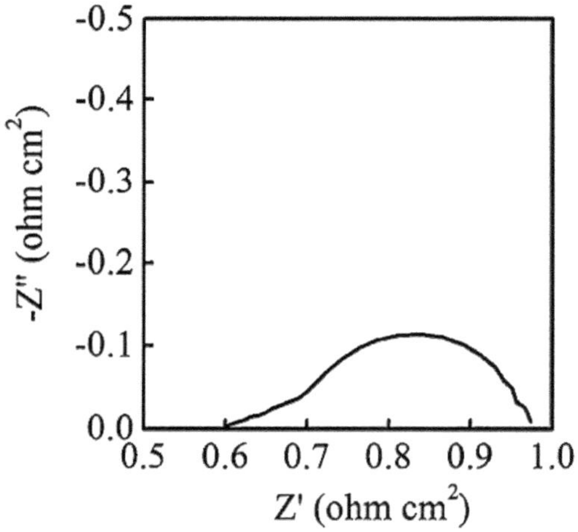

Figure 3.52 Impedance spectra of fuel cells measured at 0.4 V, which are based on the NPS glass membranes with a thickness of 500 μm [59]. Copyright 2012 Elsevier.

(NPS/SPEEK) composite membrane will be introduced, including its preparation and the evaluation of its performance.

3.6.2.1 Preparation of the NPS/SPEEK composite membrane

The preparation procedures, materials used, and the characterizations of the composite membrane can be found in Ref. [76].

3.6.2.2 Measurement of methanol permeability

Methanol permeability of the membranes was determined by using a two-chamber method. The membrane was clamped between the two chambers, the contents of which were stirred continuously during the experiments. One chamber contained pure deionized water (600 ml) and the other contained an 8 vol.% aqueous solution of methanol (600 ml). Samples were taken from the water chamber at intervals, and a gas chromatograph was used to determine their methanol concentrations. The methanol permeability was calculated by the following equation:

$$C_B(t)=ADKC_A(t-t_0)/V_BL \tag{3.1}$$

where C_A and C_B are the methanol concentrations in the methanol chamber and in the water chamber; A and L are the effective area and thickness of the membrane; V_B is the volume of the water chamber; t is the permeation time; and D and K are the methanol diffusivity and partition coefficient, respectively. The product DK is the methanol permeability.

3.6.2.3 Preparation and characterization of DMFC

A single DMFC was prepared by using an NPS/SPEEK composite membrane with a thickness of 96 mm. GDEs with a Pt-Ru loading of 1.75 mg/cm^2 as anode catalyst and a Pt loading of 1.75 mg/cm^2 as cathode catalyst deposited on wetproofed carbon papers were attached on both sides of the NPS/SPEEK composite membrane to form a sandwich structure. A membrane-electrode assembly (MEA) with an active area of 1.69 cm^2 (1.3cm*1.3 cm) was prepared by hot-pressing the sandwich structure under 0.3 MPa at 135°C. The respective graphite blocks on the anode and cathode sides had a serpentine flow field with a width of 0.7 mm and a depth of 1 mm. Gold-coated copper plates were used as current collectors. The respective heating sheets were attached to aluminum end plates.

The DMFC assembled with the above MEA was fed with methanol solution supplied to the anode and oxygen supplied to the cathode. The performance of the DMFC under various operating conditions was characterized by means of a fuel-cell testing system with a methanol supply unit. The layout of the DMFC testing apparatus is illustrated in Fig. 3.53. Details of the testing conditions are listed in Table 3.3.

Table 3.3 Experimental conditions

Cell temperrature	55, 65, 75, 85°C
Anode flow rate	0.5, 2, 5 mL min^1
Cathode backpressure	0.05, 0.1, 0.15, 0.2 MPa

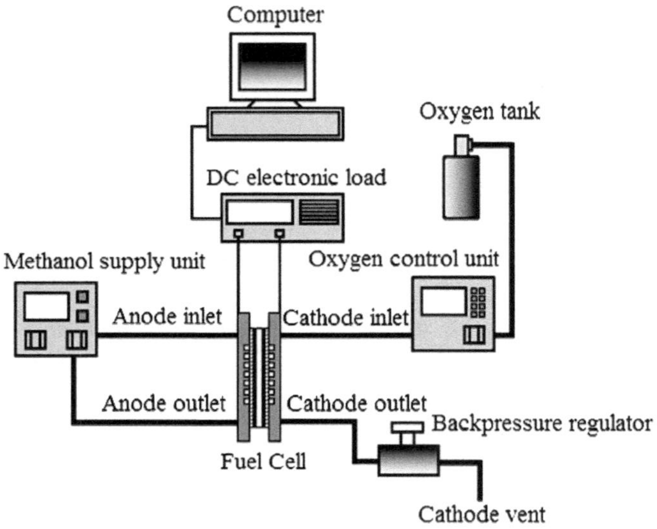

Figure 3.53 Schematic layout of the DMFC testing apparatus [82]. Copyright 2013 Elsevier.

3.6.2.4 Characterization, analysis, and discussion

3.6.2.4.1 *The NPS/SPEEK composite membrane and its methanol permeation properties*

An NPS/SPEEK composite membrane was prepared by mixing NPS glass powder and SPEEK polymer by a mechanical ball-milling approach. The NPS glass used had been synthesized by a sol-gel method from TEOS, H_3PO_4, and Nafion® solutions with a subsequent low-temperature hydrothermal treatment (below 150°C), which prevented thermal degradation of the organic component. The obtained glass exhibited proton conductivities of the order of 10^{-2}–10^{-1} S/cm [39] comparable to those of commercially available Nafion® membranes. For another thing, it is known that a high sulfonated degree (DS) with many hydrophilic groups leads to high conductivity, but is associated with a greater susceptibility to swelling and fragility, whereas a low DS gives low conductivity, but high mechanical strength and thermal stability [44]. SPEEK with a DS of 65% was used for the preparation of the composite membrane. In this study, we focused on the 6NPS/4SPEEK membrane consisting of NPS glass and SPEEK polymer in a weight ratio of 6:4, which has been reported to possess a proton conductivity of ca. 1×10^{-2} S/cm and a tensile strength of 18 MPa [76].

Figure 3.54a,b shows photographs of a 6NPS/4SPEEK composite membrane and an MEA with the NPS/SPEEK composite membrane, respectively. As can be seen in Fig. 3.54a, the composite membrane had a yellowish, opaque appearance, and was smooth and crackfree, indicating that mechanical ball-milling is an efficient approach for preparing composite membranes consisting of glass powder and a polymer, even if the proportion of the glass component is as high as 60 wt.%. Figure 3.54b shows that the MEA based on the 6NPS/4SPEEK composite membrane had a square active area of 1.3 cm × 1.3 cm.

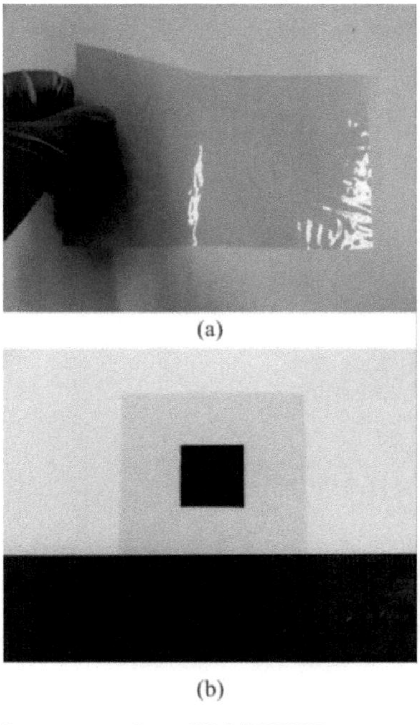

(a)

(b)

Figure 3.54 Optical images of an NPS/SPEEK composite membrane (a), and an MEA with the NPS/SP [82]. Copyright 2013 Elsevier.

In DMFCs, the permeation of methanol should be suppressed as much as possible, because if methanol fuel at the anode diffuses through the membrane to the cathode, it will react there, resulting in a mixed potential without generation of electricity. In an NPS/SPEEK composite membrane, the SPEEK component is promising for reducing methanol cross-over through the membrane, due to its narrower hydrophilic channels and greater branching than Nafion® [77]. Figure 3.55 shows the methanol concentration that permeated through the NPS/SPEEK composite membrane as a function of time. The slope obtained from the linear relationship was used to calculate the methanol permeability according to Equation (3.1). The methanol permeability of the membrane was determined to be 7.5×10^{-7} cm^2/s, which is lower than that of 2×10^{-6} cm^2/s of a Nafion® membrane [78]. This may be attributed to the presence of the relatively impermeable SPEEK within the composite membrane adopted as a polymer matrix, and the incorporation of NPS glass powder particles as nano-scale reinforcements and fixtures that serve to restrain the swelling of the membrane body. Indeed, as previously reported, an NPS/SPEEK membrane has a small swelling ratio of 0.5% [76].

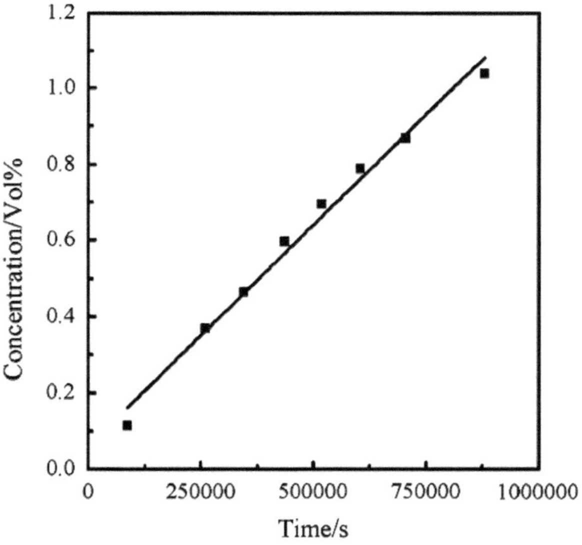

Figure 3.55 Methanol concentration diffusing through the NPS/SPEEK composite membrane as a function of time [82]. Copyright 2013 Elsevier.

3.6.2.4.2 *Effect of cell temperature on the DMFC performance*

A single DMFC was assembled by using an MEA with an active area of 1.69 cm^2, which was prepared by hot-pressing of the NPS/SPEEK composite membrane, together with anode and cathode GDEs. This composite membrane could endure sufficient hot-pressing to assemble the MEA due to its good flexibility, so that good interfacial contacts between the electrolyte membrane and electrodes were formed.

The performance of the DMFC was characterized, and polarization and power output curves were recorded. To assess the effect of cell temperature on the performance of the DMFC, the experimental cell temperature was increased from 55°C to 85°C in increments of 10°C. As shown in Fig. 3.56, the DMFC had an OCV of ca. 0.6 V at the various temperatures. As expected, the output power of the DMFC increased markedly with increasing cell temperature, with the peak power density increasing from 39.4 mW/cm^2 at 55°C to 73.1 mW/cm^2 at 85°C. This improvement of the DMFC performance with increasing cell temperature may be attributed to decreases of ohmic loss and activation loss. Ohmic loss can mainly be ascribed to the ohmic resistance of the composite membrane, and it decreases with increasing temperature. This may be attributed to an increase in membrane conductivity. For example, proton conductivities of ca. 4×10^{-3} S/cm at 30°C and ca. 1×10^{-2} S/cm at 80°C under 90% relative humidity have been reported in the literature [76]. Activation loss is derived from both anode and cathode polarizations. According to electrochemical reaction kinetics, increasing the temperature could increase the rate of the electrode reaction, and thereby decrease the activation loss [75]. In addition, an increased operation temperature could also improve the mass-transfer process, resulting in improved performance.

3.6.2.4.3 *Effect of the methanol flow rate on DFMC performance*

The effect of the methanol flow rate on the DMFC performance was evaluated. As shown in Fig. 3.57, the cell performance increased on increasing the methanol flow rate from 0.5 ml/min to 2 ml/min, but higher methanol flow rates had a negative effect. This can be

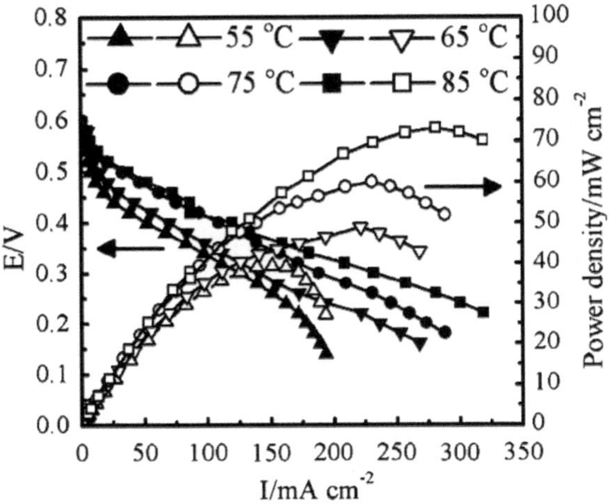

Figure 3.56 The performances of a DMFC at 55°C (a), 65°C (b), 75°C (c), and 85°C (d) under operating conditions of a methanol concentration of 2 M, a methanol flow rate of 1 ml/min, an oxygen flow rate of 200 ml/min, and an oxygen backpressure of 0.2 MPa [82]. Copyright 2013 Elsevier.

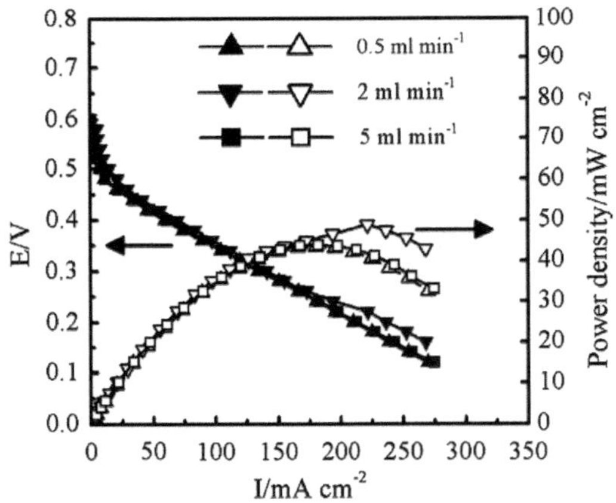

Figure 3.57 The performances of a DMFC at various methanol flow rates of 0.5 ml/min (a), 2 ml/min (b), and 5 ml/min (c) at a cell temperature of 65°C, a methanol concentration of 2 M, an oxygen flow rate of 200 ml/min, and a cathode backpressure of 0.1 MPa [82]. Copyright 2013 Elsevier.

rationalized in terms of increasing methanol flow rate enhancing the methanol transportation and providing a sufficient supply of methanol to the anode catalyst layer, but a higher methanol flow rate leading to an increase of methanol crossover through the composite membrane. In this study, the optimum value of the methanol flow rate was identified as 2 ml/min.

3.6.2.4.4 *Effect of the cathode backpressure on the DMFC performance*

Figure 3.58 shows the effect of the oxygen backpressure on the DMFC performance at a cell temperature of 85°C, as the oxygen backpressure was increased from 0.05 MPa to 0.2 MPa in 0.05 MPa intervals. It can be seen that the peak power density of the DMFC was 33.2 mW/cm^2 at 0.05 MPa, which is much lower than the value at 0.2 MPa. The DMFC performance was obviously improved with increasing oxygen backpressure. It is suggested that methanol crossover was impeded by a decrease in electro-osmotic drag within the membrane under the increased oxygen backpressure, and that the reduction reaction at the cathode was enhanced by an increase in oxygen partial pressure [79]. In addition, increasing the oxygen backpressure may have enhanced the transportation of oxygen to the cathode catalyst layer, resulting in improvement of the DMFC performance.

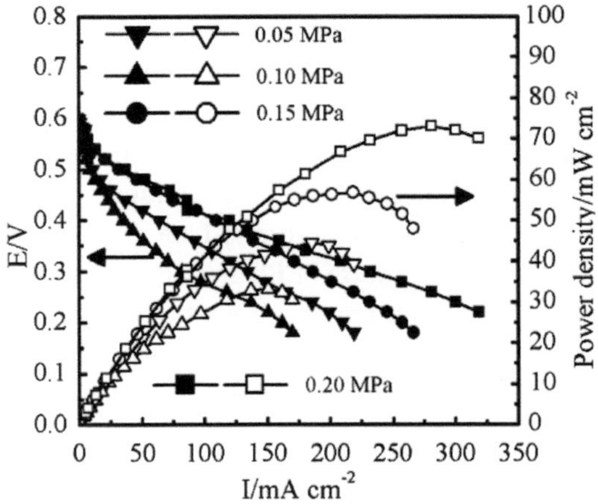

Figure 3.58 Effect of the cathode backpressure on the performances of a DMFC at various oxygen backpressures of 0.05 MPa (a), 0.10 MPa (b), 0.15 MPa (c), and 0.20 MPa (d) at a cell temperature of 85°C, a methanol concentration of 2 M, a methanol flow rate of 1 ml/min, and an oxygen flow rate of 200 ml/min [82]. Copyright 2013 Elsevier.

3.7 Conclusion and Outlook

From the above experiment and discussion, it is easy to find that the sol-gel method along with the hydrothermal treatment at temperature 150°C is conducive to boosting the performance of membranes and the production of large-size and crack-free membranes. On the basis of the method, different kinds of membranes, consisting of various materials and possessing promising performance, were fabricated step by step, which can be concluded succinctly as the following:

The phosphosilicate membranes (PSMs) synthesized from TEOS and H_3PO_4 exhibit the outstanding proton conductivities of the order of 10^{-1} S/cm comparable to those of commercially available Nafion® membranes. The membrane by incorporating Nafion into PSMs, short for NPSMs, not only keeps the high proton conductivity, but it also shows strong mechanical properties and good thermal stability as well as a peak power density of 207 mW/cm^2 in a single H_2/O_2 fuel cell with a 500 μm thick NPSM by optimized assembling method. For the sake of ameliorating the fragility and inflexibility of glass membrane, and thereby improving the performance of membrane, SPEEK is incorporated into the NPS. The experiment results show that the NPS/SPEEK composite membranes exhibit excellent flexibility and promising proton conductivity of over 10^{-3} S/cm^2; simultaneously, the peak power density of a single H_2/O_2 fuel cell equipped with the composite membranes is elevated to 322 mW/cm^2. Moreover, the NPS/SPEEK composite membrane also exhibits good performance in the direct methanol fuel cells. For another thing, the perfluorosulfonate-free composite membranes based on NPS (SiO_2-P_2O_5) and SPEEK also shows fabulous properties such as high proton conductivity more than 10^{-2} S/cm^2, excellent thermal stability, and sufficient mechanical strength combined with flexibility; meanwhile, the peak power density of a single H_2/O_2 fuel cell equipped with the membrane is as high as 355.6 mW/cm^2, which ensures it as a potential alternative membrane for applications in PEMFCs and DMFCs. In addition, the hybrid glass membrane based on the [dema][TfO]/SiO_2 for application in intermediate-temperature range was also studied. It shows a good ionic conductivity exceeding 10^{-2} S/cm^2 and excellent stability at 120°C.

Despite the membranes with outstanding property obtained, there is still great space of improvement of membrane properties. It is imperative to develop the membranes possessing high proton conductivity, thermal stability, and outstanding flexibility, via optimizing the synthesis process and utilizing novel materials, which can be used for low, intermediate, and high temperature range. Certainly, more importantly, all the membranes developed can keep high properties when put into practice.

Acknowledgements

This work was supported by the National Natural Science Foundation of China (Grant No. 50672058), the National High-Tech R&D Program (863 Program) (Grant No. 2009AA05Z113), the International Science & Technology Cooperation Program of Ministry of Science & Technology (Grant 2008DFA51200).

References

1. M Winter, RJ Brodd, *Chem Rev* 104 (2004) 4245.
2. BCH Steele, A Heinzel, *Nature* 414 (2001) 345.
3. L Carrette, KA Friedrich, U Stimming *Chem Phys Chem* 1 (2000) 162.
4. JH Kim, YY Jo, EA Cho, JH Jang, HJ Kim, TH Lim, IH Oh, JJ Ko, IJ Son *J Electrochem Soc* 157 (2010) 633.
5. M Nogami, L Gandham *Adv Sol-Gel Deriv Mater Technol* (2012) 37–58.
6. MW Verbrugge, RF Hill, Analysis of promising perfluorosulfonic acid membranes for fuel-cell electrolytes, *J Electrochem Soc* 137 (1990) 3770–3777.
7. KD Kreuer, *Solid State Ionics* 97 (1997) 1–15.
8. M Nogami, H Matsushita, Y Goto, T Kasuga, *Adv Mater* 12 (2000) 1370–1372.
9. M Nogami, R Nagao, C Wong, T Kasuga, T Hayakawa, *J Phys Chem B* 103 (1999) 9468–9472.
10. T Uma, M Nogami, *Chem Phys Chem* 8 (2007) 2227–2234.
11. T Kasuga, M Nakano, M Nogami, *Adv Mater* 14 (2002) 1490–1492.
12. H Li, M Nogami, *Adv Mater* 14(2002) 912–914.
13. M Nogami, Y Usui, T Kasuga, *Fuel Cells* 1 (2002) 181–185.
14. K Daiko, T Kasuag, MNogami, *J Ceram Soc Jpn* 109 (2001) 815–817.

15. SJ Huang, HK Lee, WH Kang, *Bull Korean Chem Soc* 26 (2005) 241.
16. CW Clavier, DL Rodman, JF Sinski, LR Allain, HJ Im, Y Yang, JC Clark, ZL Xue, *J Mater Chem* 15 (2005) 2356.
17. I Vasiliu, M Gartner, M Anastasescu, L Todan, L Predoana, M Elisa, C Grigorescu, C Negrila, C Logofatu, M Enculescu, A Moldovan, G Pavelescu, M Zaharescu, *Opt Quant Electron* 39 (2007) 511.
18. H Li, T Kunitake, *Micropor Mesopor Mater* 97 (2006) 42–48.
19. Y Jin, S Qiao, JCD da Costa, BJ Wood, BR Ladewig, GQ Lu, *Adv Funct Mater* 17 (2007) 3304.
20. SP Tung, BJ Hwang, *J Memb Sci* 241 (2004) 315.
21. A Aronne, M Turco, G Bagnasco, P Pernice, M Di Serio, NJ Clayden, E Marenna, E Fanelli, *Chem Mater* 17 (2005) 2081.
22. MTN Colomer, *Adv Mater* 18 (2006) 371.
23. KD Kreuer, *Chem Mater* 8 (1996) 610.
24. H Li, M Nogami, *Adv Mater* 14 (2002) 912.
25. Y Daiko, T Kasuga, M Nogami, *Micropor Mesopor Mater* 69 (2004) 149.
26. N Koone, Y Shao, TW Zerda, *J Phy Chem B* 99 (1995) 16976.
27. CH Lee, HB Park, YM Lee, RD Lee, *Ind Eng Chem Res* 44 (2005) 7617.
28. F Lufrano, I Gatto, P Staiti, V Antonucci, E Passalacqua, *Solid State Ionics* 145 (2001) 47.
29. J Halim, FN Büchi, O Haas, M Stamma, GG Scherer, *Electrochim Acta* 39 (1994) 1303.
30. M Yamada, DL Li, I Honma, HS Zhou, *J Am Chem Soc* 127 (2005) 13092.
31. Y Daiko, T Akai, T Kasuga, M Nogami, *J Ceram Soc Jap* 109 (2001) 815.
32. SP Tung, BJ Hwang, *J Mater Chem* 15 (2005) 3532.
33. A Matsuda, T Kanzaki, M Tatsumisago, T Minami, *Solid State Ionics* 145 (2001) 161.
34. Q Deng, CA Wilkie, RB Moore, KA Mauritz, *Polymer* 39 (1998) 5961–5972.
35. MTN Colomer, *Adv Mater* 18 (2006) 371–374.
36. F Jiang, A Kaltbeitzel, B Fassbender, G Brunklaus, WH Meyer, HW Spiess, G Wegner, *Macromol Chem Phys* 209 (2008) 2494–2503.
37. T Uma, M Nogami, *Anal Chem* 80 (2008) 506–508.
38. H Tang, M Pan, S Lu, J Lu, S Jiang, *Chem Commun* (2010) 4351–4353.

39. F Jiang, Z Di, H Li, H Tu, Q Yu, *J Power Sources* 196 (2011) 1048–1054.
40. AF Ismail, NH Othman, A Mustafa, *J Membr Sci* 329 (2009) 18–29.
41. S Lu, D Wang, S Jiang, Y Xiang, J Lu, J Zeng, *Adv Mater* 22 (2010) 971–976.
42. P Xing, GP Robertson, MD Guiver, SD Mikhailenko, K Wang, S Kaliaguine, *J Membr Sci* 229 (2004) 95–106.
43. SMJ Zaidi, SD Mikhailenko, GP Robertson, MD Guiver, S Kaliaguine, *J Membr Sci* 173 (2000) 17–34.
44. D Marani, A D'Epifanio, E Traversa, M Miyayama, S Licoccia, *Chem Mater* 22 (2010) 1126–1133.
45. P Innocenzi, *J Non-Cryst Solids* 316 (2003) 309–319.
46. H Lim, M Yarmo, N Huang, P Khiew, W Chiu, *J Phys Sci* 20 (2009) 23–36.
47. P Krishnan, JS Park, CS Kim, *J Membr Sci* 279 (2006) 220–229.
48. N Intaraprasit, P Kongkachuichay, *J Taiwan Inst Chem Eng* 42 (2011) 190–195.
49. J Cao, W Su, Z Wu, T Kitayama, K Hatada, *Polymer* 35 (1994) 3549–3556.
50. MT Colomer, *Adv Mater* 18 (2006) 371–374.
51. H Li, D Jin, X Kong, H Tu, Q Yu, F Jiang, *Microporous Mesoporous Mater* 138 (2011) 63–67.
52. R Jiang, HR Kunz, JM Fenton, *J Power Sources* 150 (2005) 120–128.
53. T Kobayashi, M Rikukawa, K Sanui, N Ogata, *Solid State Ionics* 106 (1998) 219–225.
54. GP Robertson, SD Mikhailenko, K Wang, P Xing, MD Guiver, S Kaliaguine, *J Membr Sci* 219 (2003) 113–121.
55. DH. Kang, D Kim, *Kor J Chem Eng* 24 (2007) 1101–1105.
56. HK Kim, HT Ju, JW Hong, *Eur Polym J* 39 (2003) 2235.
57. P Krishnan, JS Park, TH Yang, WY Lee, CS Kim, *J Power Sources* 163 (2006) 2.
58. S Werner, L Joerissen, U Heider, *Ionics* 2 (1996) 19.
59. ZG Di, HB Li, M Li, DL Mao, XJ Chen, M Xiao, J Gu, *J Power Sources* 207 (2012) 86.
60. HB Li, DL Jin, QC Yu, HY Tu, *J Power Sources* 196 (2011) 3836.
61. F Jiang, H Pu, WH Meyer, *Electrochim Acta* 53 (2008) 4495.
62. J Britz, WH Meyer, G Wegner, *Adv Mater* 22 (2010) E72.

63. D Cheddie, N Munroe, *J Power Sources* 156 (2006) 414.
64. K Jiao, X Li, *Fuel Cells* 10 (2010) 351.
65. BS Lalia, N Yoshimoto, M Egashira, M Mprota, *J Power Sources* 195 (2010) 7426.
66. W Lu, LT Qu, K Henry, LM Dai, *J Power Sources* 189 (2009) 1270.
67. L Guo, X Pan, CN Zhang, WQ Liu, M Wang, XQ Fang, SY Dai, *Solar Energy* 84 (2010) 373.
68. H Nakamoto, M Watanabe, *Chem Commun* 24 (2007) 2539.
69. SY Lee, T Yasuda, M Watanabe, *J Power Sources* 195 (2010) 5909.
70. SY Lee, A Ogawa, M Kanno, H Nakamoto, T Yasuda, M Watanabe, *J Am Chem Soc* 132 (2010) 9764.
71. G Lakshminarayana, M Nogami, IV Kityk, *J Electrochem Soc* 157 (2010) B892.
72. G Lakshminarayana, M Nogami, *Electrochim Acta* 55 (2010) 1160.
73. K Mori, S Hashimoto, T Yuzuri, K Sakakibara, *Bull Chem Soc Jpn* 83 (2010) 328.
74. A Kaltbeitzel, S Schauff, H Steininger, B Bingoel, G Brunklaus, WH Meyer, HW Spiess, *Solid State Ionics* 178 (2007) 469.
75. Y Tang, J Zhang, C Song, H Liu, J Zhang, H Wang, S Mackinnon, T Peckham, J Li, S McDermid, P Kozak, *J Electrochem Soc* 153 (2006) A2036–A2043.
76. H Li, X Chen, F Jiang, M Ai, Z Di, J Gu, *J. Power Sources* 199 (2012) 61–67.
77. KD Kreuer, *J Membr Sci* 185 (2001) 29–39.
78. YA Elabd, E Napadensky, JM Sloan, DM Crawford, CW Walker, *J Membr Sci* 217 (2003) 227–242.
79. SH Seo, CS Lee, *Energy Fuels* 22 (2008) 1212–1219.
80. Z Di, H Li, M Li, D Mao, X Chen, M Xiao, *Fuel Cells* 13 (2013), 126–133.
81. H Li, F Jiang, Z Di, J Gu, *Electrochim Acta* 59 (2012) 86–90.
82. M Xiao, H Li, Z Di, et al., *J Power Sources* 239 (2013) 356–360.

Chapter 4

Silicon Nanowire Arrays: Fabrication, Properties and Energy Applications

Yang Cao and Junhui He
Functional Nanomaterials Laboratory and Key Laboratory of Photochemical Conversion and Optoelectronic Materials, Technical Institute of Physics and Chemistry, Chinese Academy of Sciences, Beijing 100190, China
jhhe@mail.ipc.ac.cn

4.1 Introduction

Among 1D nanostructures, Si nanowires (SiNWs) are widely considered as an important class of nanoscale building blocks for high-performance solar energy conversion [1–5] due to their unique structural, electrical, and optical properties in addition to their compatibility with current Si-based microelectronics.

Controllable fabrication of SiNWs is a prerequisite for their device applications. Varied methods have been developed to fabricate SiNWs via top-down or bottom-up approaches, such as vapor-liquid-solid (VLS) growth, reactive ion etching (RIE), electrochemical etching, and metal-catalyzed electroless etching (MCEE), all of which aim

Nanomaterials in Energy and Environmental Applications
Edited by Junhui He

ISBN 978-981-4463-78-2 (Hardcover), 978-981- 4463-79-9 (eBook)
www.panstanford.com

to control various parameters of the SiNWs. Among these methods, MCEE has attracted increasing attention in recent years for several reasons. First, MCEE is a simple and low-cost method for fabricating various Si nanostructures with the ability to control various parameters (e.g., diameter, length, orientation, doping type, and doping level). Almost all procedures can be accomplished in a chemical lab without expensive equipment [6-13]. Second, the MCEE SiNWs inherit electrical properties from the mother Si substrate and thus require no further doping processing for conductivity. Third, the MCEE SiNWs are an integral part of the Si substrate and are thus mechanically robust. Fourth, the MCEE SiNWs have rough surfaces making them highly antireflective due to strong light scattering and absorption. Fifth, combined with lithography techniques, MCEE allows the patterned fabrication of vertical SiNWs on Si wafers with desired geometrical features. For example, VLS-based methods can only be used to grow wires with circular cross-sections, while MCEE is much more flexible and can be used to make wires with various shape cross-sections [14, 15]. These salient attributes make MCEE a highly attractive method to fabricate SiNWs. MCEE has become increasingly important in the last decade. The Si nanostructures fabricated by MCEE have demonstrated their application potentials in fields ranging from solar energy conversion [16–19], thermal power conversion [20], and energy storage [21], to chemical and biological sensing [22, 23] and biomimic superhydrophobicity [24–26].

4.2 Fabrication Process and Formation Mechanism of SiNW Arrays

In 2002, Peng et al. [6] reported that a wafer-scale SiNW array could be readily produced via MCEE at room temperature by simply immersing a Si wafer into HF–$AgNO_3$ solution. The overall fabrication process is as follows: A whole piece of commercial Si wafer was ultrasonically washed, and was then immersed in 5% HF aqueous solution, in which the Si surface was H-terminated. Secondly, the Si wafer was immediately immersed into a solution containing HF and $AgNO_3$ mixture to coat a layer of uniform Ag nanoparticles on the Si surface. Thirdly, the Si wafer was washed with deionized water to remove the excess Ag^+ ions, and was placed in an etching solution containing HF and H_2O_2 in dark ambience. Fourthly, the Si wafer was rinsed thoroughly in deionized water, and then put into diluted aqueous HNO_3 to dissolve the Ag nanoparticles. Finally, the wafer was washed with 5% HF again to remove the oxide layer and then cleaned with deionized water and dried by N_2 flow.

The formation mechanism of the MCEE SiNWs is galvanic displacement reaction. The reduction of metal ions (cathodic process) and the oxidation of Si atoms (anodic process) occur simultaneously on the Si surface, while the charge is exchanged through the Si substrate. The schematic diagram of the etching mechanism and TEM images of the SiNWs are illustrated in Fig. 4.1. The Si/$AgNO_3$/HF system is composed of a corrosion-type redox couple: the cathodic reduction of Ag^+ ions and its counterpart, the anodic oxidation and dissolution of silicon, which occurs locally beneath the Ag deposits. Therefore, the Ag particles trap into the pits vertically, resulting in the formation of SiNWs [11]. The electrochemistry reactions of the cathode and anode could be expressed as follows:

Cathode (Ag surface facing the electrolyte):

$$H_2O_2 + 2H^+ + 2e^- \rightarrow 2H_2O,\ E^0 = 1.77V$$

Anode (Si in close contact with Ag):

$$Si + 2H_2O \rightarrow SiO_2 + 4H^+ + 4e^-$$

$$SiO_2 + 6HF \rightarrow H_2SiF_6 + 2H_2O$$

$$Si + 6HF \rightarrow H_2SiF_6 + 4H^+ + 4e^-,\ E^0 = 1.2V$$

$$\text{Overall: } Si + 2H_2O_2 + 6HF \rightarrow H_2SiF_6 + 4H_2O$$

Generally, the used metals in the galvanic displacement reaction are Ag, Au, and Pt salts with highly positive equilibrium reduction potential [12]. Peng et al. [27] concluded that the autonomous motion of metal particles (Ag and Au) in Si is highly uniform, yet directional and preferential along the (100) crystallographic orientation of Si, rather than always being normal to the silicon surface.

MCEE allows control over the diameter, length, density [28] and crystallographic orientation of SiNWs [10, 29]. For example, single-crystal zigzag SiNWs with different turning angles in wafer can be fabricated via selecting the crystallographic orientation of Si wafer, reaction temperature, and etchant concentration [30]. The morphology of SiNWs is tunable from solid nonporous nanowires, non-porous/nanoporous core/shell nanowires, to entirely nanoporous nanowires by controlling the hydrogen peroxide concentration in the etching solution [31].

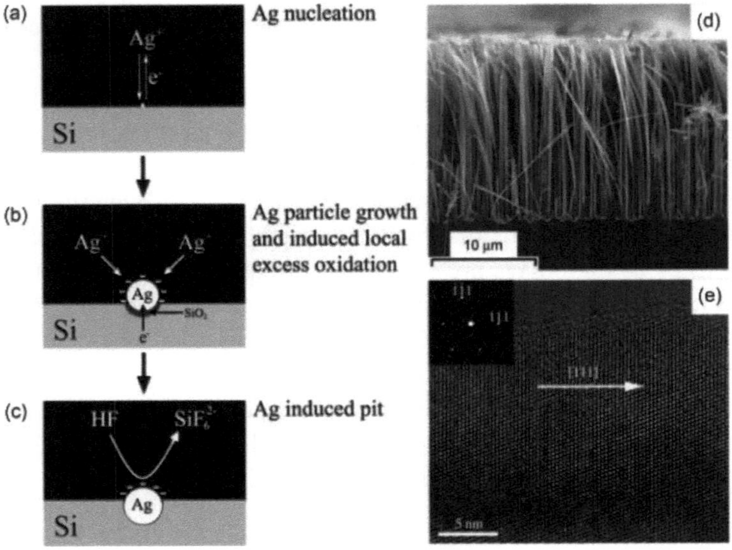

Figure 4.1 Mechanism of electroless Ag deposition on a Si substrate in $HF/AgNO_3$ solution. (a) Ag^+ ions in the vicinity of the silicon surface capture electrons from Si and deposite in the form of metallic Ag nuclei on a nanoscopic scale. (b) Electron exchange between Ag^+ ions and Si catalyze the subsequent reduction of Ag ions and oxidation of Si, inducing etching of SiO_2 by HF solution. (c) Ag particles trapped in these pits vertically. (d) TEM image showing a cross-sectional view of SiNWs arrays on an etched silicon wafer. (e) HRTEM image of a single silicon nanowire and its SAED pattern [11, 12]. Copyright © 2006 WILEY-VCH.

4.3 Properties of SiNW Arrays

Compared to bulk silicon, SiNW arrays show some special physical properties, including optical and electrical properties.

4.3.1 Optical Properties: Improved Anti-Reflection and Broadband Absorption

One of the most important features of aligned SiNW arrays is the significant suppression of reflection over a wide spectrum range from visible to near-infrared. Peng et al. reported that SiNW arrays prepared by the MCEE method exhibit excellent antireflection properties [1, 10–12]. Figure 4.2 shows the hemispherical reflectance of MCEE SiNW arrays on single-crystal and polycrystalline Si wafers, porous Si, and polished single-crystal Si wafers. Clearly, SiNW arrays drastically suppress light reflection over a wide spectral range. The reflectance is less than 1.4% over the range of 300–600 nm for the SiNWs fabricated on single-crystal Si wafer. This remarkably low reflectance of the SiNW array is attributed to several distinct advantages associated with the wire geometry: (1) the ultrahigh surface areas of high-density SiNWs; (2) the subwavelength light-trapping effects of SiNW arrays; and (3) the collective light scattering interactions among SiNWs, which trap light and make it travel many turns over distances much longer than the array thickness.

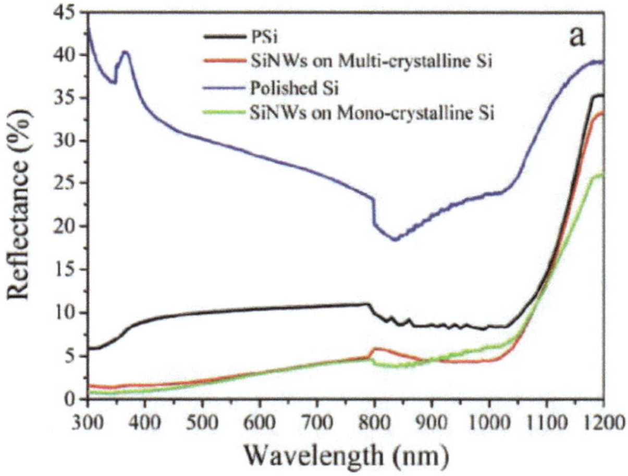

Figure 4.2 Hemispherical reflectance measurements of SiNW arrays, porous Si (PSi), and polished crystalline Si. The PSi was prepared by conventional stain etching in HF/HNO_3 solution [1]. Copyright © 2005 WILEY-VCH.

Furthermore, compared to thin films, SiNW arrays with moderate filling ratio have a much lower reflectance in a wide spectrum range and wide incidence angle range simultaneously which can be achieved without specially designed antireflection coating [32, 33].

However, efficient antireflection is not equivalent to effective light absorption because light transmission can be significant if the dimensions of the SiNW arrays are not optimized. Yu's group found that light transmission is very sensitive to the array periodicity, and it can be effectively suppressed when the array periodicity is around 600 nm, as shown in Fig. 4.3c [34].

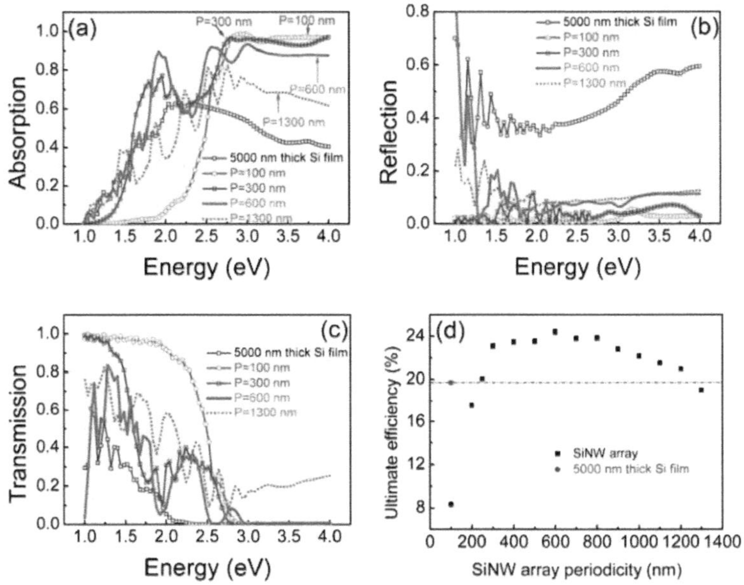

Figure 4.3 Calculated (a) absorption, (b) reflection, and (c) transmission spectra of SiNW arrays as a function of the array periodicity (P). The SiNWs are squarely arranged with a length of 5 μm. The ratio of the wire diameter to P is 0.5. (d) summarizes the ultimate efficiency of SiNW arrays for different periodicity [34]. Copyright © 2009 American Institute of Physics.

In the meanwhile, effective antireflection can be well maintained (see Fig. 4.3b). Figure 4.3a shows that the enhanced light absorption can be realized in the full energy range from 1 to 4 eV, covering the major energy region of the solar spectrum [35]. Figure 4.3d summarizes the calculated ultimate efficiencies [36] with various array periodicities, for the sake of quantitative comparison of the light trapping capability. When the array periodicity is >300 nm, the total light trapping capability of SiNW arrays becomes superior to the Si thin film with the same thickness. There is a wide "window" of periodicity variation till 1200 nm for light trapping enhancement, which offers flexibility to manufacture the related high efficient solar cells. In the meanwhile, the optimized ratio of wire diameter to array periodicity is found to be around 0.8. Yu's group briefly discussed the mechanism governing the interaction between nanostructures incorporated system and incident light. From the point of view of wave optics, if the feature size of the object is much smaller than the light wavelength, light could directly bypass the object. If the wavelength of incident light is comparable with the interacted object, the light will be strongly scattered, resulting in the elongated optical path length and thus the enhanced light trapping. However, when the light wavelength is much shorter than the object feature size, the interaction between the incident light and the objects would be dominated by the reflection process [36].

4.3.2 Electrical Properties: Enhanced Carrier Collection Efficiency

In the top-down process, SiNWs inherit the electrical characteristics of the mother Si wafers and do not need further doping processing for conductivity. The most significant electrical property of the SiNW arrays potentially beneficial to the solar cell applications is the capability of forming radial (or core–shell) p-n junction to facilitate carrier collection along a short collection path, i.e., the radial direction [37].

It is known that photo-carrier collection and light absorption are in competition for planar structured solar cells. Efficient carrier collection requires thin materials to shorten the minority carrier travel distance, and efficient light absorption requires thick materials. This conflict may not be as severe in single crystalline Si as it has a minority carrier diffusion length in hundreds of micrometers. However, the conflict is significant for poly-crystalline and even nanocrystalline materials which have relatively short carrier diffusion lengths [38]. To greatly decrease the competition between carrier collection and light absorption, a core-shell array with a radial p-n junction configuration for solar cells has been proposed. Compared to planar p-n junction solar cells, the core-shell nanowire arrays are superior: on one hand, short collection length along the radial direction; on the other hand, carrier collection occurs perpendicular to light absorption, resulting in decrease of the competition between carrier collection and light absorption.

The impact of SiNWs surface defects on electrical characteristics is worth noting [39]. Figure 4.4 depicts the relationship between the calculated power conversion efficiency (PCE) and the surface recombination velocity (SRV) [40]. With increasing SRV from 0 to 10^3 cm s^{-1}, the predicted PCE has a relatively gradual decrease from 17.3% to 15.2%. However, further increase of the surface recombination velocity to 10^5 cm s^{-1} would significantly reduce PCE to only 6.1%. Recently, several groups have started to tackle the issues of SiNWs' surface passivation. Gunawan and Guha investigated effects of Al_2O_3 passivation on the performance of SiNW based solar cells. They demonstrate that after adding a 10-nm-thick Al_2O_3 conformal passivation layer onto the SiNWs surfaces, the PCE is improved from 1% to 1.8%, with an increment of 47 mV and 5 mA cm^{-2} in V_{oc} and J_{sc}, respectively [41]. Demichel *et al.* demonstrate that SiO_2 formed by thermal oxidation and subsequent forming gas (N_2 /H_2—95%/5%) annealing has excellent passivation function to SiNWs, in which surface recombination velocity can reach as low as 20 cm s^{-1} [42]. By using a thin a-Si layer passiviated SiNWs, Dan et al. reduced SRV by nearly 2 orders of magnitude, i.e., from ~3×10^5 to 4.5 ×010^3 cm s^{-1} [43]. In addition, Kim et al introduced a-SiN:H to passivate Si microwire based solar cells, and the cell efficiency was enhanced from 8.9% to 10.6% [44].

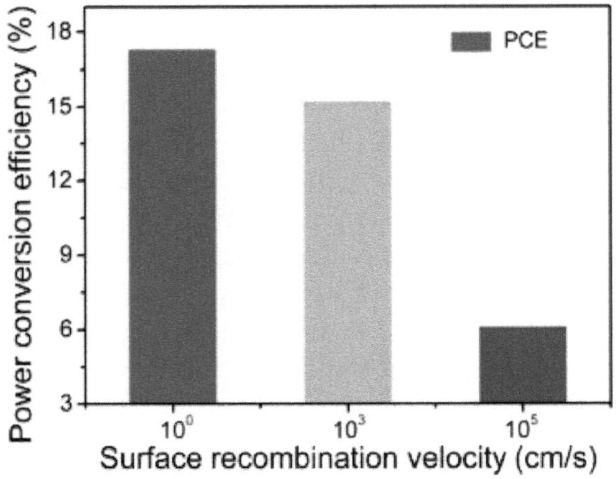

Figure 4.4 Predicted power conversion efficiency (PCE) of the SiNW arrays incorporated solar cell vs. surface recombination velocity (SRV) [40]. Copyright © 2010, IEEE.

4.4 Energy Applications of SiNW Arrays

Due to the band-gap matching the solar spectrum and the unique optical and electrical properties, the SiNW arrays are attracting intense interest as a promising material for cost-effective and high-performance solar energy conversion. Current applications of SiNW arrays in the solar energy conversion mainly include solar cells, photoelectrochemical solar cells and photocatalytic water splitting.

4.4.1 Solar Cells

Silicon is the most popular material in commercial solar cell modules. The success of Si photovoltaics (PVs) is due to several beneficial features of Si [45]: (1) Si is plentiful and ubiquitous, being the second most abundant element on Earth; (2) Si is generally stable and nontoxic; and (3) the 1.12 eV band-gap of Si is almost ideally matched to the terrestrial solar spectrum. Conventional Si solar cells are made of planar junctions of p-type and n-type Si. Such PVs require relatively large amounts of Si to fully absorb the incident sunlight because Si has an indirect gap and low absorption in the visible and near-infrared spectrum. In addition, since charge carriers are collected along the light path, Si material must be of high-purity solar grade and have long minority-carrier diffusion lengths for minimal carrier recombination and enhanced carrier collection. The requirements for large amounts of high-purity Si materials translate into the high cost of Si PVs. Cost and efficiency are the two major challenges facing the PV industry.

For low-cost solar energy conversion, SiNW array based solar cells are an extremely attractive candidate due to their unique geometrical features [1–5, 37]. First, SiNW arrays have strong optical absorption in the solar spectrum, i.e., less than 1% equivalent of Si materials in SiNW arrays can achieve the same amount of absorption as traditional planar wafer-based PV devices. Second, SiNW solar cells with radial p-n junctions offer a short collection path for charge carriers, thus allowing the use of lower-quality Si materials. These advantages may substantially reduce the production cost of SiNW-based solar cells while retaining competitive efficiencies [46].

4.4.1.1 Solar cells based on radial p-n junction SiNW arrays

A large number of works have been carried out on radial p-n Junction SiNW array solar cells. In 2008, Garnett and Yang reported the first SiNW array solar cell with a radial p-n junction [4]. The starting material is phosphorus doped n-type Si wafers with a resistivity of 0.6 Ωcm. After obtaining the MCEE SiNW array, a p-type a-Si layer was deposited using low pressure chemical vapor deposition (LPCVD) at 450°C with disilane (silicon source) and boron trichloride (dopant gas), followed by thermal annealing at 1000°C for 10 s to crystallize the p-type Si shell.

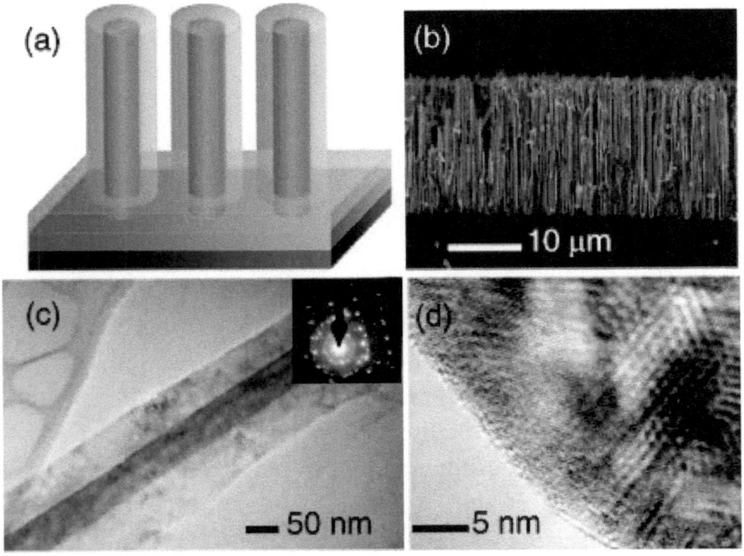

Figure 4.5 SiNW array solar cell structure. (a) Schematic cell design with the single crystalline n-SiNW core in brown, the polycrystalline p-Si shell in blue, and the back contact in black. (b) Cross-sectional SEM of a completed device demonstrating excellent vertical alignment and dense wire packing. (c) TEM image showing the single crystalline n-Si core and polycrystalline p-Si shell. The inset is the selected area electron diffraction pattern. (d) TEM image from the edge of the core-shell nanowire showing nanocrystalline domains [4]. Copyright © 2008, American Chemical Society.

The schematic of the radial p-n junction SiNW array based cell configuration and the cross-sectional SEM image of a completed cell are shown in Fig. 4.5a,b respectively. The SiNWs have lengths of 18 μm, with a packing density of ~50%. The typical wire diameter is 350–400 nm, including the 150-nm-thick poly-Si shell and 50–100 nm SiNWs core. Figure 4.5c shows the single crystalline n-Si core and polycrystalline p-Si shell. The TEM image taken near the edge of the nanowire (Fig. 4.5d) shows roughly 5 nm domains, indicating a nanocrystalline film. *I–V* characteristics (Fig. 4.6) for the solar cells with the size of 1 ×1 cm^2 show that the device has an open circuit voltage (V_{oc}) of 0.29 V, a short circuit current (J_{sc}) of 4.28 mA/cm^2, and a fill factor (FF) of 0.33 for an overall efficiency of 0.46% under AM 1.5G illumination.

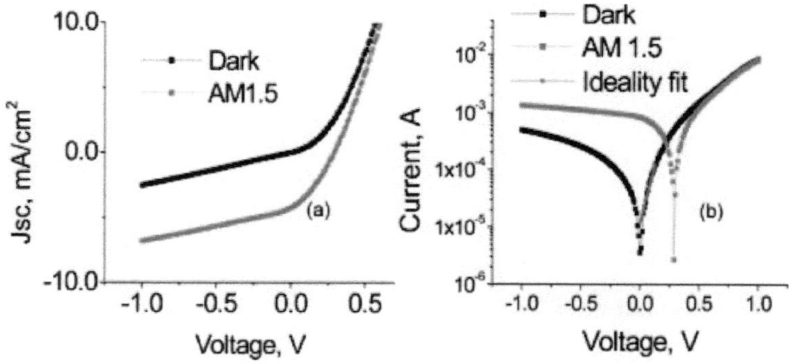

Figure 4.6 Silicon nanowire solar cell electrical performance. (a) Current-voltage behavior in the dark and under AM1.5 simulated sunlight irradiation. (b) Semilog plot of the same data used to extract the diode ideality factor of 2.1 [4]. Copyright © 2008, American Chemical Society.

The low V_{oc} stems from interfacial recombination as evidenced by the large dark current and high diode ideality factor (Fig. 4.6b) [47].

In 2011, Lee's group [48] demonstrated that the organic-inorganic solar cell based on hybrid composites of conjugated molecules and SiNW array yielded excellent power conversion efficiency (PCE) of 9.70% under AM 1.5 solar spectrum irradiation at 100 mW cm^{-2}. The high efficiency was ascribed to two aspects: one was the improvement of the light absorption by SiNWs structure on the planar components; the other was the enhancement of charge extraction efficiency, resulting from the novel top contact by forming a thin organic layer shell around the individual silicon nanowire.

The SiNWs were coated with spiro-OMe-TAD (chemical structure shown in Fig. 4.7a) as the hole transport layer to form a core-shell structure (Fig. 4.7b) or embedded structure (Fig. 4.7c) on the planar component by controlling the thickness of the organic layer.

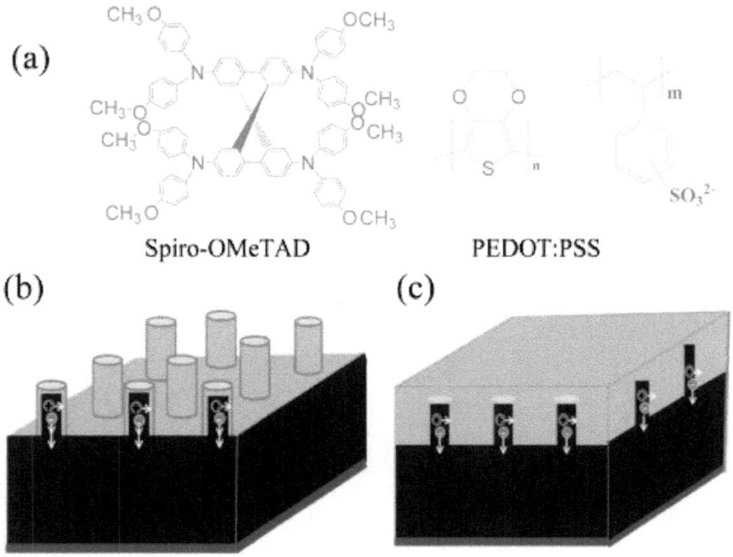

Figure 4.7 (a) Chemical structures of spiro-OMeTAD and PEDOT:PSS; (b) SiNWs solar cell with core-shell architectures, schematic cross-section view of the inorganic silicon nanowire as core on bulk silicon substrate (deep blue) and organic spiro-OMeTAD as shell (light blue); and (c) SiNWs solar cell with embedded architectures, schematic cross-section view of the embedded structure with SiNWs on bulk silicon substrate (deep blue) embedded in spiro-OMeTAD (light blue) [48]. Copyright © 2011, American Chemical Society.

The external quantum efficiency (EQE) spectra of device 1 (In:Ga/SiNWs/spiro-OMe-TAD/PEDOT:PSS/Cu(core-shell)), device2 (In:Ga/SiNWs/spiro-OMeTAD/Cu(core-shell)) and device 3 (In:Ga/SiNWs/spiro-OMe-TAD/Cu (embedded structure)) as shown in Fig. 4.8a indicate that the carrier transfer and collection should be rather efficient in the present inorganic-organic core-shell structure [49]. Figure 4.8b shows electrical output characteristics of device with the core-shell structure is also better than that of device with embedded structure and their parameters were summarized in Table 4.1.

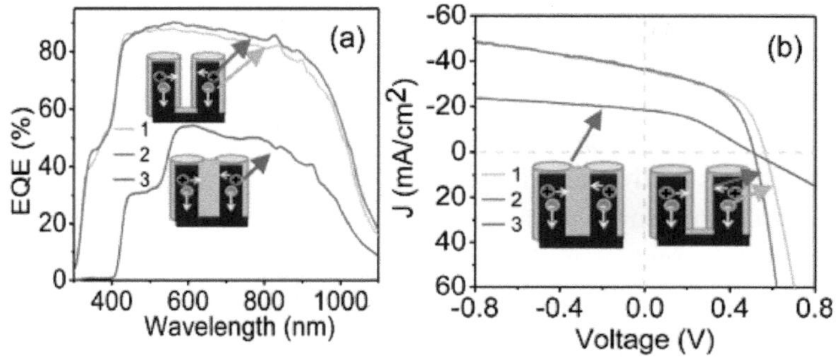

Figure 4.8 Output characteristics of hybrid devices with different device structures. (a) EQE spectra of device 1 (In:Ga/SiNWs/spiro-OMe-TAD/PEDOT:PSS/Cu (core- shell)), device 2 (In:Ga/SiNWs/spiro-OMeTAD/Cu (core-shell)), and device 3 (In:Ga/SiNWs/spiro-OMe-TAD/Cu (embedded structure)). (b) Current density voltage behavior of devices 1, 2, and 3 under simulated AM 1.5 solar irradiation at 100 mW/cm^2. The inset images stand for the structures of the devices [48]. Copyright © 2011, American Chemical Society.

Table 4.1 Response of hybrid solar cells with different device structures under simulated AM 1.5 solar irradiation at 100 mW/cm^2 [48]. Copyright © 2011, American Chemical Society

Device	V_{oc} (V)	J_{sc} (mA/cm^2)	FF	PCE(%)
1	0.568	32.5	0.504	9.30
2	0.512	33.7	0.500	8.63
3	0.504	16.7	0.371	3.12

Moreover, Hwang et al. demonstrated that a rapid formation of radial p-n junctions on MCEE SiNWs [50]. With a low-cost objective, the radial p-n junctions were prepared using a homemade nonhazardous diffusion source of high phosphor concentration annealed at a small thermal budget.

They demonstrated the existence of the radial p-n junctions within the NWs of the fabricated MCEE SiNW solar cell using a conducting AFM scanning over the fractured cross sections of the NWs as shown in Fig. 4.9a.

It can be understood that the zero-bias photocurrent can be detected on top of the MCEE SiNW cross section only if a radial p-n junction exists inside the NWs. Figure 4.9c shows a zero-biased photocurrent mapping of Fig. 4.9b measured with the configuration of Fig. 4.9a. The clear white dots within the bundled nanowires in Fig. 4.9c provided solid evidence that the radial p-n junctions within the SiNWs really existed. Figure 9d shows a current profile of Fig. 4.9c. The currents were all unidirectional and positive, which means that the currents were flowing upward from the scanned cross sections to the AFM tip, due to the existence of the radial junctions, and were photogenerated.

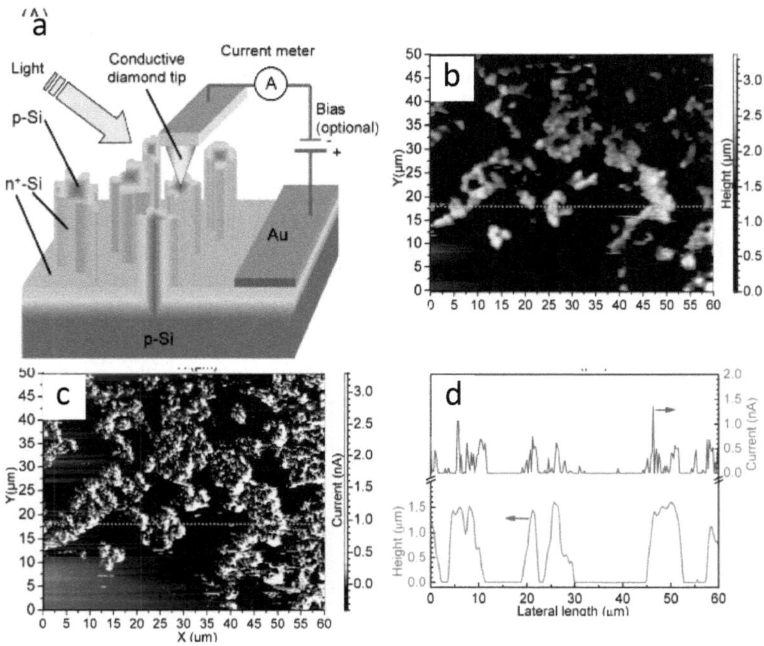

Figure 4.9 (a) Schematic diagram of a conductive AFM (cAFM) mapping the photocurrents generated from the fractured SiNWs with radial p-n junctions. (b) Results of cAFM current mapping (b-d). The morphology (b) and the cAFM current mapping image (c) on the fractured SiNWs evidenced the photocurrents generated from radial p-n junctions. Profiles of (b) and (c) (specified by the dotted lines) are shown in (d) [50]. Copyright © 2011, American Chemical Society.

The performance of the planar and the MCEE-SiNW solar cells is shown in Fig. 4.10a,b, respectively. The η of the MCEE-SiNWs reaches 8.41%, which is about 30% higher than that of its planar counterpart

(η = 6.45%). Likewise, the short-circuit current density (J_{sc}) of the MCEE-SiNW solar cell (24.75 mA/cm^2) is 33% larger than that of the planar one, while the two cells have similar fill factors (FF) and open-circuit voltages (V_{oc}). The enhanced η may be attributed to the improved anti-reflection, in combination with the efficient carrier collection in the radial junction NWs that are passivated by the remnant oxide.

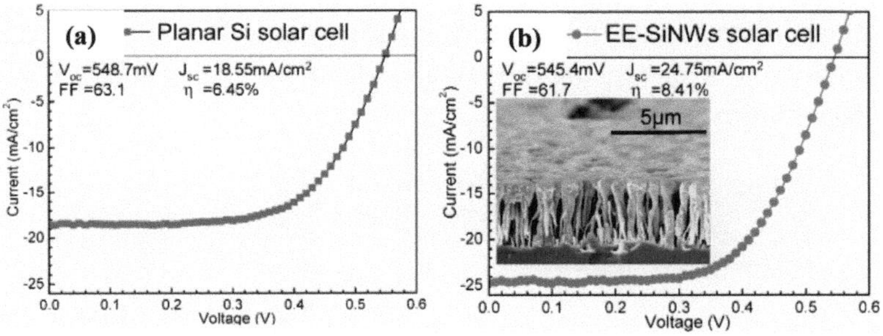

Figure 4.10 Performance of solar cells made with planar Si (a) and with MCEE-SiNWs (b). The corresponding solar cell parameters are both indicated. A cross-sectional SEM image revealing the gold contact above an MCEE-SiNW solar cell is shown in the inset of (b) [50]. Copyright © 2011, American Chemical Society.

4.4.1.2 Solar cells based on graphene film/SiNW array Schottky junction

Schottky junction solar cells based on SiNW array have also been widely studied. Fan et al. [51] assembled Schottky junction solar cells by directly coating graphene films on n-type SiNW arrays. The graphene/SiNW junction showed enhanced light trapping and faster carrier transport compared to the graphene/planar Si structure. SiNW-based solar cells showed energy conversion efficiencies of up to 2.86% under AM1.5 conditions, opening a possibility of using graphene/semiconductor nanostructures in photovoltaic applications.

Jie's group [52] constructed Schottky junction solar cells by combining the monolayer graphene (MLG) films and the SiNW arrays. The step-wise process for the device construction is illustrated in Fig. 4.11. They investigated the photovoltaic characteristics of devices with both p-MLG/n-SiNWs and n-MLG/p-SiNWs structures, and found that due to the balance between light absorption and surface carrier recombination, devices made of SiNW arrays with a medium length showed better performance.The conversion efficiency continues to increase to a maximum value of 0.40% for the device made of 20 min etched SiNW array, nearly ten times of the initial value for Si wafer. Nevertheless, further prolongation of the etching time (40 min) results in the degeneration of the device performance. Besides, the performance of the Schottky junction solar cells could be further improved by enhancing the MLG conductivity via appropriate surface treatment or doping.

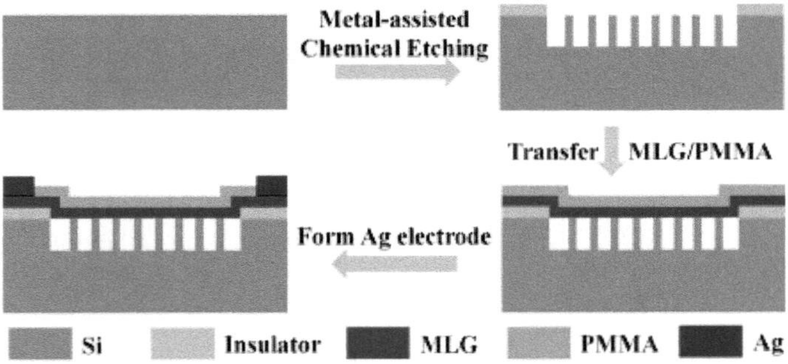

Figure 4.11 Schematic illustration of the fabrication process for the MLG film/SiNW array Schottky junction solar cell [52]. Copyright © 2011, AIP Publishing LLC.

4.4.2 Photoelectrochemical Solar Cells

Although solid-state radial p-n junction SiNW array solar cells are theoretically and experimentally shown to be an attractive low-cost alternative to traditional planar Si-based solar cells, they have to date exhibited poor performance primarily due to excessive interfacial recombination losses and large shunting. As a prototypical system to explore new concepts for solar energy conversion, liquid-junction SiNW photoelectrochemical solar cells (PEC) have attracted intense attention owing to their potential low-cost and performance advantages over their solid-state junction counterparts [53–58].

Peng et al. fabricated metallic nanoparticles (MNPs)-decorated SiNW electrode geometry for PEC applications [45]. MNPs-decorated

SiNW electrode provided additional advantages: First, the conductive SiNWs provide direct 1D pathway for efficient charge-carrier transport to the Si wafer substrate. Second, MNPs serve as catalysts for electrochemical reactions on SiNW electrodes. Third, MNPs scatter impinging light among the SiNWs, which leads to increased light absorption. Thus, MNPs were coated on n-SiNWs to enhance solar energy conversion [53]. High-density aligned n-type SiNW arrays decorated with discrete 5–10 nm platinum nanoparticles (PtNPs) have been fabricated by MCEE of Si wafer followed by electroless platinum deposition. Coating of PtNPs on SiNW sidewalls yielded a substantial enhancement in photoconversion efficiency and an apparent energy conversion efficiency of up to 8.14% for the PtNP-decorated SiNW-based photoelectrochemical solar cell using a liquid electrolyte containing Br^-/Br_2 redox couple. The results demonstrate PtNP-decorated SiNWs to be a promising hybrid system for solar energy conversion.

However, stability of the SiNW PECs is a critical issue. Suffering from photocorrosion, the SiNW PECs undergo rapid performance degradation. Hence, reducing photocorrosion by surface passivation is crucial for the realization of high-performance and stable SiNW PECs. Recently, Peng et al. [57] reported a novel design of platinum nanoparticle (PtNP)-decorated carbon/silicon core/shell nanowire (PtNPs@C@SiNW) arrays for high-performance and stable PECs. Ultrathin carbon film was coated onto SiNWs to form a core/shell structure by microwave plasma-enhanced chemical vapor deposition (MPECVD), followed by decoration with PtNPs by electroless metal deposition (EMD) as shown in Fig. 4.12.

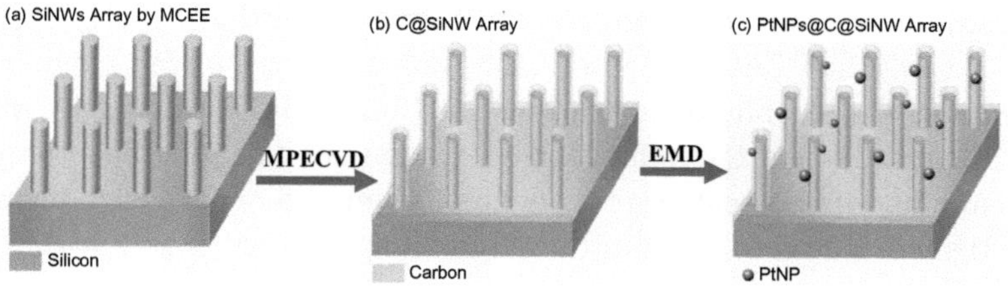

Figure 4.12 Schematic illustration of the fabrication process of a PtNPs@C@SiNW array. a) SiNW array prepared by the MCEE method. b) C/Si core/shell nanowire array after carbon MPECVD. c) PtNPs@C@SiNW array after PtNP EMD [57]. Copyright © 2011 WILEY-VCH.

Heterostructured PtNPs@C@SiNWs PEC exhibited high power conversion efficiency of 10.86% and excellent operational stability (see Fig. 4.13b,c) in aggressive aqueous HBr/Br_2 electrolyte under 1-sun AM 1.5G illumination. Comparison of photovoltaic output characteristics and time-dependent photocurrent density of SiNWs samples showed that carbon coating is a highly effective means of passivating SiNW surfaces, and PtNPs can significantly enhance interfacial charge transfer. Consequently, PtNPs@C@SiNW arrays are highly promising for high-performance and stable PECs.

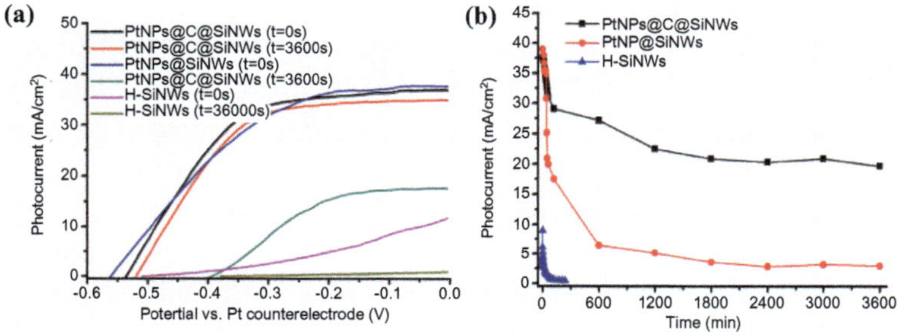

Figure 4.13 (a) Photocurrent versus potential of H-SiNWs, PtNPs@SiNWs, and PtNPs@C@SiNWs under continuous light illumination. (b) Photocurrent densities of H-SiNWs, PtNPs@SiNWs, and PtNPs@C@SiNWs as a function of measurement time [57]. Copyright © 2011 WILEY-VCH.

4.4.3 Photocatalytic Water Splitting

Using solar energy for converting water to hydrogen has been considered a promising route for clean, low-cost and environmentally friendly fuel [59, 60]. Since Fujishima and Honda [59] reported a TiO_2-anode photo-electrochemical (PEC) cell that can split water into H_2 and O_2 under ultraviolet-light (UV) irradiation, a large number of photocatalysts for hydrogen generation have been investigated. However, most of these photocatalysts work only with UV irradiation given their large bandgaps. UV light accounts for only approximately 4% of solar energy, whereas visible light accounts for approximately 43% [61]. Therefore, the development of stable photocatalysts that can work with visible light or even the entire solar spectrum is highly desirable. So far a few materials have been reported to be able to work under visible light [62].

Silicon has a conduction band which is about 0.07 eV higher than the potential of hydrogen generation from water (pH = 7), and an excellent optical absorbing property over near IR and visible light [5]. Moreover, the optical absorbing property could be further improved in SiNW arrays [63]. Although the bandgap of Si (1.1 eV) is slightly smaller than the energy required for water splitting (1.23 eV), combined photocatalysts for H_2 and O_2 evolution can be used to drive the water oxidation/reduction process separately via a redox shuttle, such as I^- /IO^{3-} or Fe^{3+} /Fe^{2+} [64]. In a PEC solar hydrogen-generation system, the reduction reaction takes place at the photocathode, whereas the oxidation reaction occurs at the photoanode. One challenge that remains for the utilization of silicon as a photoanode for hydrogen generation in solution is its unstable surface [65].

Decorating the surface of SiNW with a suitable semiconductor layer may both stabilize its surface and form a hybrid junction to enhance the separation of photogenerated charge carriers [66]. Zhang's group demonstrated polymer modified SiNW core-shell arrays as an efficient photocatalyst for solar hydrogen generation [67, 68]. They reported a controllable electropolymerization method for coating a dense layer of poly ((2,9,16,23-tetraaminophthalocyanine) Cu)) (CuTAPc) polymer on the surface of a SiNW array to form a Si/poly-CuTAPc coaxial core–shell nanowire array. Figure 4.14a shows a schematic diagram of the hydrogen generation cell in which the Si/poly-CuTAPc core–shell nanowire array was employed as the anode, a Pt net as the cathode and a saturated calomel electrode as the reference electrode. The three electrodes were immersed in a 3.0 mol L^{-1} KI aqueous solution (the pH was adjusted to 1 with sulfuric acid). Figure 4.14b,c shows TEM images of typical poly-CuTAPc coated SiNW. It can be seen that the ultrathin and nonporous polymer layer is uniform on the surface of the SiNW.

The mechanism of hydrogen production is shown in Fig. 4.14a: light is absorbed by the Si/poly-CuTAPc coaxial core–shell nanowire along the axial direction and the photogenerated electon–hole pairs are separated by the space charge layer in the radial direction. The holes react with I^- on the surface of poly-CuTAPc film and the electrons are transferred to the Pt counter electrode and then reduce

H^+ to H_2. The poly-CuTAPc serves as a holes transport material and a protective coating here.

The modified SiNW array demonstrated significantly enhanced photocatalytic hydrogen generation performance (Fig. 4.14d) and stability (Fig. 4.14e). The PEC employing this Si/poly-CuTAPc coaxial core–shell nanowire array as the photoanode was shown to effectively produce hydrogen at a rate of 83.5 μmol cm^{-2} h^{-1} under 100 mW cm^{-2} irradiation with 0 V applied potential (vs. SCE). The Si/poly-CuTAPc coaxial core–shell nanowire array could be a promising visible photocatalyst for solar hydrogen generation.

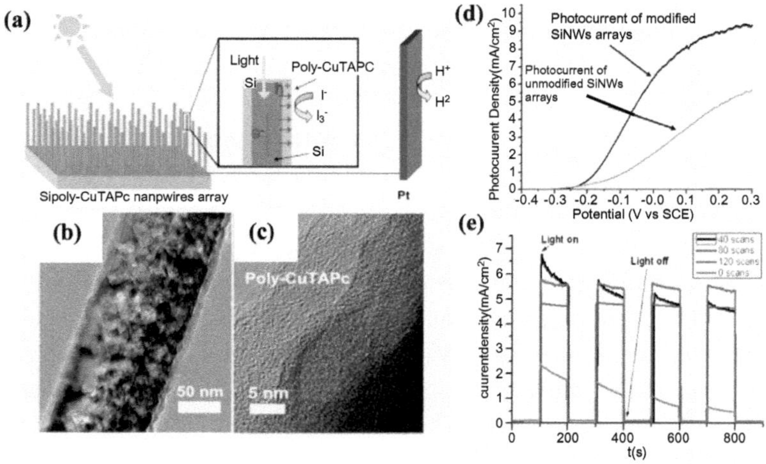

Figure 4.14 Schematic of the hydrogen production solar cell (a) and TEM images (b and c) of the SiNWs modified with poly-CuTAPc. (d) Dark (blue and red) and photocurrent (100 mW cm^{-2} xenon lamp) densities for Si/poly-CuTAPc coaxial core-shell nanowire array (black) and H-terminated n-type SiNW array photoanodes (green), collected at a scan rate of 50 mVs^{-1}. (e) The short-circuit photocurrent response of the Si/poly-CuTAPc coaxial core–shell nanowire array with different number of electropolymerizing scan cycles (zero scan corresponds to the SiNW array with no polymer coating). All currents were measured with three-electrode photoelectrolysis cells at zero applied potential (vs. SCE). The electrolyte was 3.0 M KI (pH=1) and was irradiated with a xenon lamp at an intensity of 100 mW cm^{-2} [67]. Copyright © 2012, Royal Society of Chemistry.

To further improve the efficiency of hydrogen production, Zhang's group later developed a PEC cell, in which an iodine-doped PEDOT/CuPc modified SiNW core-shell array was used as the photoanode (Fig. 4.15a) [68]. A water-insoluble conductive polymer PEDOT was used as a stable protective layer of SiNWs due to its good stability in harsh solvent environments and under relatively high temperatures. The PEDOT film was electropolymerized on the SiNW arrays. After polymerization, the PEDOT film was doped by I_2 vapor. I_2 is commonly used as an oxidizing agent for polymer doping, and the oxidized doping process results in the formation of the I^- ions in the PEDOT film. Copper phthalocyanine (CuPc) was deposited onto the PEDOT:I-modified SiNW arrays as a cocatalyst to accelerate I^- oxidation in the bulk electrolyte solution. The counter electrode was a platinum foil, and the electrolyte solution used was 0.2 M hydriodic acid (0.5 M H_2SO_4, pH= 0). Upon illumination with 100 mW cm^{-2}, the SiNW arrays absorbed light to produce photoelectrons and photoholes; then, iodine ions in the PEDOT film were oxidized by the photoholes immediately. The photoelectrons that transferred to the counter electrode reduced $2H^+$ to form H_2. To verify the nanowires photoactivity, they compared the *J–V* characteristics of the photoelectrochemical performance of the silicon wafer and the SiNW both coated with PEDOT:I and CuPc cocatalyst under the same experimental conditions. As shown in Fig. 4.15b, the maximum photocurrent of the coated SiNWs photoanode is nearly 30 mA cm^{-2} (when the potential is 0.598 V (vs. SCE)) in the H^+ reduction process. This is more than one order higher than that of the coated silicon-wafer photoanode. The obvious advantage of the nanowire-array photoanode is caused by the nanowires unique photoelectrical properties. Firstly, the light-trapping effect by the nanowire arrays enhances light absorption over the planar silicon wafer [5]; secondly, nanostructures show high surface-to-volume ratios, which enhances the charge-carrier separation and collection efficiencies [69]; Thirdly, the radial separation of the charge-carrier of the nanowires facilitates the use of a low diffusion-length material; thus, the short diffusion length for the minority carriers improves the carrier separation and collection efficiencies [70].

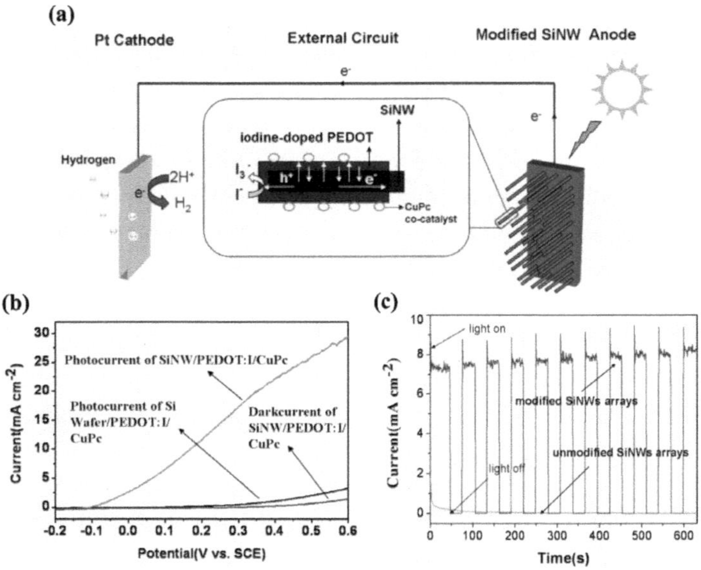

Figure 4.15 (a) Schematic diagram of the hydrogen-production PEC cell. (b) Si/iodine-doped PEDOT/CuPc coaxial core-shell nanowire-array anode and Si planar wafer/iodine-doped PEDOT/CuPc anode *J* – *V* data. The data were collected as a single sweep from negative to positive potentials at 0.1 V s^{-1}. The anode was irradiated using a xenon lamp at an intensity of 100 mW cm^{-2} in a three-electrode system in 0.2 M hydriodic acid solution. (c) The short-circuit current response of the Si/iodine-doped PEDOT/CuPc coaxial core-shell nanowire-array as well as the unmodified SiNW arrays anode during on-off cycles of illumination in a two-electrode photoelectrolysis cell [68]. Copyright © 2012 WILEY-VCH.

Figure 4.15c shows that comparison of the modified SiNW array with an unmodified SiNW array. The good stability shown by the modified SiNW array indicates that it can be used as a stable photocatalyst for hydrogen generation. The hydrogen-generation rate in the PEC cell is shown to be 112.39 μ mol cm^{-2} h^{-1} at 0.1 V applied bias (vs. SCE). The solar-to-chemical energy-conversion efficiency is 3.0%, which is among the best reported for a stable visible-light photoanode for hydrogen generation [71–73].

In addition, He's group demonstrated inorganic nanoparticles modified core-shell SiNW arrays as a photoanode for solar hydrogen generation. They tried to make full use of the advantages of TiO_2, Au nanoparticles and SiNW arrays, and designed and developed a novel low-cost solution approach followed by annealing in Ar and air atmosphere, respectively, to prepare the unique structure of Si/TiO_2 core-shell nanowire arrays modified with Au nanoparticles (SiNWs/TiO_2@AuNPs) (see Fig. 4.16) as photoanode for PEC water splitting [74].

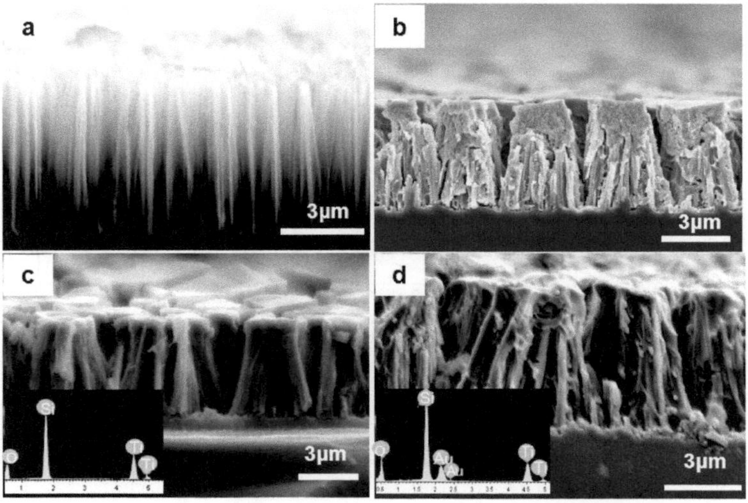

Figure 4.16 Cross-section SEM images of (a) SiNWs array, (b) SiNWs/TiO_2-air, (c) SiNWs/TiO_2-Ar, (inset: EDS) and (d) SiNWs/TiO_2@AuNPs-Ar (inset: EDS) [74]. Copyright © 2013 American Scientific Publishers.

They found that the core-shell arrays annealing in inert atmosphere could significantly improve the performance for water splitting, as shown in Fig. 4.17a. At a 0.5 V vs. SCE bias voltage, the photocurrent density of SiNWs/TiO_2-air sample was only 0.04 mA/cm^2, while the photocurrent density of SiNWs/TiO_2-Ar sample could reach 0.25 mA/cm^2, which is comparable to that of n-Si/n-TiO_2 core/shell nanowire arrays prepared by atomic layer deposition (ALD) [18] (0.28mA/cm^2) under otherwise identical measurement conditions. The SiNWs/TiO_2-Ar sample enhanced the photocurrent by 6.3 times compared to the SiNWs/TiO_2-air sample. The reason why the Ar ambience is more helpful improving the photocurrent than the air atmosphere may be the significantly lower reflectance of SiNWs/TiO_2-Ar than that of SiNWs/TiO_2-air (Fig. 4.17c), and the lower reflectance would lead to absorbing sunlight more efficiently. Another reason may be the existence of silicon dioxide (SiO_2) at the interface between the SiNW and TiO_2 of the SiNW/TiO_2-air sample annealed in the oxygen-rich environment, and the presence of SiO_2 can impede the charge flow in the n-Si/n-TiO_2 junctions, leading to the decrease of photocurrent.

Decoration of Au nanoparticles on the SiNWs/TiO_2-Ar sample (SiNWs/TiO_2@AuNPs-Ar) can further enhance the photocatalytic activity, as shown in Fig. 4.17a

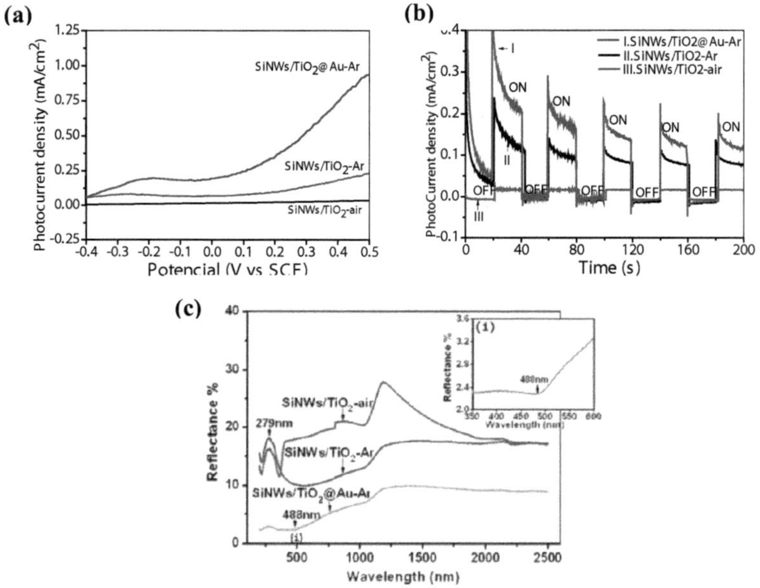

Figure 4.17 (a) Photocurrent density versus bias potential (vs SCE) of varied SiNW-based heterojunctions: SiNWs/TiO_2@Au-Ar, SiNWs/TiO_2-Ar, SiNWs/TiO_2-air. (b) Photocurrent density as a function of time for SiNWs/TiO_2@AuNPs-Ar (I), SiNWs/TiO_2-Ar (II), and SiNWs/TiO_2-air (III), respectively. The photocurrent measurements of solar water splitting were carried out in a 1 M KOH electrolyte using a three electrode setup, in which the core/shell nanowire arrays served as the working electrode, a Pt foil as counter electrode, and a standard saturated calomel electrode (SCE) as the reference electrode, respectively. A 500 W Xe lamp passing through an AM 1.5 G filter was used as the light source to simulate solar light. The light intensity about 100mW cm^{-2}. (c) UV-Vis reflection spectra of SiNWs/TiO_2-air, SiNWs/TiO_2-Ar, and SiNWs/TiO_2@AuNPs-Ar, the inset highlight the absorption peak of Au element at 488 nm [74]. Copyright © 2013 American Scientific Publishers.

The unique heterostructure of SiNWs/TiO_2@AuNPs–Ar gave an even higher photocurrent density of 0.9 mA/cm^2 at 0.5 V vs. SCE bias voltage, which is about 3.6 times as high as that of the SiNWs/TiO_2-Ar without Au nanoparticles. The reason can be attributed to the fact that the reflectance of the SiNWs/TiO_2@AuNPs-Ar is significantly (10%) lower than that of the sample without Au nanoparticles in the wavelength range of 200–2500 nm (Fig. 4.17c). Another reason may be the effect of surface Plasmon resonance (SPR) of Au nanoparticles on the TiO_2 surface, which can enhance the optical near-field that would facilitate the generation of exitons at the TiO_2 surface, thus increasing the amount of photogenerated charges contributing to solar water splitting [75]. However, the SiNWs/TiO_2@AuNPs-Ar sample showed a declining tendency of photocurrent density in the light-on region (see Fig. 4.17b), which was probably caused by the imperfect TiO_2 shell that exposed part of SiNW arrays to the electrolyte solution, leading to the reaction between SiNW and the electrolyte solution.

4.5 Conclusions and Outlook

In summary, this review described the fabrication procedures and mechanisms of SiNW arrays by the MCEE method. Compared to the VLS and oxide-assisted growth (OAG) growth, the MCEE method offers several unique advantages: simple, cost-effective, low-temperature, and wafer-scale production. Particularly, the MCEE SiNW arrays inherit electrical properties from the mother Si substrate and thus require no further doping processing for conductivity. Based on the unique optical and electrical properties of the SiNW arrays, the SiNW arrays have been shown to be an important and promising class of nanomaterial for cost-effective and high-performance solar energy conversion including solar cells, photoelectrochemical solar cells and photocatalytic water splitting.

However, there are still many important issues remaining to be addressed. In particular, the excessive recombination at the high surface and interface areas of SiNWs must be reduced and kept comparable to that of the bulk in order to capitalize on the advantages of SiNWs in light absorption and charge collection. Besides optimal fabrication of high-quality p-n junctions and effective surface passivation are needed to improve device performance. By surface modification of SiNW arrays to form radial SiNW heterojunctions may solve some of the above problems. Recent reports showed that SiNW array-based device performance could be significantly improved by surface modification of SiNW arrays [57, 67, 68, 74]. Therefore, it is believed that modification methods and materials used needs further study in future.

Acknowledgements

This work was financially supported by the National Natural Science Foundation of China (Grant No. 11104283), the National Basic Research Program of China (Grant No. 2010CB934103) and Open Research Fund Program of the State Key Laboratory of Low-Dimensional Quantum Physics (Grant No. KF201207).

References

1. Peng, K. Q., Xu, Y., Wu, Y., Yan, Y. J., Lee. S. T., Zhu, J. *Small.* **2005**, *1*, 1062.
2. Tian, B. Z., Zheng, X. L., Kempa, T. J., Fang, Y. N., Yu, F., Yu, G. H., Huang, J. L., Lieber, C. M. *Nature* **2007**, *449*, 885.
3. Tsakalakos, L., Balch, J., Fronheiser, J., Korevaar, B. A., Sulima, O., Rand, *J.Appl. Phys. Lett.* **2007**, *91*, 233117.
4. Garnett, E. C., Yang, P. D. *J. Am. Chem. Soc.* **2008**, *130*, 9224.
5. Kelzenberg, M. D., Boettcher, S. W., Petykiewicz, J. A., Turner-Evans, D. B., Putnam, M. C., Warren, E. L., Spurgeon, J. M., Briggs, R. M., Lewis, N. S., Atwater, H. A. *Nat. Mater.* **2010**, *9*, 239.
6. Peng, K. Q., Yan, Y. J., Gao, S. P., Zhu, J. *Adv. Mater.* **2002**, *14*, 1164.
7. Peng, K. Q., Yan, Y. J., Gao, S. P., Zhu, J. *Adv. Funct. Mater.* **2003**, *13*, 127.
8. Peng, K. Q., Zhu, *J. Electroanal. Chem.* **2003**, *55*, 35.
9. Peng, K. Q., Zhu, *J. Electrochem. Acta.* **2004**, *49*, 2563.
10. Peng, K. Q., Wu, Y., Fang, H., Zhong, X. Y., Xu, Y., Zhu, J. *Angew. Chem. Int. Ed.* **2005**, *44*, 2737.
11. Peng, K. Q., Fang, H., Hu, J. J., Wu, Y., Zhu, J., Yan, Y. J., Lee, S. T. *Chem.-Eur. J.* **2006**, *12*, 7942.
12. Peng, K. Q., Hu, J. J., Yan, Y. J., Wu, Y., Fang, H., Xu, Y., Lee, S. T., Zhu, J. *Adv. Funct. Mater.* **2006**, *16*, 387.

13. Peng, K. Q., Zhang, M. L., Lu, A. J., Wong, N. B., Zhang, R. Q., Lee, S. T. *Appl. Phys. Lett.* **2007**, *90*, 163123.

14. Chern, W., Hsu, K., Chun, I., de Azeredo, B. P., Ahmed, N., Kim, K. H. *Nano Lett.* **2010**, *10*, 1582.

15. Choi, W. K., Liew, T. H., Dawood, M. K., Smith, H. I., Thompson, C. V., Hong, M. H. *Nano Lett.* **2008**, *8*, 3799.

16. Fang, H., Li, X. D., Song, S., Xu, Y., Zhu, J. *Nanotechnology*, **2008**, *19*, 255703.

17. Peng, K. Q., Wang, X., Lee, S. T. *Appl. Phys. Lett.* **2008**, *92*, 163103.

18. Hwang, Y. J., Boukai, A., Yang, P. D. *Nano Lett.* **2009**, *9*, 410.

19. Sivakov, V., Andra, G., Gawlik, A., Berger, A., Plentz, J., Falk, F., Christiansen, S. H. *Nano Lett.* **2009**, *9*, 1549.

20. Hochbaum, A. I., Chen, R. K., Delgado, R. D., Liang, W. J., Garnett, E. C., Najarian, M., Majumdar, A., Yang, P. D. *Nature* **2008**, *451*,163.

21. Peng, K., Jie, J., Zhang, W., Lee, S. T. *Appl. Phys. Lett.* **2008**, *93*, 033105.

22. Zhang, B. H., Wang, H. S., Lu, L. H., Ai, K. L., Zhang, G., Cheng, X. L. *Adv. Funct. Mater.* **2008**, *18*, 2348.

23. Zhang, M. L., Yi, C. Q., Fan, X., Peng, K. Q., Wong, N. B., Yang, M. S., Zhang, R. Q., Lee, S. T. *Appl.Phys. Lett.* **2008**, *92*, 043116.

24. Cao, M. W., Song, X. Y., Zhai, J., Wang, J. B., Wang, Y. L. *J. Phys. Chem. B.* **2006**, *110*, 13072.

25. Shi, F., Song, Y. Y., Niu, H., Xia, X. H., Wang, Z. Q., Zhang, X. *Chem. Mater.* **2006**, *18*, 1365.

26. Xiu, Y., Zhu, L., Hess, D. W., Wong, C. P. *Nano Lett.* **2007**, *7*, 3388.

27. Peng, K. Q., Lu, A. J., Zhang, R. Q., Lee, S. T. *Adv. Funct. Mater.* **2008**, *18*, 3026.

28. Huang, Z. P., Zhang, X. X., Reiche, M., Liu, L. F., Lee, W., Shimizu, T., Senz, S., G¨osele, U. *Nano Lett.* **2008**, *8*, 3046.

29. Zhang, M. L., Peng, K. Q., Fan, X., Jie, J. S., Zhang, R. Q., Lee, S. T., Wong, N. B. *J. Phys. Chem. C* **2008**, *112*, 4444.

30. Chen, H., Wang, H., Zhang, X. H., Lee, C. S., Lee, S. T. *Nano Lett.* **2010**, *10*, 864.

31. Qu, Y. Q., Liao, L., Li, Y. J., Zhang, H., Huang, Y., Duan, X. F. *Nano Lett.* **2009**, *9*, 4539.

32. Zhu, J., Yu, Z. F., Burkhard, G. F., Hsu, C. M., Connor, S. T., Xu, Y. Q., Wang, Q., McGehee, M., Fan, S. H., Cui, Y. *Nano Lett.* **2009**, *9*, 279.

33. Hu, L., Chen, G. *Nano Lett.* **2007**, *7*, 3249.

34. Li, J. S., Yu, H. Y., Wong, S. M., Li, X. C., Zhang, G., Kwong, D. L. *Appl. Phys. Lett.* **2009**, *95*, 243113.
35. Air Mass 1.5 Spectra, American Society for Testing and Materials, http://rredc.nrel.gov/solar/spectra/am1.5/#1962.
36. Li, J. S., Yu, H. Y., Wong, S. M., Zhang, G., Sun, X. W., Lo, P. G. Q., Kwong, D. L. *Appl. Phys. Lett.* **2009**, *95*, 033102.
37. Kayes, B. M., Atwater, H. A., Lewis, N. S. *J. Appl. Phys.* **2005**, *97*, 114302.
38. Yu, M., Long, Y. Z., Sun, B., Fan, Z. Y. *Nanoscale*, **2012**, *4*, 2783
39. Allen, J. E., Hemesath, E. R., Perea, D. E., Lensch-Falk, J. L., Li, Z. Y., Yin, F., Gass, M. H., Wang, P., Bleloch, A. L., Palmer, R. E., Lauhon, L. *J. Nat. Nanotechnol.* **2008**, *3*, 168.
40. Li, J. S., Wong, S. M., Li, Y. L., Yu, H. Y. *IEEE Photovoltaic Spec. Conf., 35th*, **2010**, 1469.
41. Gunawan, O., Guha, S. *Sol. Energy Mater. Sol. Cells*, **2009**, *93*, 1388.
42. Demichel, O., Calvo, V., Besson, A., Noé, P., Salem, B., Pauc, N., Oehler, F., Gentile, P., Magnea, N. *Nano Lett.* **2010**, 10, 2323.
43. Dan, Y., Seo, K., Takei, K., Meza, J. H., Javey, A., Crozier, K. B. *Nano Lett.* **2011**, *11*, 2527.
44. Kim, D. R., Lee, C. H., Rao, P. M., Cho, I. S., Zheng, X. *Nano Lett.* **2011**, *11*, 2704.
45. Peng, K. Q., Lee, S. T. *Adv. Mater.* **2011**, *23*, 198.
46. Lu, Y. R., AmitLal, *Nano Lett.* **2010**, 10, 4651.
47. Green, M. A. In Solar Cells: Operating Principles, Technology and SystemApplications, Prentice-Hall, Inc.: Englewood Cliffs, NJ, **1982**, p 274.
48. Shen, X. J., Sun, B. Q., Liu, D., Lee, S. T., *J. Am. Chem. Soc.* **2011**, *133*, 19408
49. Snaith, H. J., Moule, A. J., Klein, C., Meerholz, K., Friend, R. H., Gratzel, M. *Nano Lett.* **2007**, *7*, 3372.
50. Hwang, J. S., Kao, M. C., Shiu, J. M., Fan, C. N., Ye, S. C., Yu, W. S., Lin, H. M., Lin, T. Y. *J. Phys. Chem. C* **2011**, *115*, 21981.
51. Fan, G. F., Zhu, H. W., Wang, K. L., Wei, J. Q., Li, X. M., Shu, Q. K., Guo, N., Wu, D. H. *ACS Appl. Mater. Interfaces*, **2011**, *3*, 721.
52. Xie, C., Lv, P., Nie, B., Jie, J. S., Zhang, X. W., Wang, Z., Jiang, P., Hu, Z. Z., Luo, L. B. *Appl. Phys. Lett.* **2011**, *99*, 133113.
53. Peng, K. Q., Wang, X., Wu, X. L., Lee, S. T. *Nano Lett.* **2009**, *9*, 3704.

54. Dalchiele, E. A., Martin, F., Leinen, D., Marotti, R. E., Ramos-Barrado, J. R. *J. Electrochem. Soc.* **2009**, *156*, K77.

55. Shu, Q. K., Wei, J. Q., Wang, K. L., Zhu, H. W., Li, Z., Jia, Y., Gui, X. C., Guo, N., Li, X., Ma, C., Wu, D. H. *Nano Lett.* **2009**, *9*, 4338.

56. Shen, X. J., Sun, B. Q., Yan, F., Zhao, J., Zhang, F., Wang, S. D., Zhu, X. L., Lee, S. T. *ACS Nano* **2010**, *4*, 5869.

57. Wang, X., Peng, K. Q., Pan, X. J., Chen, X., Yang, Y., Li, L., Lee, S. T. *Angew. Chem. Int. Ed.* **2011**, *50*, 9861.

58. Huang, Z. P., Zhong, P., Wang, C. F., Zhang, X. X., Zhang, C. *ACS Appl. Mater. Interfaces*, **2013**, *5*, 1961.

59. Fujishima, A., Honda, K. *Nature*, **1972**, *238*, 37.

60. Seger, B., Kamat, P. V. *J. Phys. Chem. C* **2009**, *113*, 18946.

61. Levinson, R., Berdahl, P., Akbari, H. *Sol. Energy Mater. Sol. Cells*, **2005**, *89*, 319.

62. Lee, Y. L., Chi, C. F., Liau, S. Y. *Chem. Mater.* **2010**, *22*, 922.

63. Boettcher, S. W., Spurgeon, J. M., Putnam, M. C., Warren, E. L., Turner-Evans, D. B., Kelzenberg, M. D., Maiolo, J. R., Atwater, H. A., Lewis, N. S. *Science*, **2010**, *327*, 185.

64. Abe, R., Sayama, K., Sugihara, H. *J. Phys. Chem. B* **2005**, *109*, 16052.

65. Shao, M. W., Cheng, L., Zhang, X. H., Ma, D. D. D., Lee, S. T. *J. Am. Chem. Soc.* **2009**, *131*, 17738.

66. Shiu, S. C., Chao, J. J., Hung, S. C., Yeh, C. L., Lin, C. F. *Chem. Mater.* **2010**, *22*, 3108.

67. Liu, Z., Wang, H., Ou, X. M., Lee, C. S., Zhang, X. H. *Chem. Commun.* **2012**,*48*, 2815.

68. Yang, T., Wang, H., Ou, X. M., Lee, C. S., Zhang, X. H. *Adv. Mater.* **2012**, *24*, 6199.

69. Cho, I. S., Chen, Z. B., Forman, A. J., Kim, D. R., Rao, P. M., Jaramillo, T. F., Zheng, X. L. *Nano Lett.* **2011**, *11*, 4978.

70. Maiolo III, J. R., Kayes, B. M., Filler, M. A., Putnam, M. C., Kelzenberg, M. D., Atwater, H. A., Lewis, N. S. *J. Am. Chem. Soc.* **2007**, *129*, 2346.

71. Wang, G. M., Wang, H. Y., Ling, Y. C., Tang, Y. C., Yang, X. Y., Fitzmorris, R. C., Wang, C. C., Zhang, J. Z., Li, Y. *Nano Lett.* **2011**, *11*, 3026.

72. Park, J. H., Park, O. O., Kim, S. *Appl. Phys. Lett.* **2006**, *89*, 163106.

73. Shi, J., Hara, Y., Sun, C. L., Anderson, M. A., Wang, X. D. *Nano Lett.* **2011**, *11*, 3413.

74. Liu, L., Cao, Y., He, J., Qi, X., Shi, W., Yang, Q. *J. Nanosci. Nanotechnol.* **2013**, *13*, 6835.

75. Liu, Z. W., Hou, W. B., Pavaskar, P., Aykol, M., Stephen, B. *Nano.Lett.* **2011**, *11*, 1111.

Chapter 5

Positron-Positronium Annihilation Studies of Energy- and Environment-Related Nanomaterials

Kiminori Sato
Department of Environmental Sciences,
Tokyo Gakugei University,
4-1-1 Koganei, Tokyo 184-8501, Japan
sato-k@u-gakugei.ac.jp

5.1 Introduction

Nanometer-sized open spaces (open nanospaces) are ubiquitous in artificially and naturally formed substances. In intermetallic compounds and compound semiconductors, open nanospaces called as vacancies on element-specific sublattice govern atomic diffusion, plastic deformation, and structural phase transformation [1]. Polymers and inorganic glasses possesses free volumes with definite affinity, which play an important role in gas permeation and ionic conduction [2–4] such as Li^+ diffusion associated with oxygen atoms in polymer electrolytes [5]. There also exists free volumes in geological environment, which offer an evolutionary pathway

Nanomaterials in Energy and Environmental Applications
Edited by Junhui He

ISBN 978-981-4463-78-2 (Hardcover), 978-981- 4463-79-9 (eBook)
www.panstanford.com

of uptaking elements together with underground water [6]. The open nanospace is expected to be crucial to the molecular design of energy- and environment-related nanomaterials as well as a number of important environmental issues as, e.g., the dispersion of radioactive nuclide, but has not been considered so far.

Mercury intrusion porosimetry is known to be suitable for the macroscopic open space larger than 50 nm in the diameter called as macropore, because quite large molecules of mercury are employed for probing the open space. In this porosimetric technique, the open space connected to the surface, so-called open-type pore, can be detected because mercury molecules have to be penetrated into the open spaces from the surface. Gas absorption porosimetry can be conducted for investigating further smaller open space in the range from 2 nm to 50 nm, which is called as mesopore. Similarly to the mercury intrusion porosimetry, the gas absorption porosimetry provides the information on the open-type pore since gas molecules adsorbed at the pore surface is analyzed with BET theory. Small angle x-ray scattering can be applied for the mesopore, of which the pore is completely closed inside the materials. Transmission electron microscopy (TEM) observation could be promising to identify the open space smaller than 2 nm, called as micropore. It is however difficult to obtain clear contrast in TEM images for the materials consisting of light elements as, e.g., polymer.

Positron is the anti-particle of electron, positively-charged electron. Positron-positronium annihilation spectroscopy has great potential for the characterization of open nanospaces in materials [7]. In metals, the vacancy size and elemental environment around the vacancy can be explored by the measurements of positron lifetime and element-specific momentum distribution of positron-electron annihilation photons, respectively [8–15]. In insulating materials as, e.g., polymers, glasses, and minerals, a fraction of energetic positrons forms the bound state with an electron, positronium (Ps) [16]. Triplet *ortho*-Ps (*o*-Ps) with antiparallel spins of an electron and positron undergoes two-photon annihilation with one of the bound electrons with opposite spin in free volumes [17–18]. This process is known as *o*-Ps pick-off annihilation and its lifetime gives the information on the size of open nanospace through quantum mechanical model [19–22]. The momentum distribution of *o*-Ps pick-off annihilation is influenced by the electrons bound to the surrounding molecules [23–24], thus enabling the investigation of elemental environment around open nanospaces [25–26].

Here, recent advances in the studies of energy- and environment-related nanomaterials are reviewed from the viewpoint of open nanospaces. In the following section, the principle of positron-positronium annihilation spectroscopy is described after the introduction of positron and positronium. Analytical procedures for investigating the size of open nanospace and its elemental environment are particularly focused. And then the following topics are discussed based on the results of open nanospace analyses: 1. mechanism of Bi precipitation in environmentally friendly Sn-Bi eutectic system [27–28], 2. densification dynamics of gadolinium-doped ceria (GDC), one of the most promising candidates for the electrolytes of solid-oxide fuel cell [29], 3. formation mechanism of biogenic silica quartz in the steady-state environment [30–31], and 4. elemental migration through geological environment [32].

5.2 Positron-Positronium Annihilation Spectroscopy

In this section, a series of positron-positronium annihilation spectroscopy is introduced after describing how positrons behave in materials. Analytical procedures for investing the size of open nanospace and its surrounding environment are particularly focused.

5.2.1 Positron and Positronium in Materials

When energetic positrons are implanted into materials, they rapidly lose their energy and reach thermal equilibrium within a few ps. After thermalization, positrons begin to diffuse in materials. The average positron diffusion length L_+ is typically of the order of 100 nm in the case of metals [7]. During diffusion positrons interact with surrounding atoms and finally annihilate with their electrons. Positrons are Bloch delocalized in the perfectly ordered matrix in crystalline materials, characterizing the annihilation rate λ_b of given materials. At the open nanospace such as monovacancies, the potential felt by positrons is lowered due to reduction in the Coulomb repulsion, which results in a lower energy level than for delocalized-free state. The transition from the delocalized state to the localized one is called as positron trapping. Since local electron density at the vacancy site is lower than in defect-free region, positron lifetime τ_d (the reciprocal of annihilation rate λ_d) of the trapped positrons becomes longer than in defect-free region τ_b (the reciprocal of λ_b). This enables to investigate the size of open nanospaces in metal, intermetallic compound, semiconductor, quasicrystal, and ceramic and so on.

In insulating materials as, e.g., polymers, glass, and minerals, a fraction of energetic positrons injected into samples often forms Ps. Singlet *para*-Ps (*p*-Ps) with the spins of the positron and electron antiparallel and triplet *o*-Ps with parallel spins are formed at a ratio of 1 : 3. Hence, three states of positrons: *p*-Ps, *o*-Ps, and free positrons exist in Ps-formaing materials. The annihilation of *p*-Ps results in the emission of two γ-ray photons of 511 keV with lifetime ~ 125 ps. Free positrons are trapped by negatively-charged parts such as polar elements and annihilated into two photons with lifetime ~ 450 ps. The positron in *o*-Ps undergoes two-photon annihilation with one of the bound electrons with a lifetime of a few ns after localization in angstrom-scale pores. The last process is known as *o*-Ps pick-off annihilation and provides the information on the size of open nanospaces in Ps-forming materials as polymer through quantum mechanical model.

Annihilation of a positron-electron pair produces γ-ray photons, which have a total energy of $2m_0c^2$. Here, m_0 and c are the electron rest mass and the velocity of light, respectively. Energy-momentum conservation requires the release of two or more photons. In the center-of-mass frame of positron-electron pair, two annihilation γ-rays are emitted in the opposite directions to each others. In the laboratory frame, in which positron is considered to be in the rest, γ-ray energies are Doppler shifted and deviate from 180°, depending on the momentum P_L of the annihilating pair. The momentum distributions of positron and *o*-Ps pick-off annihilation are significantly influenced by the electron momenta participating in annihilation, which can be utilized for investigating the elemental environment around open nanospaces.

5.2.2 Positron-Positronium Lifetime Spectroscopy

Positron-positronium lifetime spectroscopy is well documented as a powerful tool for examining the size of open nanospaces in materials. ^{22}Na radioisotope, generally employed as positron source in the laboratory frame, emits a positron together with a 1.27 MeV birth γ-ray. The positron implanted into the sample annihilates with the electron, providing 511 keV annihilation γ-rays. The average lifetimes at individual positron states are measured as the time difference between the birth and 511 keV annihilation γ-rays. The positron lifetime spectrum is recorded with a fast-fast coincidence or digital oscilloscope-based system employing a photomultiplier tube with a scintillator. The time resolution of the system is 180–220 ps full width at the half maximum (FWHM). For each spectrum at least 1.0×10^6 coincident counts are corrected. The validity of our lifetime measurements and data analysis was confirmed with a certified reference materials (NMIJ CRM 5601-a and 5602-a) provided by National Metrology Institute of Japan, National Institute of Advanced Industrial Science and Technology (AIST) [33–34]. After subtracting the background, positron lifetime spectra are numerically analyzed by POSITRONFIT code [35]. In the case of Ps-forming materials such as polymer, the longest-lived component and its relative intensity are attributed to pick-off annihilation of *o*-Ps localized in the open nanospace. The size of open nanospace R can be evaluated through the lifetime of *o*-Ps pick-off annihilation $\tau_{o\text{-Ps}}$ based on the Tao-Eldrup model [20–21]:

$$\tau_{o-Ps} = 0.5\left[1 - \frac{R}{R_0} + \frac{1}{2\pi}\sin\left(\frac{2\pi R}{R_0}\right)\right]^{-1} \tag{5.1}$$

where $R_0 = R + \Delta R$, and $\Delta R = 0.166$ nm is the thickness of homogeneous electron layer in which the positron in *o*-Ps annihilates. Lifetime distributions were obtained by using the MELT code [36].

5.2.3 Coincident Doppler Broadening (CDB) Spectroscopy

Coincident Doppler broadening (CDB) spectroscopy is suitable for investigating elemental environment around the open nanospace in materials, in which no Ps formation occurs as in metal. In the

Doppler broadening spectrum, the positron annihilation events with valence electrons contribute to the low-momentum part of the momentum distribution, whereas tightly bound core electrons contribute to the high momentum part. One-detector Doppler broadening measurements have been employed to study line-shape variations in the low-momentum region. However, line-shape variations due to the core electrons in the high-momentum region cannot be measured satisfactorily by one-detector Doppler broadening spectroscopy because of the poor signal-to-noise (S/N) ratio. Coincident measurements of the energies of two positron-electron annihilation photons gives the possibility to increase the S/N ratio up to 10^5 and extends the range of measurements of the Doppler broadened line up to 50×10^{-3} m_0c, which can be used for a local chemical analysis of elements [37].

The coincident measurements are performed by measuring the energies of the two annihilation quanta E_1 and E_2 with a collinear set-up of two high-purity Ge detectors as shown in Fig. 5.1. The Doppler broadening spectra are obtained by cutting the E_1 and E_2 spectra along the energy conservation line E_1+E_2 = (1022±1) keV,

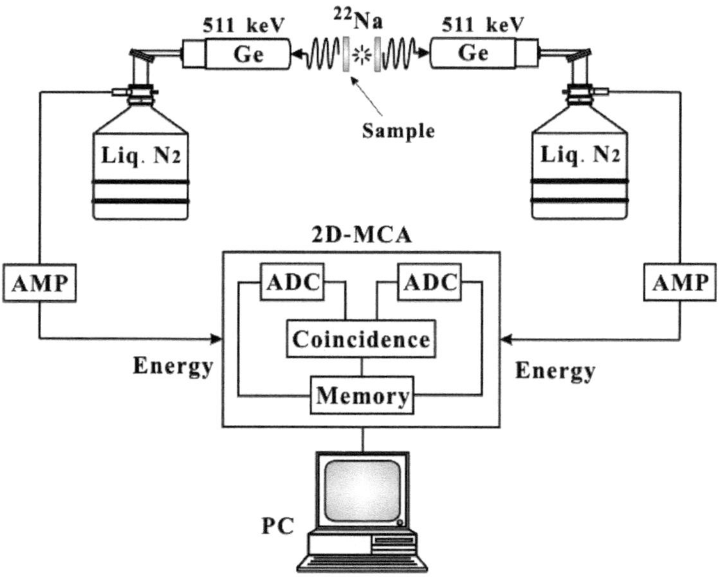

Figure 5.1 Schematic diagram of coincidence Doppler broadening (CDB) system. Reprinted with permission from Ref. [25]. Copyright 2009 American Chemical Society.

taking into account the annihilation events within a strip of a certain energy width. For convenience, Doppler broadening spectrum is often analyzed by introducing two kinds of shaping parameters, *S* and *W* parameters. *S* parameter is defined as the ratio of integrated central counts around 511 keV peak (low-momentum part) to the total integrated counts. On the other hand, *W* parameter is defined as the ratio of integrated counts corresponding to wing area (high-momentum part) to the total counts.

5.2.4 Positron-Age-Momentum Correlation (AMOC) Spectroscopy

In Ps-forming materials, there exist three states of positrons: *p*-Ps, *o*-Ps, and a free positron. Each of them individually contributes to the annihilation photons in its own manner so that the elemental information involved in their momentum distributions is obscured. Furthermore, Ps-forming materials are generally composed of light atoms compared with common metals, the momentum distribution being less element specific. In order to obtain information specific to the open nanospace, the momentum distribution of *o*-Ps pick-off annihilation has to be extracted by time resolving the annihilation photons with positron-age-momentum correlation (AMOC) spectroscopy [25].

Figure 5.2 shows a schematic diagram of our AMOC spectrometer. In AMOC spectroscopy, the Doppler broadening data correlated with positron lifetimes are recorded. The 1.27 MeV positron birth γ-ray from a ^{22}Na source and one of the 511 keV γ-rays emitted as a result of positron annihilation in the samples are detected by BaF_2 scintillators with 2" diameter × 1" thickness coupled with photomultiplier tubes (PMT1 and PMT2). PMT1 and PMT2 are squarely aligned to each other. The 1.27 MeV positron birth γ-ray and 511 keV annihilation γ-ray are energy-selected by constant-fraction-differential discriminators (CFDD1 and CFDD2), and the timing pulse from CFDD1 is delayed by a time-delay module (DELAY). Subsequently, a time-to-amplitude converter (TAC) produces an analog output, whose height is proportional to the time interval between 1.27 MeV and 511 keV γ-rays. Analog signals from TAC are transferred to a two-parameter multi-channel analyzer (2D-MCA) with a built-in analog-to-digital-convertor (ADC). For the simultaneous measurement of the energy of the second annihilation γ-rays, a high purity Ge detector is placed collinearly to PMT2. Analog pulses from the Ge detector are transferred to the 2D-MCA with the built-in ADC through an amplifier (AMP). A pair of correlated digital outputs of positron lifetime and Doppler broadening data within a certain coincident time is stored in a random-access-memory, controlled by a personal computer (PC).

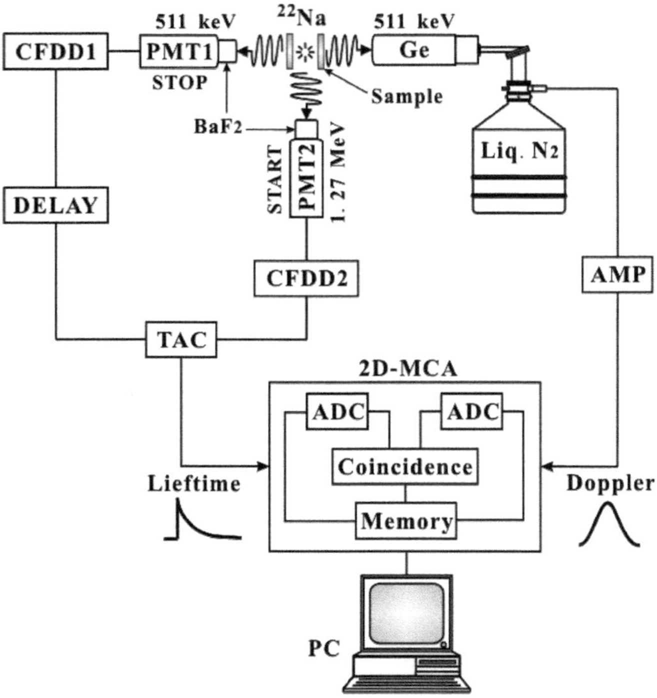

Figure 5.2 Schematic diagram of the positron-age-momentum correlation (AMOC) system. Reprinted with permission from Ref. [25]. Copyright 2009 American Chemical Society.

Figure 5.3 shows a typical AMOC spectrum for amorphous SiO_2, in which coincidence counts are plotted in logarithmic scale as functions of both the positron age and momentum of the annihilation pairs. Positron-age dependent Doppler broadening spectra were parameterized as the shaping *S* parameter, which was defined as the ratio of integrated central counts around 511 keV peak to the total integrated counts after subtracting the background.

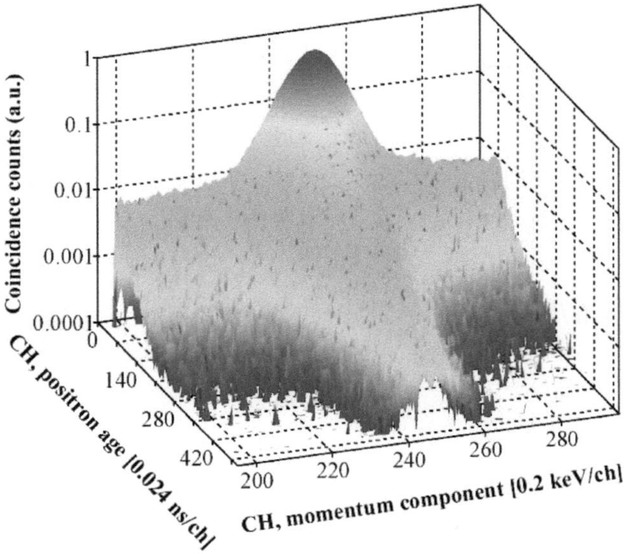

Figure 5.3 Positron-age-momentum correlation (AMOC) spectrum observed for amorphous SiO_2. The coincidence counts are plotted in logarithmic scale as functions of both the positron age and momentum component of annihilation photons. Reprinted with permission from Ref. [25]. Copyright 2009 American Chemical Society.

Figure 5.4a shows the *S* parameters as a function of positron age (*S*-*t* plot) for undoped and Cs-doped amorphous SiO_2, as one of examples. The *S* parameters indicate high values around $t = 0$ ns. They rapidly decrease with increasing positron age up to 0.5 ns and then level off above 2.0 ns. In Fig. 5.4a, the average lifetimes of *p*-Ps, free positrons, and *o*-Ps are indicated with thick arrows pointing to the corresponding positron ages of the horizontal axis. At younger positron age, *p*-Ps, *o*-Ps, and positrons all contribute to the momentum distributions of positron-electron annihilation. Around $t = 0$ ns, high values of *S* parameters are observed due to the narrow momentum distribution of *p*-Ps self annihilation. At the positron age above 0.4 ns, more than 90 % of the shortest-lived *p*-Ps has already disappeared. Above 1.2 ns, the positrons have mostly disappeared as well and, as a consequence, *o*-Ps dominantly contributes to the momentum distribution. The saturated values of *S* parameter at older positron age thus reflect the momentum distributions of *o*-Ps pick-off annihilation.

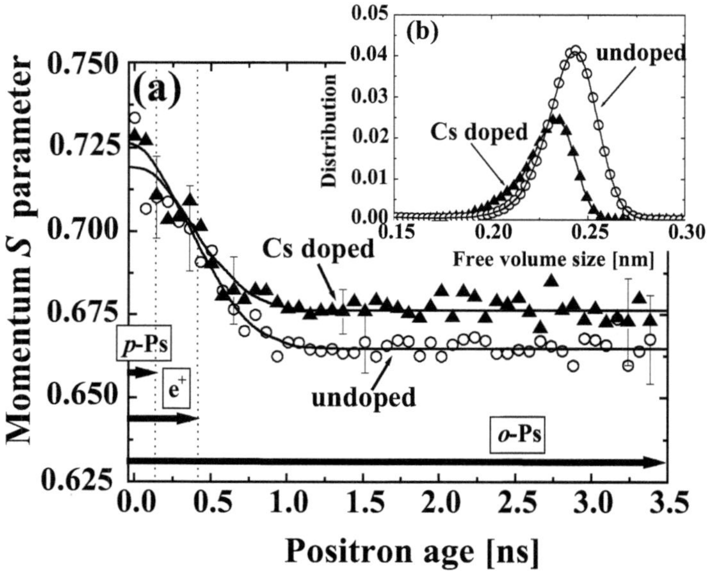

Figure 5.4 (a) Positron age dependence of the *S* parameters obtained by AMOC spectroscopy for undoped (open circles) and Cs-doped (solid triangles) amorphous SiO_2. Solid lines are drawn for guiding the eye. Time scales of lifetimes of *p*-Ps, free positron, and *o*-Ps are illustrated with thick arrows. (b) Free-volume size distributions for undoped (open circles) and Cs-doped (solid triangles) amorphous SiO_2. Reprinted with permission from Ref. [27]. Copyright 2009 American Chemical Society.

The saturated *S* parameters at older positron age are significantly higher for Cs-doped amorphous SiO_2 than for undoped one. This demonstrates narrowing of the momentum distribution of *o*-Ps pick-off annihilation in the former. Upon Cs doping, open nanospaces in amorphous SiO_2 mostly surrounded by O atoms [38] get partially occupied by Cs atoms, which is seen in the variation of the free-volume size distribution obtained from $\tau_{o\text{-Ps}}$ determined by conventional lifetime spectroscopy with the Tao-Eldrup model (see Fig. 5.4b). Higher *S* parameters after Cs doping in the older positron age region indicate the narrower momentum distribution of *o*-Ps pick-off annihilation in open nanospaces partially surrounded by Cs atoms. *S* parameters at older positron age are thus sensitive to the elemental environment around open nanospaces.

5.2.5 Positron Diffusion Experiment

Positron diffusion experiments are conducted by Doppler broadening measurements with a slow positron beam. The Doppler broadening spectra obtained at each positron incident energies are analyzed by taking the S parameter. The average positron-diffusion length L_+ can be generally estimated from the analysis of the positron-incident energy variation of S parameter (S-E plot). After implantation at a positron incident energy E, positrons rapidly thermalize. The spatial distribution of the thermalized positrons $P(E, z)$ along the incident direction (z) can be described by a scaling function [39–40]

$$P(E,z)=N_{lm}\left(\frac{u}{C_{lm}}\right)\exp\left[-\left(\frac{u}{C_{lm}}\right)^m\right], \tag{5.2}$$

where N_{lm} is a normalization constant, and C_{lm}, l, and m are parameters which depend on material in consideration [40–41], and u is defined as,

$$u=\frac{z}{\langle z(E)\rangle}, \tag{5.3}$$

where $<z(E)>$ is the mean implantation depth. $<z(E)>$ is assumed to be given by

$$\langle z(E)\rangle=\frac{A_i}{\rho_i}E^{\alpha_i+\beta_i\ln E}, \tag{5.4}$$

where E and ρ_i are the incident energy of the positron beam and the density of the sample, respectively. The parameters A_i, α_i, and β_i depend on the atomic number of the constituent elements as confirmed by Monte Carlo simulation [42].

After thermalization, positrons begin to diffuse in the sample. The one-dimensional positron diffusion is described by the following equation:

$$D_+\frac{\partial^2 N(z,t)}{\partial z^2}-\frac{\partial N(z,t)}{\partial t}-\Gamma N(z,t)=0, \tag{5.5}$$

where D_+ is the positron diffusion coefficient, $N(z, t)$ is the positron density as a function of both time and position, and Γ is the effective annihilation rate of the positron. The observed S parameter is given by a linear combination of contributions from different annihilation sites

$$S(E) = \sum_i S_i F_i(E), \tag{5.6}$$

where $F_i(E)$ is the fraction of positrons annihilating in the i-th state characterized by the S_i parameter. The fraction $F_i(E)$ can be obtained by solving the diffusion equation, subjected to the positron implantation profile and boundary conditions. By fitting the measured S parameters to Equation (5.6), one can obtain the average positron-diffusion length L_+.

5.3 Application of Positron-Positronium Annihilation Spectroscopy to Energy- and Environment-Related Nanomaterials

In this section, the examples of application of positron-positronium annihilation spectroscopy to energy- and environment-related nanomaterials are described. The following topics are discussed: 1. mechanism of Bi precipitation in environmentally friendly Sn-Bi eutectic system [27–28], 2. densification dynamics of gadolinium-doped ceria (GDC), one of the most promising candidates for the electrolytes of solid-oxide fuel cell [29], 3. formation mechanism of biogenic silica quartz in the steady-state environment [30–31], and 4. elemental migration through geological environment [32].

5.3.1 Mechanism of Bi Precipitation in Environmentally Friendly Sn-Bi Eutectic System

The Sn-Bi eutectic alloy system is one of typical heterogeneous materials. Owing to its low melting temperature, the Sn-Bi eutectic alloy without any toxic elements has been accepted as the most potential candidate for environmentally friendly solders [43–44]. In Bi-containing eutectic alloys there often appears Bi precipitation during cooling, which is macroscopically visible [45]. Bi particles grow with decreasing the cooling rate [46–47]. The Bi precipitation considerably deteriorates the mechanical property of the alloy [43,48]. In this subsection, the mechanism of Bi precipitation in an eutectic alloy of Sn-Bi is discussed.

Pure elements of Sn and Bi with nominal composition of $Sn_{65}Bi_{35}$ at. % were melted and several specimens were cut from the same ingot. Electron probe microanalysis (EPMA) revealed chemical composition of $Sn_{65.4}Bi_{34.6}$. A melting temperature of 412 K was determined by differential scanning calorimetry (DSC). The specimens were heated up to 398 K and then cooled at different cooling rates of 10^{-1} Ks^{-1}, 10^{-3} Ks^{-1}, and 10^{-5} Ks^{-1}.

Figure 5.5 shows backscattering electron (BE) images of $Sn_{65.4}Bi_{34.6}$ cooled at different cooling rates (1) 10^{-1} Ks^{-1}, (2) 10^{-3} Ks^{-1}, and (3) 10^{-5} Ks^{-1}. There appear bright parts arising from the electron backscattering effect from particles of higher atomic numbers (see arrows in Fig. 5.5). Since the atomic number of Bi (83) is much higher than Sn (50), we reasonably ascribe bright parts to Bi-rich particles. They substantially grow with decreasing the cooling rate. Extremely slow cooling with 10^{-5} Ks^{-1} yields surprisingly huge Bi-rich particles exceeding 1 μm in size. The sizes of Bi-rich particles estimated from BE images are listed in Table 5.1.

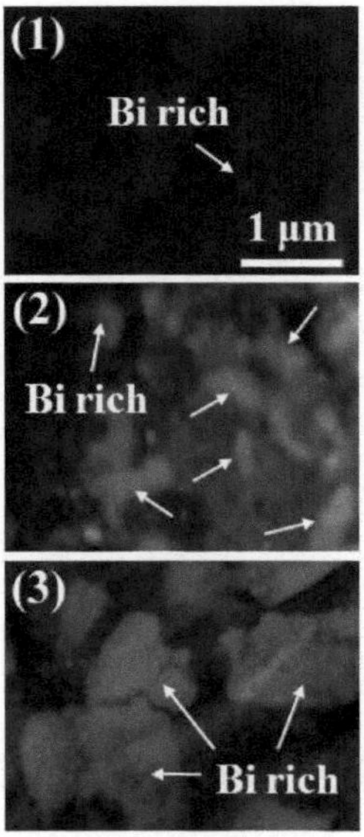

Figure 5.5 Backscattering electron (BE) images of $Sn_{65.4}Bi_{34.6}$ cooled at the cooling rates of (1) 10^{-1} Ks^{-1}, (2) 10^{-3} Ks^{-1}, and (3) 10^{-5} Ks^{-1}. The magnification factor is × 40,000 for all images. Bright parts indicated by arrows are resultant from Bi precipitation. Reprinted with permission from Ref. [27]. Copyright 2008 the Institute of Physics.

Table 5.1 Average size of Bi-rich particles in $Sn_{65.4}Bi_{34.6}$ cooled at the cooling rates of (1) 10^{-1} Ks^{-1}, (2) 10^{-3} Ks^{-1}, and (3) 10^{-5} Ks^{-1} evaluated from the data of backscattering electron (BE) images

Cooling rate [Ks^{-1}]	BE imaging [nm]	XRD [nm]
10^{-1}	100 ± 30	34 ± 3
10^{-3}	600 ± 50	48 ± 4
10^{-5}	1000 ± 80	55 ± 6

Note: The table includes Bi crystallite size evaluated from x-ray diffraction (XRD).

Figure 5.6 shows small angle x-ray scattering (SAXS) profiles obtained for $Sn_{65.4}Bi_{34.6}$ cooled at (1) 10^{-1} Ks^{-1}, (2) 10^{-3} Ks^{-1}, and (3) 10^{-5} Ks^{-1}. Scattered x-ray intensities $I(Q)$ are presented as a function of scattering vector $Q = 4\pi\sin(\theta/2)/\lambda$ defined by scattering angle θ and x-ray wavelength λ. With decreasing the cooling rate, SAXS intensity in the low Q region attributable to the scattering from Bi-rich particles is strongly enhanced, in agreement with the results of BE imaging (see Table 5.1).

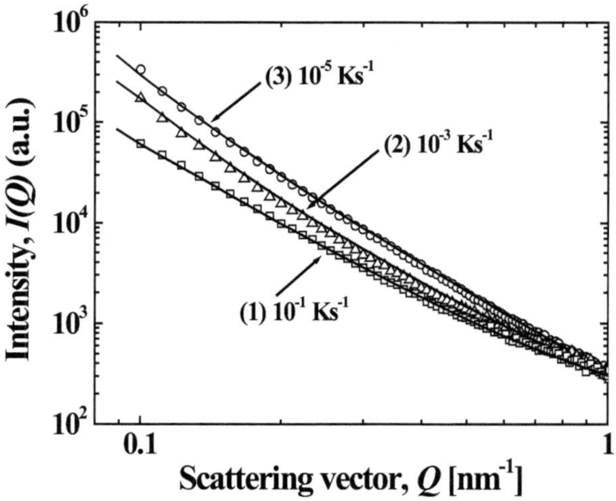

Figure 5.6 Small-angle x-ray scattering (SAXS) profiles of $Sn_{65.4}Bi_{34.6}$ cooled at the cooling rates of (1) 10^{-1} Ks^{-1}, (2) 10^{-3} Ks^{-1}, and (3) 10^{-5} Ks^{-1}. Solid lines are drawn for guiding the eye. Reprinted with permission from Ref. [27]. Copyright 2008 the Institute of Physics.

Figure 5.7 shows x-ray diffraction (XRD) patterns measured for $Sn_{65.4}Bi_{34.6}$ cooled at the cooling rates of (1) 10^{-1} Ks^{-1}, (2) 10^{-3} Ks^{-1}, and (3) 10^{-5} Ks^{-1} together with the calculated intensities. Diffraction peaks are well indexed with those of a body-centered tetragonal β-Sn and rhombohedral Bi phases, signifying an eutectic structure of β-Sn and Bi. We note here that Bi(012) peaks are narrower in the order of (3) 10^{-5} Ks^{-1}, (2) 10^{-3} Ks^{-1}, and (1) 10^{-1} Ks^{-1}. This demonstrates that Bi crystallite grows with decreasing the cooling rate (see Fig. 5.3c). Average sizes of Bi crystallites evaluated from the Bi(012) peak broadening by employing Scherrer's formula are listed in Table 5.1.

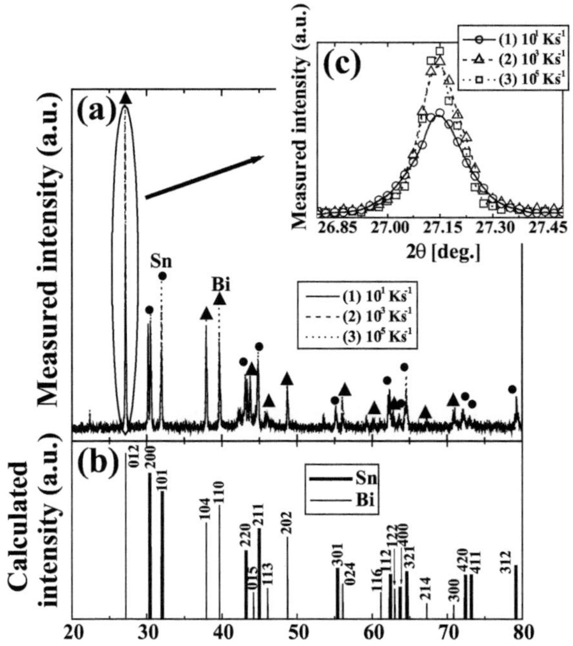

Figure 5.7 (a) Measured x-ray diffraction (XRD) patterns of $Sn_{65.4}Bi_{34.6}$ cooled at the cooling rates of (1) 10^{-1} Ks^{-1} (solid line), (2) 10^{-3} Ks^{-1} (dashed line), and (3) 10^{-5} Ks^{-1} (dotted line). XRD peaks indexed to the body-centered tetragonal β-Sn and rhombohedral Bi phases are marked with solid circles and triangles, respectively. (b) Calculated XRD patterns. Thick and thin lines correspond to the diffraction from Sn and Bi, respectively. (c) Blown-up section of Bi(012) peaks. Reprinted with permission from Ref. [27]. Copyright 2008 the Institute of Physics.

Positron lifetime spectroscopy yields a single component over the entire temperature range studied, as shown in Fig. 5.8. Observed positron lifetimes are much longer than the defect-free matrix of β-Sn (205 ps) [49] and Bi (225 ps) [50], and rather close to the values of β-Sn monovacancy (242 ps) [49] and Bi monovacancy (258 ps) [51]. We therefore conclude the presence of vacancy-sized free volumes in concentration over 10^{-4}, giving rise to positron saturation trapping [52]. The positron lifetime decreases from ~ 275 ps at 398 K close to the melting point (T_m) down to ~ 255 ps with slow cooling the sample at 10^{-5} Ks^{-1}. Obviously the decreasing tendency is less pronounced for $Sn_{65.4}Bi_{34.6}$ at the cooling rate of 10^{-3} Ks^{-1} and disappears at 10^{-1} Ks^{-1}. This clearly demonstrates that the free volumes change in their states depending on the cooling rate, which should be further studied.

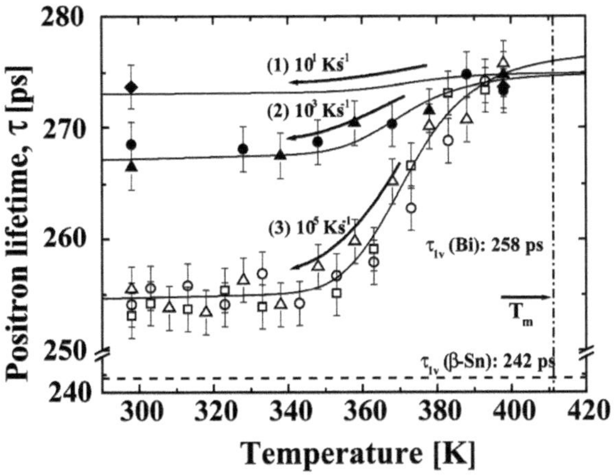

Figure 5.8 Temperature variation of positron lifetime of $Sn_{65.4}Bi_{34.6}$ cooled at the cooling rates of (1) 10^{-1} Ks^{-1}: solid diamonds, (2) 10^{-3} Ks^{-1}: solid circles, solid triangles, and (3) 10^{-5} Ks^{-1}: open circles, open triangles, open squares. The solid lines are drawn for guiding the eye. Dotted and dashed lines indicate positron lifetimes corresponding to Bi and β-Sn monovacancies. The dash-dot-dashed line indicates the melting temperature. Reprinted with permission from Ref. [28]. Copyright 2009 the Japan Society of Applied Physics.

Figure 5.9 shows the correlation between *S* and *W* parameters (*S*-*W* correlation) taken from coincident Doppler broadening spectra. In the presence of one kind of free volume, the *S* parameter linearly varies with the *W* parameter. The data at 398 and 298 K for 10^{-1} Ks^{-1} are similar to each other. This behavior is typical for quenched alloys [12], suggesting that the microstructure at 398 K is maintained after cooling. The data at 398 and 298 K for 10^{-1} Ks^{-1} are located on the straight line for pure Bi (Bi line), indicating that the chemical surroundings around free volumes in $Sn_{65.4}Bi_{34.6}$ are similar to those around free volumes in Bi. The data of $Sn_{65.4}Bi_{34.6}$ cooled at (2) 10^{-3} Ks^{-1} begin to deviate from the Bi line and approach the straight line for pure Sn (Sn line). The data of $Sn_{65.4}Bi_{34.6}$ cooled at (3) 10^{-5} Ks^{-1} are steadily located on the Sn line below 343 K. The above results reveal that Bi-rich environment of the free volumes varies to Sn-rich environment with decreasing the cooling rate.

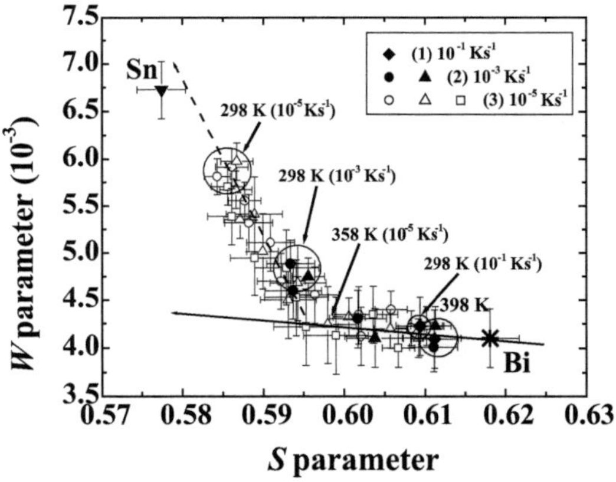

Figure 5.9 The correlation between *S* and *W* parameters taken from Doppler broadening spectra of $Sn_{65.4}Bi_{34.6}$ at various temperatures during cooling at the cooling rates of (1) 10^{-1} Ks^{-1}, (2) 10^{-3} Ks^{-1}, and (3) 10^{-5} Ks^{-1} together with those of pure Sn and pure Bi. The straight solid and dashed lines with the data of pure Bi (Bi line) and pure Sn (Sn line) are drawn for guiding the eye. Reprinted with permission from Ref. [28]. Copyright 2009 the japan society of applied physics.

Figure 5.10 shows the *S* parameter as a function of positron incident energy for $Sn_{65.4}Bi_{34.6}$ cooled at (1) 10^{-1} Ks^{-1}, (2) 10^{-3} Ks^{-1}, and (3) 10^{-5} Ks^{-1}. The *S* parameters rapidly increase with increasing positron incident energy and become saturated above 10 keV. The average positron-diffusion lengths evaluated based on a one-dimensional positron diffusion model [8,53] for the simple surface-bulk structure are 35, 28, and 14 nm, respectively, for $Sn_{65.4}Bi_{34.6}$ cooled at (1) 10^{-1} Ks^{-1}, (2) 10^{-3} Ks^{-1}, and (3) 10^{-5} Ks^{-1}. Positron diffusion lengths obtained here are much shorter than in pure Sn and pure Bi [54–55]. The considerably short positron diffusion lengths indicate that the free volumes are concentrated at interfaces among heterogeneously segregated short-range ordered phases in the $Sn_{65.4}Bi_{34.6}$ alloy. This is also supported by the observation that the peak counting rate of the 511 keV annihilation peak for $Sn_{60}Bi_{40}$ eutectic alloys dramatically decreases upon the disappearance of the interfaces [56]. It should be noted here that the positron diffusion lengths shorten with decreasing the cooling rates. This indicates that the concentration of free volumes at the interfaces increase along with the decrease of cooling rate.

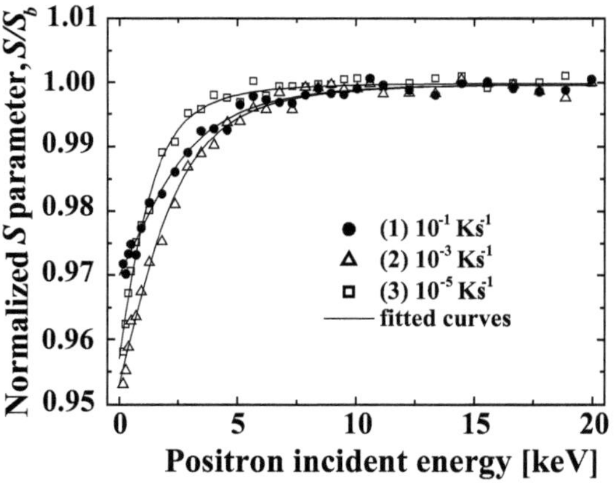

Figure 5.10 Positron diffusion data of $Sn_{65.4}Bi_{34.6}$ cooled at the cooling rates of (1) 10^{-1} Ks^{-1} (solid circles), (2) 10^{-3} Ks^{-1} (open triangles), and (3) 10^{-5} Ks^{-1} (open squares) obtained by a slow positron beam at room temperature. Measured *S* parameters are shown as a function of positron incident energy. *S* parameters are normalized to those for the bulk (S_b). The solid lines are results of fits. Reprinted with permission from Ref. [28]. Copyright 2009 the Japan Society of Applied Physics.

It is accepted that a cooling rate has influence on the microstructure and mechanical properties of Bi-containing eutectic alloys [46–47]. The growth of Bi particles at a lower cooling rate deteriorates the mechanical properties. In fact, indentation creep experiments revealed that Sn-Bi alloys cooled at a high cooling rate of 8 Ks^{-1} have more creep resistance than those slowly cooled at 10^{-2} Ks^{-1} with massive precipitation of Bi particles [57]. In the present study we observed the increase of the particle size upon lowering the cooling rate for $Sn_{65.4}Bi_{34.6}$ by BE imaging and SAXS. On the other hand, the average size of Bi crystallites evaluated from XRD data is one order of magnitude smaller than the particle size obtained by BE imaging. This provides evidence that the segregation of Bi nanocrystallites occurs inside the Bi-rich particles. The increase of the crystallite size with decreasing cooling rate shows that Bi nanocrystallites and Bi-rich particles intergrow with each other. Furthermore, we found that the concentration of free volumes at the interfaces is enhanced with Bi precipitation caused by the decreased cooling rate for a Sn-Bi eutectic alloy. Therefore, besides the Bi precipitation, increased concentration of free volumes at the interfaces may contribute to the deterioration in mechanical properties.

We now discuss the mechanism of macroscopic Bi precipitation based on nanoscopic positron data obtained in the present work. There exists high concentration of free volumes at interfaces in the $Sn_{65.4}Bi_{34.6}$ eutectic system, which is supposed to play an important role in Bi precipitation. At lower cooling rate, Bi atoms sufficiently diffuse through the free volumes at the interfaces leading to the growth of Bi particles. This is evidenced by the change of chemical surroundings around the free volumes from Bi- to Sn-rich environment. In addition to that, the free volume concentration increases at lower cooling rate. The increase of free volume concentration accelerates atomic diffusion, which generates massive Bi precipitation. We therefore conclude that the free volume-intermediated atomic diffusion drives macroscopic Bi precipitation.

5.3.2 Densification Dynamics of Gadolinium-Doped Ceria Upon Sintering

Gadolinium-doped ceria (GDC) is one of the most promising candidates for the electrolytes of solid oxide fuel cell (SOFC) owing to the highest oxygen ionic conductivity at intermediate temperature in the range from 773 K to 973 K [58]. For a practical application of GDC to the electrolyte layer of SOFC, direct reactions between fuels and oxygen gasses have to be avoided [59]. It is thus required for GDC to achieve a dense structure without micro cracks and pores. Densification of GDC has been extensively attempted by the solid-state reaction, so-called sintering technique. GDC is however known as a typical material, which exhibits highly reducible nature of Ce ions. This often causes the release of gaseous phase resulting in incomplete densification with crack formation as well as a faulty matrix [60–61].

In a general picture of sintering, agglomeration of particles occurs and then the particles grow by atomic diffusion at boundaries among the particles and inside the particles [62–65]. It is thus of significance to investigate the variation of local atomic structures at the particle boundaries and inside the particles caused by sintering. In principle, ceramics powders typically employed as starting materials are composed of the particles of small agglomerates separated by the boundaries. Furthermore, the particle is composed of crystallites with high concentration of interfaces, which is the most microscopic structural unit. Current two-particle sintering theory does not address the presence of crystallite interface.

In this subsection, the densification mechanism of GDC upon sintering is comprehensively discussed from both macroscopic and microscopic points of view based on the results of Archimedes method, XRD, high-resolution dilatometry (DLT), and a series of positron annihilation spectroscopy. We focus on the structural evolution of the particle boundaries as well as the crystallite interfaces to highlight the densification dynamics of GDC by introducing time-resolved DLT and positron annihilation spectroscopy.

GDC was prepared using oxalate coprecipitation method [66]. The Ce-Gd nitrate solution mixed at a Gd molar ratio of ~ 0.2 was dropped into stirred oxalate acid solution to produce oxalate precipitates. The precipitates were calcined at 873 K for 1 h under the atmospheric condition to form oxides. XRD studies together with the Gd solubility on the basis of empirical relation by Hong et al. [67] revealed that Ce^{+3} sites are nearly perfectly replaced by Gd^{+3}. The precipitates were compacted into pellets by uniaxial pressing with a pressure of 20 MPa and subjected to the sintering experiments in the temperature range from 873 K to 1273 K in the air.

Figure 5.11 shows temperature-dependent length change (solid circles), relative density (open circles), mean positron lifetime (solid squares), and crystallite size (open squares) for GDC samples sintered in the temperature range from 873 K to 1273 K. The crystallite sizes of GDC samples were evaluated from (111) XRD peaks following Scherrer's formula. The bulk densities of sintered samples were measured by Archimedes method with ion-exchanged water. Both the relative density and length change increase with increasing temperature indicating macroscopic densification for GDC upon sintering (see Fig. 5.11a). The crystallite size increases together with temperature, whereas the mean positron lifetime decreases (see Fig. 5.11b). In Fig. 5.11b, the positron lifetime obtained for Yttrium-Stabilized Zirconia (YSZ) single crystal of similar oxide ceramic compound is also added with a dotted line for comparison, in accordance with that reported earlier [68]. The positron lifetimes for GDC obtained here are consistently higher than that of YSZ single crystal in the whole temperature region. This indicates the presence of vacancy-like nano-defects in the GDC samples, which will be discussed more in detail later. The results of XRD together with positron lifetime spectroscopy reveal microscopic densification, in which the GDC crystallites grow and the vacancy-like nano-defects simultaneously disappear.

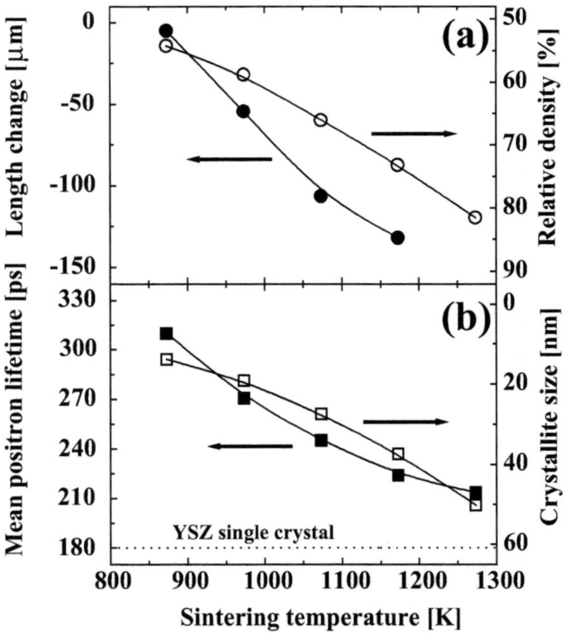

Figure 5.11 (a) Temperature-dependent length-change (solid circles) and relative density (open circles). (b) Temperature-dependent data of positron lifetime (solid squares) and crystallite sizes (open squares). Positron lifetime of YSZ single crystal is indicated by dashed line for comparison. The GDC samples were sintered at each sintering temperature for 6 h. Reprinted with permission from Ref. [29]. Copyright 2012 the Japan Society of Applied Physics.

Figure 5.12 shows time-resolved length changes for GDC samples measured at 873 K (open circles), 973 K (open triangles), 1073 K (open squares), and 1173 K (open diamonds) after up-quenching at a heating rate of 20 K/min. The data indicates drastic shrinkage up to 1 h and gradual change thereafter, as indicated by thick arrows (see Fig. 5.12a). The decreasing tendency with sintering time becomes more prominent with sintering temperature up to 1173 K, which indicates an acceleration of densification upon sintering. It is generally accepted that the rearrangement of particles occurs for ceramics powders when the relative density is lower than 65 % [69]. We thus expect the occurrence of particle rearrangement yielding drastic shrinkage around 1 h in the present GDC samples, where the relative density is ~ 55 % (see Fig. 5.11a).

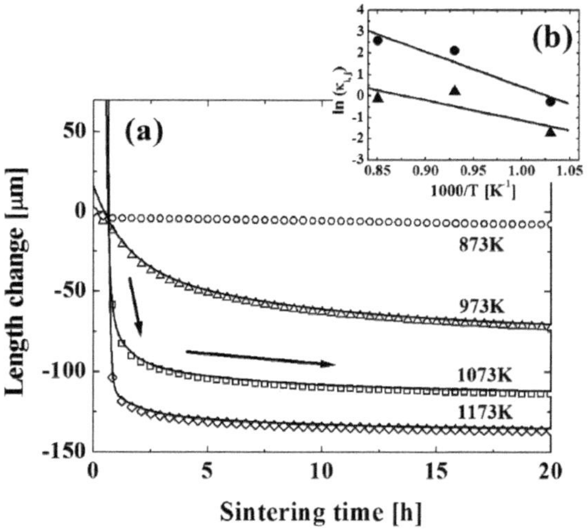

Figure 5.12 (a) Time-dependent length-change data of GDC samples sintered at 873 K (open circlees), 973 K (open triangles), 1073 K (open squares), and 1173 K (open diamonds). Solid lines denote the results of the fits by three superimposed exponentials. (b) Logarithms of first (solid circles) and second (solid triangles) decay constants as a function of inverse temperature. Reprinted with permission from Ref. [29]. Copyright 2012 the Japan Society of Applied Physics.

Figure 5.13 shows time-dependent mean positron lifetimes for GDC samples at 773 K, 873 K, 973 K, 1073 K, 1173 K, and 1273 K. In contrast to the length-change data, the positron lifetimes gradually decrease along with sintering time (see Fig. 5.13a) signifying the sensitivity of positrons on densification after the particle rearrangement. Similarly to the time-dependent DLT data, the decreasing tendency with sintering time becomes prominent due to acceleration of densification upon sintering when the temperature is raised up to 1273 K. The average positron diffusion length in YSZ single crystal is reported as ~ 60 nm [70]. As revealed by XRD experiments (see Fig. 5.11b), the GDC crystallite grows up to ~ 50 nm at the highest sintering temperature of 1273 K, which is significantly smaller than the positron diffusion length in YSZ single crystal. Positrons implanted in the GDC crystallites is thus expected to sufficiently diffuse out to reach the crystallite interfaces among the GDC crystallites in the whole temperature region, indicating that the nano-defects are located at the crystallite interfaces in the present GDC samples. We reasonably infer that the nano-defects are formed at the interfaces inside the particles as a consequence of misfit inclusion or triple point junction among the crystallites.

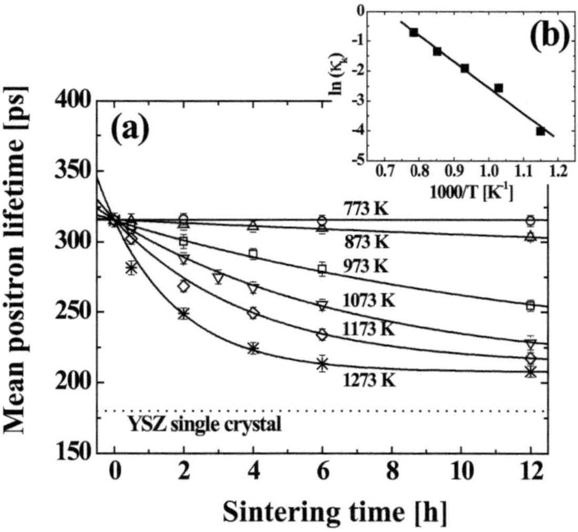

Figure 5.13 (a) Time-dependent positron lifetime data of GDC sintered at 773 K (open circles), 873 K (open triangles), 973 K (open squares), 1073 K (open inverse triangles), 1173 K (open diamonds), and 1273 K (stars). Solid lines denote the results of the fits by an exponential. Positron lifetime of YSZ single crystal is indicated by dotted line for comparison. (b) Logarithms of decay constants as a function of inverse temperature. Reprinted with permission from Ref. [29]. Copyright 2012 the Japan Society of Applied Physics.

Figure 5.14 shows the correlation between *S* and *W* parameters (*S*-*W* correlation) taken from coincident Doppler broadening spectra. The data in the correlation are located on the straight line of Gd_2O_3 (Gd_2O_3 line) in the temperature range from 773 K to 973 K, indicating that surrounding atoms of the vacancy-like nano-defects are similar to those of Gd_2O_3. The data deviate from the Gd_2O_3 line and are in turn located on the straight line of CeO_2 (CeO_2 line) above 1073 K, which indicates the chemical surroundings are getting toward to those of CeO_2. The above element-specific results demonstrate that Gd atoms are transferred from the crystallite interfaces to the GDC matrix. Similar observations on atomic diffusion between the interface to the matrix were reported for Fe-B-Si nanocrystalline [15,71] and Sn-Bi eutectic [27–28] materials.

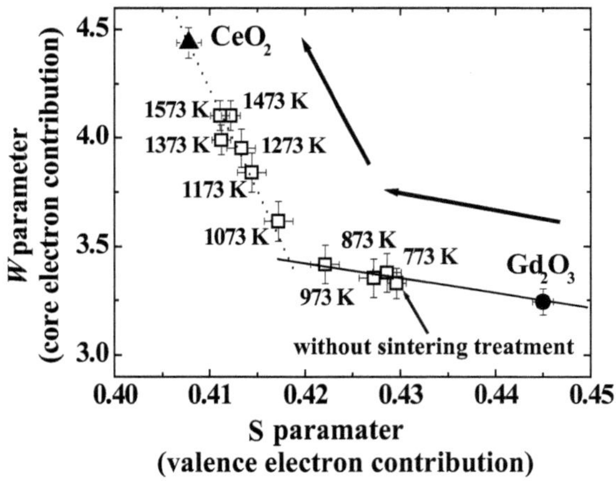

Figure 5.14 The correlation between *S* and *W* parameters taken from coincidence Doppler broadening spectra of the GDC samples without sintering treatment and sintered at the temperatures of 773, 873, 973, 1073, 1173, 1273, 1373, 1473, and 1573 K (open squares). The data for Gd_2O_3 and CeO_2 are presented for comparison as solid circle and triangle, respectively. The solid and dotted lines with data of Gd_2O_3 and CeO_2 are drawn for guiding the eye, respectively. Reprinted with permission from Ref. [29]. Copyright 2012 The Japan Society of Applied Physics.

It is of high interest that Gd atoms are transferable during densification upon sintering for the calcined GDC samples, in sharp contrast with earlier results for YSZ by CDB spectroscopy [68,72]. As mentioned above, XRD suggests that Ce^{+3} sites are nearly perfectly replaced by Gd^{+3} in the GDC samples. Presumably, Gd_2O_3 is still residual at the crystallite interfaces with below the detection limit of XRD. Such a small amount of Gd_2O_3 heterogeneously distributed at the interfaces cannot be observed by transmission electron microscopy (TEM) as well. We suppose that difficulties in densification upon sintering often observed for GDC is relevant to the microscopic precipitation of Gd_2O_3 at the interfaces.

It should be noted here that positrons are dominantly annihilated at the vacancy-like nano-defects at the crystallite interfaces, providing the microscopic information on densification inside the particles. On the other hand, macroscopic shrinkage observed by DLT corresponds to overall densification (see Fig. 5.12a). Below, we discuss densification at the particle boundaries relevant to a sintering neck and in the particles. Good fits of isotherms are obtained for the DLT data at 973 K, 1073 K, and 1173 K by three superimposed exponentials with decay constants κ_i, κ_j, and κ_k, as indicated with solid lines in Fig. 5.12a. The logarithms of 1st and 2nd decay constants κi and κ_j plotted as a function of inverse temperature follow the Arrhenius law indicating the thermally activated reaction (Fig. 5.12b). From temperature variation, the activation energies of overall densification E_{t1} and E_{t2} are derived as 138 and 79 kJmol^{-1}. The activation energy of overall densification largely decreases together with sintering time. This signifies that different mechanisms are involved in densification at different time scales. In contrast to the DLT data, a single exponential with a decay constant κ_l reproduces the time-dependent positron lifetimes, as indicated with solid lines in Fig. 5.13a. The logarithms of decay constants obtained from exponential fits are plotted as a function of inverse temperature in Fig. 5.13b. From temperature variation, the activation energy of densification inside the particle grain is derived as 73 kJmol^{-1}.

Here, we model time-dependent overall densification that is contributed from those inside the particle and at the particle boundary. The activation energies of overall densification E_t is then described as $E_t = E_p + E_b$, where E_p and E_b are the activation energies of densification inside the particle and at the particle boundary, respectively. Applying above the model to the Arrhenius plots in Figs. 5.12b and 5.13b, the activation energies E_t, E_i, and E_b are determined as listed in Table 5.2. At the initial stage of sintering, similar activation energies are obtained for densification inside the particles and at the particle boundary. The high activation energy E_b would be reasonable because the particle rearrangement effectively occurs at this stage. Along with sintering, the particle boundary becomes less contributable to densification than at the initial stage while the contribution of particle stays constant. The particle thus gets to be responsible for densification above the sintering time of 1 h.

Table 5.2 Activation energies of overall densification E_t, inside the particle grain E_p, and at the grain boundary E_b

	E_p [kJ mol^{-1}]	E_b [kJ mol^{-1}]	E_t [kJ mol^{-1}]
Up to 1 h	73	65	138
Above 1 h	73	6	79

Based on the data of XRD, DLT, and element-specific positron annihilation spectroscopy, the following densification dynamics can be drawn for GDC. At the initial stage of sintering, the temperature-induced particle rearrangement effectively occurs to agglomerate due to the loose particle packing, which triggers off densification upon sintering. The boundary among particles relevant to the sintering neck is thus responsible for densification. This initial sintering stage completes within 1 h and is then shifted to the next sintering stage with different mechanism, in which the interior of the particle rather plays an important role in densification. The densification of particle interior is caused by the growth of the crystallites followed by the transfer of Gd atoms from the interfaces to the crystallites. The present results furnish direct evidence that the crystallite interfaces as well as the particle boundaries contribute to densification of ceramics powder upon sintering.

5.3.3 Formation Mechanism of Biogenic Silica Quartz in the Steady-State Environment

Biogenic silica has attracted much attention due to its pivotal importance for advancing earth sciences as well as the strong impact on ecological systems. As is well known, silica is one of the most common minerals in earth's crust, thus preserving an environmental history on its formation processes. Amorphous silica transforms to crystalline quartz through biochemical and geophysical processes on geological time scales in the steady-state environment as, e.g., low temperature and low pressure. This widespread phenomenon has stimulated a long history of diagenetic investigation for sequential stratigraphy and marine sediments [73–76]. In addition, biogenic silica plays a crucial role in the global cycling of carbon, influencing on atmospheric carbon dioxide concentrations [77–78]. A wealth of phytoplankton productivity of biogenic silica increases bacterial metabolism and diversity offering potential reservoirs in deep subsurface environment as ecosystems [79].

Biogenic silica ubiquitously present in nature originates from a wide variety of living organisms as, e.g., diatom, radiolarian, silicoflagellate, siliceous sponge, and plant [80–82]. Diatom, unicellular algae found in marine environments, may be largely responsible for the global scale for biogenic silica formation. Rice husk is an abundant agricultural waste, in which biogenic silica is easily produced. Upon the production of biogenic silica, organisms take up silicon to form well-defined silica structures in their skeletons, which is of particular importance in understanding global cycling of silicon [83]. In spite of an interest in the formation mechanism dependent on the organism, detailed studies have not been performed due to the difficulty in finding the characteristic signature of organism-dependent origin for structurally complex silica composed of contaminated light elements. In this subsection, the steady-state formation of biogenic quartz arising from diatom algae and rice husks is discussed focusing on open nanospaces. The mechanism of organism-dependent formation is highlighted along with the data of quartz formation on diatom algae and rice husk.

The cell wall of diatom algae and rice husk ash are known to contain silica. We thus employed amorphous silica originated from algae frustules collected from Mariana Trench and rice husk ash originated from Niigata Koshihikari, as starting materials. Hydrothermal reaction on a 70-day time scale was performed in the temperature range from 383 K to 443 K in a cylindrical Teflon vessel.

Figure 5.15 shows XRD patterns observed for (a) rice husk- and (b) diatom-originated materials reacted at 423 K for 0 h (0 week), 336 h (2 weeks), 672 h (4 weeks), 1176 h (7 weeks), and 1680 h (10 weeks). Structural evolution toward crystalline quartz with the reaction time can be seen in the variation for both the materials. The diffraction pattern exhibits a typical amorphous structure for the rice husk-originated material at 0 h. XRD peaks arising from the quartz phase begin to appear together with those from the cristobalite phase at 336 h. The peaks of cristobalite phase disappear at 672 h and the quartz phase is dominantly formed thereafter. The same sequential reactions take place for the diatom-originated materials, but starts at 1176 h largely later than for the rice husk-originated materials. This demonstrates that biogenic quartz formation for rice husk-originated materials occurs more easily than for the diatom-originated materials.

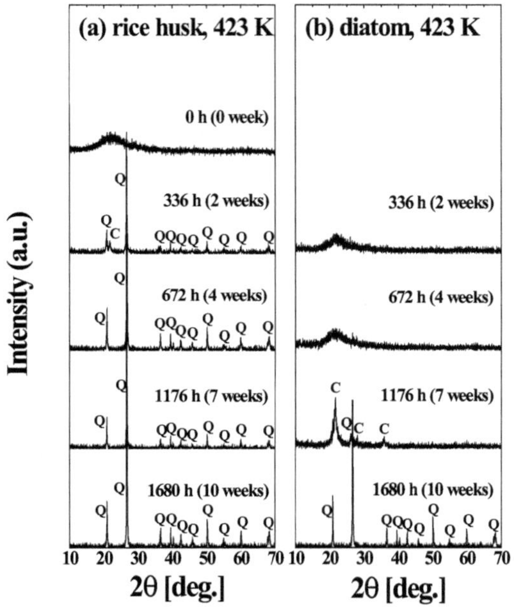

Figure 5.15 (a) X-ray diffraction (XRD) patterns observed for (a) rice husk- and (b) diatom-originated materials hydrothermally reacted at 423 K for 0 h (0 week), 336 h (2 weeks), 672 h (4 weeks), 1176 h (7 weeks), and 1680 h (10 weeks). XRD peaks indexed to cristobalite and quartz are marked with C and Q, respectively. Reprinted with permission from Ref. [31]. Copyright 2011 American Chemical Society.

Positron-positronium lifetime spectroscopy for the diatom- and rice husk-originated amorphous silica yields three components of lifetimes, of which the longest-lived component is attributable to the pick-off annihilation of *o*-Ps localized in the angstrom-scale pores of the amorphous matrix. The *o*-Ps lifetimes obtained for the diatom- and rice husk-originated amorphous silica are 1.6 ns and 3.1 ns corresponding to the pore sizes of 2.4 Å and 3.7 Å in radius, respectively. Note that the pore size for the rice husk-originated amorphous silica is slightly larger by 1.3 Å than for the diatom-originated amorphous silica. No systematic changes are observed in the lifetime τ_3 for both the materials when the quartz formation proceeds at longer reaction time. In light of the fact that no long lifetime component τ_3 is observed in silica polymorphs with high crystallinity as, e.g., crystalline quartz [84], angstrom-scale pores are located not in the quartz phase but in the intergranular amorphous regions.

Figure 5.16 shows reaction-time variations of pore intensity at 383 K, 403 K, 423 K, and 443 K for (a) rice husk- and (b) diatom-

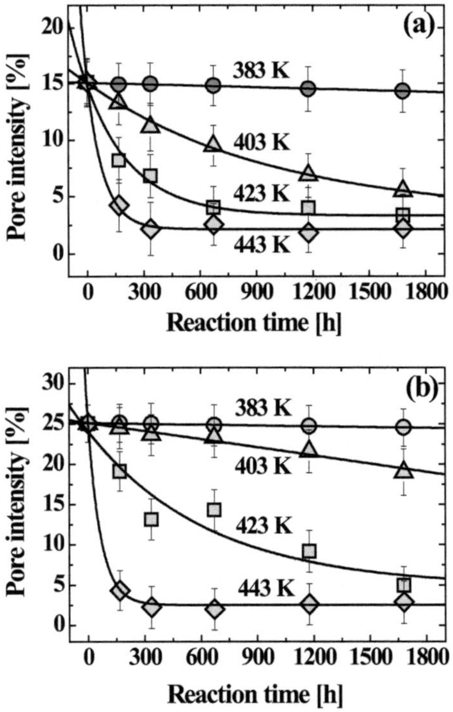

Figure 5.16 Reaction-time variations of pore intensity at 383 K (open circles), 403 K (open triangles), 423 K (open squares), and 443 K (open diamonds) for (a) rice husk- and (b) diatom-originated materials. Solid lines denote the results of the fits by an exponential with a decay constant k_f.

originated materials. The pore intensities decrease with increasing the reaction time up to 1680 h (10 weeks) for both materials. The decreasing tendency with the reaction time becomes increasingly prominent when the temperature is raised up to 443 K. The quartz formation in the amorphous matrix creates the intergranular amorphous regions and its growth reduces the volume fraction of intergranular amorphous regions. It is thus reasonable that angstrom-scale pores in the intergranular amorphous regions decrease together with the quartz formation.

The quartz formation can be now visible with respect to the structural modification of silica itself apart from aqueous solution. Good fits of isotherms are obtained by an exponential with a decay constant k_f, as indicated with solid lines in Fig. 5.16. In Fig. 5.17a, the

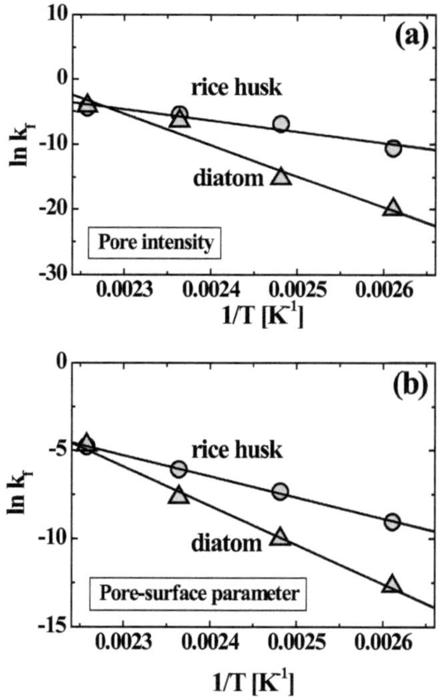

Figure 5.17 (b) Logarithms of decay constants obtained from (a) reaction-time variations of pore intensity at different temperatures in Fig. 5.16 and (b) reaction-time variations of pore-surface parameter at different temperatures in Fig. 5.17 as a function of inverse temperature for rice husk- (open circles) and diatom-originated (open triangles) materials.

logarithms of decay constants are plotted as a function of inverse temperature. They follow the Arrhenius law demonstrating that the quartz formation is governed by the thermally activated reaction. The data of diatom-originated materials indicate a steeper gradient than for the rice husk-originated materials. From the temperature variations, the activation energies of quartz formation are derived to be 1.5 eV and 4.2 eV for the rice husk- and diatom-originated materials, respectively, in consistent with the XRD data of Fig. 5.15.

Figure 5.18 shows the results of AMOC spectroscopy for diatom-originated materials, in which reaction-time variations of *S*-*t* plot at 423 K are presented. It is well seen that the momentum *S* parameter at older positron age is significantly influenced by elemental environment around free volumes being sensitive to the surface state of open nanospace wall. We thus call the momentum *S* parameter pore-surface parameter. Pore-surface parameters at 383 K, 403 K, 423 K, and 443 K for (a) rice husk- and (b) diatom-originated materials are shown in Fig. 5.19 as a function of reaction time. The decrease of pore-surface parameters with increasing reaction time are a consequence resultant from increased oxygen atoms incorporated with pore surfaces. The thermally activated

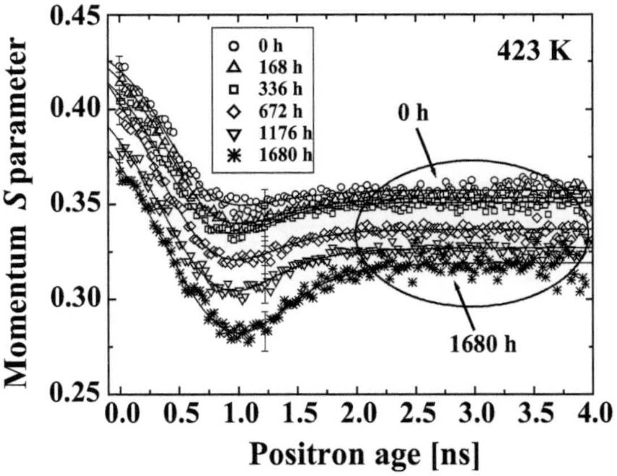

Figure 5.18 Results of AMOC spectroscopy for diatom-originated materials obtained at different reaction times at 423 K. Momentum *S* parameter at the reaction times of 0 h (open circles), 168 h (open triangles), 336 h (open squares), 672 h (open diamonds), 1176 h (open inverse triangles), and 1680 h (stars) are plotted as a function of positron age.

reactions of water molecules at pore surfaces are well seen. The logarithms of the decay constants obtained from an exponential fit are plotted as a function of inverse temperature together with those of diatom-originated materials in Fig. 5.17b. Similarly to the case of quartz formation in Fig. 5.17a, Arrhenius plots for the diatom-originated materials indicate a steeper gradient than for the rice husk-originated materials, in which the activation energies of water reactions at pore surfaces are derived to be 1.0 eV and 1.9 eV for the rice husk- and diatom-originated materials, respectively.

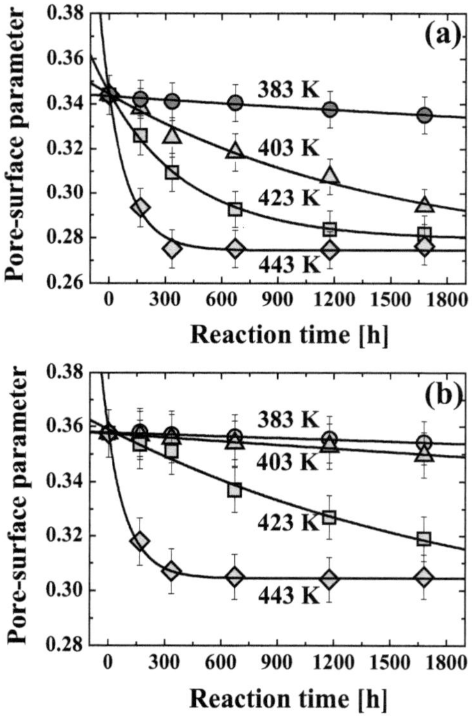

Figure 5.19 Reaction-time variations of pore-surface parameter at 383 K (open circles), 403 K (open triangles), 423 K (open squares), and 443 K (open diamonds) for (a) rice husk- and (b) diatom-originated materials. Solid lines denote the results of the fits by an exponential with a decay constant k_r.

Following our recent model that explains the steady-state formation of biogenic quartz by the mechanistic theory of nucleation taking account of diffusion and reactions of water molecules in angstrom-scale pores [30–31], the temperature-dependent quartz formation rate J is described as

$$J(T) = A\exp\left(-\frac{E_f}{kT}\right), \tag{5.7}$$

where A, T, E_f, and k are a pre-exponential factor, temperature, the activation energy of quartz formation, and the Boltzmann constant, respectively. Being diffusion of water molecules through angstrom-scale pores and reactions of water molecules at pore surfaces involved in the quartz formation [30,85–87], Equation (5.7) is rewritten as

$$J(T) = A\exp\left(-\frac{E_d}{kT}\right)\exp\left(-\frac{E_r}{kT}\right), \tag{5.8}$$

where E_d and E_r are the activation energies of water diffusion through angstrom-scale pores and water reaction at pore surfaces, respectively. Table 5.3 lists E_f, E_d, and E_r values determined by applying eqs (2) and (3) to the Arrhenius plots in Figs. 5.17 and 5.19.

Table 5.3 Activation energies of quartz formation E_f, water diffusion through angstrom-scale pores E_d, and water reaction at pore surfaces E_r

Organism	r [Å]	E_f [eV]	E_d [eV]	E_r [eV]
Diatom	2.4	4.2	2.3	1.9
Rice husk	3.7	1.5	0.5	1.0

Note: The average pore sizes in radius r are presented as well.

The activation energy of water diffusion E_d for the rice husk-originated materials is much lower than for the diatom-originated materials. As mentioned above, the pore size for the rice husk-originated materials is larger by 1.3 Å than for diatom-originated materials. The results thus demonstrate that water molecules more easily diffuse through larger pores in the rice husk-originated materials. According to first principle calculations on amorphous silica [88], the activation energy of water diffusion through angstrom-scale pores is 2.2 eV at the pore radius of 2 Å. It decreases down to 0.8 eV when the pore size is raised up to 3.5 Å. The activation energies for both the materials evaluated in the present study well coincide with the above calculation, thus ascertaining the validity of our model [30–31].

The activation energy of water reaction E_r for the rice husk-originated materials is almost half of that of diatom-originated materials. This is in contrast with our speculation that the value of E_r for the rice husk-originated materials with larger pores is higher than for the diatom-originated materials owing to less interaction of water molecules with the silica network. First principle calculation predicts that the concentration of water molecules becomes high in the pores larger than 6 Å and water molecules significantly cluster together [89]. Water molecules thus react with each other forming OH^{-1} and H_3O^+ at considerably low energy in the vicinity of pore surfaces, which exactly coincides with our case. We most probably expect that the low activation energy E_r obtained for the rice husk-originated materials is the consequence of mutual interaction among water molecules in the 3.7 Å pores.

Here, we describe the reason why the biogenic-quartz formation for the rice husk-originated materials occurs more easily than for the diatom-originated materials. The rice husk- and diatom-originated amorphous silica intrinsically possesses angstrom-scale pores with the sizes of 3.7 Å and 2.4 Å in amorphous matrices, respectively. Water molecules diffuse through the larger pores of rice husk-originated materials with the low activation energy of 0.5 eV. Water molecules easily reach pore surfaces to react thereon. In the larger pores of rice husk-originated materials, the concentration of water molecules becomes high and clusters of water molecules are significantly formed. The reaction among water molecules at pore surfaces is thus rather significant than that between water molecules and the silica network resulting in the lower activation energy of 1.0 eV than expected. This triggers off the nucleation of biogenic quartz involving the totally lower energy cost of 1.5 eV than for the diatom-originated materials. The present findings evidence that angstrom-scale pores intrinsically residing in the amorphous matrix is the organism-dependent origin of steady-state biogenic quartz formation.

5.3.4 Elemental Migration in Geological Environment

Following the nuclear disaster at the Fukushima reactors caused by the recent earthquake and tsunami, migration of radioactive materials released to geological environment has come to the forefront of worldwide concern [90–92]. In geological environment, nuclides dissolved in underground water [93–94] are transported by the combined processes of advection in the rock fractures [95] and molecular diffusion in the rock matrix [96]. Nuclide migration by advection through continuously fractured rock aquifers proceeds dynamically. On the contrary, molecular diffusion is appreciated as a static process due to solute transport with virtually stagnant water within the rock matrix. This process, called as matrix diffusion, has been expected to cause the retardation of nuclides being acknowledged in the context of geological disposal of radioactive waste [97–98].

The matrix diffusion has been extensively investigated by means of tracer diffusion experiment for different types of rocks at a number of underground test sites as e.g., Mirror Lake, USA [99], Äspö, Sweden [100–101], and Yucca Mountain, USA [102–103]. The results have shown that matrix diffusion coefficients observed on a field scale are 30 to 50 times higher than of laboratory-scale experiments [99–104]. There have thus been made microscopic speculations on rock matrix as e.g., advective region [99], fast transport path [102], and many small-scale fracture [104].

In this subsection, migration of underground saline water at Horonobe, Hokkaido, Japan is studied on a field scale up to 1 km. Positron-positronium annihilation spectroscopy is applied to probe angstrom-scale pores that has never observed by conventional porosimetric techniques as e.g., mercury intrusion porosimetry, gas absorption porosimetry, and high resolution transmission electron microscopy (TEM). We focus on the matrix diffusion of saline water based on the slight variation of angstrom-scale pores influenced by saline water diffusion.

Horonobe area is located on the eastern margin of late Pliocene sedimentary basin of northern Hokkaido, Japan. The geology comprises siliceous marine sediments [105]. The mineralogical composition of sedimentary rocks examined by x-ray diffraction (XRD) shows mainly quartz, opal, and plagioclase. In this area, the uplift of sea floor together with marine sediments occurred following maximum burial ~ 1.5 million years ago [106–107]. Infiltration of rain water into marine sediments decreases the heavy saline water beneath, by which the boundary between fresh and saline water is formed following Ghyben-Herzberg relation [108]. In practice, the boundary exhibits a transition zone with a certain thickness arising from the geological-time scale of molecular diffusion, which is driven by a concentration gradient between fresh and saline water. The present samples were taken from a borehole (HDB-11) at following drilling depths, 55 m, 110 m, 273 m, 385 m, 440 m, 625 m, 700 m, 809 m, 875 m, and 951 m for estimation of matrix diffusivity. Diatomaceous mud of the Koetoi formation is distributed above 460 m drilling depth overlying hard shale of the Wakkanai formation in HDB-11 borehole [106–107]. The water squeezed from the samples indicates the transition between fresh and saline water at 273 m in depth. Water seepage was eliminated from the samples by baking at 150 deg. for 24 h.

Figure 5.20 shows size distributions of angstrom-scale pores for the sample taken at 273 m in depth evaluated by mercury intrusion porosimetry, N_2-gas absorption porosimetry, small-angle x-ray scattering, and *o*-Ps lifetime spectroscopy. A variety of size distribution ranging from subnanometer to micrometer is seen indicating complex pore structures of natural rock samples. In contrast with previous reports where the pores with the lower limit of several 100 nm were detected [109], *o*-Ps lifetime spectroscopy sensitively probes angstrom-scale pores. As mentioned above, the present rock samples are mainly composed of quartz, opal, and plagioclase. We thus conducted *o*-Ps lifetime spectroscopy for each mineral taken at different fields to identify the origin of the angstrom-scale pores in the rock samples. No significant angstrom-scale pores were observed for the minerals regardless of the fields, in agreement with our earlier studies [84]. It is therefore concluded that the angstrom-scale pores detected in the rock samples are located at interfaces among minerals. This observation is similar to the results of *o*-Ps lifetime spectroscopy for quartz and chalcedony, in which angstrom-scale pores are not found in quartz but in chalcedony possessing interfaces among quartz particles [84].

Figure 5.21 shows the average sizes of angstrom-scale pores R and their fractions f as a function of depth. Both the size and fraction exhibit the highest and lowest peaks at the depth of 273 m, which coincides with the boundary between fresh and saline water. In the

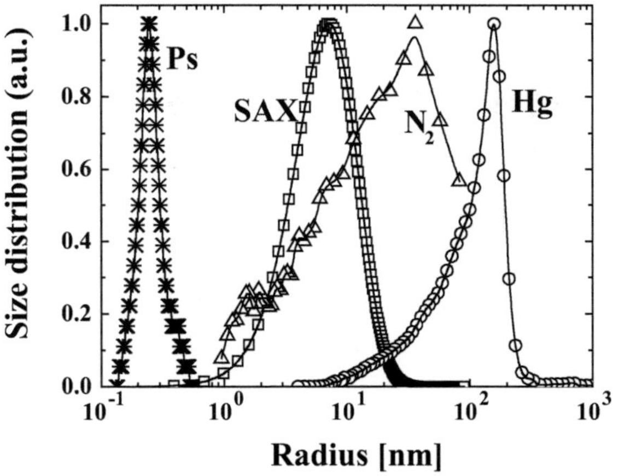

Figure 5.20 Pore size distributions evaluated for the sample taken at 273 m in depth by mercury intrusion porosimetry (Hg, open circles), N_2-gas absorption porosimetry (N_2, open triangles), small-angle x-ray scattering (SAX, open squares), and ortho-positronium lifetime spectroscopy (Ps, stars). Reprinted with permission from Ref. [32]. Copyright 2013 The Physical Society of Japan.

fresh water region up to 273 m, the pore size decreases with increasing depth whereas the pore fraction increases. This is typical for the fractionation of pores due to pressurization in deep underground environment, known as compaction effect [110]. It is of interest that the angstrom-scale pores located at interfaces among minerals are significantly influenced by the compaction effect in the fresh water region. In contrast, both the size and fraction indicate opposite depth dependence in the saline water region that cannot be explained by the compaction effect.

The depth dependence of angstrom-scale pores is more informatively seen in the size distributions (see Fig. 5.22). The microscopic pores gradually appear with increasing the depth in accordance with the compaction effect in the fresh water region up to 273 m. They in turn decrease in the saline water region up to 624 m, thus resulting in the increase of effective pore sizes and the decrease of fraction in the saline water region (see Fig. 5.21). We now recognize that angstrom-scale pores at interfaces among minerals are occupied by certain molecules in the saline water region.

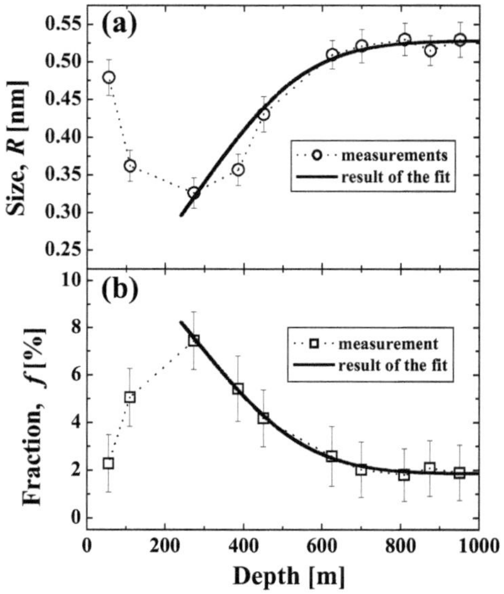

Figure 5.21 (a) Average sizes of angstrom-scale pores, R (open circles) and (b) their fractions, f (open squares) as a function of depth. Solid lines are the results of the fit with Fick's second law. Reprinted with permission from Ref. [32]. Copyright 2013 The Physical Society of Japan.

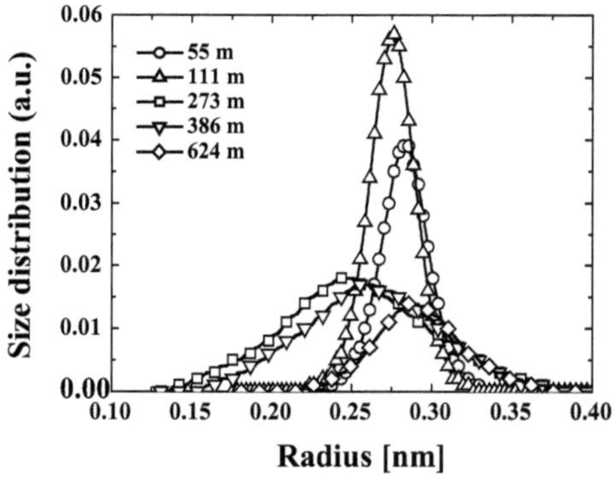

Figure 5.22 Size distributions of angstrom-scale pores for samples taken at 55 m (open circles), 111 m (open triangles), 273 m (open squares), 386 m (open inverse triangles), and 624 m (open diamonds). Reprinted with permission from Ref. [32]. Copyright 2013 The Physical Society of Japan.

Figure 5.23 shows the result of the analysis of pore interior by CDB momentum spectroscopy, which addresses the issue of which kinds of molecules occupy angstrom-scale pores in the saline water region. Here, all the spectra are normalized to that of NaCl crystal to highlight the slight difference of spectra influenced by saline water diffusion. It is clearly seen in the spectra that the element-specific high momentum parts are gradually shifted toward to that of NaCl crystal with increasing the depth. This demonstrates that more population of NaCl molecules resides in the angstrom-scale pores at interfaces among minerals in deeper region.

As mentioned above, the fresh-saline water boundary exhibits a transition zone arising from water mixture due to long-term molecular diffusion between fresh and saline water. The transition is clearly visible in the successive depth-variation of NaCl molecules residual in the angstrom-scale pores. It is of great surprise that the diffusion process of saline water through angstrom-scale pores is visible over ~ 1 km scale. This evidences the presence of widely interconnected angstrom-scale pores in the rock matrix.

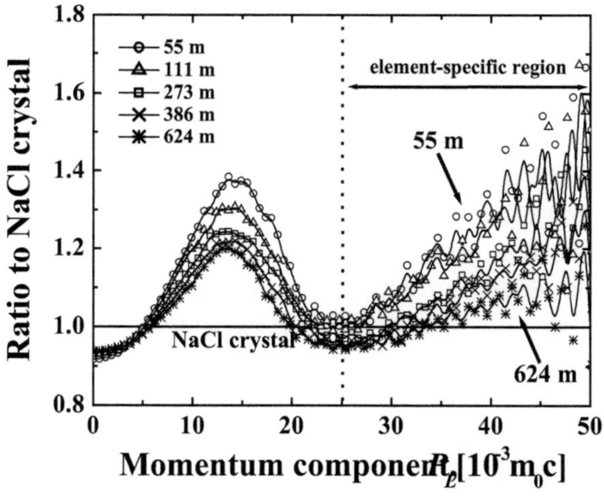

Figure 5.23 Coincident Doppler broadening (CDB) spectra for samples taken at 55 m (open circles), 111 m (open triangles), 273 m (open squares), 386 m (crosses), and 624 m (stars). Each spectrum is normalized to that of NaCl crystal (Y=1). Reprinted with permission from Ref. [32]. Copyright 2013 The Physical Society of Japan.

It is now worthwhile evaluating the effective matrix diffusion coefficient of saline water applying the Fick's second law:

$$\frac{\partial C(x,t)}{\partial t} = D\frac{\partial^2 C(x,t)}{\partial^2 x}, \tag{5.9}$$

where x, t, c, and D are depth, time, concentration of water molecules, and effective diffusion coefficient, respectively. We obtain the analytical solution:

$$C(x,t) = C_0\left[1 - erf\left(\frac{x}{2\sqrt{Dt}}\right)\right], \tag{5.10}$$

under the boundary and initial conditions:

$$t > 0, \quad C(0, t) = C_0 = r_{273\text{m}}(f_{273\text{m}}) \approx const., \tag{5.11}$$

$$t = 0, \quad C(x, 0) = 0, 0 < x < \infty. \tag{5.12}$$

Here, t of 1.5 million years is employed according to the inception of sea floor uplift. The solid lines in the saline water region of Figs. 5.21a, b represent the results of calculation. It can be seen that the depth variation of angstrom-scale pores at interfaces among minerals well follows the Fick's 2nd law, evidencing that the migration of saline water through angstrom-scale pores is driven by molecular diffusion. Fits with Equation (5.10) to the data in Figs. 5.21a, b yield the exactly the same effective matrix diffusion coefficient of 4×10^{-6} cm^2s^{-1}, verifying the detection of angstrom-scale pore with complete accuracy in the present work.

The effective matrix diffusion coefficient evaluated here is in agreement with those obtained from field-scale tracer data at other test sites as for example, Mirror Lake, USA [99], Äspö, Sweden [100,101], and Yucca Mountain, USA [102,103]. It is reasonably inferred that field-scale matrix diffusion is enhanced by the interconnected angstrom-scale pores clarified in this work. We suppose that there also exist the interconnected angstrom-scale pores enhancing matrix diffusion at other test sites. The microscopic structures as for example, "advective region", "many small-scale fracture", and "fast transport path" speculated so far presumably correspond to the angstrom-scale pores as identified in the present study. Such microscopic pores could not be observed by conventional techniques as for example, mercury intrusion porosimetry, gas absorption porosimetry, and even high-resolution TEM.

The present matrix diffusion coefficient yields the migration distance more than 100 m within a million. In light of long halflives of nuclear fission products as for example, ^{135}Cs (2.3 million years) and ^{129}I (15.7 million years), above migration distance is long enough to reach the depth supposed for the reference case of underground disposal of radioactive wastes. The model studies within a framework of laboratory-scale experiments may significantly underestimate the extent of elemental migration by matrix diffusion. The present results demonstrate that matrix diffusion rather plays an important role in transforming element species in geological environment.

5.4 Conclusion

Recent advances in the studies of energy and environment-related nanomaterials by using positron-positronum annihilation spectroscopy have been reviewed. Positron-positronium annihilation spectroscopy is found to be a powerful tool for characterizing open nanospaces in materials. It should be emphasized that the open nanospace plays an important role in the molecular design of energy and environment-related nanomaterials as well as environmental issues, as for example, elemental migration through geological environment.

References

1. K. Sato, F. Baier, W. Sprengel, R. Würschum, H.-E. Schaefer, *Phys. Rev. Lett.* 92, 127403 (2004).
2. Ho Bum Park, Chul Ho Jung, Young Moo Lee, Anita J. Hill, Steven J. Pas, Stephen T. Mudie, Elizabeth Van Wagner, Benny D. Freeman, and David J. Cookson, Science 318, 254 (2007).
3. N. A. Stolwijk and Sh. Obeidi, *Phys. Rev. Lett.* 93, 125901 (2004).
4. A. Maitra and A Heuer, *Phys. Rev. Lett.* 98, 227802 (2007).
5. O. Borodin, G.D. Smith, and R. Douglas, *J. Phys. Chem. B* 107, 6824 (2003).
6. A. B. Kersting, D. W. Efurd, D. L. Finnegan, D. J. Rokop, D. K. Smith, J. L. Thompson, *Nature* 397, 56 (1999).
7. for example, Positron Solid State Physics, edited by W. Brandt and A. Dupasquier (North-Holland, Amsterdam, 1983).

8. K. Sato, Y. Takahashi, H. Uchiyama, I. Kanazawa, R. Tamura, K. Kimura, F. Komori, R. Suzuki, T. Ohdaira, and S. Takeuchi, *Phys. Rev. B* 59, 6712 (1999).
9. K. Sato, H. Uchiyama, Y. Takahashi, I. Kanazawa, R. Suzuki, T. Ohdaira, T. Takeuchi, T. Mizuno, and U. Mizutani, *Phys. Rev. B* 64, 024202 (2001).
10. K. Sato, F. Baier, W. Sprengel, and H.-E. Schaefer, *Phys. Rev. B* 66, 092201 (2002).
11. K. Sato, H. Uchiyama, K. Arinuma, I. Kanazawa, R. Tamura, T. Shibuya, and S. Takeuchi, *Phys. Rev. B* 66, 052201 (2002)
12. K. Sato, F. Baier, A.A. Rempel, W. Sprengel, and H.-E. Schaefer, *Phys. Rev. B* 68, 214203 (2003).
13. K. Sato, Y. Kobayashi, I. Kanazawa, R. Tamura, and S. Takeuchi, *Phys. Rev. B*, 70, 184204 (2004).
14. K. Sato, Y. Kobayashi, K. Arinuma, I. Kanazawa, R. Tamura, T. Shibuya, and S. Takeuchi, *Phys. Rev. B*, 70, 094107 (2004).
15. K. Sato, H. Murakami, W. Sprengel, H.-E. Schaefer, and Y. Kobayashi, *Appl. Phys. Lett.* 94, 1719041 (2009).
16. J. Algers, P. Sperr, W. Egger, G. Kögel, and F. H. Maurer, *Phys. Rev. B* 67, 125404 (2003).
17. C. Wästlund and F. H. J. Maurer, *Macromolecules* 30, 5870 (1997).
18. M. Schmidt and F. H. J. Maurer, *Macromolecules* 33, 3879 (2000).
19. Y. Kobayashi, K. Haraya, S. Hattori, and T. Sasuga, Polymer 35, 925 (1994).
20. S. J. Tao, J. *Chem. Phys.* 56, 5499 (1972).
21. M. Eldrup, D. Lightbody, J. N. Sherwood, *Chem. Phys.* 63, 51 (1981).
22. K. Ito, H. Nakanishi, Y. Ujihira, *J. Phys. Chem.* 103, 4555 (1999).
23. K. Sato, K. Ito, K. Hirata, R.S. Yu, and Y. Kobayashi, *Phys. Rev. B*, 71, 012201 (2005).
24. K. Sato, D. Shanai, Y. Hotani, T. Ougizawa, K. Ito, K. Hirata, and Y. Kobayashi, *Phys. Rev. Lett.*, 96, 228302 (2006).
25. K. Sato, H. Murakami, K. Ito, K. Hirata, and Y. Kobayashi, *Macromolecules* 42, 4853 (2009).
26. K. Sato and W. Sprengel, *J. Chem. Phys.* 137, 104906 (2012).
27. K. Sato, H. Murakami, K. Fujimoto, M. Nakata, T. Oka, and Y. Kobayashi, *J. Phys.: Condensed Matter* 20, 3952341 (2008).
28. K. Sato, H. Murakami, K. Fujimoto, M. Nakata, and Y. Kobayashi, *Jpn. J. Appl. Phys.* 48, 0202211 (2009).

29. Kiminori Sato, *Jpn. J. Appl. Phys.* 51, 0773011 (2012).
30. K. Sato, K. Fujimoto, M. Nakata, and T. Hatta, *J. Phys. Chem. C* 115, 18131 (2011).
31. K. Sato, *J. Phys. Chem. B* 115, 14874 (2011).
32. Kiminori Sato, Koichiro Fujimoto, Masataka Nakata, and Naotatsu Shikazono, *J. Phys. Soc. Jpn.* 82, 014901 (2013).
33. Interlaboratory Comparison of Positron Annihilation Lifetime Measurements for Synthetic Fused Silica and Polycarbonate, Kenji Ito, Toshitaka Oka, Yoshinori Kobayashi, Yasuharu Shirai, Haruo Saito, Yoshihide Honda, Yasuyoshi Nagai, Masanori Fujinami, Akira Uedono, Kiminori Sato, Tetsuya Hirade, Akira Shimazu, Hiroyuki Hosomi, and Koji Sakaki, *J. Applied Phys.* 104, 0261021 (2008).
34. Interlaboratory Comparison of Positron Annihilation Lifetime Measurements, K. Ito, T. Oka, Y. Kobayashi, Y. Shirai, K. Wada, M. Matsumoto, M. Fujinami, T. Hirade, Y. Honda, H. Hosomi, Y. Nagai, K. Inoue, H. Saito, K. Sakaki, K. Sato, A. Shimazu, and A. Uedono, *Mater. Sci. Forum* 607, 248 (2009).
35. P. Kirkegaard and M. Eldrup, *Comput. Phys. Commun.* 7, 401 (1974).
36. A. Shukla, L. Hoffmann, A. A. Manuel, M. Peter, *Mater. Sci. Forum* 255–257, 233 (1997).
37. P. Asoka-Kumar, M. Alatalo, V. J. Ghosh, A. C. Kruseman, B. Nielsen, and K. G. Lynn, *Phys. Rev. Lett.* 77, 2097 (1996).
38. W. H. Zachariasen, *J. Am. Chem. Soc.* 54, 3841 (1932).
39. V. J. Ghosh, D. O. Welch, and K. G. Lynn, in Proc. 5th Inter. Workshop on Slow Positron Beams for Solids and Surface, edited by E. H. Ottewitte (AIP, New York, 1992), p. 937.
40. V. J. Ghosh, *Appl. Surf. Sci.* 85, 187 (1995).
41. G. C. Aers, P. A. Marshall, T. C. Leung, and R. D. Goldberg, *Appl. Surf. Sci.* 85, 196 (1995).
42. G. C. Aers, *J. Appl. Phys.* 76, 1622 (1994).
43. D. Mitlin, C. H. Raeder, and R. W. Messler, *Metall. Mater. Trans.* 30A, 115 (1999).
44. T. Reinikainen and J. Kivilahti, *Metall. Mater. Trans.* 30A, 123 (1999).
45. Y. S. Kim, K. S. Kim, C. W. Hwang, and K. Suganuma, *J. Alloy Compd.* 352, 273 (2003).
46. A. F. da Silveria, W. B. de Castro, B. A. Luciano, and C. S. Kiminami, *Mater. Sci. Eng. A* 375–377, 473 (2004).

47. B. A. Khalifa, M. R. Nagy, and R. Afify, *Phys. Status Solidi A* 202, 55 (2005).
48. C. M. Chuang, T. S. Lui, and L. H. Chen, *J. Mater. Sci.* 37, 191 (2002).
49. D. Herlach, H. Stoll, W. Trost, H. Metz, T. E. Jackman, K. Maier, H. -E. Schaefer, and A. Seeger, *Appl. Phys.* 12, 59 (1977).
50. J. M. Clayton, S. G. Usmar, H. M. Fretwell, I. K. MacKenzie, and M. A. Alam, *J. Phys.: Condens. Matter* 7, 3181 (1995).
51. I. Lemahieu, L. Dorikens-Vanpraet, D. Segers, M. Dorikens, P. Moser, C. Corbel, and P. Bois, *Phys. Status Solidi A* 102, 659 (1987).
52. M. J. Puska and R. M. Nieminen, *J. Phys.* F 13, 333 (1983).
53. P. J. Schultz and K. G. Lynn, *Rev. Mod. Phys.* 60, 701 (1988).
54. B. Bergersen, E. Pajanne, P. Kubica, M. J. Stott, and C. H. Hodges, *Solid State Commun.* 15, 1377 (1974).
55. E. Gramsch, K. G. Lynn, J. Throwe, and I. Kanazawa, *Phys. Rev. B* 59, 14282 (1999).
56. St. Chabik, W. Rudzińska, M. Szuszkiewicz, and Cz. Szymański, *Acta Phys. Pol. A* 95, 479 (1999).
57. R. Mahmudi, A. R. Geranmayeh, S. R. Mahmoodi, and A. Khalatbari, *J. Mater. Sci: Mater. Electron* 18, 1071 (2007).
58. J. A. Kilner and C. D. Waters, *Solid State Ionics* 6, 253 (1982).
59. K. R. Reddy and K. Karan, *J. Electroceram.* 15, 45 (2005).
60. H. Yoshida and T. Inagaki, *J. Alloys Compd.* 408–412, 632 (2006).
61. E. C. C. Souza and E. N. S. Muccillo, *J. Alloys Compd.* 473, 560 (2009).
62. F. Wakai, K. Chihara and M. Yoshida, *Acta Mater.* 55, 4553 (2007).
63. G. C. Kuczynski, *Trans. Am. Inst. Mining Met. Eng.* 185, 169 (1949).
64. W. D. Kingery and M. Berg, *J. Appl. Phys.* 26, 1205 (1955).
65. R. L. Coble, *J. Appl. Phys.* 32, 787 (1961).
66. K. Higashi, K. Sonoda, H. Ono, S. Sameshima, and Y. Hirata, *J. Mater. Res.* 14, 957 (1999).
67. S. J. Hong and A.V. Virkar, *J. Am. Ceram. Soc.* 78, 433 (1995).
68. J. Čížek, O. Melikhova, I. Procházka, J. Kuriplach, R. Kužel, G. Brauer, W. Anwand, T. E. Konstantinova and I. A. Danilenko, *Phys. Rev. B* 81, 024116 (2010).
69. P. L. Chen and I. W. Chen, *J. Am. Ceram. Soc.* 80, 637 (1997).
70. R. I. Grynszpan, S. Saudé, L. Mazerolles, G. Brauer, and W. Anwand, *Radiat. Phys. Chem.* 76, 333 (2007).

71. K. Sato, H. Murakami, K. Fujimoto, M. Nakata, W. Sprengel, H.-E. Schaefer, Y. Ueji, Y. Amemiya, and Y. Kobayashi, *Jpn. J. Appl. Phys.* 48, 0855051 (2009).

72. J. E. Garay, S. C. Glade, P. Asoka-Kumar, U. Anselmi-Tamburini, and Z. A. Munir, *J. Appl. Phys.* 99, 024313 (2006).

73. M. Kellermeier, E. Melero-Garcia, F. Glaab, R. Klein, M. Drechsler, R. Rachel, J. M. García-Ruiz, and W. Kunz, *J. Am. Chem. Soc.* 132, 17859 (2010).

74. J. S. Vorhies and R. Gaines, *Nat. Geosci.* 2, 221 (2009).

75. J. Diaz, E. Ingall, C. B. Nelson, D. Paterson, M. D. de Jonge, I. McNulty, and J. A. Brandes, *Science* 320, 652 (2008).

76. A. C. Allwood, J. P. Grotzinger, A. H. Knoll, I. W. Burch, M. S. Anderson, M. L. Coleman, and I. Kanik, *Proc. Natl. Acad. Sci. U S A* 106, 9548 (2009).

77. K. E. Kohfeld, C. L. Quéré, S. P. Harrison, and R. F. Anderson, *Science* 308, 74 (2005).

78. D. L. Rabosky and U. Sorhannus, *Nature* 457, 183 (2009).

79. C. Hallmann, L. Schwark, and K. Grice, *Nat. Geosci.* 1, 588 (2008).

80. H. Lowenstam, *Science* 211, 1126 (1981).

81. S. Mann and G. Ozin, *Nature* 382, 313 (1996).

82. V. Bansal, A. Ahmad, and M. Sastry, *J. Am. Chem. Soc.* 128, 14059 (2006).

83. M. J. Leng, G. E. A. Swann, M. J. Hodson, J. J. Tyler, S. V. Patwardhan, and H. Sloane, *Silicon* 1, 65 (2009).

84. K. Yoshizawa, K. Sato, H. Murakami, N. Shikazono, K. Fujimoto, and M. Nakata, *Mater Sci Forum* 607, 189 (2009).

85. P. M. Dove, N. Han, A. F. Wallace, and J. J. De Yoreo, *Proc. Natl. Acad. Sci. U S A* 105, 9903 (2008).

86. P. M. Dove, N. Han, and J. J. De Yoreo, *Proc. Natl. Acad. Sci. U S A* 102, 15357 (2005).

87. A. F. Wallace, G. V. Gibbs, and P. M. Dove, *J. Phys. Chem.* 114, 2534 (2010).

88. T. Bakos, S. N. Rashkeev, and S. T. Pantelides, *Phys. Rev. Lett.* 88, 0555081 (2002).

89. T. Bakos, S. N. Rashkeev, and S. T. Pantelides, Phys. Rev. B 69, 1952061 (2004).

90. G. Brumfiel, D. Cryanoski, *Nature* 474, 135 (2011).

91. M. Simons, S. E. Minson, A. Sladen, F. Ortega, J. Jiang, S. E. Owen, L. Meng, J. P. Ampuero, S. Wei, R. Chu, D. V. Helmberger, H. Kanamori, E. Hetland, A. W. Moore, and F. H. Webb, *Science* 332, 1421 (2011).
92. D. Butler, *Nature* 471, 555 (2011).
93. A. B. Kersting, D. W. Efurd, D. L. Finnegan, D. J. Rokop, D. K. Smith, and J. L. Thompson, *Nature* 397, 56 (1999).
94. P. A. Witherspoon, N. G. W. Cook, and J. E. Gale, *Science* 211, 894 (1981).
95. P. Maloszewski and A. Zuber, *Water Resour. Res.* 29, 2723 (1993).
96. I. Neretnieks, *J. Geophys. Res.* 85, 4379 (1980).
97. S. Colley and J. Thomson, *Nature* 346, 260 (1980).
98. E. A. Kwicklis, A. V. Wolfsber, P. H. Stauffer, M. A. Walvoord, and M. J. Sully, *Vadose Zone J.* 5, 934 (1980).
99. A. M. Shapiro, *Water Resour. Res.* 37, 507 (2001).
100. I. Neretnieks, *J. Contam. Hydrol.* 55, 175 (1980).
101. P. Andersson, J. Byegård, E. Tullborg, T. Doe, J. Hermanson, and A. Winberg, *J. Contam. Hydrol.* 70, 271 (2004).
102. H. H. Liu, Y. Q. Zhang, Q. Zhou, and F. J. Molz, *J. Contam. Hydrol.* 90, 41 (2007).
103. G. Lu, H. H. Liu, and R. Salve, *J. Hydrol.* 396, 33 (2011).
104. Y. S. Wu, H. H. Liu, G. S. Bodvarsson, and K. E. Zellmer, *J. Contam. Hydrol.* 73, 145 (2004).
105. K. Hama, T. Kunimaru, R. Metcalfe, and A. J. Martin, *Phys. Chem. Earth* 32, 170 (2007).
106. E. Ishii, K. Yasue, H. Ohira, A. Furusawa, T. Hasegawa, and M. Nakagawa, *J. Geol. Soc. Japan* 114, 286 (2008).
107. E. Ishii, H. Funaki, T. Tokiwa, and K. Ota, *J. Struct Geology* 32, 1792 (2010).
108. T. R. E. Chidley and J. W. Lloyd, *Ground Water* 15, 215 (1977).
109. T. Hirono, S. Nakashima, and C. J. Spiers, *Inter. J. Rock Mech. Min. Sci.* 45, 450 (2008).
110. K. Kiho, T. Oyama, and Y. Mahara, *J. Jpn Soc. Eng. Geology* 40, 260 (1999).

Chapter 6

Graphene: A New Star Nanomaterial in Energy and Environment Applications

Jiayi Zhu[a,b] and Junhui He[a]
[a]*Functional Nanomaterials Laboratory and Key Laboratory of Photochemical Conversion and Optoelectronic Materials,*
Technical Institute of Physics and Chemistry,
Chinese Academy of Sciences, Zhongguancundonglu 29,
Haidianqu, Beijing 100190, China
[b]*University of Chinese Academy of Sciences, Beijing 100049, China*
jhhe@mail.ipc.ac.cn

6.1 Introduction

The exponential growth of graphene research in both the scientific and engineering communities has taken place after the Geim group isolated "free" and "perfect" graphene sheets and demonstrated the unprecedented electronic properties of graphene in 2004 (Graphene, 2010 Nobel Prize for Physics) [1]. Graphene, a single layer of two-dimensional carbon lattice, has shown many unique properties, such as the quantum hall effect (QHE), high carrier mobility at room tem-

Nanomaterials in Energy and Environmental Applications
Edited by Junhui He

ISBN 978-981-4463-78-2 (Hardcover), 978-981- 4463-79-9 (eBook)
www.panstanford.com

perature (~10,000 $cm^2 V^{-1} s^{-1}$) [2], large theoretical specific surface area (~2630 $m^2 g^{-1}$) [3], good optical transparency (~97.7% per layer) [4], high Young's modulus (~1 TPa) [5], and excellent thermal conductivity (~3000–5000 $W m^{-1} K^{-1}$) [6]. To exploit these properties in various kinds of applications, several synthetic routes have been developed for the preparation of graphene and its derivatives, ranging from the bottom-up epitaxial growth to the top-down exfoliation of graphite. In particular, chemical exfoliation and reduction starting from the oxidation of graphite is an efficient process to produce graphene sheets in a low-cost, scalable, controllable, and reproducible manner [7]. Owing to the highly versatile and tunable properties, graphene has attracted a great deal of attention in many important applications, such as optoelectronic devices [8], energy storage materials [9], photo-catalysis [10], chemical and biological sensors [11], and polymer composites [12].

In the 21st century, aggravating energy and environmental problems such as fossil fuel depletion, global warming, and pollution are ringing the alarm bell to the human society. Thus, green energy and environment technologies have been the urgent and important areas. Among several possible alternatives for fossil energy, eventually solar energy is probably the only one that can meet the multifold-demand for the long-term human needs. The utilization of solar energy consists of two steps: First, solar energy can be effectively converted to applicable forms (electricity or fuel) from solar power to suppress energy crisis and global warming. Aiming at this goal, solar cells and photo-catalysts for production of H_2 and reduction of CO_2 are mostly concerned [13–17]. Second, high performance energy storage devices are also required. This is mainly due to the intermittent characteristics of solar energy and other renewable energy sources. Supercapacitor is one of the promising devices for this purpose [18–20]. Due to highly remarkable properties, graphene has been useful in various energy and environment applications, such as transparent conductive electrodes or active materials in thin film solar cells, counter electrodes in dye-sensitized solar cells, high-performance electrodes in supercapacitors, and photo-catalysts for reduction of CO_2 and degradation of organic pollutants. Therefore, this review mainly focuses on recent advances in the synthesis of graphene and graphene-based materials and their applications in the energy and environment related systems described above.

6.2 Preparations of Graphene-Based Materials

6.2.1 Graphene

The first piece of single-layer graphene sheet was prepared by mechanical exfoliation, in which highly oriented pyrolitic graphite (HOPG) was peeled using scotch tape and the subsequent release of the graphene flake on Si/SiO_2 after the tape was removed [1]. The exfoliated graphene manifests a unique structure and superior properties, although this production method is not applicable on a large scale. Inspired by the pioneering work, several alternative techniques have been developed for fabricating graphene materials. These methods for fabricating graphene materials used in energy and environment applications can be generally classified into the bottom-up and top-down approaches.

The bottom-up approach involves the direct synthesis of graphene materials from the carbon sources, such as the chemical vapor deposition (CVD), which is a typical method used to grow large-area, single, and few-layer graphene sheets on metal substrates. When the metal surfaces are heated, hydrocarbon (or carbon oxide) decomposes into carbon atoms and hydrogen gas (or oxygen gas), and the carbon atoms then form a graphene monolayer (Fig. 6.1a). Furthermore, as seen from Fig. 6.1b, the obtained graphene films on metal surface can be transferred to other target substrates via metal etching, which is very important for device applications [21]. Recently, this approach was employed to prepare large-area, continuous, few-layered graphene as photo-anodes for organic photovoltaic devices [14]. Moreover, the epitaxial growth process has also been exploited to prepare single-layer graphene via the sublimation of SiC. This process can provide a higher yield with much less defects, but cannot easily fabricate a large-area graphene [22]. In addition to the above methods based on the solid-phase deposition, graphene is also obtainable via the wet chemical reaction of ethanol and sodium followed by pyrolysis [23], or through the organic synthesis to give graphene-like polyaromatic hydrocarbons (Fig. 6.1c) [24].

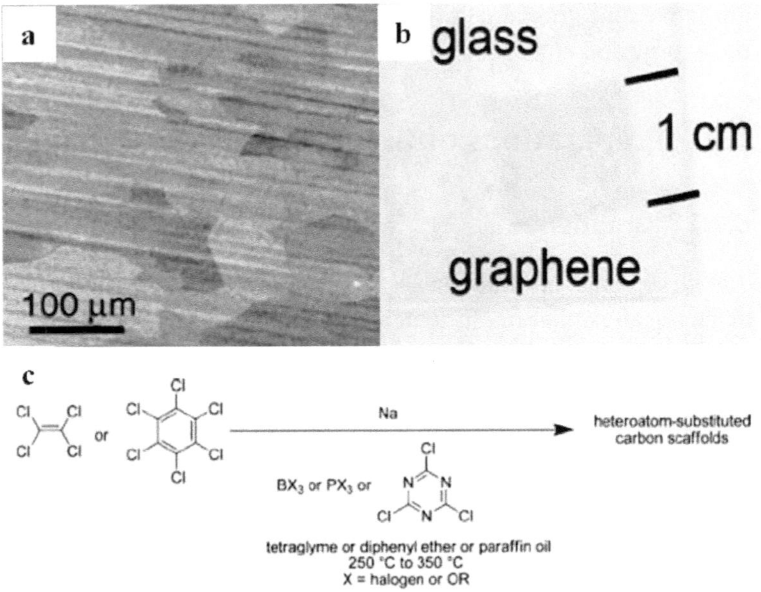

Figure 6.1 (a) SEM image of graphene sheets on a Cu foil with a growth time of 30 min. (b) Graphene films transferred onto a glass plate. (c) Schematic of bottom-up preparation of heteroatom-substituted carbon scaffolds by a solution-phase process. Reproduced from Refs. [21], Copyright 2009 American Association for the Advancement of Science; and [24], Copyright 2010 American Chemical Society.

Different from the bottom-up approaches, the top-down approaches are advantageous in terms of high yield, solution-based processability, and ease of implementation. Chemical exfoliation and reduction starting from the oxidation of graphite is an efficient process to produce graphene sheets in a low-cost, controlled, scalable, and reproducible manner [25]. The procedure is through three steps: oxidation of graphite, exfoliation of graphite oxide into graphene oxide (GO), and reduction of graphene oxide. The graphite oxide is frequently obtained by the Hummers method via the reaction of graphite with a mixture of potassium permanganate ($KMnO_4$) and concentrated sulfuric acid (H_2SO_4) [26]. A variety of thermal and mechanical approaches can be used to exfoliate graphite oxide to single-layer graphene oxide, but sonicating graphite oxide in a solvent is most commonly used. GO can be regarded as a functionalized graphene containing hydroxyl, epoxy, and carboxylic groups. So far, various types of reduction methods have been reported to obtain reduced graphene oxide (RGO) sheets, such as chemical reagent reduction [7], photochemical reduction [27], thermal reduction [28], photothermal reduction [29], microwave-assisted reduction [30], and electrochemical reduction [31]. Among these methods, the chemical reduction is the most versatile one with many reduction agents being usable, such as hydrazine [7], $NaBH_4$ [32], supercritical water [33], strong alkaline media [34], vitamin C or ascorbic acid [35], transition metal powder (Fe or Zn) [36], and hydriodic acid [37]. Despite the fact that the conductivity of RGO sheets is orders lower than the pristine graphene due to the presence of residual oxygen-containing groups and defects, the reactive surfaces of GO and RGO sheets provide the tunability in electronic and optoelectronic properties via chemical reactions, and the feasibility for composite incorporation with good dispersion (Fig. 6.2a,b). Other approaches

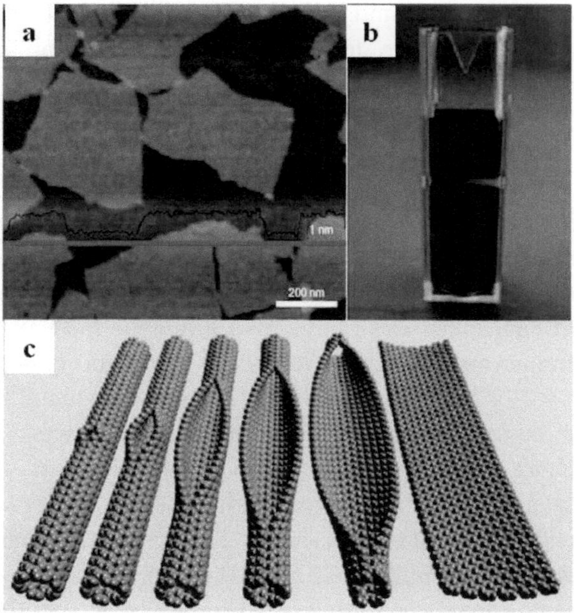

Figure 6.2 (a) Tapping mode AFM image of graphene sheets with a height profile (blue curve; scale bar, 1 nm) taken along the red line. (b) The Tyndall effect confirming the colloidal nature of the graphene dispersions. (c) Representation of the gradual unzipping of one wall of a carbon nanotube to form a nanoribbon. Reproduced from Refs. [7], Copyright 2008 Nature Publishing Group; and [41], Copyright 2009 Nature Publishing Group.

directly exfoliate graphite via sonication and intercalation. The resulting graphene materials have much fewer defects than those of RGO sheets. However, these methods are not widely used because of their unsatisfactory yields of single layer graphene or sonication-induced scissoring of graphene sheets into pieces of very small sizes [38–40]. Last, unzipping carbon nanotubes (CNTs) can offer the possibility of large-scale production of narrow graphene nanoribbons (GNRs) with well-controlled widths. As seen from Fig. 6.2c, some unzipping methods, such as oxidative treatment of CNTs and cutting of CNTs induced by transition metal nanoparticles (Ni or Co), have been developed to effectively unzip CNTs to form GNRs [41–43].

Although significant progress has been witnessed in developing the methods of fabricating graphene materials, the controllable production of graphene materials with desirable size, shape, quality in a low-cost, scalable, and reproducible manner is still an essential technological challenge for graphene potential applications in different fields.

6.2.2 Graphene-Based Thin Films

Assembling 2D architectures using nanoscale materials as functional building blocks is a very exciting research topic. Graphene shows great potential as an advanced building block for developing different functional films. First, large area graphene films of both high transmittance and high conductivity can be produced on metal substrates by CVD methods. As initially deposited on metal foils, these films can be transferred to transparent glass or polymer substrates after etching away the metal supports [44–48]. Figure 6.3a shows an industrial applicable "roll-to-roll" technique for large area and continuous fabrication of transparent CVD graphene electrodes [49].

Second, the assembly of GO or RGO sheets prepared by chemical routes can be an efficient process to produce large-area thin films in a low-cost, controlled, scalable, and reproducible manner. A number of approaches have been used to assemble graphene sheets into thin films with tailorable properties, including dip-coating [50], vacuum filtration [51], spray-coating [52], spin-coating [53], Langmuir–Blodgett assembly [54], and LbL assembly [55]. Among them, the LbL assembly of charged nanosized graphene sheets is a controllable and reproducible route for generating large-scale but still uniform graphene thin films in an aqueous environment, and is considered as a promising candidate for creating function-embedded graphene thin films. Considerable efforts have been made to fabricate different graphene thin films through LbL techniques. For instance, Wallace and coworkers reported on the usage of negatively charged RGO sheets combined with positively charged polyelectrolytes to fabricate multilayered graphene films [7]. Recently, we reported a new approach to fabrication of flexible transparent conductive thin films via the LbL assembly of oppositely charged reduced RGOs and benign step-by-step post-treatment (Fig. 6.3b) [56]. Figure 6.3c shows the dependence of the transmittance at 550 nm on the number of RGO–PDDA$^+$/RGO–O$^-$ bilayers. The linear relationship between the transmittance and the number of RGO–PDDA$^+$/RGO–O$^-$ bilayers suggests that LbL assembly is a simple and controllable method to fabricate uniform thin films. Moreover, the sheet resistance as well as the transmittance decreases with an increase in the number of LbL cycles, which is attributed to the increase in the number of conductive graphene sheets on the substrate after each LbL cycle.

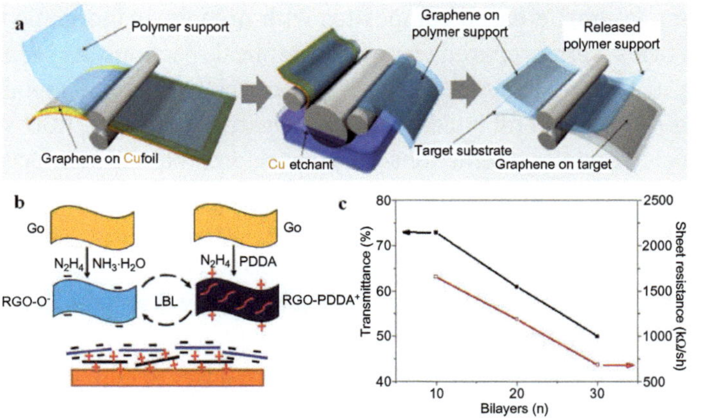

Figure 6.3 (a) Schematic of the roll-based production of graphene films grown on a copper foil. (b) A schematic illustration of the fabrication of graphene thin films by LbL assembly. (c) Transmittance (550 nm) and sheet resistance of the $(RGO–PDDA^+/RGO–O^-)_{10}$, $(RGO–PDDA^+/RGO–O^-)_{20}$, and $(RGO–PDDA^+/RGO–O^-)_{30}$ films. Reproduced from Refs. [49], Copyright 2010 Nature Publishing Group; and [56], Copyright 2012 Royal Society Chemistry.

6.2.3 Graphene-Based Composites

With the development of nanoscience and nanotechnology, appropriate combination of different nanoscale materials with different morphologies or sizes or components may lead to the development of multifunctional nano-assembled systems that simultaneously exhibit novel optical, electronic, magnetic and mechanical properties, or even have enhanced performance. In graphene-based composites, graphene generally acts either as a functional component or a substrate for immobilizing other components. The high surface area and the conductive robust structure of graphene often facilitate charge transfer, redox reaction as well as enforce the mechanical strengths of composites. Therefore, anchoring redox active materials, and photo-catalysts on graphene would improve the performances of the composites for energy conversion and storage devices and/or degradation of organic pollutants [57–60].

Methods for fabricating graphene-based composites used in energy and environment applications can be classified into two general strategies: in situ reaction and ex situ hybridization. For example, many graphene-based composites with organic or inorganic nanoparticles were prepared via in situ chemical reactions. As seen from Fig. 6.4a,b, in chemical syntheses, positively charged metal ions (e.g., Pd^{2+}) preferably adsorbed on negatively charged RGO sheets via electrostatic interactions [61]. Therefore, metalization occurred preferentially on RGO sheets to form the corresponding composite materials [62]. Negatively charged ions, such as $PtCl_4^{2-}$, can also be reduced to metal nanoparticles on RGO sheets [63]. Moreover, if graphene is a reactant (e.g., reducing MnO_4^- by graphene), the reaction is self-limited on the surface of graphene sheets [64]. On the contrary, the ex situ hybridization involves the mixing of graphene sheets and pre-synthesized or commercially available nanocrystals in solutions. RGO sheets are usually dispersible in water or various organic solvents [7]. Furthermore, graphene can be easily functionalized into either positively or negatively charged derivatives. Thus, they can form composites with other charged components via electrostatic assembly. Recently, in our work, we reported a facile approach to fabricate novel graphene wrapped–MnO_2 nanocomposites by co-assembly between honeycomb MnO_2 nanospheres and graphene sheets via electrostatic interaction [9,65–67]. As seen from Fig. 6.4c,d, the graphene wrapped–MnO_2 nanocomposites show crinkled and rough textures, which is associated with the presence of flexible and ultrathin graphene sheets. The honeycomb MnO_2 nanospheres are firmly attached to the graphene sheets. On the basis of the above techniques, graphene-based composites with various functions have been successfully prepared.

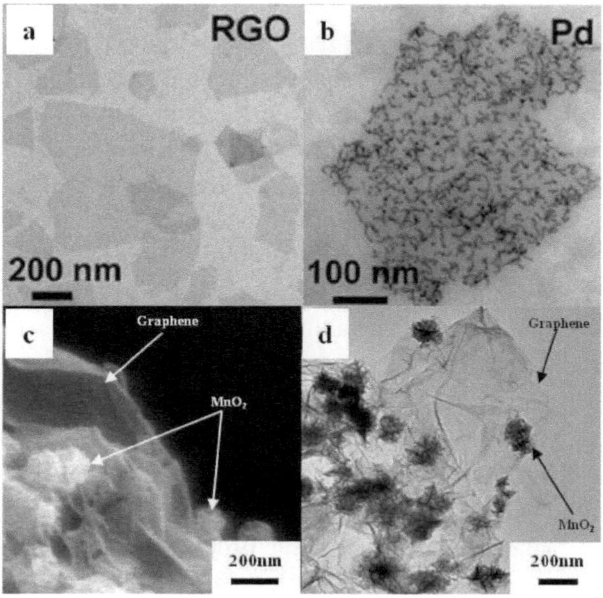

Figure 6.4 TEM images of (a) graphene sheets. (b) graphene sheets with Pd nanoparticles with a worm-like shape ~4 nm in width. (c,d) SEM and TEM images of graphene-wrapped MnO_2 nanocomposites. Reproduced from Refs. [61], Copyright 2010 American Chemical Society; and [9], Copyright 2012 American Chemical Society.

6.3 Applications of Graphene-Based Materials in Energy and Environment-Related Systems

6.3.1 Solar Cells

Solar cell can directly convert solar energy to electrical power. Thus, it is one of the most promising devices satisfying the global energy requirements. As a novel and unique star among carbon nanomaterials, graphene manifests attractive application potentials in thin film, dye-sensitized as well as heterojunction solar cells.

6.3.1.1 Thin film solar cells

6.3.1.1.1 *Transparent electrodes*

Organic and inorganic thin film solar cells are being pursued as viable alternatives to silicon solar cells. A critical aspect of these solar cells is the transparent, conductive illuminated side (photo-anode). Traditional materials, such as cadmium oxide and indium tin oxide (ITO), have been in use for many years. ITO thin films are the most frequently used because of their low electrical resistance and high visible light transmittance [68]. However, transparent conductive metal oxides suffer from considerable drawbacks, including the scarcity of indium reserves, costs associated with complicated vacuum deposition, the instability in the presence of acid or base, and especially highly brittle nature, making them unsuitable for flexible substrates. Recently, several materials are currently being developed for use as transparent electrodes to replace ITO, such as conductive polymers [69] and carbon-based materials [70–72].

Graphene, a single layer of two-dimensional carbon lattice, exhibits remarkable optoelectronic properties as well as excellent chemical resistance and mechanical flexibility, and is considered to be an ideal candidate to replace ITO for transparent electrodes. Several methods have been developed for the preparation of large-area graphene films to satisfy optoelectronic applications. Generally, there are mainly two strategies to prepare graphene-based transparent electrodes: solution-based assembly of RGO sheets into thin films and CVD fabrication of large-size continuous graphene films.

The solution-based assembly of RGO sheets into thin films has several advantages. On one hand, the starting material of RGO is usually low-cost and abundant graphite; on the other hand, it is facile to process RGO sheets into a large area film via wet processing techniques (e.g., spin coating, dip-coating, LbL assembly, or vacuum filtration) [50–55]. Very recently, we reported a new approach to fabrication of flexible, transparent conductive thin films via LbL assembly of oppositely charged RGOs, and benign step-by-step post-treatment [56]. The graphene thin films showed remarkable optical and electronic properties as well as good electrical conductivities even under excessively multiple-cycle bending conditions, as seen from Fig. 6.5a,b. This could be attributed to the superior structural properties of graphene sheets, such as flexibility, mechanical stability, and high tendency to stack together via π-π stacking. In contrast, the ITO on flexible substrate showed significant loss of electrical conductivity after multiple cycles of bending, which should be due to an increased number of cracks in the rigid inorganic structure of ITO under identical bending conditions. However, the conductivity and transparency of solution-based assembled RGO thin films are still far from the requirements of practical applications as transparent conductive electrodes for solar cells (sheet resistance <100 Ω sq^{-1}, transmittance >90%).

However, the production of large-area continuous graphene films may circumvent the above problem. As seen from Fig. 6.5c,d, Wang et al. fabricated a large-area graphene film using a CVD process [14]. For graphene film with transmittance of 91–72% in the visible light wavelength range, the average sheet resistance varied from 1350 to 210 Ω sq^{-1}, which is lower than that of the solution-based assembled RGO thin films by a factor of 2–3. Furthermore, they exploited the graphene film as a photo-anode for the organic photovoltaic device. The performance of the photovoltaic cell was measured and the graphene anode exhibited excellent performance characteristics (V_{oc}=0.55 V, J_{sc}= 6.05 mA cm^{-2}, FF =51.3%, and PCE =1.71%). Under the same conditions, the photovoltaic performance of a reference device made with ITO anode showed V_{oc}, J_{sc}, FF, and PCE of 0.56 V, 9.03 mA cm^{-2}, 61.1%, and 3.10%, respectively. This indicates that the solar cell with a graphene anode has reached 55.2% PCE of a structurally identical cell with ITO anode.

Graphene-based transparent conductive films have proved effective photo-anodes for solar cells, which have high chemical and mechanical stabilities as well as excellent transparency. However, the power conversion efficiency of graphene-based cells, which is still lower than that of ITO based cells, needs to be improved in future researches.

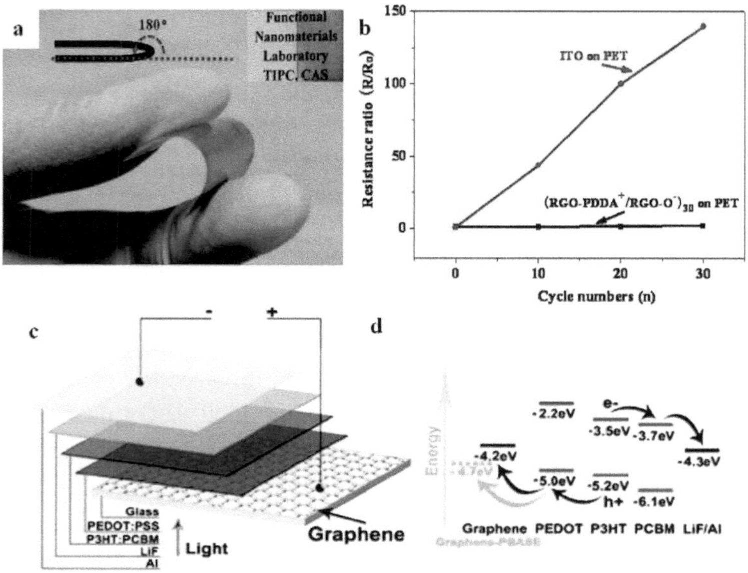

Figure 6.5 (a) Image of the $(RGO-PDDA^{+}/RGO-O^{-})_{30}$ film on a PET substrate during the bending test (inset: left is blank PET and right is PET with the $(RGO-PDDA^{+}/RGO-O^{-})_{30}$ film). (b) The change in the resistance ratio (R/R_0) of $(RGO-PDDA^{+}/RGO-O^{-})_{30}$ on PET (dark line) and ITO on PET (red line) with the number of bending cycles at an angle of 180°. (c,d) Schematic and energy diagram of the fabricated device with the structure of graphene/PEDOT:PSS/P3HT:PCBM/LiF/Al. Reproduced from Refs. [56], Copyright 2012 Royal Society Society; and [9], Copyright 2009 AIP Publishing LLC.

6.3.1.1.2 *Electron transport/acceptors*

The electron donor/acceptor layer is one of the most important components in a polymer-based thin film solar cell, which contains a conjugated polymer [e.g., poly(3-hexyl)thiophene, P3HT] to generate electron–hole pairs upon photon absorption, and an acceptor with a relatively high electron affinity to dissociate the electron–hole pairs into separate charges. Fullerene and its derivatives (e.g., phenyl-C_{61}-butyric acid methyl ester, PCBM) are the mostly employed acceptor materials due to their high electron affinity and charge mobility. The combination of P3HT and PCBM has been studied intensively in various configurations and the cell efficiencies of up to nearly 6% have been achieved [73].

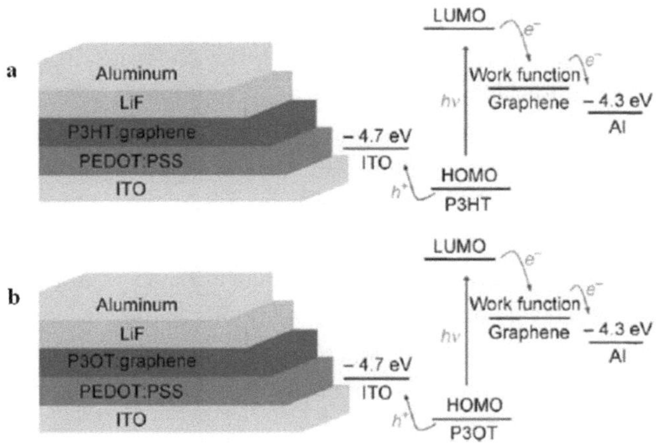

Figure 6.6 Representation of devices a) with P3HT:graphene thin film as the active layer and the structure ITO/PEDOT:PSS/P3HT:graphene/LiF/Al, and b) with P3OT:graphene thin film as the active layer and the structure ITO/PEDOT:PSS/P3OT:graphene/LiF/Al. Reproduced from Refs. [74], Copyright 2009 WILEY-VCH Verlag GmbH Co. KGaA; and [75], Copyright 2008 WILEY-VCH Verlag GmbH Co. KGaA.

Owing to its large surface area for donor/acceptor interfaces and continuous pathway for electron transfer, graphene-based materials are anticipated to be used for the effective electron–hole separation and charge transport when blended with conjugated polymers. Recently, functionalized graphene has been explored as an electron acceptor for OPV devices. As shown in Fig. 6.6a,b Chen and his co-workers found that the strong photoluminescence of P3HT solution was remarkably reduced after the functionalized graphene was introduced [74,75]. This efficient quenching of the photoluminescence emission indicates that functionalized graphene can be used as an effective electron-acceptor material for organic photovoltaic applications. Indeed, the overall performance of the P3HT/functionalized graphene-based photovoltaic device was much higher than that of the control device based on pristine P3HT. This indicates that there was a charge (electron) transfer from the P3HT donor to the functionalized graphene acceptor.

6.3.1.1.3 *Hole transport layers*

Direct connection between the cathode and anode by the electron donor/acceptor layer will result in the fast recombination of charge carriers and current leakage. Therefore, a hole transport layer is usually incorporated between the anode and the electron donor/acceptor layer, such as the commonly used poly(3,4-ethylenedioxythiophene) : poly(styrenesulfonate) (PEDOT : PSS). As seen from Fig. 6.7a,b, GO film could be used as a simple and effective alternative to PEDOT : PSS in polymer-based thin film solar cells [13,76]. In the configuration of glass/PBASE-ITO/GO/P3HT:PCBM/Al, the GO film with a band gap of nearly 3.6 eV is able to hinder the electron transport from the PCBM LUMO to the ITO anode, while bridging the holes to the anode. Moreover, the investigation into the effects of film thickness on the performance of polymer-based thin film solar cells was also discussed.

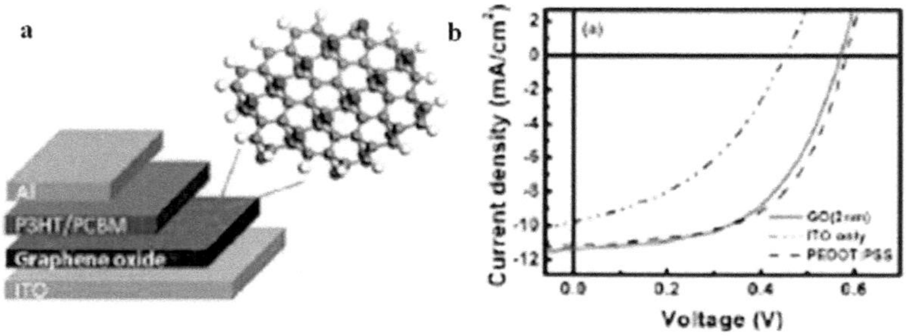

Figure 6.7 (a) Schematic illustration of GO used as the hole transport layer in a polymer solar cell. (b) Current–voltage characteristics of photovoltaic devices with no hole transport layer (curve labeled as ITO), with a 30 nm PEDOT:PSS layer, and a 2 nm thick GO film. Reproduced from Ref. [76] Copyright 2010 American Chemical Society.

6.3.1.2 Dye sensitized solar cells

6.3.1.2.1 *Electron transport promoter for TiO_2 layer*

Dye-sensitized solar cells (DSSCs) are attractive because of their low cost, convenient manufacturing process, and comparable efficiencies to those of amorphous silicon solar cells. A DSSC consists of a transparent electrode as photo-anode, a mesoporous semiconductor (usually TiO_2) layer, a monolayer of dye, an electrolyte containing redox pairs (usually I_2/I_3^-), and a counter electrode. Upon light irradiation, the dye molecules are excited to inject electrons into the conduction band of the semiconductor layer in DSSCs. Thus, it is important to improve the conductivity of electrons at the TiO_2 layers to compete with charge recombination [77,78]. Recently, Tang et al. reported that RGO is an attractive material for application in the above aspect (Fig. 6.8a,b). The benefit of RGO sheets incorporation in TiO_2 nanostructures is that the work function of RGO sheets is between the conduction band of TiO_2 and the work function of ITO, which allows for the fast collection of electrons at the anode to suppress electron–hole pairs recombination. Apart from the enhanced conductivity, the large surface area of RGO sheets improves the loading and dispersion of the dye molecules to enhance the photon absorption [79–81].

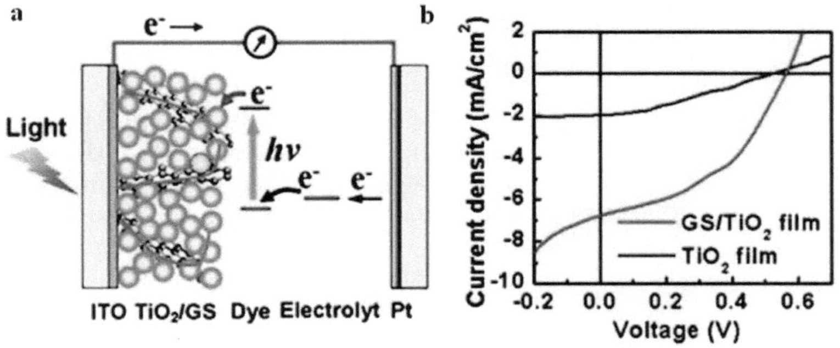

Figure 6.8 (a) Schematic structure of the DSSCs with an active cell area of 0.5×0.5 cm^2. (b) Current density–voltage (*J-V*) characteristics of DSSCs based on graphene/TiO_2 and TiO_2 nanoparticles films under 1 sun illumination. Reproduced from Ref. [79] Copyright 2010 American Chemical Society.

6.3.1.2.2 *Counter electrodes*

The duty of the counter electrode in a DSSC is to catalyze the reduction of I_2 after electron injection. In a DSSC, the reductions of I_2 at both the counter electrode and TiO_2 surface are competing; therefore, the electrocatalytic efficiency of the counter electrode should be as high as possible to inhibit the backward reaction. Platinum cathode, as the commonly used counter electrode, should be replaced by inexpensive materials for reducing cost and highly conductive graphene-based materials are promising candidates. Recently, Shi and his co-workers used graphene functionalized by pyrenebutyrate (PB^-) as the counter electrode material of DSSC (Fig. 6.9a,b) [82,83].

The performance of PB$^-$-graphene/FTO electrode greatly surpassed that of bare FTO. Furthermore, the composite film is transparent and flexible, and manifests an application potential in both flexible solar cells and power-producing windows.

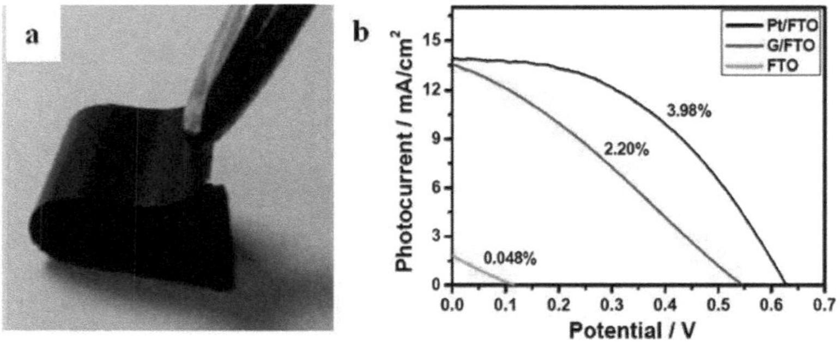

Figure 6.9 (a) Photograph of a 6 μm thick PB$^-$-graphene film (1×3 cm). (b) Photocurrent density-voltage curves of the DSSCs under 1.5 AM, 100 mW cm^{-2} illumination. Reproduced from Ref. [83] Copyright 2008 American Chemical Society.

6.3.1.3 Heterojunction solar cells

In order to partially replace silicon, thus reducing cost in solar cell devices, carbon based materials have been attempted in the p-type amorphous carbon/n-type silicon (p-AC/n-Si) heterojunctions and CNT/Si heterojunctions. Recently, we fabricated heterojunction structures (Fig. 6.10a,b), which consisted of a double-walled carbon nanotube (DWCNT) thin film coated either on an n-type silicon wafer or an n-type silicon nanowires (SiNW) array with varied lengths. We found that the photoresponse of the heterojunctions dramatically depends on the length of SiNWs. The heterojunction with a SiNW length of ca. 600 nm has the highest photoresponse value of 10.72. The heterojunction also showed fast photocurrent response (<10 ms) with good reproducibility [84].

Moreover, graphene-based films can be prepared with controlled thickness, good surface continuity, and tunable properties via variable functionalizations. As seen from Fig. 6.10c,d, CVD grown graphene sheets had been deposited on n-Si with 100% coverage to make the Schottky junction solar cell, which showed an efficiency of up to nearly 1.5% with a filling factor of nearly 56%. In addition, the graphene film served as a semitransparent electrode for the graphene/n-Si solar cells [85,86].

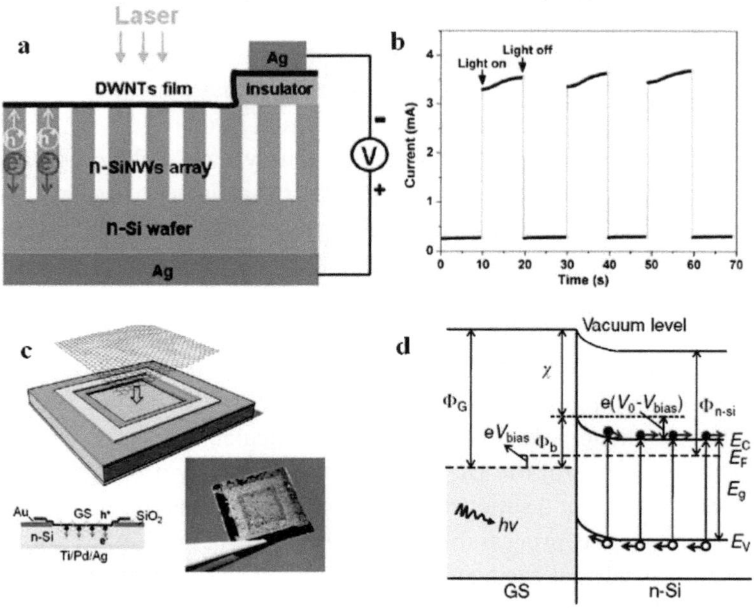

Figure 6.10 (a) Schematic diagram of *I–V* curve measurements of heterojunctionsfabricated with DWCNT film and SiNW array. (b) Current on and off as a function of time for DWCNT/SiNWs (600 nm) under 532 nm laser and at V_{bias} = 1 V. (c) Schematic illustration of the device configuration. Bottom-left inset: cross-sectional view, photogenerated holes (h+) and electrons (e−) are driven into the graphene and n-Si, respectively, by the built-in electric field. Bottom-right inset: photograph of a graphene/n-Si Schottky cell with a 0.1 cm^2 junction area. (d) Energy diagram of the forward-biased graphene/n-Si Schottky junction upon illumination. Reproduced from Refs. [84], Copyright 2010 Elsevier B.V.; and [85], Copyright 2010 WILEY-VCH Verlag GmbH Co. KGaA.

6.3.2 Supercapacitors

Supercapacitors are charge-storage devices that have been attracting tremendous attention due to their high power density, excellent reversibility and long cycle life. A high performance supercapacitor should have high energy density (~1–10 Wh/kg, determined by its capacitance and voltage), high power density (~10^3–10^5 W/kg, determined by its voltage and internal resistance), and ultra-long cycling life (>100,000 cycles) [87–89]. Thus, supercapacitors are considered to be promising power supplies for versatile applications such as environmentally friendly automobiles, artificial organs, high-performance portable electronics, and so on.

Generally, supercapacitors can be simply classified by their storage mechanisms into two groups. One is the electrical double layer capacitor (EDLC), which stores energy via an electrostatic process, that is, the charges are accumulated at the electrode/electrolyte interface through polarization. Hence, it is essentially important to use electrode materials with good conductivity and large specific surface areas in EDLCs, such as the activated carbon, CNTs, carbon nanofibers, and the emerging graphene-based materials [90–93]. The graphene-based materials, in particular RGO, are advantageous in terms of the chemically active surface with large specific surface area, good conductivity, low cost, and mass production with solution-based processability as mentioned before. In addition, it is suggested that the aggregated graphene sheets exhibit an open-pore structure, which allows for the easy access of electrolyte ions to form electric double layers. Ruoff and his co-workers reported the first EDLC based on RGO. A specific capacitance of 135 F g^{-1} in the aqueous electrolyte was obtained, which is comparable with the traditional carbon-based electrode materials [3].

The other type of supercapacitor is so called the pseudocapacitor, which is based on the rapid redox reactions of the chemical species present in the electrode. Commonly used electrode materials are metal oxides (e.g., RuO_2, NiO, and MnO_2) and conducting polymers [e.g., polyaniline (PANi) and polypryrrole (PPy)]. This type of electrode affords higher specific capacitance per unit surface area compared to the porous carbon based electrode materials [94–96]. Among all the materials, MnO_2 is considered as a good candidate for supercapacitors because of its electrochemical behavior, low cost, and environmental compatibility. Varied methods have been reported to synthesize MnO_2-based materials with different structures and properties for supercapacitors [97–101]. Moreover, it is highly desired to develop a rational design to maximize their electrochemically active sites for redox reactions through obtaining "opened" structures to further increase their energy storage density. Recently, we synthesized monodisperse honeycomb MnO_2 nanospheres via micro-emulsion method and studied the effects of reaction conditions on the structure, morphology and size of honeycomb MnO_2 nanospheres [100–102]. Nevertheless, MnO_2 does not usually deliver ideal specific capacitance due to its poor electrical conductivity and electrochemical dissolution during cycling. To circumvent these obstacles, carbonaceous materials with high electrical conductivity and buffer matrix have been widely chosen as matrices for MnO_2-based materials to improve their conductivity and stability. Compared with other carbon matrices such as graphite, carbon black, and carbon nanotubes, graphene is emerging as one of the most appealing carbon materials due to its unique properties such as superior electrical conductivity, excellent mechanical flexibility, and high thermal and chemical stability [103–107].

Recently, we synthesized novel graphene wrapped-MnO_2 nanocomposites by self-assembly of honeycomb MnO_2 nanospheres and graphene sheets via an electrostatic co-precipitation method [9,64,95]. As shown in Fig. 6.11a,b, the graphene wrapped–MnO_2 nanocomposites exhibited enhanced capacitive performance than bare honeycomb MnO_2 nanospheres, which is attributed to the synergistic effect between the pseudocapacitance of honeycomb MnO_2 nanospheres and good electrical conductivity of graphene sheets. The honeycomb MnO_2 nanospheres contribute to the energy storage via the redox reaction between the III and IV oxidation states of Mn (reaction (6.1)), which involves the intercalation of alkali metal ions present in the electrolyte, for example, Na^+.

$$MnO_2 + C^+ + e^- \Rightarrow MnOOC\ (C^+ = H^+, Na^+) \tag{6.1}$$

The graphene sheets, act as an electronic conductive channel in the nanocomposites, and are beneficial to the fast transfer of electrons throughout the whole electrode, while they may also provide the capacitance through the electron double layer at the carbon surface. Besides, a graphene/Co_3O_4 supercapacitor displays a maximum specific capacity of 243.2 F g^{-1} at a scan rate of 10 mV s^{-1} in KOH aqueous solution, and about 95.6% of the specific capacitance is retained after 2000 cycles [108].

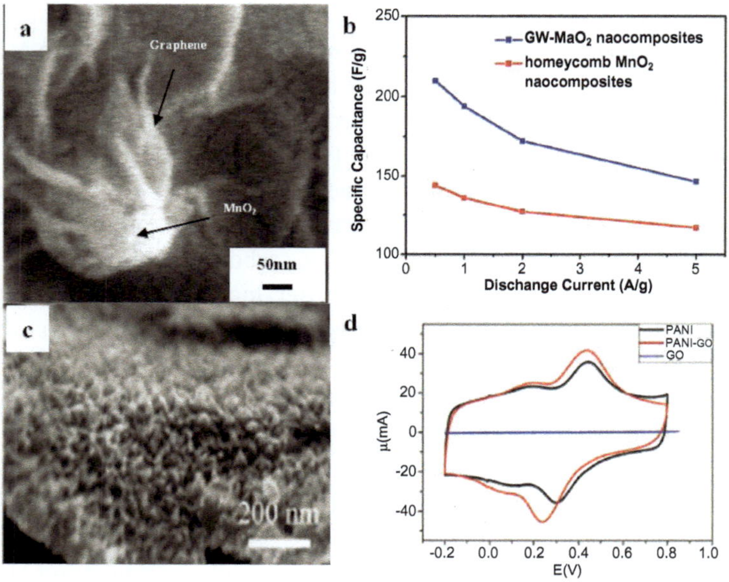

Figure 6.11 (a) SEM image of graphene wrapped-MnO_2 nanocomposites. (b) Specific capacitances of graphene wrapped-MnO_2 nanocomposites and honeycomb MnO_2 nanospheres at different discharge currents (0.5–5A g^{-1}) in 1 M Na_2SO_4 solution. (c) SEM images of PANi-GO nanocomposites at the concentration of aniline of 0.05 M. (d) CV curves of pristine GO, random connected PANi nanowires, and PANi-GO nanocomposite at a scan rate of 20 mV s^{-1}. Reproduced from Refs. [9], Copyright 2012 American Chemical Society; and [111], Copyright 2010 American Chemical Society.

Conducting polymers are also attractive materials for supercapacitor electrodes, because they have moderate conductivity, fast charge–discharge kinetics and doping–undoping processes, and flexibility for thin film based fabrications. Till now, PANi is one of the most employed polymers to be incorporated with GO or RGO sheets for supercapacitors [62,109]. As seen from Fig. 6.11c,d, Wei and his co-workers reported at the suitable concentration of aniline (0.05 M), uniform aligned PANi nanowire arrays on GO sheets was obtained with a high yield. In the CV curves (Fig. 6.11d), there was nearly no peaks for GO and two pairs of redox peaks appeared for PANi-GO and pristine PANi, which were attributed to two redox transitions of PANi. Therefore, the capacitance of PANi-GO mainly comes from

Faradaic reactions of PANi at the electrode/electrolyte surface, which is different from that of the electric double layer capacitance of carbon-based materials. Besides the enhanced conductivity and capacitance, GO is also able to improve the cycling stability of the composite electrode. The hierarchical composite electrode of PANi–GO can retain 92% of the initial capacitance after 2000 cycles, whereas the initial capacitance of pure PANi electrode drops to 74% [110,111].

6.3.3 Photo-Catalysts for Reduction of CO_2 and Degradation of Organic Pollutants

After the global climate conference in Copenhagen in 2009, how to reduce emissions of the greenhouse gas CO_2 causing global warming and transform it into useful chemical products has become the hot spot in the global research. A great deal of effort has been expended to reduce CO_2 emissions from the industries where the largest percentages of fossil fuels are used. Recently, we reviewed the study on CO_2 capture, storage and transformation so far, where the principle and characteristic of each CO_2 utilization technology were shown, their advantages and disadvantages were summarized and an outlook of CO_2 utilization was finally presented [112,113]. To date, photo-catalytic conversion of CO_2 to valuable hydrocarbons using solar energy has attracted a great deal of attention because it is one of the best solutions to both the energy shortage and the global warming problems, two birds with one stone for energy and environment issues. In a photo-catalysis reaction, the light-induced electron-hole pairs formed on the catalyst surfaces (e.g., TiO_2) are directly scavenged by chemical reactions instead of being collected by an electrode. However, effective electron-hole pairs can be recombined and dissipated as heat easily before they arrive at the photo-catalyst surface, which limits the efficiency of photo-catalytic activity [60,81,114,115]. Recently, Zhou and his co-workers reported about utilizing graphene-based TiO_2 nanocomposites for high photo-catalytic CO_2 conversion into renewable fuel (Fig. 6.12a,b). The enhanced performances are attributed to the large surface area of graphene sheets for loading of catalysts, and their excellent conductivity for electron capture and transport [10,113].

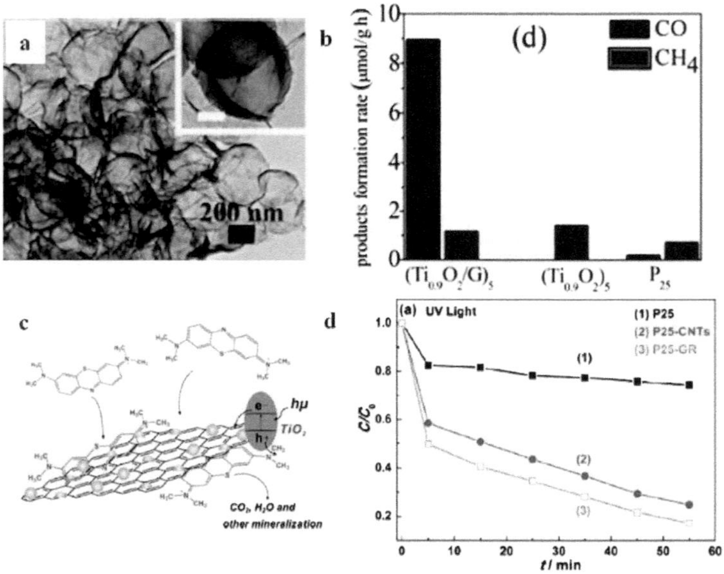

Figure 6.12 (a) (Graphene-$Ti_{0.91}O_2$)$_5$ hollow spheres. Scale bar of the inset in (a) 100 nm. (b) Comparation of the average product formation rates. (c) Schematic structure of P25-graphene and tentative processes of the photo-degradation of MB over P25-graphene. (d) Photo-degradation of MB under UV light over different photo-catalysts, respectively. Reproduced from Refs. [10], Copyright 2012 WILEY-VCH Verlag GmbH Co. KGaA; and [17], Copyright 2010 American Chemical Society.

Moreover, photo-degradation process of organic pollutants has attracted increasing attention during the past decades [116–122]. A chemically bonded TiO_2-graphene nanocomposite photo-catalyst was fabricated by a facile one-step hydrothermal method [17]. As shown in Fig. 6.12c,d, the TiO_2-graphene photo-catalyst possessed great adsorptivity of dyes, extended light absorption range, and efficient charge separation properties, and thus contributed to high photo-degradation of methylene blue (MB).

6.4 Conclusion and Outlook

Graphene is attractive because of its unique atom-thick 2D structure and excellent electrical, thermal, mechanical and optical properties. Furthermore, it has high chemical stability, and can be produced in a low-cost, controlled, scalable, and reproducible manner. Thus, graphene is regarded as a promising material for energy and environment applications. In the current review, we have systematically reviewed the synthesis of graphene-based materials and their applications as photo-electrodes in solar cells, high-performance electrodes in supercapacitors, and photo-catalysts for reduction of CO_2 and degradation of organic pollutants.

Although considerable progress has been achieved, the studies of graphene-based materials in the above fields are in their primary stages. And at least, the following challenges still remain. First, different applications require large scale syntheses of different grades of graphene, where different structures (layer number, size, edge group, and so on) and quantities must be considered. For example, as photo-anodes in solar cells, a CVD-grown large-area continuous graphene film is considered to be an ideal candidate to replace ITO for transparent electrodes. However, water-soluble RGO sheets are beneficial to the loading of catalysts or active materials for photo-catalysts or supercapacitors. Second, apart from experimental studies, it is highly desired to understand and develop new mechanisms of graphene-based materials working in energy and environment related systems. An advanced theory may help with understanding and predicting new properties associated with graphene-based materials. Rational design that combines theoretical modeling with state of the art experimental techniques would significantly facilitate graphene researches in energy and environment applications. Third, despite the fact that scientific results show superb advantages of graphene for different application potentials, there is still a long way to go for possible exploitation of graphene-based commercial products and a clear timetable in how soon the applications will reach the user or consumer may be needed. Nevertheless, so far, all the tremendous efforts for graphene have brought us a few real and significant perspectives. It could be convinced that the development of graphene-based materials will certainly lead to future advances in energy and environment applications.

Acknowledgments

This work was supported by the National High Technology Research and Development Program ("863" Program) of China (Grant No.

2011AA050525), the Knowledge Innovation Program of the Chinese Academy of Sciences (CAS) (Grant Nos. KGCX2-YW-370, KGCX2-EW-304-2), and the National Basic Research Program of China (Grant No. 2010CB934103). It was also partially supported by Key Laboratory of Photochemical Conversion and Optoelectronic Materials, Technical Institute of Physics and Chemistry (TIPC), CAS.

References

1. Novoselov, K. S., Geim, A. K., Morozov, S. V., Jiang, D., Zhang, Y., Dubonos, S. V., Grigorieva, I. V., Firsov, A. A., Electric field effect in atomically thin carbon films. *Science* **2004,** *306* (5696), 666–669.
2. Mayorov, A. S., Gorbachev, R. V., Morozov, S. V., Britnell, L., Jalil, R., Ponomarenko, L. A., Blake, P., Novoselov, K. S., Watanabe, K., Taniguchi, T., Geim, A. K., Micrometer-scale ballistic transport in encapsulated graphene at room temperature. *Nano Letters* **2011,** *11* (6), 2396–2399.
3. Stoller, M. D., Park, S., Zhu, Y., An, J., Ruoff, R. S., Graphene-based ultracapacitors. *Nano Letters* **2008,** *8* (10), 3498–3502.
4. Nair, R. R., Blake, P., Grigorenko, A. N., Novoselov, K. S., Booth, T. J., Stauber, T., Peres, N. M. R., Geim, A. K., Fine structure constant defines visual transparency of graphene. *Science* **2008,** *320* (5881), 1308.
5. Lee, C., Wei, X., Kysar, J. W., Hone, J., Measurement of the elastic properties and intrinsic strength of monolayer graphene. *Science* **2008,** *321* (5887), 385–388.
6. Balandin, A. A., Thermal properties of graphene and nanostructured carbon materials. *Nat Mater* **2011,** *10* (8), 569–581.
7. Li, D., Muller, M. B., Gilje, S., Kaner, R. B., Wallace, G. G., Processable aqueous dispersions of graphene nanosheets. *Nat Nano* **2008,** *3* (2), 101–105.
8. Lee, D. W., Hong, T.-K., Kang, D., Lee, J., Heo, M., Kim, J. Y., Kim, B.-S., Shin, H. S., Highly controllable transparent and conducting thin films using layer-by-layer assembly of oppositely charged reduced graphene oxides. *Journal of Materials Chemistry* **2011,** *21* (10), 3438–3442.
9. Zhu, J., He, J., Facile synthesis of graphene-wrapped honeycomb MnO_2 nanospheres and their application in supercapacitors. *ACS Applied Materials & Interfaces* **2012,** *4* (3), 1770–1776.
10. Tu, W., Zhou, Y., Liu, Q., Tian, Z., Gao, J., Chen, X., Zhang, H., Liu, J., Zou, Z., Robust hollow spheres consisting of alternating titania nanosheets

and graphene nanosheets with high photocatalytic activity for CO_2 conversion into renewable fuels. *Advanced Functional Materials* **2012,** *22* (6), 1215–1221.

11. Liu, Y., Yu, D., Zeng, C., Miao, Z., Dai, L., Biocompatible graphene oxide-based glucose biosensors. *Langmuir* **2010,** *26* (9), 6158–6160.
12. Weng, Z., Su, Y., Wang, D.-W., Li, F., Du, J., Cheng, H.-M., Graphene–cellulose paper flexible supercapacitors. *Advanced Energy Materials* **2011,** *1* (5), 917–922.
13. Yun, J.-M., Yeo, J.-S., Kim, J., Jeong, H.-G., Kim, D.-Y., Noh, Y.-J., Kim, S.-S., Ku, B.-C., Na, S.-I., Solution-processable reduced graphene oxide as a novel alternative to PEDOT:PSS hole transport layers for highly efficient and stable polymer solar cells. *Advanced Materials* **2011,** *23* (42), 4923–4928.
14. Wang, Y., Chen, X., Zhong, Y., Zhu, F., Loh, K. P., Large area, continuous, few-layered graphene as anodes in organic photovoltaic devices. *Applied Physics Letters* **2009,** *95* (6), 063302.
15. Krantz, J., Richter, M., Spallek, S., Spiecker, E., Brabec, C. J., Solution-processed metallic nanowire electrodes as indium tin oxide replacement for thin-film solar cells. *Advanced Functional Materials* **2011,** *21* (24), 4784–4787.
16. Wang, X., Zhi, L., Tsao, N., Tomović, Ž., Li, J., Müllen, K., Transparent carbon films as electrodes in organic solar cells. *Angewandte Chemie International Edition* **2008,** *47* (16), 2990–2992.
17. Zhang, H., Lv, X., Li, Y., Wang, Y., Li, J., P25-graphene composite as a high performance photocatalyst. *ACS Nano* **2009,** *4* (1), 380–386.
18. Chen, Y., Zhang, X., Zhang, D., Yu, P., Ma, Y., High performance supercapacitors based on reduced graphene oxide in aqueous and ionic liquid electrolytes. *Carbon* **2011,** *49* (2), 573–580.
19. Wang, L., Wang, D., Dong, X. Y., Zhang, Z. J., Pei, X. F., Chen, X. J., Chen, B., Jin, J., Layered assembly of graphene oxide and Co-Al layered double hydroxide nanosheets as electrode materials for supercapacitors. *Chemical Communications* **2011,** *47* (12), 3556–3558.
20. Chen, S., Zhu, J., Wu, X., Han, Q., Wang, X., Graphene oxide–MnO_2 nanocomposites for supercapacitors. *ACS Nano* **2010,** *4* (5), 2822–2830.
21. Li, X., Cai, W., An, J., Kim, S., Nah, J., Yang, D., Piner, R., Velamakanni, A., Jung, I., Tutuc, E., Banerjee, S. K., Colombo, L., Ruoff, R. S., Large-area synthesis of high-quality and uniform graphene films on copper foils. *Science* **2009,** *324* (5932), 1312–1314.

22. Emtsev, K. V., Bostwick, A., Horn, K., Jobst, J., Kellogg, G. L., Ley, L., McChesney, J. L., Ohta, T., Reshanov, S. A., Rohrl, J., Rotenberg, E., Schmid, A. K., Waldmann, D., Weber, H. B., Seyller, T., Towards wafer-size graphene layers by atmospheric pressure graphitization of silicon carbide. *Nat Mater* **2009,** *8* (3), 203–207.

23. Choucair, M., Thordarson, P., Stride, J. A., Gram-scale production of graphene based on solvothermal synthesis and sonication. *Nat Nano* **2009,** *4* (1), 30–33.

24. Jin, Z., Sun, Z., Simpson, L. J., O'Neill, K. J., Parilla, P. A., Li, Y., Stadie, N. P., Ahn, C. C., Kittrell, C., Tour, J. M., Solution-phase synthesis of heteroatom-substituted carbon scaffolds for hydrogen storage. *Journal of the American Chemical Society* **2010,** *132* (43), 15246–15251.

25. Yang, D., Velamakanni, A., Bozoklu, G., Park, S., Stoller, M., Piner, R. D., Stankovich, S., Jung, I., Field, D. A., Ventrice Jr, C. A., Ruoff, R. S., Chemical analysis of graphene oxide films after heat and chemical treatments by X-ray photoelectron and Micro-Raman spectroscopy. *Carbon* **2009,** *47* (1), 145-152.

26. Hummers, W. S., Offeman, R. E., Preparation of graphitic oxide. *Journal of the American Chemical Society* **1958,** *80* (6), 1339–1339.

27. Cote, L. J., Cruz-Silva, R., Huang, J., Flash reduction and patterning of graphite oxide and its polymer composite. *Journal of the American Chemical Society* **2009,** *131* (31), 11027–11032.

28. Lin, Z., Yao, Y., Li, Z., Liu, Y., Li, Z., Wong, C.-P., Solvent-assisted thermal reduction of graphite oxide. *The Journal of Physical Chemistry C* **2010,** *114* (35), 14819–14825.

29. Wu, Z.-S., Ren, W., Gao, L., Zhao, J., Chen, Z., Liu, B., Tang, D., Yu, B., Jiang, C., Cheng, H.-M., Synthesis of graphene sheets with high electrical conductivity and good thermal stability by hydrogen arc discharge exfoliation. *ACS Nano* **2009,** *3* (2), 411–417.

30. Hassan, H. M. A., Abdelsayed, V., Khder, A. E. R. S., AbouZeid, K. M., Terner, J., El-Shall, M. S., Al-Resayes, S. I., El-Azhary, A. A., Microwave synthesis of graphene sheets supporting metal nanocrystals in aqueous and organic media. *Journal of Materials Chemistry* **2009,** *19* (23), 3832–3837.

31. Liu, N., Luo, F., Wu, H., Liu, Y., Zhang, C., Chen, J., One-step ionic-liquid-assisted electrochemical synthesis of ionic-liquid-functionalized graphene sheets directly from graphite. *Advanced Functional Materials* **2008,** *18* (10), 1518–1525.

32. Shin, H.-J., Kim, K. K., Benayad, A., Yoon, S.-M., Park, H. K., Jung, I.-S., Jin, M. H., Jeong, H.-K., Kim, J. M., Choi, J.-Y., Lee, Y. H., Efficient reduction

of graphite oxide by sodium borohydride and its effect on electrical conductance. *Advanced Functional Materials* **2009,** *19* (12), 1987–1992.

33. Zhou, Y., Bao, Q., Tang, L. A. L., Zhong, Y., Loh, K. P., Hydrothermal dehydration for the "green" reduction of exfoliated graphene oxide to graphene and demonstration of tunable optical limiting properties. *Chemistry of Materials* **2009,** *21* (13), 2950–2956.
34. Fan, X., Peng, W., Li, Y., Li, X., Wang, S., Zhang, G., Zhang, F., Deoxygenation of exfoliated graphite oxide under alkaline conditions: a green route to graphene preparation. *Advanced Materials* **2008,** *20* (23), 4490–4493.
35. Zhang, J., Yang, H., Shen, G., Cheng, P., Zhang, J., Guo, S., Reduction of graphene oxide vial-ascorbic acid. *Chemical Communications* **2010,** *46* (7), 1112–1114.
36. Fan, Z.-J., Kai, W., Yan, J., Wei, T., Zhi, L.-J., Feng, J., Ren, Y.-m., Song, L.-P., Wei, F., Facile synthesis of graphene nanosheets via fe reduction of exfoliated graphite oxide. *ACS Nano* **2010,** *5* (1), 191–198.
37. Pei, S., Zhao, J., Du, J., Ren, W., Cheng, H.-M., Direct reduction of graphene oxide films into highly conductive and flexible graphene films by hydrohalic acids. *Carbon* **2010,** *48* (15), 4466–4474.
38. O'Neill, A., Khan, U., Nirmalraj, P. N., Boland, J., Coleman, J. N., Graphene dispersion and exfoliation in low boiling point solvents. *The Journal of Physical Chemistry C* **2011,** *115* (13), 5422–5428.
39. Coleman, J. N., Liquid exfoliation of defect-free graphene. *Accounts of Chemical Research* **2012,** *46* (1), 14–22.
40. Lotya, M., Hernandez, Y., King, P. J., Smith, R. J., Nicolosi, V., Karlsson, L. S., Blighe, F. M., De, S., Wang, Z., McGovern, I. T., Duesberg, G. S., Coleman, J. N., Liquid phase production of graphene by exfoliation of graphite in surfactant/water solutions. *Journal of the American Chemical Society* **2009,** *131* (10), 3611–3620.
41. Kosynkin, D. V., Higginbotham, A. L., Sinitskii, A., Lomeda, J. R., Dimiev, A., Price, B. K., Tour, J. M., Longitudinal unzipping of carbon nanotubes to form graphene nanoribbons. *Nature* **2009,** *458* (7240), 872–876.
42. Jiao, L., Zhang, L., Wang, X., Diankov, G., Dai, H., Narrow graphene nanoribbons from carbon nanotubes. *Nature* **2009,** *458* (7240), 877–880.
43. Elías, A. L., Botello-Méndez, A. s. R., Meneses-Rodríguez, D., Jehová González, V., Ramírez-González, D., Ci, L., Muñoz-Sandoval, E., Ajayan, P. M., Terrones, H., Terrones, M., Longitudinal cutting of pure and doped

carbon nanotubes to form graphitic nanoribbons using metal clusters as nanoscalpels. *Nano Letters* **2009,** *10* (2), 366–372.

44. Vlassiouk, I., Regmi, M., Fulvio, P., Dai, S., Datskos, P., Eres, G., Smirnov, S., Role of hydrogen in chemical vapor deposition growth of large single-crystal graphene. *ACS Nano* **2011,** *5* (7), 6069–6076.

45. Su, C.-Y., Lu, A.-Y., Wu, C.-Y., Li, Y.-T., Liu, K.-K., Zhang, W., Lin, S.-Y., Juang, Z.-Y., Zhong, Y.-L., Chen, F.-R., Li, L.-J., Direct formation of wafer scale graphene thin layers on insulating substrates by chemical vapor deposition. *Nano Letters* **2011,** *11* (9), 3612–3616.

46. Lee, W. H., Suk, J. W., Lee, J., Hao, Y., Park, J., Yang, J. W., Ha, H.-W., Murali, S., Chou, H., Akinwande, D., Kim, K. S., Ruoff, R. S., Simultaneous transfer and doping of CVD-grown graphene by fluoropolymer for transparent conductive films on plastic. *ACS Nano* **2012,** *6* (2), 1284–1290.

47. Wang, Y., Zheng, Y., Xu, X., Dubuisson, E., Bao, Q., Lu, J., Loh, K. P., Electrochemical delamination of CVD-grown graphene film: toward the recyclable use of copper catalyst. *ACS Nano* **2011,** *5* (12), 9927–9933.

48. Suk, J. W., Kitt, A., Magnuson, C. W., Hao, Y., Ahmed, S., An, J., Swan, A. K., Goldberg, B. B., Ruoff, R. S., Transfer of CVD-grown monolayer graphene onto arbitrary substrates. *ACS Nano* **2011,** *5* (9), 6916–6924.

49. Bae, S., Kim, H., Lee, Y., Xu, X., Park, J.-S., Zheng, Y., Balakrishnan, J., Lei, T., Ri Kim, H., Song, Y. I., Kim, Y.-J., Kim, K. S., Ozyilmaz, B., Ahn, J.-H., Hong, B. H., Iijima, S., Roll-to-roll production of 30-inch graphene films for transparent electrodes. *Nat Nano* **2010,** *5* (8), 574–578.

50. Park, J. S., Cho, S. M., Kim, W.-J., Park, J., Yoo, P. J., Fabrication of graphene thin films based on layer-by-layer self-assembly of functionalized graphene nanosheets. *ACS Applied Materials & Interfaces* **2011,** *3* (2), 360–368.

51. Geng, J., Jung, H.-T., Porphyrin functionalized graphene sheets in aqueous suspensions: from the preparation of graphene sheets to highly conductive graphene films. *The Journal of Physical Chemistry C* **2010,** *114* (18), 8227–8234.

52. Shim, D., Jung, S.-H., Han, S. Y., Shin, K., Lee, K.-H., Han, J. H., Improvement of SWCNT transparent conductive films via transition metal doping. *Chemical Communications* **2011,** *47* (18), 5202–5204.

53. Hong, J.-Y., Shin, K.-Y., Kwon, O. S., Kang, H., Jang, J., A strategy for fabricating single layer graphene sheets based on a layer-by-layer self-assembly. *Chemical Communications* **2011,** *47* (25), 7182–7184.

54. Zheng, Q., Ip, W. H., Lin, X., Yousefi, N., Yeung, K. K., Li, Z., Kim, J.-K., Transparent conductive films consisting of ultralarge graphene sheets

produced by Langmuir–Blodgett assembly. *ACS Nano* **2011,** *5* (7), 6039–6051.

55. Li, Z., Wang, J., Liu, X., Liu, S., Ou, J., Yang, S., Electrostatic layer-by-layer self-assembly multilayer films based on graphene and manganese dioxide sheets as novel electrode materials for supercapacitors. *Journal of Materials Chemistry* **2011,** *21* (10), 3397–3403.

56. Zhu, J., He, J., Assembly and benign step-by-step post-treatment of oppositely charged reduced graphene oxides for transparent conductive thin films with multiple applications. *Nanoscale* **2012,** *4* (11), 3558–3566.

57. Zhao, B., Song, J., Liu, P., Xu, W., Fang, T., Jiao, Z., Zhang, H., Jiang, Y., Monolayer graphene/NiO nanosheets with two-dimension structure for supercapacitors. *Journal of Materials Chemistry* **2011,** *21* (46), 18792–18798.

58. Huang, Z.-D., Zhang, B., Oh, S.-W., Zheng, Q.-B., Lin, X.-Y., Yousefi, N., Kim, J.-K., Self-assembled reduced graphene oxide/carbon nanotube thin films as electrodes for supercapacitors. *Journal of Materials Chemistry* **2012,** *22* (8), 3591–3599.

59. Huang, H., Wang, X., Graphene nanoplate-MnO_2 composites for supercapacitors: a controllable oxidation approach. *Nanoscale* **2011,** *3* (8), 3185–3191.

60. Min, S., Lu, G., Dye-sensitized reduced graphene oxide photocatalysts for highly efficient visible-light-driven water reduction. *The Journal of Physical Chemistry C* **2011,** *115* (28), 13938–13945.

61. Liu, J., Fu, S., Yuan, B., Li, Y., Deng, Z., Toward a universal "adhesive nanosheet" for the assembly of multiple nanoparticles based on a protein-induced reduction/decoration of graphene oxide. *Journal of the American Chemical Society* **2010,** *132* (21), 7279–7281.

62. Zhang, D., Zhang, X., Chen, Y., Yu, P., Wang, C., Ma, Y., Enhanced capacitance and rate capability of graphene/polypyrrole composite as electrode material for supercapacitors. *Journal of Power Sources* **2011,** *196* (14), 5990–5996.

63. Zhang, S., Shao, Y., Liao, H., Engelhard, M. H., Yin, G., Lin, Y., Polyelectrolyte-induced reduction of exfoliated graphite oxide: a facile route to synthesis of soluble graphene nanosheets. *ACS Nano* **2011,** *5* (3), 1785–1791.

64. Yan, J., Fan, Z., Wei, T., Qian, W., Zhang, M., Wei, F., Fast and reversible surface redox reaction of graphene-MnO_2 composites as supercapacitor electrodes. *Carbon* **2010,** *48* (13), 3825–3833.

65. Yang, S., Feng, X., Ivanovici, S., Müllen, K., Fabrication of graphene-encapsulated oxide nanoparticles: towards high-performance anode materials for lithium storage. *Angewandte Chemie International Edition* **2010,** *49* (45), 8408–8411.

66. Zhou, G., Wang, D.-W., Li, F., Zhang, L., Li, N., Wu, Z.-S., Wen, L., Lu, G. Q., Cheng, H.-M., Graphene-wrapped Fe_3O_4 anode material with improved reversible capacity and cyclic stability for lithium ion batteries. *Chemistry of Materials* **2010,** *22* (18), 5306–5313.

67. Zhou, W., Zhu, J., Cheng, C., Liu, J., Yang, H., Cong, C., Guan, C., Jia, X., Fan, H. J., Yan, Q., Li, C. M., Yu, T., A general strategy toward graphene@metal oxide core-shell nanostructures for high-performance lithium storage. *Energy & Environmental Science* **2011,** *4* (12), 4954–4961.

68. De, S., Coleman, J. N., Are there fundamental limitations on the sheet resistance and transmittance of thin graphene films. *ACS Nano* **2010,** *4* (5), 2713–2720.

69. Crispin, X., Jakobsson, F. L. E., Crispin, A., Grim, P. C. M., Andersson, P., Volodin, A., van Haesendonck, C., Van der Auweraer, M., Salaneck, W. R., Berggren, M., The origin of the high conductivity of poly(3,4-e thylenedioxythiophene)–poly(styrenesulfonate) (PEDOT–PSS) plastic electrodes. *Chemistry of Materials* **2006,** *18* (18), 4354–4360.

70. Tenent, R. C., Barnes, T. M., Bergeson, J. D., Ferguson, A. J., To, B., Gedvilas, L. M., Heben, M. J., Blackburn, J. L., Ultrasmooth, large-area, high-uniformity, conductive transparent single-walled-carbon-nanotube films for photovoltaics produced by ultrasonic spraying. *Advanced Materials* **2009,** *21* (31), 3210–3216.

71. Zhao, J., Pei, S., Ren, W., Gao, L., Cheng, H.-M., Efficient preparation of large-area graphene oxide sheets for transparent conductive films. *ACS Nano* **2010,** *4* (9), 5245–5252.

72. Zhu, Y., Sun, Z., Yan, Z., Jin, Z., Tour, J. M., Rational design of hybrid graphene films for high-performance transparent electrodes. *ACS Nano* **2011,** *5* (8), 6472–6479.

73. Sun, Y., Welch, G. C., Leong, W. L., Takacs, C. J., Bazan, G. C., Heeger, A. J., Solution-processed small-molecule solar cells with 6.7% efficiency. *Nat Mater* **2012,** *11* (1), 44–48.

74. Liu, Q., Liu, Z., Zhang, X., Yang, L., Zhang, N., Pan, G., Yin, S., Chen, Y., Wei, J., Polymer photovoltaic cells based on solution-processable graphene and P3HT. *Advanced Functional Materials* **2009,** *19* (6), 894–904.

75. Liu, Z., Liu, Q., Huang, Y., Ma, Y., Yin, S., Zhang, X., Sun, W., Chen, Y., Organic photovoltaic devices based on a novel acceptor material: graphene. *Advanced Materials* **2008,** *20* (20), 3924–3930.

76. Li, S.-S., Tu, K.-H., Lin, C.-C., Chen, C.-W., Chhowalla, M., Solution-processable graphene oxide as an efficient hole transport layer in polymer solar cells. *ACS Nano* **2010,** *4* (6), 3169–3174.

77. Krishnamoorthy, T., Tang, M. Z., Verma, A., Nair, A. S., Pliszka, D., Mhaisalkar, S. G., Ramakrishna, S., A facile route to vertically aligned electrospun SnO_2 nanowires on a transparent conducting oxide substrate for dye-sensitized solar cells. *Journal of Materials Chemistry* **2012,** *22* (5), 2166–2172.

78. Wang, X., Zhi, L., Mullen, K., Transparent, conductive graphene electrodes for dye-sensitized solar cells. *Nano Letters* **2007,** *8* (1), 323–327.

79. Tang, Y.-B., Lee, C.-S., Xu, J., Liu, Z.-T., Chen, Z.-H., He, Z., Cao, Y.-L., Yuan, G., Song, H., Chen, L., Luo, L., Cheng, H.-M., Zhang, W.-J., Bello, I., Lee, S.-T., Incorporation of graphenes in nanostructured TiO_2 films via molecular grafting for dye-sensitized solar cell application. *ACS Nano* **2010,** *4* (6), 3482–3488.

80. Lightcap, I. V., Kosel, T. H., Kamat, P. V., Anchoring semiconductor and metal nanoparticles on a two-dimensional catalyst mat. Storing and shuttling electrons with reduced graphene oxide. *Nano Letters* **2010,** *10* (2), 577–583.

81. Ng, Y. H., Lightcap, I. V., Goodwin, K., Matsumura, M., Kamat, P. V., To what extent do graphene scaffolds improve the photovoltaic and photocatalytic response of TiO_2 nanostructured films. *The Journal of Physical Chemistry Letters* **2010,** *1* (15), 2222–2227.

82. Hong, W., Xu, Y., Lu, G., Li, C., Shi, G., Transparent graphene/PEDOT–PSS composite films as counter electrodes of dye-sensitized solar cells. *Electrochemistry Communications* **2008,** *10* (10), 1555–1558.

83. Xu, Y., Bai, H., Lu, G., Li, C., Shi, G., Flexible graphene films via the filtration of water-soluble noncovalent functionalized graphene sheets. *Journal of the American Chemical Society* **2008,** *130* (18), 5856–5857.

84. Cao, Y., He, J.-H., Zhu, J.-L., Sun, J.-L., Fabrication of carbon nanotube/silicon nanowire array heterojunctions and their silicon nanowire length dependent photoresponses. *Chemical Physics Letters* **2011,** *501* (4–6), 461–465.

85. Li, X., Zhu, H., Wang, K., Cao, A., Wei, J., Li, C., Jia, Y., Li, Z., Li, X., Wu, D., Graphene-on-silicon Schottky junction solar cells. *Advanced Materials* **2010,** *22* (25), 2743–2748.

86. Hu, Y. H., Wang, H., Hu, B., Thinnest two-dimensional nanomaterial—graphene for solar energy. *ChemSusChem* **2010,** *3* (7), 782–796.

87. Sun, Y., Wu, Q., Shi, G., Graphene based new energy materials. *Energy & Environmental Science* **2011,** *4* (4), 1113–1132.

88. Huang, X., Qi, X., Boey, F., Zhang, H., Graphene-based composites. *Chemical Society Reviews* **2012,** *41* (2), 666–686.

89. Sahoo, N. G., Pan, Y., Li, L., Chan, S. H., Graphene-based materials for energy conversion. *Advanced Materials* **2012,** *24* (30), 4203–4210.

90. Yoo, J. J., Balakrishnan, K., Huang, J., Meunier, V., Sumpter, B. G., Srivastava, A., Conway, M., Mohana Reddy, A. L., Yu, J., Vajtai, R., Ajayan, P. M., Ultrathin planar graphene supercapacitors. *Nano Letters* **2011,** *11* (4), 1423–1427.

91. Mishra, A. K., Ramaprabhu, S., Functionalized graphene-based nanocomposites for supercapacitor application. *The Journal of Physical Chemistry C* **2011,** *115* (29), 14006–14013.

92. Li, W., Zhang, F., Dou, Y., Wu, Z., Liu, H., Qian, X., Gu, D., Xia, Y., Tu, B., Zhao, D., A self-template strategy for the synthesis of mesoporous carbon nanofibers as advanced supercapacitor electrodes. *Advanced Energy Materials* **2011,** *1* (3), 382–386.

93. Bose, S., Kuila, T., Mishra, A. K., Rajasekar, R., Kim, N. H., Lee, J. H., Carbon-based nanostructured materials and their composites as supercapacitor electrodes. *Journal of Materials Chemistry* **2012,** *22* (3), 767–784.

94. Yu, D., Dai, L., Self-assembled graphene/carbon nanotube hybrid films for supercapacitors. *The Journal of Physical Chemistry Letters* **2009,** *1* (2), 467–470.

95. Wei, W., Cui, X., Chen, W., Ivey, D. G., Manganese oxide-based materials as electrochemical supercapacitor electrodes. *Chemical Society Reviews* **2011,** *40* (3), 1697–1721.

96. Mu, J., Chen, B., Guo, Z., Zhang, M., Zhang, Z., Zhang, P., Shao, C., Liu, Y., Highly dispersed Fe_3O_4 nanosheets on one-dimensional carbon nanofibers: Synthesis, formation mechanism, and electrochemical performance as supercapacitor electrode materials. *Nanoscale* **2011,** *3* (12), 5034–5040.

97. Wang, Y.-T., Lu, A.-H., Zhang, H.-L., Li, W.-C., Synthesis of nanostructured mesoporous manganese oxides with three-dimensional frameworks and their application in supercapacitors. *The Journal of Physical Chemistry C* **2011,** *115* (13), 5413–5421.

98. Prasad, K. R., Miura, N., Potentiodynamically deposited nanostructured manganese dioxide as electrode material for electrochemical redox supercapacitors. *Journal of Power Sources* **2004,** *135* (1-2), 354-360.

99. Brousse, T., Toupin, M., Dugas, R., Athouel, L., Crosnier, O., Belanger, D., Crystalline MnO as possible alternatives to amorphous compounds in electrochemical supercapacitors. *Journal of the Electrochemical Society* **2006,** *153* (12), A2171–A2180.

100. Chen, H., He, J., Facile synthesis of monodisperse manganese oxide nanostructures and their application in water treatment. *The Journal of Physical Chemistry C* **2008,** *112* (45), 17540–17545.

101. Chen, H., He, J., Zhang, C., He, H., Self-assembly of novel mesoporous manganese oxide nanostructures and their application in oxidative decomposition of formaldehyde. *The Journal of Physical Chemistry C* **2007,** *111* (49), 18033–18038.

102. Zhu, J., He, J., Effects of reaction conditions on the structure, morphology and size of manganese oxide nanoparticles. *Acta Chimica Sinica* **2010,** *68* (10), 961–968.

103. Lee, H., Kang, J., Cho, M. S., Choi, J.-B., Lee, Y., MnO_2/graphene composite electrodes for supercapacitors: the effect of graphene intercalation on capacitance. *Journal of Materials Chemistry* **2011,** *21* (45), 18215–18219.

104. Lei, Z., Shi, F., Lu, L., Incorporation of MnO_2-coated carbon nanotubes between graphene sheets as supercapacitor electrode. *ACS Applied Materials & Interfaces* **2012,** *4* (2), 1058–1064.

105. Li, Z., Mi, Y., Liu, X., Liu, S., Yang, S., Wang, J., Flexible graphene/MnO_2 composite papers for supercapacitor electrodes. *Journal of Materials Chemistry* **2011,** *21* (38), 14706–14711.

106. Lin, Y.-H., Wei, T.-Y., Chien, H.-C., Lu, S.-Y., Manganese oxide/carbon aerogel composite: an outstanding supercapacitor electrode material. *Advanced Energy Materials* **2011,** *1* (5), 901–907.

107. Mao, L., Zhang, K., On Chan, H. S., Wu, J., Nanostructured MnO_2/graphene composites for supercapacitor electrodes: the effect of morphology, crystallinity and composition. *Journal of Materials Chemistry* **2012,** *22* (5), 1845–1851.

108. Yan, J., Wei, T., Qiao, W., Shao, B., Zhao, Q., Zhang, L., Fan, Z., Rapid microwave-assisted synthesis of graphene nanosheet/Co_3O_4 composite for supercapacitors. *Electrochimica Acta* **2010,** *55* (23), 6973–6978.

109. Sathish, M., Mitani, S., Tomai, T., Honma, I., MnO_2 assisted oxidative polymerization of aniline on graphene sheets: superior nanocomposite electrodes for electrochemical supercapacitors. *Journal of Materials Chemistry* **2011,** *21* (40), 16216–16222.

110. Wang, H., Hao, Q., Yang, X., Lu, L., Wang, X., Graphene oxide doped polyaniline for supercapacitors. *Electrochemistry Communications* **2009,** *11* (6), 1158–1161.

111. Xu, J., Wang, K., Zu, S.-Z., Han, B.-H., Wei, Z., Hierarchical nanocomposites of polyaniline nanowire arrays on graphene oxide sheets with synergistic effect for energy storage. *ACS Nano* **2010,** *4* (9), 5019–5026.

112. Zhu, J., He, J., Capture, Storage and transformation of carbon dioxide. *Imaging Science and Photochemistry* **2011,** *29* (3), 168–180.

113. Liang, Y. T., Vijayan, B. K., Gray, K. A., Hersam, M. C., Minimizing graphene defects enhances titania nanocomposite-based photocatalytic reduction of CO_2 for improved solar fuel production. *Nano Letters* **2011,** *11* (7), 2865–2870.

114. Yao, H.-B., Wu, L.-H., Cui, C.-H., Fang, H.-Y., Yu, S.-H., Direct fabrication of photoconductive patterns on LBL assembled graphene oxide/PDDA/titania hybrid films by photothermal and photocatalytic reduction. *Journal of Materials Chemistry* **2010,** *20* (25), 5190–5195.

115. Lee, J. S., You, K. H., Park, C. B., Highly photoactive, low bandgap TiO_2 nanoparticles wrapped by graphene. *Advanced Materials* **2012,** *24* (8), 1084–1088.

116. Guo, S., Dong, S., Graphene nanosheet: synthesis, molecular engineering, thin film, hybrids, and energy and analytical applications. *Chemical Society Reviews* **2011,** *40* (5), 2644–2672.

117. Zhu, Y., James, D. K., Tour, J. M., New routes to graphene, graphene oxide and their related applications. *Advanced Materials* **2012,** *24* (36), 4924–4955.

118. Wan, X., Huang, Y., Chen, Y., Focusing on energy and optoelectronic applications: a journey for graphene and graphene oxide at large scale. *Accounts of Chemical Research* **2012,** *45* (4), 598–607.

119. Zhu, J., Xu, L., He, J., Assembly of graphene nanosheets and SiO_2 nanoparticles towards transparent, antireflective, conductive, and superhydrophilic multifunctional hybrid films. *Chemistry – A European Journal* **2012,** *18* (51), 16393–16401.

120. Zhu, J., He, J., Fabrication and its transient optical properties of graphene thin films. *Materials Science Forum* **2013,** *743*, 892–902.

121. Zhu, J., He, J., Self-assembled fabrication of graphene based materials with optical-electronic, transient optical and electrochemical properties. *International Journal of Nanoscience* **2012,** *11* (6), 1240032.

122. Novoselov, K. S., Falko, V. I., Colombo, L., Gellert, P. R., Schwab, M. G., Kim, K., A roadmap for graphene. *Nature* **2012,** *490* (7419), 192–200.

Chapter 7

Recent Advances in Synthesis and Applications of Metal-Added Carbon Nanotubes and Graphenes

K. P. Annamalai and Yousheng Tao
CAS Key Laboratory of Design and Assembly of Functional Nanostructures, FJIRSM, Chinese Academy of Sciences (CAS) and Fujian Provincial Key Laboratory of Nanomaterials, Haixi Institutes of CAS, Fuzhou, Fujian 350002, China
taoys@tom.com

7.1 Introduction

Carbon nanotubes (CNTs) and graphenes have unique nanostructures with large surface area, high thermal, electrical, mechanical, and optical properties [1–4]. They are promising for various industrial applications. For example, their large surface area can be utilized to support metals for the applications in gas adsorption, energy storage, and catalysis. But these large surface areas are restricted by stacking behavior due to Vander walls force and π-π interaction. To overcome these drawbacks, CNTs and graphene/polymer composite and metal-added CNTs and graphene are explored [5–8].

Nanomaterials in Energy and Environmental Applications
Edited by Junhui He

ISBN 978-981-4463-78-2 (Hardcover), 978-981- 4463-79-9 (eBook)
www.panstanford.com

CNTs are tubular structure having nanometer scale diameter but micrometer scale length [9]. CNTs could be prepared by means of arc discharge, laser ablation, and chemical vapor deposition (CVD). Arc discharge and laser ablation techniques are difficult to control the size, shape, and alignment due to higher amorphous carbon impurities and high probability to obtain MWCNTs [10]. CVD is an efficient method to synthesize CNTs with controlled size, shape, alignment, high yield, and low impurities.

The surface of pure CNTs are inert, while uniform dispersion of metal nanoparticle can be achieved by introducing some functional group on CNT or surface oxidation. Without surface modification, CNTs lacks sufficient binding sites for anchoring metal nanoparticles, which further leads to poor dispersion and nanoparticle aggregation [11].

Graphene is a one-atom thick 2D material, which was prepared first by mechanical exfoliation, that is, scotch tape method [12]. Now several methods were invented to synthesize graphenes. Thermal decomposition of SiC wafer under ultrahigh vacuum (UHV) condition, CVD growth on metal substrate, and metal-free CVD growth of graphene are mass production routes to synthesize graphene. Bulk production technique to produce graphene is chemical exfoliation of graphite oxide [13], of which producing graphene from graphene oxide (GO) reduction normally gives some functional groups as well as defects in it.

To avoid the stacking behavior of carbon nanotubes and graphene layers, metals, metal nanoparticles, organic compounds, and other nanocarbons are inserted between the graphene nano-sheets or the carbon nanotubes. Among them, metal nanoparticle intercalation is a facile route because the nanoparticle intercalation and decoration prevent the restacking of nano sheets [14]. Furthermore, it avoids the nanoparticle aggregation. In other words, intercalated nanoparticles prevent these nanocarbons stacking; nanocarbon support prevents nanoparticle aggregation. Note that metal nanoparticles have large surface area, high energy storage capability, good photochemical and electrochemical properties. They assist and improve the properties of nanocarbons. This synergistic effect of these metal-added nanocarbon materials has promising applications in energy and environmental engineering [15]. This chapter discusses about the synthesis of metal-added nanocarbon materials and their promising applications such as hydrogen adsorption, electrochemical energy storage, catalysis, and many other aspects.

7.2 Synthesis of Metal-Added Carbon Nanotubes and Graphenes

Metal-added CNTs and graphenes can be synthesized by wet chemical method and dry method, of which wet chemical methods are used to prepare different nanoparticles with uniform sizes. Synthesizing metal nanoparticles of small sizes and with narrow size distribution is of interest in the field.

7.2.1 Metal-Added Carbon Nanotubes

Initially, nanoparticles were synthesized with lack of uniformity because of uncontrolled agglomeration due to attractive Vander Walls forces between the nanoparticles; it can give rise to micro structural particles. So, researchers used polymers to achieve uniform nanoparticles. It was reported that nanoparticle sizes can be controlled by polymer concentration [16,17]. Because polymers are adsorbed on the surface of nanoparticles, it is a drawback that they block the active sites of the nanoparticles, which restrict the free access of fuels in fuel cell application and decrease the catalytic activity. Schmidt et al. synthesized uniform Pt nanoparticles using poly acrylic acid (PAA) [18]. XPS results showed that the Pt nanoparticles contained significant amount of carbon; it indirectly indicated that PAA capping agent was present on the nanoparticles even after several wash. Post-treatment (heat treatment) could release the active site from surface blocking, but burning out these polymers by heat treatments in air or oxygen causes aggregation of nanoparticles [19].

In recent years, nanocarbon-assisted nanoparticles synthesis techniques have been of great interest due to extraordinary properties of nanocarbons and nanoparticles. New physical and chemical properties emerged as a result of synergetic effect of nanocarbon-nanoparticle interaction. Although nanocarbons have unique features, surface of pristine nanocarbons is inert. These inert surfaces of nanocarbons could not strongly bind nanoparticles, which results in that prepared nanoparticles tend to agglomerate easily [11]. Functionalization of nanocarbon surface is efficient for nanocarbons to anchore nanoparticles and high surface area facilitates ease of nanoparticles adsorption. These nanocarbon-supported nanoparticles have shown a promising building block for various potential applications.

CNTs are normally functionalized with harsh oxidation agent, for example, concentrated acids (HNO_3:H_2SO_4) reflux. It also gives defects and oxygenated functional groups on CNTs, which act as anchoring agent for nanoparticle deposition. Functionalized CNTs further mixed into metal ion solutions and then reducing agent was added to convert this metal ion into metallic nanoparticles. Although acid-treated CNTs produce active surface and provide better nanoparticles dispersion, harsh oxidation process harms the CNTs intrinsic properties [20]. In order to withstand the original properties of CNTs, non-covalent functionalization is emerged [21,22]. Figure 7.1 shows that schematic illustration of non-covalent functionalization followed by nanoparticle deposition.

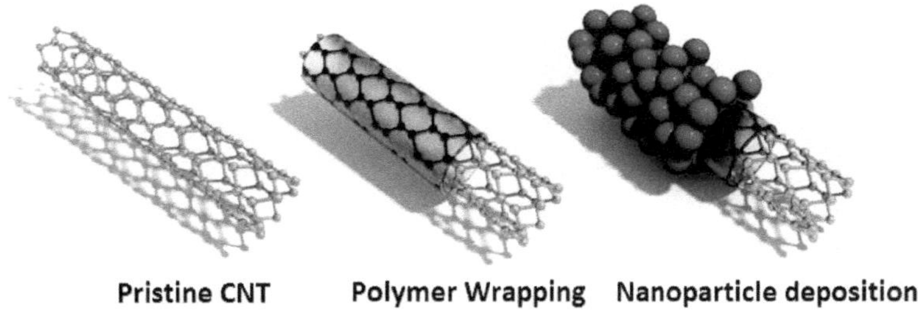

Figure 7.1 Schematic diagram of polymer wrapping and deposition of nanoparticles on CNT surface. Reproduced from *J. Mater. Chem.* 2006, **16**, 22–25. Copyright © 2006, The Royal Society of Chemistry [22].

In the early stages, nanoparticle–CNT hybrids synthesized by ex situ method, that is, already-prepared (polymer coated) nanoparticles mixed into CNTs under constant stirring. For example, Huang et al. synthesized Au-decorated CNTs using 1-pyrenemethylamine linker [23]. Au nanoparticles are surrounded by pyrenemethylamine stabilizer; further, this stabilizer was attached with CNTs non-covalently through π-π interaction. Figure 7.2 shows the uniform distribution of metal nanoparticles on CNTs.

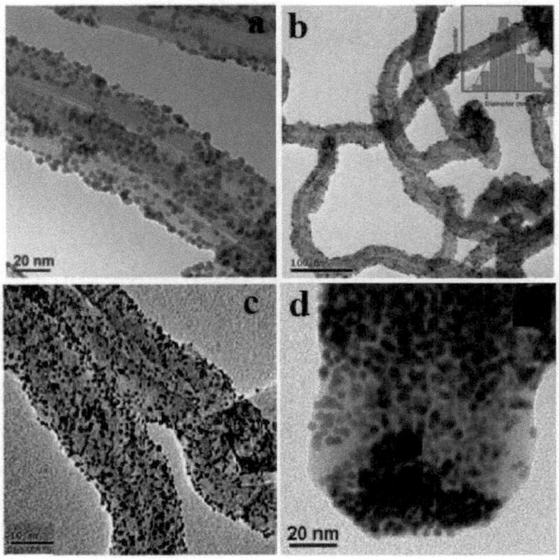

Figure 7.2 TEM images of nanoparticles on polymer-blended CNTs: (a) MWCNT-PyPBI-Pt nanoparticles (reproduced from *Adv. Mater.* 2013, **25**, 1666–1681. Copyright © 2013, Wiley-VCH Verlag GmbH & Co. KGaA [24]), (b) MWCNT-PDOP-Pt nanoparticles (reproduced from *Nanotechnology* 2013, **24**, 065501. Copyright © 2013, IOP Publishing Ltd [25]), (c) MWCNT-1-AP-PtRu nanoparticles (reproduced from *Langmuir* 2008, **24**, 10505–10512. Copyright © 2008, American Chemical Society [26]) and (d) Py-CH_2NH_2-MWCNT-Au nanoparticles (reproduced from J. Phys. Chem. B 2006, **110**, 2031–2036. Copyright © 2006, American Chemical Society [23]). [PyPBI: polybenzimidazole; PDOP: Polydopamine; 1-AP: 1-Aminopyrene; Py-CH_2NH_2: 1-pyrenemethylamine.]

Fujigaya et al. reported that poly benzimidazole (PBI) wrapped CNTs induce uniform and strong binding of Pt nanoparticles [24]. Synthesis method involved that CNTs were simply blend with PBI; subsequently, Pt decoration was done by conventional polyol method. Interestingly, different sizes of nanoparticles were observed with varying the concentration of ethylene glycol, as shown in Fig. 7.3.

Polydopamine-coated CNTs enhance the nanoparticles–CNTs interaction that leads to uniform distribution of nanoparticles [25]. Jiang et al. [26] stated uniform nanoparticles deposition with the aid of 1-aminopyrene as a linker. Negative charge of the Pt ions attracted by positively charged amino group binded on CNT surface.

In addition, Ru ions bind in the same place and result in bimetal nanoparticles successfully loaded on CNTs. Bimetal nanoparticles were investigated to boost up the monometallic nanoparticles performance. Recently, Annamalai et al. discussed CNTs-supported PtRu bimetallic nanoparticles and suggested the foremost industrial applications [27]. Synthesis of bimetal nanoparticles is the same as conventional preparation of nanoparticles except second metal ion addition. Cheng et al. investigated poly ethyleneimine (PEI) functionalized CNTs for bimetal nanoparticles support [28]. Metal nanoparticles deposited on PEI blended-CNT by means of polyol method combined with microwave treatment. Resultant PtRu-PEI-CNTs exhibited higher performance in comparison to conventional acid treated CNTs. Chen et al. synthesized PtPd/MWCNTs with aqueous H_2PtCl_6 and Na_2PdCl_4 solutions and MWCNTs as starting materials by using the colloidal reduction method [29].

Although wet chemical methods are facile and easy to practice for the preparation of nanoparticles, dry methods such as sputtering method are easy to control the sizes of nanoparticles on nanocarbons without need of any surface modifier. Precursors were sprayed through the nebulizers and the droplets were reduced at high temperature. Here, nanoparticle loading was controlled by the concentration of metal precursors. Electrostatic force directed assembly is one of the dry methods processed at room temperature. Nanoparticles such as SnO_2 and Ag were successfully incorporated on sidewalls of CNTs by this method [30].

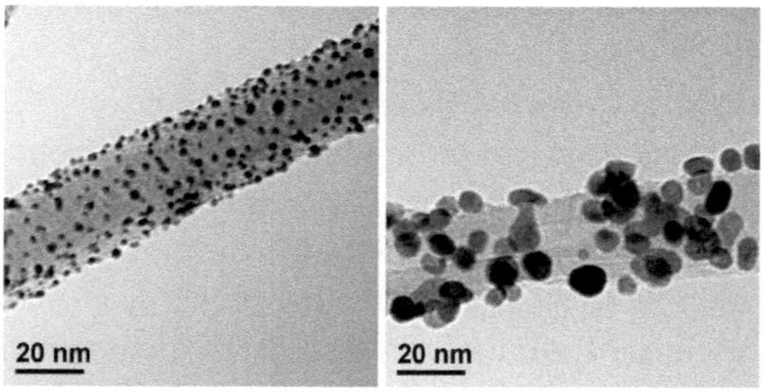

Figure 7.3 TEM images of Pt nanoparticle added CNT synthesized using 60% aq. solution (left) and 30% aq. solution (right) of ethylene glycol. Reproduced from *Adv. Mater.* 2013, **25**, 1666–1681. Copyright © 2013, Wiley-VCH Verlag GmbH & Co. KGaA [24].

7.2.2 Metal-Added Graphene

In order to add metals onto graphene, pristine graphene is usually charged with some functionalization agents. For example, graphene surface can be non-covalently modified with water-dispersible polymers and ionic liquids in order to affirm uniform nanoparticles on it [31]. Chemically exfoliated graphene derivatives intrinsically produce defects and functional groups on it, and its high surface area facilitates nanoparticle adsorption.

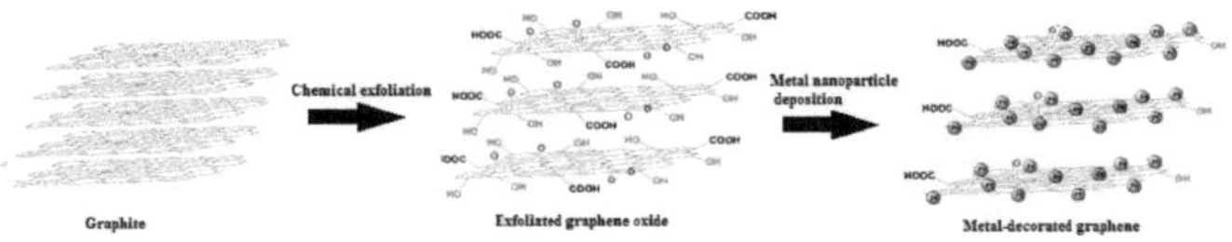

Figure 7.4 The schematic diagram of metal nanoparticle decoration on graphene surface. Reproduced from *Carbon* 2010, 48, 1124–1130. Copyright © 2010, Elsevier Ltd [32b].

Samulski et al. recently showed that nanoparticles could act as a spacer to prevent the reversible aggregation of graphene nanosheets [32]. When the sizes of nanoparticles are 3–4 nm, they can intercalate between the graphene nanosheets, consequently producing space between the nanosheets. Siburian et al. reported that graphene mixed and stirred with an ethanol solution of $H_2PtCl_6.H_2O$ for 3 h and reduced by H_2 at 400°C gave a Pt/graphene composite [33]. According to this method, Pt size was controlled by varying the concentration of the Pt precursor, but higher loading of Pt lead to bigger nanoparticles, as shown in Fig. 7.5. It was found that sub-nanoclusters of 0.8 nm Pt-nanoparticles with high surface area of 170 m^2/g could be formed on graphene nanosheets.

Ye et al. synthesized RuO_2-decorated graphene using 40% PDDA-assisted hydrothermal reaction; here, the surfactant can act as a stabilizer and as a reducing agent [34]. Gopiraman et al. synthesized graphene-supported RuO_2 nanorods by a simple mechanochemical method followed by calcinations at 300°C [35]. Recently, Tan et al. proved that DNA could be an efficient binder for nanoparticles [36]. Homogeneously distributed Au nanoparticles and Ag nanoparticles on graphene surface were obtained with double-stranded DNA (dsDNA) at room temperature. Figure 7.6 shows the different metal nanoparticles on graphene surface using dsDNA. Gu et al. prepared single noble metal and bimetal nanoparticles on graphene support at room temperature: nanoparticles and graphene oxides were reduced by Vitamin C [37].

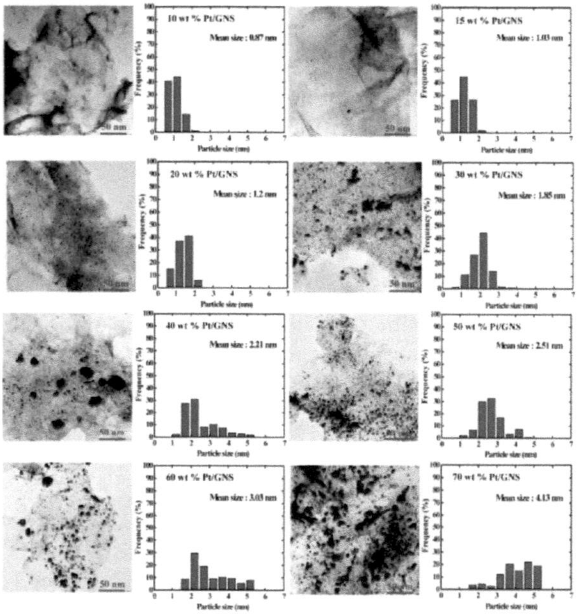

Figure 7.5 TEM image of Pt-decorated graphene sheets with different Pt weight percentage. Reproduced from *J. Phys. Chem. C* 2003, **117**, 3635–3645. Copyright © 2003, American Chemical Society [33].

Zhai et al. proposed a novel method to obtain 1.8 nm Pt nanoparticles deposited graphene by γ-ray irradiated: graphene oxides and Pt ion solutions are mixed, saturated with Ar gas and exposed to γ-ray using ^{60}Co at room temperature [38]. Sattayasamitsathit et al. reported the electrodeposition of Pd and Au nanoparticles on 3D graphene [39]. The synthesis procedure involved that 4 mM $PdCl_2$ and 2 mM $HAuCl_4$ solutions were mixed with 1:1 solution of $HClO_4$:acetonitrile and H_2SO_4:acetonitirile, respectively. In addition, the size of nanoparticles could be controlled by changing the deposition time and potential. Raston et al. reported a novel approach using P-calix [8] arene-mediated graphene for deposition of Pt and Pd nanoparticles. Metal ions are reduced through H_2 gas bubbling [40].

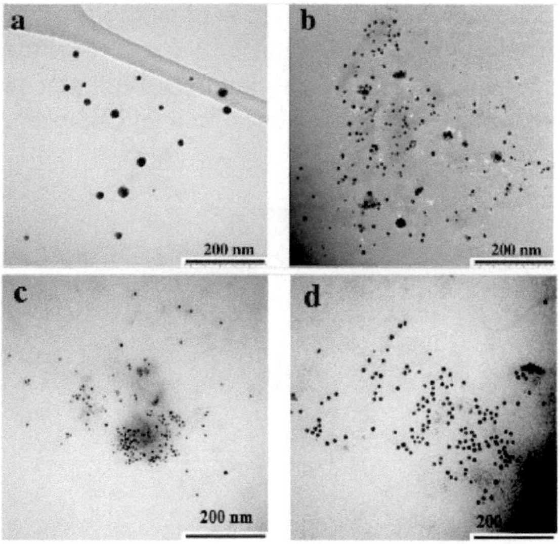

Figure 7.6 TEM images of DNA-guided nanoparticles-decorated graphene: (a) ~16 nm Au nanoparticles, (b) ~12 nm Ag nanoparticles, (c) ~13 nm Cu nanoparticles, and (d) ~10 nm Pt nanoparticles. Reproduced from *Adv. Mater.* 2013, **25**, 2319–2325. Copyright © 2013, Wiley-VCH Verlag GmbH & Co. KGaA [36].

Dry methods are also attempted for the preparation of metal-added graphene. Pt nanoparticles were decorated in graphene sheets by aerosol spray pyrolysis [41]. Fluorine-doped tin oxide (FTO) supported graphene sheets were decorated with Pt nanoparticles by pulsed laser deposition method. After 10 s of laser ablation, particle sizes were ca. 5 nm and after 30 s of laser ablation, particles become larger, which indicated that the sizes of nanoparticles were easily tuned by varying deposition time [42]. Magnetron sputtering deposition was used to deposit bimetallic Ag@Pd and Au@Pd nanoparticles on graphene layer, as shown in Fig. 7.7 [43].

Dry methods of high energy methods are highly defect-induced techniques, which could import undesirable properties to nanoparticles-granphene hybrid. Qiu et al. investigated the defect-free synthesis of metal-loaded graphene using flip sputtering at high temperature [44]. Compared with direct sputtering, flip sputtering (flipping the sample surface into opposite side) decreased the energy when the nanoparticles arrived near the graphene, resulting in minimum disorder or defects to graphene layers. Co, Fe, and Al nanoparticles were effectively deposited by this flip technique with minimum defects. Another method for defect-free synthesis of metal loaded-graphene is thermal evaporation. With the technique of thermal evaporation, Kim et al. decorated Pd nanoparticles on CVD grown graphene sheets [45].

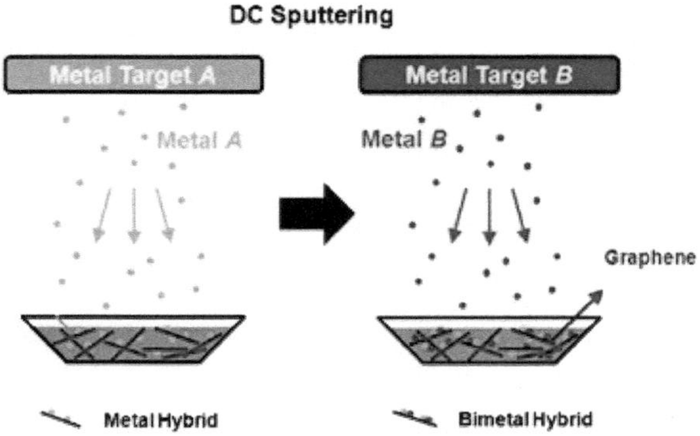

Figure 7.7 The sputtering decoration of mono and bi-metal nanoparticles on graphene support. Reproduced from *ACS Appl. Mater. Interfaces* 2013, **5**, 5072–5079. Copyright © 2013, American Chemical Society [43].

7.3 Applications of Metal-Added Carbon Nanotubes and Graphenes

Applications of metals and metal nanoparticles become broaden when it combines with nanocarbons. Extraordinary properties are evolved due to metal–nanocarbon interaction. High surface area and high capacitance of nanocarbons combined with high catalytic ability of nanoparticles can solve the energy and environmental demands. To boost up the performance, metal alloys nanoparticles are pinned with nanocarbons. Theoretical and experimental researches prove that heteroatom-doping also offers uniform distribution of nanoparticles by constructive heteroatom-metal bonds.

7.3.1 Hydrogen Adsorption

Hydrogen storage is one of the world's primary energy demands. Physical storage of hydrogen requires high energy and safety at the same time. Chemical storage thus is highly favorable. Nanocarbons of higher surface area alone can store hydrogen. It was reported that single-layer graphene could store maximum 0.2 wt. % hydrogen at room temperature [46]. Rao et al. described that multilayer graphene had higher hydrogen storage capability of 5 wt. % than single-layer graphene [47]. Hydrogen adsorption affinity toward the adsorbent improves when the metal nanoparticles are loaded on it. It was suggested theoretically that Al adsorbed on both sides of graphene increases the surface for hydrogen adsorption; thus, Al-adsorbed graphene gave an excellent hydrogen storage capacity of 13.79 wt. % [48]. Metal nanoparticles have a higher capability to catalyse the molecular hydrogen and spread the adsorbed hydrogen atom into nanocarbon backbones. This process is called spillover mechanism, as shown in Fig. 7.8. Pd deposited on functionalized hydrogen exfoliated graphene gave 1.76 wt. % hydrogen adsorption at room temperature [49]. Simple mechano-chemical preparation of CNT–Pd composites gave 3.98 wt. % of hydrogen storage [50].

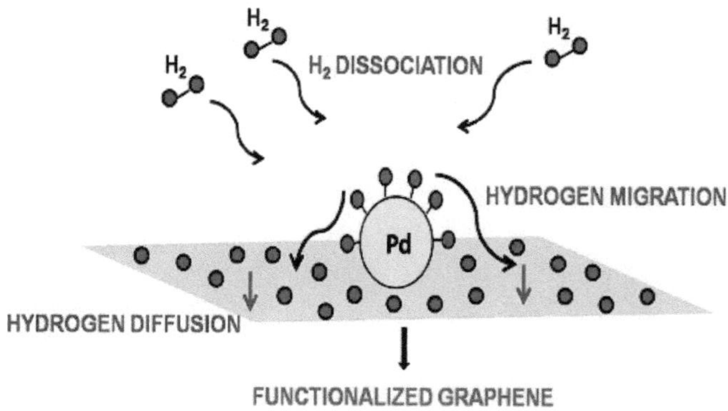

Figure 7.8 Schematic illustration of the spillover mechanism.

Tsai et al. supercritically prepared Pt nanoparticles–MWCNT nano-composites which had a higher hydrogen adsorption 2.7 wt. % than pure MWCNT [51]. Figure 7.9 shows that Pt decoration increased the hydrogen adsorption due to formation of active sites by metal nanoparticle addition, and further hydrogen adsorption was enhanced by acid etching.

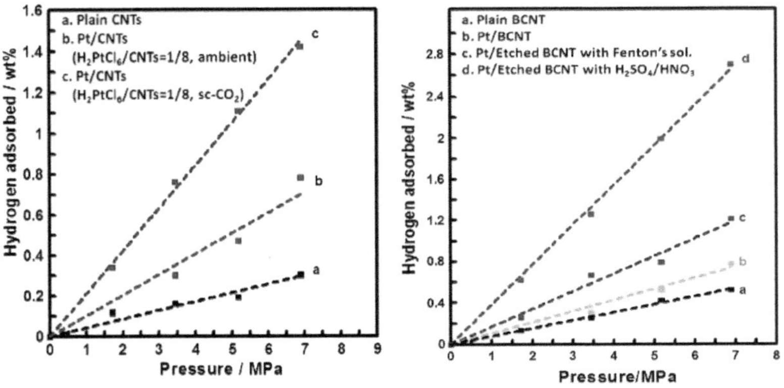

Figure 7.9 Influence of hydrogen adsorption on metal nanoparticle impregnation under ambient and $scCO_2$ atmosphere (left) and acid etching (right). Reproduced from *Int. J. Hydrogen Energy* 2012, **37**, 6714–6720. Copyright © 2012, Hydrogen Energy Publications, LLC. Published by Elsevier Ltd. [51].

On the contrary, Ni-nanoparticles impregnated MWCNT surface shows significant hydrogen uptake of 2.8 wt.% [52]. Nahm prepared Pt-decorated MWCNTs that exhibited approximately 40 times greater adsorption than pristine MWCNTs at room temperature [53].

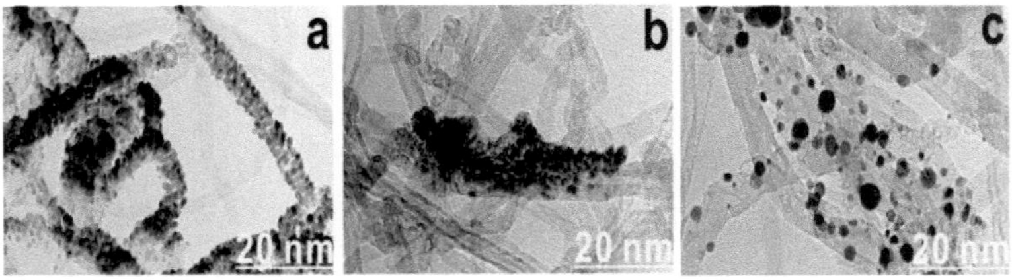

Figure 7.10 TEM images of (a) Pt-Pd coated CNTs, (b) adsorption, and (c) desorption of hydrogen after several cycles. Reproduced from *J. Alloys Compounds* 2009, **480**, L20–L24. Copyright © 2009, Elsevier B.V. [54].

But after several cycles, these composites' hydrogen uptake decreased rapidly caused by formation of unwanted agglomerated nanoparticles, as shown in Fig. 7.10 [54]. This agglomeration can be overcome by nanocarbon-heteroatom doping. Ramaprabhu et al. described that nitrogen-doped graphene had a great interaction with metal nanoparticles compared with pristine graphene. This type of strong interaction decreases the hydrogen dissociation energy, which further improves the hydrogen uptake. Moreover, this strong interaction causes least aggregation of nanoparticles during repeating adsorption and desorption [55].

7.3.2 Electrochemical Energy Storage

7.3.2.1 Applications in Li-ion battery

Li ion batteries are primary candidates for most of the electronic devices. In commercial Li ion batteries, graphite serves as anode. To develop electrodes in low cost with high performance, metal oxides and nanocarbons are utilized. Most recently, $NaBH_4$ reduced Sn nanoparticles-reduced graphene oxides (Sn-RGO) were reported as superior anode material for Li ion batteries, in which graphene plays an important role to decrease the Sn nanoparticles expansion. Highly conductive 2D-network simplifies the Li ion insertion. Sn-RGO composite shows high capacity of 885 mAh/g and 97% capacity preserved, even above 50 cycles [56]. Formation of thick solid electrolyte interface (SEI) film on electrode due to irreversible capacity suppresses electrolyte accession and it lowers the Li ion diffusion [57]. Wu et al. established LiF nanoparticle modified graphene electrode, which minimized the SEI layer formation due to porous nature of LiF-modified graphene [58]. Thin and porous film ease the Li ion transportation between electrode–electrolyte interfaces, which can increase the charge capacity. Co_3O_4-loaded graphene sheets also show higher performance for Li ion battery [59].

Xue et al. improved the Li ion battery performance by double protection of Ge nanoparticles [60]. Initially, Ge nanoparticles were coated with carbon, which was further impregnated on 2D graphene network. It prevented the direct contact of nanoparticles to electrolyte and avoided the nanoparticle aggregation. The double-protected Ge nanoparticles showed higher charge capacity of 938 mAh/g, which was much higher than Ge–carbon coated nanoparticles.

Recently, Guo et al. examined SnO_2 nanocrystals (SnO_2NC) located on nitrogen doped-RGO composites for Li-ion battery application [61]. The capacity reached to 1352 mAh/g, which was very close to the theoretical value (1494 mAh/g) of SnO_2. Figure 7.11 shows the STEM images of uniformly distributed SnO_2 NC even after 500 cycles. The extreme high capacity caused by stronger interaction between N and SnO_2 nanocrystals prevents SnO_2NC from the aggregation after cycling. Moreover, the formation of micropores between SnO_2 and graphene sheets enhances the activity of SnO_2 and is beneficial to the Li diffusion.

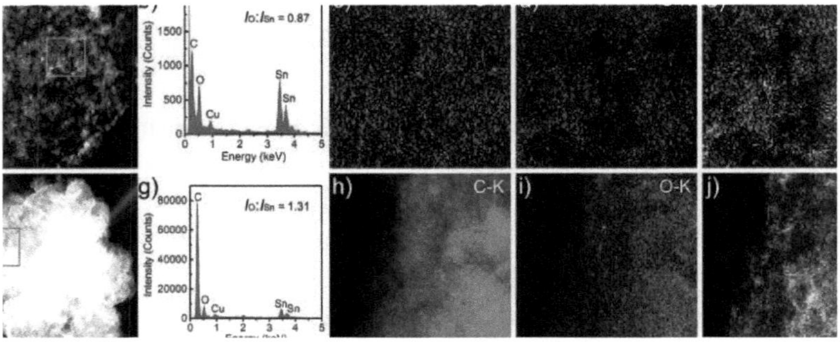

Figure 7.11 (a) STEM image, (b) EDX spectrum, (c–e) elemental mapping of SnO_2NC@NGO, (f) STEM image, (g) EDX spectrum, (h–j) corresponding elemental mapping after 500 charge/discharge cycles. Reproduced from *Adv. Mater.* 2013, **25**, 2152–2157. Copyright © 2013, Wiley-VCH Verlag GmbH & Co. KGaA [61].

Lee et al. recently reported a novel route through microwave-irradiated in situ growth of CNTs on Fe nanoparticles loaded graphene (G-CNT-Fe), which led to 3D hierarchical nanostructures of the composite [62]. Due to large open pores and high surface area, this 3D material achieved higher reversible capacity of 1024 mAh/g. This result was comparable to Fe_3O_4-decorated graphene [63].

Shin proposed that small nanoparticles provided greater Li ion capacity than larger nanoparticles [64]. Around 3 nm NiO deposited nitrogen-doped CNTs (NiO_{3nm}-NCNT) showed higher Li capacity of 3500 mAh/g, while 10 nm NiO–NCNTs (NiO_{10nm}-NCN) showed decreased capacity of 3000 mAh/g. Figure 7.13 describes role of nanoparticle size for efficient Li battery performance.

Zeng et al. reported that various metals and metal oxides could decorate MWCNTs via self-assembly manner, and for example TiO_2 self-assembled on MWCNTs gave reversible capacity of about 225 mAh/g much higher than other crystalline TiO_2 [65].

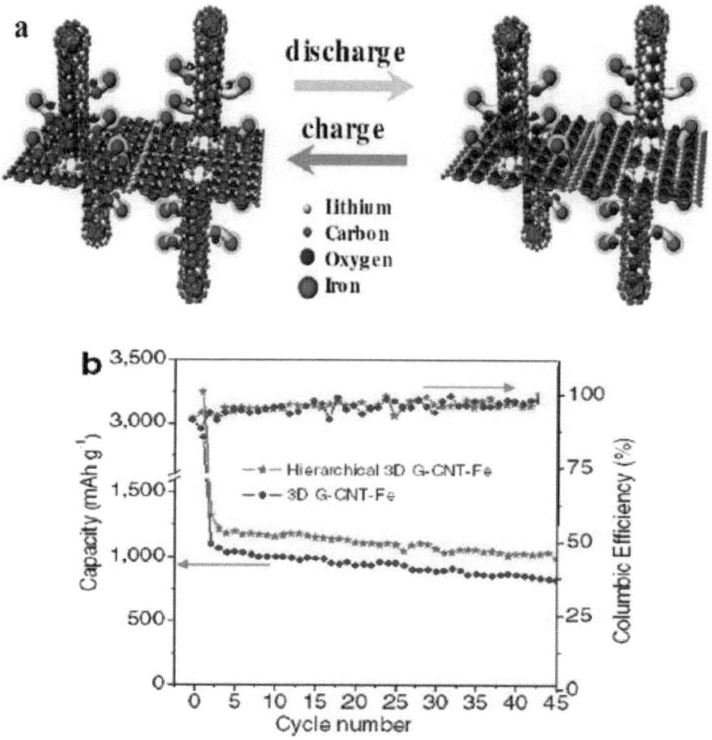

Figure 7.12 (a) 3D hierarchical structures of G-CNT-Fe and (b) specific capacity and columbic efficiency of anode electrode for both baseline and 3D G-CNT-Fe. Reproduced from *ACS Nano* 2013, **7**, 4242–4251. Copyright © 2013, American Chemical Society [62].

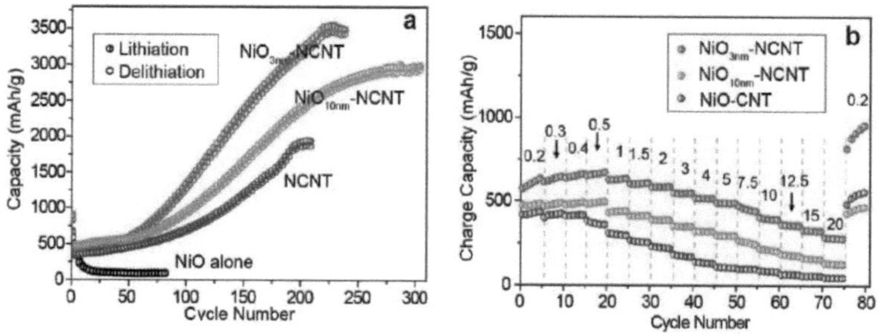

Figure 7.13 (a) Cycling performance of NiO_{3nm}-NCNT, NiO_{10nm}-NCNT and NiO alone and (b) Power capability of NiO_{3nm}, NiO_{10nm}, and NiO-CNT. Reproduced from *Nano Lett.* 2012, **12**, 2283–2288. Copyright © 2012, American Chemical Society [64].

7.3.2.2 Applications in Supercapacitors

Metal oxide nanoparticles have a vital role in supercapacitor applications. Manganese oxide (MO) nanoflowers deposited CNT arrays by one-step electrodeposition showed an excellent capacitance than pristine CNT array [66]. Although MO plays a vital role in improving capacitance, higher loading blocks the nanopores of carbon materials, inducing ion diffusion resistance. Mazloumi et al. fabricated 3D mesoporous hybrid CNT/oxide architectures with metal oxide nanoparticles such as Co_3O_4 and ZnO by microwave irradiation techniques. The prepared 3D CNT-nanoparticles had higher surface area and mesopores of 9–18 nm sizes, giving good electrochemical performance (Fig. 7.14) and a specific capacitance up to 123.94 F/g [67].

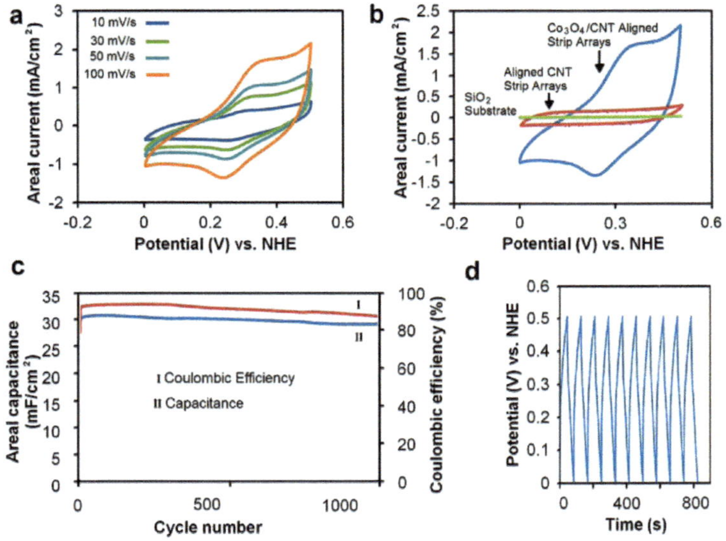

Figure 7.14 (a) CV curves at different scan rate, (b) CV curves of SiO_2 substrate, aligned CNT, and Co_3O_4/aligned CNT, (c) areal capacitance and columbic efficiency over 1000 charge-discharge cycles, and (d) galvanostatic charge-discharge vs time at 200 μA. Reproduced from *ACS Nano* 2013, **7**, 4281–4288. Copyright © 2013, American Chemical Society [67].

Li et al. used CNT as a spacer to prevent the reversible stacking of graphene sheets and then MnO_x nanometals were grown on 3D nanostructured graphene-CNT (G-CNT) composite without structural damage [68]. G-CNT-MnO_x nanopetals were prepared for a supercapacitor. Achieved 3D nanostructure exhibited higher specific capacitance of 535 F/g than pure MnO_x (233F/g) or pristine CNT-G (140F/g). The cycle stability was enhanced by adding small amount of $NaHCO_3$ or Na_2HPO_4 into electrolyte solution to avoid oxide dissolution and gas evolution.

MnO_2 nanoparticles electrodeposited on textile-supported graphene ink showed higher capacitance of 315 F/g. Textile provided flexibility to the system and graphene served as a conductive support to MnO_2 nanoparticles. Overall, device performance was measured in a cell system of SWCNT-coated fiber-negative electrode and graphene-MnO_2-fiber positive electrode. This hybrid cell had 95% capacitance retention even over 5000 cycles at 2.2A/g and 12.5 Wh/kg energy density at 1.5 V [69].

Chen et al. fabricated Fe_3O_4-graphene as anode and three-dimensional structured porous graphene as cathode and obtained ultrahigh energy density of 147 Wh/kg. The results are shown in Fig. 7.15, much higher than commercial supercapacitor and comparable with commercial Li-ion batteries [70].

7.3.3 Catalysis for Flue Gas

Nanoparticles-added nanocarbons have many applications such as the effective treatment of air and water pollutions. Performance of catalyst is based on both metal nanoparticles and nanocarbon support. Lu et al. reported that Pt/CNTs had high catalytic activity for toluene and CO conversion and conversion efficiency depended on temperature [71]. CO conversion dominated at temperature below 100°C; toluene conversion efficiency increased at high temperature and oxidation of toluene suppressed CO conversion, for example, 99% of CO conversion efficiency obtained at 90°C at the same time maximum toluene conversion of 97% at 250°C. The content of metal nanoparticles and the diameters of CNTs have an influence on the flue gas conversion [72]. Co/CNTs also exhibited good CO conversion [73].

Bandosz et al. found NH_3 catalysis with MnO_2-loaded graphene oxide at room temperature. NH_3 oxidation was easier in MnO_2/graphene composite than in pure MnO_2. Functional groups on GO activated the oxygen and then MnO_2 catalytically oxidized the NH_3 to N_2 [74].

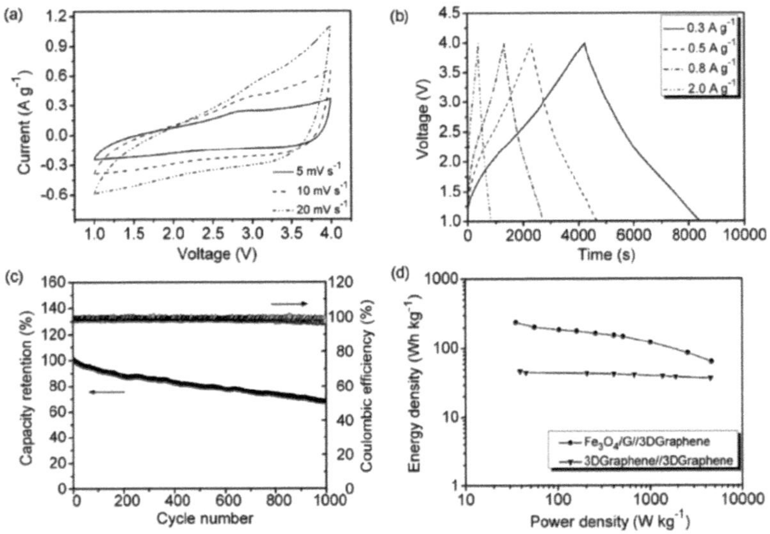

Figure 7.15 (a) CV curves of Fe_3O_4/G/3D graphene hybrid supercapacitor at various scan rate, (b) Galvanostatic charge-discharge curves at different current densities of 0.3–2.0 A/g, (c) Cycle stability Fe_3O_4/G/3D graphene hybrid supercapacitor for 1000cycles at a current density of 2 A/g, and (d) Ragone plots of the Fe_3O_4/G/3D graphene hybrid vs. Fe_3O_4/G/3D graphene symmetric supercapacitor at various charge-discharge rates. Reproduced from *Energy Environ. Sci.* 2013, **6**, 1623–1632. Copyright © 2013, American Chemical Society [70].

Song et al. reported that ZnO-decorated reduced graphene oxide had high catalytic performance against desulfurization of H_2S [75]. H_2S adsorption breakthrough test revealed that ZnO alone had high agglomeration at 350°C, causing only 15% utilization of catalyst surface, four times lower than ZnO/RGO. Additional H_2 gas supply could reduce the Zn-O and promoted the Zn^{2+} active sites for sulfur molecules, and thus further boosted up the H_2S conversion. Photocatalysis of NO_x by TiO_2/CNT could reach a conversion of 32.14% [76]. Wang et al. prepared TiO_2/graphene that had efficient photocatalysis performance of acetone degradation in air, where graphene acted as electron acceptor and transporter, reducing the charge recombination. The catalytic activity enhanced by optimal addition of 0.05 wt. % graphene, exceeding that of pure TiO_2 and Degussa P-25 by a factor of 1.7 and 1.6, respectively [77].

Apart from denitrification and desulphurization, industrial effluents can be removed on the basis of photocatalytic activity of metal, metal oxide decorated on nanocarbons. For example, Au, Ag, TiO_2 nanoparticles loaded on nanocarbons are efficient photocatalysts for industrial water soluble dyes and soluble toxic chemicals [78–80]. Their catalytic activity is greatly enhanced by nanocarbons because of the high surface area. Magnetite nanoparticles as materials of photocatalytic activity also have an important role in removing water pollutants [81–83].

7.3.4 Other Applications

Metal-added nanocarbons applications are not limited to the above-described aspects. They have applications in gas sensors and biosensors [84,85]. They can be used as an efficient catalyst for various organic reactions and water splitting [86,87]. Liu et al. and Dai et al. proposed that MoS_2-impregnated mesoporous and non-porous graphene could be efficient catalysts for hydrogen evolution reaction [88,89]. Wang et al. reported that Au@Ag core shell nanoparticles loaded on protein-coated graphene oxide improved antibacterial activity of *Escherichia coli* [90]. Sunset yellow and Tartrazine such as allergic compounds in the food could be simultaneously determined using graphene decorated with nickel nanoparticles as a sensitive substrate [91].

7.4 Conclusion and Outlook

Graphene and carbon nanotubes supported nanoparticles show promising applications in energy and environmental aspects such as hydrogen adsorption, electrochemical energy storage, catalysis, and many others. Strategies are explained to tune the nanoparticle sizes and the synergetic effects of nanocarbons and metal nanoparticles are discussed in present work. Although several methods are available, it is still highly required to prepare environment-friendly nanoparticle-carbon nanostructures in a short reaction time, at room temperature and without surfactant. Tuning nanoparticle sizes, increasing surface area, and engineering nanopores are the main challenges in field of the researches for producing the high performance materials.

Acknowledgments

This work was supported by the National Natural Science Foundation of China (21273236) and by Science and Technology Planning Projects of Fujian Province of China (2015I0008 and 2014H2008). K. P. Annamalai conveys thanks to the University of Chinese Academy of Science for the scholarship to support the research. Y. Tao appreciates the support of both "100 Talents" program of the Chinese Academy of Sciences and "100 Talents" program of Fujian Province, China.

References

1. Ning, G., Fan, Z., Wang, G., Gao, J., Qianc, W., Wei, F. *Chem. Commun.* 2011, 47, 5976–5978.
2. Balandin, A. A., Ghosh, S., Bao, W., Calizo, I., Teweldebrhan, D., Miao, F., Lau, C. *Nano Lett.* 2008, 8, 902–907.
3. Treacy, M. M., Ebbesen, T. W., Gibson, J. M. *Nature* 1996, 38, 678–680.
4. Falkovsky, L. A. *J. Phys. Conf. Ser.* 2008, 129, 012004.
5. Wang, Y., Wu, Y., Huang, Y., Zhang, F., Yang, X., Ma, Y., Chen, Y. S. *J. Phys. Chem. C* 2011, 115, 23192–23197.
6. Breuer, O., Sundararaj, U. *Polym. Compos.* 2004, 25, 630–645.
7. Nishihara, H., Kyotani, T. *Adv. Mater.* 2012, 24, 4473–4498.
8. Sun, S. T., Wu, P., *Phys. Chem. Chem. Phys.* 2011, 13, 21116–21120.
9. Iijima, S. *Nature* 1991,354, 56–58.
10. Rafique, M. M. A., Iqbal, J. *J. Encapsul. Adsorp. Sci.* 2011, 1, 29–34.
11. Ellis, A. V., Vijayamohanan, K., Goswami, R., Chakrapani, N., Ramanathan, L. S., Ajayan, P. M., Ramanath, G. *Nano Lett.* 2003, 3, 279.
12. Geim, A. K., Novoselov, K. S. *Nat. Mater.* 2007, 6, 183–191.
13. Hummer, W. S., Offeman, R. E, *J. Am. Chem. Soc.* 1958, 80, 1339–1339.
14. Pumera, M. *Chem. Record.* 2009, 9, 211–223
15. Tao, Y., Endo, M., Kaneko, K. Recent progress on the synthesis and applications of carbon nanotubes, in *Handbook of Innovative Nanomaterials*: Pan Stanford Publishing Pte. Ltd., 2012, 639–663.
16. Rozenberg, B. A., Tenne, R. *Prog. Polym. Sci.* 2008, 33, 40–112.
17. Shenhar, R., Norsten, T. B., Rotello, V. M. *Adv. Mater.* 2005, 17, 657–669.

18. Schmidt, E., Kleist, W., Krumeich, F., Mallat, T., Baiker A. *Chem. Eur. J.* 2010, 16, 2181–2192.
19. Mu, Y., Liang, H., Hu, J., Jiang, L., Wan, L. *J. Phys. Chem. B* 2005, 109, 22212–22216.
20. Z, Jin., Zou, H. L., Qing, Q., Yang, Y. L., Li, Q., Liu, Z., Guo, X., Du, Z. *J. Phys. Chem. B* 2003, 107, 3712–3718.
21. Simmons, T. J., Bult, J., Hashim, D. P., Linhardt, R. J., Ajayan, P. M. *ACS Nano* 2009, 3, 865–870.
22. Duarte, C., Miguel, A., Liz-Marzan, L. M. *J. Mater. Chem.* 2006, 16, 22–25.
23. Ou, Y. Y., Huang, M. H., *J. Phys. Chem. B* 2006, 110, 2031–2036.
24. Fujigaya, T., Nakashima, N. *Adv. Mater.* 2013, 25,1666–1681.
25. Lin, M. H., Huang, H. L., Liu, Y., Liang, C., Fei, S. D., Chen, X., Ni, C. *Nanotechnology* 2013, 24, 065501.
26. Wang, S. Y., Wang, X., Jiang, S. P. *Langmuir* 2008, 24, 10505–10512.
27. Annamalai, K. P., Tao, Y. *Recent Patents Chem. Eng.* 2013, 6, 2–7.
28. Cheng, Y., Jiang, S. P. *Electrochim. Acta* 2013,99,124–132.
29. (a) Chen, K. J., Chandrasekara Pillai, K., Rick, J., Pan, C. J., Wang, S. H., Liu, C. C., Hwang, B. J. *Biosens. Bioelectron.* 2012, 33, 120–127. (b) Watanabe, M., Uchida, M., Motoo, S. *J. Electroanal. Chem.* 1987, 229, 395–406. (c) Hwang, B. J., Chen, C. H., Sarma, L. S., Chen, J. M., Wang, G. R., Tang, M. T., Liu, D. G., Lee, J. F. *J. Phys. Chem. B* 2006, 110, 6475–6482.
30. Chen, J., Lu, G. *Nanotechnology* 2006, 17, 2891–2894.
31. Lee, J. S., Lee, T., Song, H. K., Cho, J., Kim, B. S. *Energy Environ. Sci.* 2011, 4, 4148–4154.
32. (a) Si, Y., Samulski, E. T. *Chem. Mater.* 2008, 20, 6792–6797. (b) Li, Y., Gao, W., Ci, L., Wang, C., Ajayan, P. M. *Carbon* 2010, 48, 1124–1130.
33. Siburian, R., Kondo, T., Nakamura, J. *J. Phys. Chem. C* 2003, 117, 3635–3645.
34. Shen, J., Li, T., Huang, W., Long, Y., Li, N., Ye, M. *Electrochim. Acta* 2013, 95, 155–161.
35. Gopiraman, M., Babu, S. G., Khatri, Z., Wei, K., Endo, M., Karvembu, R., Kim, I. S. *Catal. Sci. Technol.* 2013, 3, 1485.
36. Ocsoy, I., Gulbakan, B., Chen, T., Zhu, G., Chen, Z., Sari, M. M., Peng, L., Xiong, X., Fang, X., Tan, W. *Adv. Mater.* 2013, 25, 2319–2325.

37. Zhou, L., Gu, H., Wang, C., Zhang, J., Lv, M., He, R. *Coll. Surf. A: Physicochem. Eng. Aspects* 2013, 430, 103–109.

38. Zhang, Q., Zhang, Y., Gao, Z., Ma, H. L., Wang, S., Peng, J., Li, J., Zhai, M. *J. Mater. Chem.* C 2013, 1, 321.

39. Sattayasamitsathit, S., Gu, Y., Kaufmann, K., Jia, W., Xiao, X., Rodriguez, M., Minteer, S., Cha, J., Burckel, D. B., Wang, C., Polsky R., Wang, J. *J. Mater. Chem. A* 2013, 1, 1639.

40. Chen, X., Zang, W., Vimalanathan, K., Iyer, K. S., Raston, C. L. *Chem. Commun.* (Camb.) 2013, 49, 1160–1162.

41. Jang, H. D., Kim, S. K., Chang, H., Choi, J.-W., Luo, J., Huang, *J. Aerosol Sci. Technol.* 2013, 47, 93–98.

42. Bajpai, R. Roy, S., Kumar, P., Bajpai, P., Kulshrestha, N., Rafiee, J., Koratkar, N., Misra, D. S. *ACS Appl. Mater. Interfaces* 2011, 3, 3884–3889.

43. Liu, C. H., Chen, X. Q., Hu,Y. F., Sham, T. K., Sun, Q. J. Chang, J. B., Gao, X., Sun, X. H., Wang, S. D. *ACS Appl. Mater. Interfaces* 2013, 5, 5072–5079.

44. Qiu, X. P., Shin, Y. J., Niu, J., Kulothungasagaran, N., Kalon, G., Qiu, C., Yu T., Yang, H. *AIP Adv.* 2012, 2, 032121.

45. Chung, M. G., Kim, D.-H., Seo, D. K., Kim, T., Im, H. U., Lee, H. M., Yoo, J.-B., Hong, S.-H., Kang T. J., Kim, Y. H. *Sens. Actuators B: Chem.* 2012, 169, 387–392.

46. Ma, L.-P., Wu, Z.-S., Li, J., Wu, E. D., Ren, W. C., Cheng, H. M. *Int. J. Hydrogen Energy* 2009, 34, 2329–2332.

47. Subrahmanyam, K. S., Kumar, P., Maitra, U., Govindaraj, A., Hembram, K. P. S. S., Waghmare, U. V., Rao, C. N. R. *Proc. Natl. Acad. Sci.* 2011, 108,2674–2677.

48. Ao, Z. M., Peeters, F. M., *Phys. Rev.* B 2010, 81, 205406.

49. Parambhath, V. B., Nagar, R., Sethupathi, K., Ramaprabhu, S. *J. Phys. Chem. C* 2011, 115, 15679–15685.

50. Ravindra, R., Ramachandra, B. B. *Int. J. Appl. Phys. Math.* 2012, 2, 63–66.

51. Tsai, P. J., Yang, C.-H., Hsu, W.-C., Tsai, W.-T., Chang, J.-K. *Int. J. Hydrogen Energy* 2012, 37, 6714–6720.

52. Kim, H. S., Lee, H., Han, K.-S., Kim, J.-H., Song, M.-S., Park, M.-S., Lee, J.-Y., Kang, J.-K. *J. Phys. Chem. B* 2005, 109, 8983–8986.

53. Zacharia, R., Rather, S.-U., Hwang S. W., Nahm, K. S. *Chem. Phys. Lett.* 2007, 434, 286–291.

54. Hwang, S.-W., Rather, S. U., Naik, M.-U.-D., Soo, C. S., Nahm, K.-S. *J. Alloys Compounds* 2009, 480, L20–L24.

55. Parambhath, V. B., Nagar, R., Ramaprabhu, S. *Langmuir* 2012, 28, 7826–7833.

56. Nithya, C., Gopukumar, S. *Chem. Sus. Chem.* 2013, 6, 898–904.

57. Fu, L. J., Liu, H., Li, C., Wu, Y. P., Rahm E., Holze R., Wu, H. Q. *Solid State Sci.* 2006, 8, 113–128.

58. Wu, Z.-S., Xue, L., Ren, W., Li, F., Wen L., Cheng, H.-M. *Adv. Funct. Mat.* 2012, 22, 3290–3297.

59. Wu, Z.-S., Ren, W., Wen, L., Gao, L., Zhao, J., Chen, Z., Zhou, G., Li, F., Cheng, H.-M. *ACS Nano* 2010, 4, 3187–3194.

60. Xue, D. J., Xin, S., Yan, Y., Jiang, K. C., Yin, Y. X., Guo, Y. G., Wan, L. J. *J. Am. Chem. Soc.* 2012, 134, 2512–2515.

61. Zhou, X., Wan L. J., Guo, Y. G. *Adv. Mater.* 2013, 25, 2152–2157.

62. Lee, S.-H., Sridhar, V., Jung, J.-H., Karthikeyan, K., Lee, Y.-S., Mukherjee, R., Koratkar, N., Oh, I.-K. *ACS Nano* 2013, 7, 4242–4251.

63. Zhu, X., Zhu, Y., Murali, S., Stoller M. D., Ruoff, R. S. *ACS Nano* 2011, 5, 3333–3338.

64. Shin, W. H., Jeong, H. M., Kim, B. G., Kang, J. K., Choi, J. W. *Nano Lett.* 2012, 12, 2283–2288.

65. Li, J., Tang, S., Lu L., Zeng, H. C. *J. Am. Chem. Soc.* 2007, 129, 9401–9409.

66. Zhang, H., Cao, G., Wang, Z., Yang, Y., Shi Z., Gu, Z. *Nano Lett.* 2008, 8, 2664–2668.

67. Mazloumi, M., Shadmehr, S., Rangom, Y., Nazar, L. F., Tang, X. *ACS Nano* 2013, 7, 4281–4288.

68. Li, S.-M., Wang, Y.-S., Yang, S.-Y., Liu, C.-H., Chang, K.-H., Tien, H.-W., Wen, N.-T., Ma, C.-C. M., Hu, C.-C. *J. Power Sources* 2013, 225, 347–355.

69. Yu, G., Hu, L., Vosgueritchian, M., Wang, H., Xie, X., McDonough, J. R., Cui, X., Cui, Y., Bao, Z., *Nano Lett.* 2011, 11, 2905–2911.

70. Zhang, F., Zhang, T., Yang, X., Zhang, L., Leng, K., Huang Y., Chen, Y. *Energy Environ. Sci.* 2013, 6, 1623–1632.

71. Lu, C.-Y., Wei, M.-C., Chang, S.-H., Wey, M.-Y. *Appl. Catal. A: General* 2009, 354, 57–62.

72. Lu, C.-Y., Tseng, H.-H., Wey, M.-Y., Hsueh, T.-W. *Chem. Eng. J.* 2009, 145, 461–467.

73. Lu, C.-Y., Wey, M.-Y. *Fuel* 2007, 86, 1153–1161.

74. Seredych, M., Bandosz. T. J. *Micropor. Mesopor. Mater.* 2012, 150, 55–63.

75. Song, H. S., Park, M. G., Croiset, E., Chen, Z., Nam, S. C., Ryu H.-J., Yi, K. B. *Appl. Surf. Sci.* 2013, 280, 360–365.

76. Chuan-Yu, Y., Yu-Feng, L., Chih-Hung, H., Yao-Hsuan, T., Chen-Chi, M. M., Min-Chao, C., Hsin, S. *Nanotechnology* 2008, 19, 045604.

77. Wang, W., Yu, J., Xiang, Q., Cheng, B. *Appl. Catal. B: Environ.* 2012, 119–120, 109–116.

78. Woan, K., Pyrgiotakis G., and Sigmund, W. *Adv. Mater.* 2009, 21, 2233–2239.

79. Roy, P., Periasamy, A. P., Liang, C.-T., Chang, H.-T. *Environ. Sci. Technol.* 2013, 47, 6688–6695.

80. Lü, K., Zhao, G., Wang, X. *Chinese Sci. Bull.* 2012, 57, 1223–1234.

81. Fan, W., Gao, W., Zhang, C., Tjiu, W. W., Pan, J., Liu, T. *J. Mater. Chem.* 2012, 22, 25108–25115.

82. Guo, J., Wang, R., Tjiu, W. W., Pan, J., Liu, T. *J. Hazard. Mater.* 2012, 225–226, 63–73.

83. Junyong, C., Yongmei, H., Yan, L., Jiajia, G. *RSC Adv.* 2013, 3, 7254–7258.

84. Bai, J., Lai, Y., Jiang, D., Zeng, Y., Xian, Y., Xiao, F., Zhang, N., Hou, J., Jin, L. *Analyst* 2012, 137, 4349–4355.

85. Gutes, A., Hsia, B., Sussman, A., Mickelson, W., Zettl, A., Carraro, C., Maboudian, R. *Nanoscale* 2012, 4, 438–440.

86. Saberi, D., Sheykhan, M., Niknam, K., Heydari, A. *Catal. Sci. Technol.* 2013, 3, 2025–2031.

87. Ng, Y. H., Iwase, A., Kudo A., Amal, R. *J. Phys. Chem. Lett.* 2010, 1, 2607–2612.

88. Liao, L., Zhu, J., Bian, X., Zhu, L., Scanlon, M. D., Girault, H. H., Liu, B. *Adv. Funct. Mater.* 2013, 10.1002/adfm.201300318.

89. Li, Y., Wang, H., Xie, L., Liang, Y., Hong, G., and Dai, H. *J. Am. Chem. Soc.* 2011, 133, 7296–7299.

90. Wang, H., Liu, J., Wu, X., Tong, Z., Deng, Z. *Nanotechnology* 2013, 24, 205102.

91. Gan, T., Sun, J., Wu, Q., Jing, Q., Yu, S. *Electroanalysis* 2013, 25, 1505–1512.

Chapter 8

Roles of Reduced Graphene Oxide in Improving Photocatalytic Hydrogen Generation Performance over Metal Sulphide Nanocomposites

Jian Ru Gong
National Center for Nanoscience and Technology,
11 Zhongguancun Beiyitiao,
Beijng 100190, China
gongjr@nanoctr.cn

8.1 Introduction

Nowadays, global energy crisis and environmental contamination are serious issues, so seeking for renewable and clean energy is an urgent task. Solar energy has been considered as an ideal alternative, as it is the most powerful, affordable, renewable, and sustainable source of energy.[1] The economical conversion of solar energy into chemical fuels is one of the Holy Grails of 21st century science. Hydrogen is predicted to be a promising secondary energy, with the advantages of high energy density that far exceeds those of gasoline

Nanomaterials in Energy and Environmental Applications
Edited by Junhui He

ISBN 978-981-4463-78-2 (Hardcover), 978-981- 4463-79-9 (eBook)
www.panstanford.com

and coal, no carbon emission, and a useful by-product of water from combustion.[2] At present, hydrogen is mainly produced from fossil fuels such as natural gas by steam reforming, which is severely restricted by its low efficiency and high cost. Moreover, carbon dioxide is emitted in this process. As is known, nature abundantly stores hydrogen in the form of water. Therefore, hydrogen production by efficient combination of water and solar energy has enormous capacity to fulfill the present and future demand of energy around the world in an eco-friendly manner.

There are several methods for utilizing solar radiation in splitting water for hydrogen generation.[2,3] However, photovoltaic electrolysis of water uses the expensive proton exchange membranes[4] and alkaline electrolytes.[5] Solar to thermochemical water splitting needs the high temperature of 700°C–1000°C.[6] Photobiological water splitting has to overcome the natural short-term nature of biospecies hydrogen production.[7] Consequently, much attention has been focused on photocatalytic water splitting for hydrogen generation because they are cost-effective, simple and convenient, and have huge potential for further development.[8]

Honda and Fujishima reported that H_2 evolution could be observed on TiO_2 under UV-light irradiation in 1972.[9] It was the first demonstration that hydrogen could be obtained through water splitting. The concept was subsequently extended into particulate system and called heterogeneous photocatalysis by Allen J. Bard.[10] During the past 50 years, great achievements have been made in this rapidly growing field.[8,11] As the earliest H_2 production semiconductor, TiO_2 is the most investigated photocatalyst material due to its high-efficiency, low-cost, non-toxicity, and photostablility.[12] Metal oxides with ABO_3 configuration, such as $SrTiO_3$,[13] $NaTaO_3$,[14] perfectly meeting the thermodynamic requirement for water splitting, are also reported to be efficient photocatalysts. But all of them are only active under UV-light, which accounts for about 4% of the total solar spectrum. Therefore, it is necessary to design efficient photocatalysts with visible-light (~43% of the total irradiation energy of sunlight) response in order to sufficiently utilize solar energy. Compared with their wide bandgap counterparts, chalcogenides are attractive photocatalytic materials for the conversion of solar energy into chemical energy under visible-light irradiation. Specifically, these materials have a conduction band edge sufficiently more negative than the reduction potential of protons and a relatively narrow bandgap that can efficiently absorb visible light.[15,16] However, several key factors such as the low light absorption, the rapid charge recombination, and the instability of chalcogenides during chemical process are still big issues that restrict the realization of large-scale industrial application of these materials.[17]

The emergence of nanomaterials as the new building blocks to construct light energy harvesting assemblies has opened up new ways to utilize renewable energy resources because of the large surface areas, abundant surface states, and diverse morphologies compared with their corresponding bulk materials.[11f,18] Among them, graphene, a two-dimensional (2D) network of hexagonal structured sp^2-hybridized carbon atoms,[19] has stimulated tremendous research interest in various energy conversion applications.[20] Graphene exhibits many outstanding properties, such as fast room-temperature mobility of charge carriers, exceptional conductivity, large theoretical specific surface area, and excellent optical transmittance.[21] These unique properties indicate that graphene has a great potential to be an ideal construction component of photocatalytic materials for hydrogen production. Of particular interest is graphene oxide (GO), which is typically produced by chemical exfoliation of graphite through strong oxidization,[22] and widely considered as an individual sheet of graphene decorated with various oxygen-containing functional groups (such as hydroxyl, epoxy, and carboxyl) on both the basal plane and the edge.[23] The availability of solution-processable GO allows for the chemical modification with various molecules or nanoparticles via non-covalent or covalent functionalization.[22b,24,] Furthermore, GO can not only mediate the growth of nanomaterials but also help the growth of nanomaterials with good dispersion behavior because the presence of functional groups on GO sheets can serve as the nucleation sites to anchor nanoparticles. The intrinsic structural and electronic properties of graphene can be partially restored upon reduction of GO.[25] The reduced graphene oxide (RGO), in which aromatic sp^2 domains with few nanometers are surrounded by sp^3-hybridized carbon atoms, usually shows p-type semiconducting behavior.[26] During the synthetic process, GO can be reduced to RGO in chemical or thermal conditions, leading to formation of RGO-based composites.

To take the advantages of RGO, combining chalcogenides with RGO has attracted much attention recently for applications in the photocatalytic hydrogen production field, and many encouraging findings have been made in the past several years.[27] In this chapter, I will take two systems studied in our group as examples to introduce the roles of RGO in improving photocatalytic hydrogen generation performance over metal sulphide nanocomposites.

8.2 RGO/Metal Sulphides for Photocatalytic Hydrogen Generation

8.2.1 RGO/CdS Nanocomposites

Among the various visible-light driven photocatalysts, CdS is an attractive photocatalytic H_2-production material because of its narrow bandgap of 2.4 eV, which can absorb an appreciable fraction of visible light. However, there are several issues that still limit the H_2-production rate over pure CdS particles. For example, the CdS particles tend to aggregate, forming larger particles, which results in a reduced surface area and a higher recombination rate of photoinduced electron-hole pairs. To solve these problems, many approaches have been proposed to enhance the photocatalytic activity of CdS particles, including the preparation of quantum-sized CdS,[28] deposition of noble metals,[29] preparation of heterogeneous semiconductors,[30] and incorporation of semiconductor particles in the interlayer region of layered compounds.[31] For example, Bard et al.[32] introduced CdS particles into colloidal suspensions of clay; Sato et al.[33] and Hirai et al.[34] incorporated CdS and/or ZnS particles into the interlayer of hydrotalcite and mesoporous silica, respectively. The layered structure of such a supporting matrix can efficiently suppress the semiconductor particle growth as well as facilitate the transfer of the photogenerated electrons to the surface of photocatalysts. Furthermore, the recombination between the photoinduced charge carriers can be effectively suppressed, leading to the high efficiency of H_2 production. Our group proposed to combine CdS nanoparticles with RGO for improving the performance of the nanocomposite based on a few points: (1) the 2D platform structure of RGO makes it an excellent supporting matrix for photocatalyst particles, similar to the role of layer-structured matrices played in improving the efficiency of the photocatalysts as mentioned above;[35] (2) RGO can effectively inhibit the electron-hole pair recombination in the composite because of the excellent electronic conductivity imparted by its 2D planar π-conjugation structure;[36,37] (3) an appropriate amount of RGO may darken the composite and thus enhance the absorption of visible light.[38]

The CdS-cluster-decorated RGO nanosheets (RGO-CdS) were prepared by a solvothermal method.[39] In a typical synthesis of the composite, a varying amount of GO prepared by a modified Hummers' method[40] and $Cd(Ac)_2{\cdot}2H_2O$ were dispersed in DMSO. The weight ratios of GO to $Cd(Ac)_2{\cdot}2H_2O$ were 0, 0.5%, 1.0%, 2.5%, 5.0%, and 40%, and the obtained samples were labeled as GC0, GC0.5, GC1.0, GC2.5, GC5.0, and GC40, respectively. Next, the homogenous solution was transferred into a Teflon-lined autoclave and held at 180°C for 12 h after vigorous stirring and sonication. After that, the precipitates from the mixture were allowed to cool to room temperature and collected by centrifugation, and then rinsed with acetone and ethanol for several times to remove the residue of DMSO. The final product was dried in an oven at 60°C for 12 h. The bare RGO sample without any CdS clusters was prepared under the same experimental conditions for the purpose of comparison and was labeled as G.

XRD patterns were recorded for the dried RGO-CdS powder to confirm the crystallographic phase of CdS in the composite and investigate the influence of RGO on the crystallinity of CdS nanoparticles. Figure 8.1 shows the XRD patterns of RGO-CdS nanocomposites synthesized with different contents of GO compared with that of the pure CdS (i.e., GC0). The peaks at 26.5°, 44.0°, and 52.1° correspond to the diffractions of the (111), (220), and (311) planes of the cubic CdS (JCPDS 80-0019), respectively. The diffraction peaks are broad, as the crystallite sizes of CdS nanoparticles in the samples are relatively small. In general, the solubility product constant (K_{sp}) for CdS particles is quite small, leading to fast nucleation and agglomeration of CdS nanocrystals.[41] However, DMSO can regulate the nucleation rate of CdS particles by slowly releasing S^{2-} ions into solution, resulting in a much smaller crystallite size. No characteristic diffraction peaks for carbon species are observed in the patterns because of the low amount and relatively low diffraction intensity of RGO. The XRD patterns also imply that RGO may enhance the crystallinity of CdS particles. As shown in Fig. 8.1, the pure CdS particles (GC0, yellow line) have poor crystallinity, perhaps because the reactor conditions are not ideal for their nucleation. After introducing 0.5% GO, the XRD peaks of sample GC0.5 (brown line) become stronger and narrower due to the improved crystallinity of CdS particles. As GO content is increased, the intensity of XRD peaks is correspondingly enhanced. To further highlight this effect, the average crystallite sizes of different samples were calculated using Scherrer formula for the (111) facet diffraction peak. As shown in Table 8.1, the average crystallite size of CdS particles increases from 2.6 to 3.1 nm. Thus, it can be deduced that the layered structure of RGO supplies a platform on which the CdS nanoparticles can nucleate; thus, RGO can promote the crystallization of CdS nanoparticles to a certain extent.

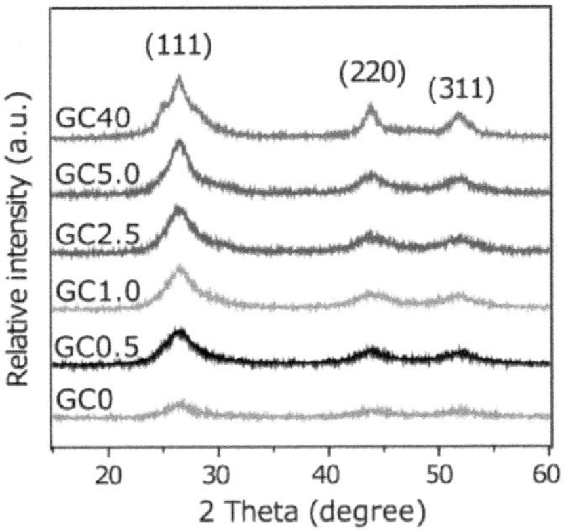

Figure 8.1 XRD patterns of samples GCx solid powders (x = 0, 0.5, 1.0, 2.5, 5.0, and 40). Reprinted with permission from Ref. [38]. Copyright 2011. American Chemical Society.

Furthermore, the morphology of samples GC0 and GC1.0 was analyzed by SEM to directly observe the structure of the RGO nanosheets decorated with CdS clusters and to specifically investigate the influence of RGO on the morphology of the CdS clusters. The SEM micrograph in Fig. 8.2a shows a significant aggregation of the CdS nanoparticles in sample GC0 and particle diameters of approximately

100 nm. However, Fig. 8.2b shows that much smaller CdS clusters spread uniformly and tightly on the RGO sheets in sample GC1.0, indicating that RGO may interact with CdS nanoparticles and inhibit their aggregation. As has been reported previously, nanoparticles may interact with RGO sheets through physisorption, electrostatic binding, or charge transfer interaction,[42] and the exact mechanism for this system is still under investigation.

Table 8.1 Effects of GO content on physicochemical properties and quantum efficiency (QE) of the RGO-CdS samples

Samples	GO content (wt%)	Crystallite size[a] (nm)	S_{BET} ($m^2 \cdot g^{-1}$)	QE (%)
GC0	0	2.6	35.5	4.6
GC0.5	0.5	2.6	40.7	7.7
GC1.0	1.0	2.7	48.2	22.5
GC2.5	2.5	2.7	58.2	11.1
GC5.0	5.0	3.0	54.2	4.6
GC40	40	3.1	96.8	0.6

[a]Average crystallite size is determined by the broadening of CdS (111) facet diffraction peak using Scherrer formula. Reprinted with permission from Ref. [38]. Copyright 2011. American Chemical Society.

Figure 8.2 SEM images of (a) sample GC0 and (b) sample GC1.0. Reprinted with permission from Ref. [38]. Copyright 2011. American Chemical Society.

The TEM image of GC1.0 (Fig. 8.3a) shows that many small CdS particles are present on the RGO sheet, which has a characteristic wrinkle on the edge. The result further confirms the combination of

RGO nanosheet and CdS clusters, which is consistent with the SEM images in Fig. 8.2b. The HRTEM image (Fig. 8.3b) shows that the size of the CdS clusters in GC1.0 is about 3 nm, which is in agreement with the value calculated by the Scherrer equation (Table 8.1). The lattice fringes of individual CdS clusters with d spacing of ca. 0.336, 0.206, and 0.177 nm can be assigned to the (111), (220), and (311) lattice planes of the cubic CdS, respectively. The selected area electron diffraction (SAED) pattern (inset in Fig. 8.3b) indicates that these nanoparticles are polycrystalline. The three inside diffraction rings correspond to the (111), (220), (311) planes of the cubic CdS, which is fully consistent with the XRD results (Fig. 8.1). In addition, the well-defined diffraction ring with six spots, five of which are indicated by the arrows in the SAED pattern, implies thin and flat RGO films.

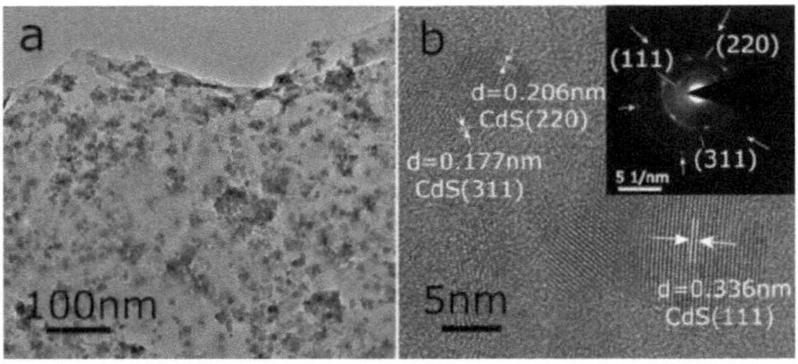

Figure 8.3 (a) TEM and (b) HRTEM images of sample GC1.0, inset of (b) showing the SAED pattern of RGO sheet decorated with CdS clusters. Reprinted with permission from Ref. [38]. Copyright 2011. American Chemical Society.

The effect of GO on the BET surface area and pore structure of the prepared samples was investigated using adsorption-desorption measurements. As shown in Table 8.1, the BET surface area (S_{BET}) of samples gradually increases with increasing GO content, from 35.5 to 96.8 m^2/g. It should be noted that the specific surface area (m^2/g) is expressed per gram of the samples, which contain some amount of carbon (graphene) with a low density. The planar density of graphene is 0.77 mg/cm^2, and the density of CdS is 4.82 g/cm^3. Consequently, the average densities of samples decrease with increasing GO, resulting in the increase of the S_{BET}. A greater specific surface area of photocatalysts can supply more surface active sites and make charge carriers transport easier, leading to an enhancement of the photocatalytic performance.[43] Thus, RGO may play a role in enhancing the photocatalytic activity.

Figure 8.4 shows the nitrogen adsorption-desorption isotherms and the corresponding curves of the pore size distribution (inset) for samples GC0, GC1.0, GC5.0, and GC40. According to the Brunauer-Deming-Deming-Teller (BDDT) classification, the majority of physisorption isotherms can be grouped into six types.[44] Typically, pure CdS (sample GC0) has an isotherm of type II, indicating the presence of large macropores, while samples GC1.0, GC5.0, and GC40 have isotherms of type IV, suggesting the presence of mesopores.[44] The shape of the hysteresis loops is of type H3, associated with slit-like pores formed by the aggregation of the plate-like particles. In other words, it appears that the prepared nanocomposite is composed of sheet-like RGO decorated with CdS clusters. The results are further confirmed by the corresponding pore-size distribution. As shown in the inset of Fig. 8.4, sample GC0 has macropores with a peak pore diameter of around 100 nm. When GO is introduced, mesopores begin to appear in samples with a typical pore diameter of around 4 nm, and the amount of macropores decreases.

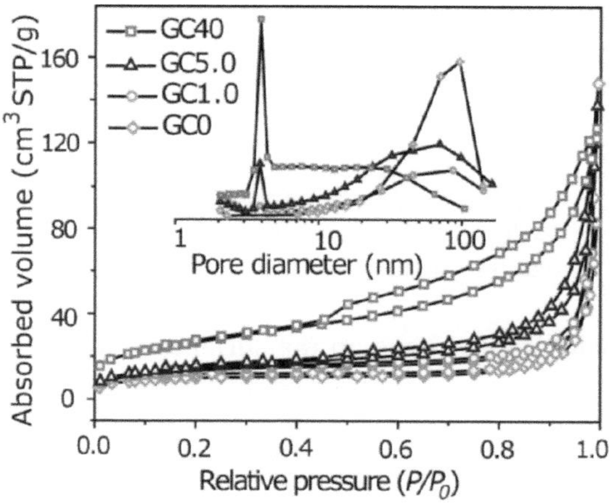

Figure 8.4 Nitrogen adsorption-desorption isotherms and corresponding pore size distribution curves (inset) of samples GCx solid powders (x = 0, 1.0, 5.0, and 40). Reprinted with permission from Ref. [38]. Copyright 2011. American Chemical Society.

A comparison of the UV-vis diffuse reflectance spectra of samples GCx (x = 0, 0.5, 1.0, 2.5, 5.0, and 40) is displayed in Fig. 8.5. There is an enhanced absorbance in the visible-light region (>500 nm) with increasing GO content. This is also observed as a color change of the samples, which become darker, that is, from pale yellow to olive, when a low amount of black GO was introduced into the pure CdS nanoparticles. The results show that the addition of GO increases the absorbance of visible light. When GO content reaches 40% (sample GC40), the absorbance is significantly higher than that of the other samples used in this study. Due to increased absorbance, a more efficient utilization of solar energy can be obtained. Therefore, we can infer that the introduction of GO in CdS particles is effective for the visible-light response of the composite.

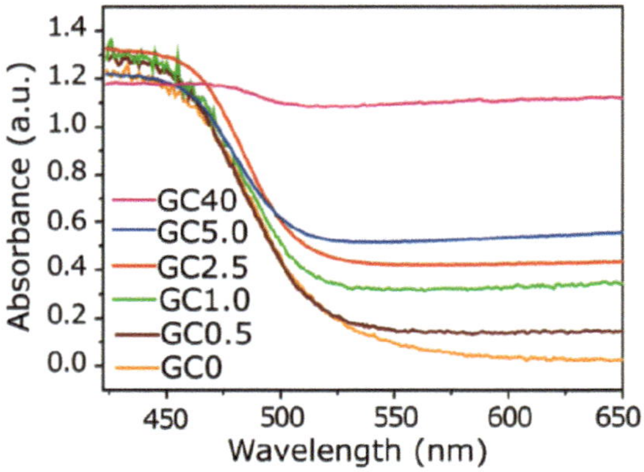

Figure 8.5 UV-vis diffuse reflectance spectra of samples GCx solid powders (x = 0, 0.5, 1.0, 2.5, 5.0, and 40). Reprinted with permission from Ref. [38]. Copyright 2011. American Chemical Society.

The hydrophilic groups on the surface of GO, such as hydroxyl and carboxyl groups, can enhance dispersion of the photocatalyst in water, providing the prerequisite for H_2 generation occurring in aqueous solution. To investigate the degree of reduction of GO in the solvothermal reduction process, high-resolution XPS spectra of C1s were collected from samples GO and GC1.0 (Fig. 8.6). The XPS spectrum of C1s from GO (Fig. 8.6a, solid line) can be deconvoluted into four smaller peaks (dashed lines), which are ascribed to the following functional groups: sp^2 bonded carbon (C-C, 284.8 eV), epoxy/hydroxyls (C-O, 286.9 eV), carbonyls (C=O, 287.8 eV), and carboxyl (O-C=O, 288.9 eV),[45,46] indicating the high percentage of oxygen-containing functional groups. In comparison, in the XPS spectrum of C1s from GC1.0 (Fig. 8.6b, solid line), the peak for C=O almost vanishes, and the peaks for C-O and O-C=O (dashed lines) still exist but with much lower intensities than those in GO, indicating the partial removal of the oxygen-containing functional groups. Furthermore, the degree of reduction of GO can be quantified by calculating the relative content of carbon in the samples. Briefly, GO has 48% graphitic carbon and 52% oxidized carbon, while 68% graphitic carbon and 32% oxidized carbon for GC1.0, respectively, showing the loss of oxygen-containing functional groups and the partial reduction of GO by the solvothermal reduction process. Thus, the RGO sheets in the composite can be dispersed in the aqueous solution to a certain extent.

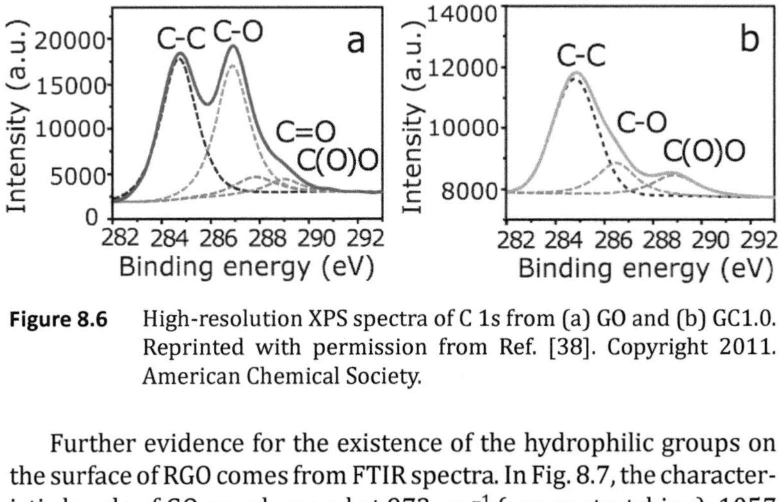

Figure 8.6 High-resolution XPS spectra of C 1s from (a) GO and (b) GC1.0. Reprinted with permission from Ref. [38]. Copyright 2011. American Chemical Society.

Further evidence for the existence of the hydrophilic groups on the surface of RGO comes from FTIR spectra. In Fig. 8.7, the characteristic bands of GO are observed at 972 cm^{-1} (epoxy stretching), 1057 cm^{-1} (alkoxy C-O stretching), 1224 cm^{-1} (phenolic C-OH stretching), 1402 cm^{-1} (carboxyl O-H stretching), and 1724 cm^{-1} (C=O stretching vibrations of carboxyl or carbonyl groups).[47] The peak at 1120 cm^{-1} is ascribed to C-O stretching vibrations of CO_2, and the broad absorption at 1624 cm^{-1} is related to H-O-H bending band of the adsorbed H_2O molecules or the in-plane vibrations of sp^2-hybridized C-C bonding.[40] Compared with the peaks of the functional groups of GO, sample GC1.0 has a similar spectrum but with much lower absorption intensity, especially for the peaks at 1224 cm^{-1}, 1402 cm^{-1}, and 1724 cm^{-1}, which are all assigned to the oxygen-containing functional groups. This result indicates the partial reduction of GO and is in good agreement with the XPS results.

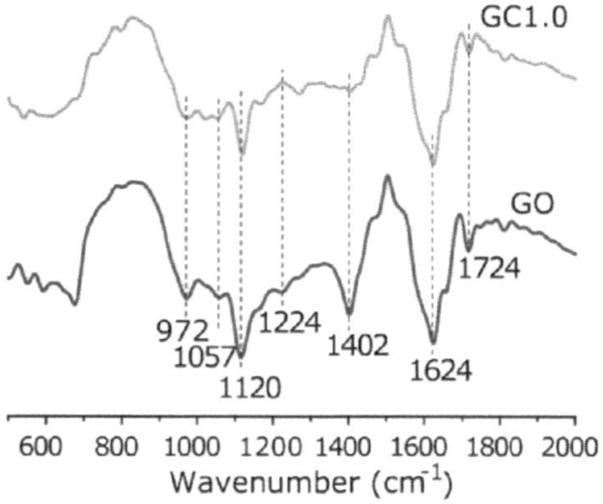

Figure 8.7 FTIR spectra of (top) GC1.0 and (bottom) GO. Reprinted with permission from Ref. [38]. Copyright 2011. American Chemical Society.

In this work, photocatalytic H_2-production activity of the prepared RGO-CdS nanocomposites was evaluated under visible-light irradiation using lactic acid as sacrificial reagent and Pt as cocatalyst. The sacrificial reagent can prevent sulfide photocatalysts from photocorrosion by providing sacrificial electron donors to consume photogenerated holes, and Pt can reduce the overpotential in the production of hydrogen from water and suppress the fast backward reaction (recombination of hydrogen and oxygen into water) as well.[48–50] GO exhibited a significant influence on the photocatalytic activity (Fig. 8.8a). Even with a small amount of GO (0.5–2.5 wt%), the H_2-production rate was noticeably increased. For CdS alone (GC0), a relatively low photocatalytic H_2-production rate (0.23 mmol/h) was observed as expected due to the rapid recombination of conduction band (CB) electrons and valence band (VB) holes. In the presence of a small amount of GO (0.5%), the activity of sample GC0.5 was slightly enhanced to 0.38 mmol/h, perhaps because the amount of RGO nanosheets was not large enough to efficiently disperse the CdS clusters. When content was 1.0% (GC1.0), the H_2-production rate reached the highest value of 1.12 mmol/h with apparent quantum efficiency of 22.5% at 420 nm. In this regard, the photocatalytic activity of sample GC1.0 exceeds that of GC0 by a factor of 4.87, and the H_2 production rate is significantly greater than that of most semiconductor photocatalysts. This is attributed to two factors: (1) compared with the pure CdS counterpart (GC0), the larger specific surface area of GC1.0 offers more active adsorption sites and photocatalytic reaction centers, which favors an enhanced photocatalytic activity; (2) in the RGO-CdS system, RGO serves as an acceptor of the electrons generated in the CdS semiconductor and effectively decreases the recombination probability of the photoexcited electron-hole pairs, leaving more charge carriers to form reactive species. Control experiments indicated that no appreciable hydrogen production was detected in the absence of either irradiation or photocatalyst, suggesting that hydrogen was produced by photocatalytic reactions on the photocatalyst. However, a further increase in GO content led to deterioration of the catalytic performance. In particular, at GO content of 40% (sample GC40), the photocatalytic activity dramatically decreased, with an H_2-production rate of only 0.02 mmol/h and quantum efficiency of 0.6% at 420 nm. It is reasonable because the introduction of a large percentage of black RGO led to screening of the active sites on the catalyst surface and also rapidly decreased the intensity of light through the depth of the reaction solution, which could be called "shielding effect."[51] As a consequence, a suitable content of GO is crucial for optimizing the photocatalytic activity of RGO-CdS nanocomposites. In comparison, no hydrogen was detected when sample G was used as the photocatalyst with Pt as a cocatalyst, suggesting that the bare RGO without CdS clusters is likely not active for photocatalytic H_2 production under the experimental conditions studied.

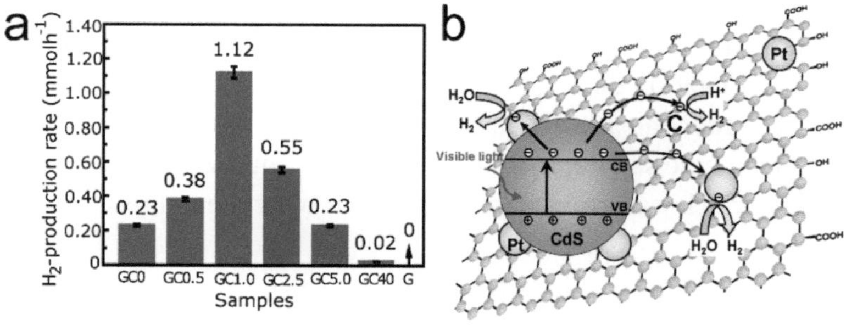

Figure 8.8 (a) Comparison of the visible-light photocatalytic activity of samples GC0, GC0.5, GC1.0, GC2.5, GC5.0, GC40, and RGO for the H_2 production using 10 vol% lactic acid aqueous solution as sacrificial reagent and 0.5 wt% Pt as cocatalyst; (b) Schematic illustration of the charge separation and transfer in the RGO/CdS system under visible light. Reprinted with permission from Ref. [38]. Copyright 2011. American Chemical Society.

On the basis of the above results, a tentative mechanism of the photocatalytic reaction is proposed as illustrated in Fig. 8.8b. Under visible-light irradiation, electrons (e^-) are excited from the VB to the CB of the CdS semiconductor and then likely transfer in one of three following ways: (1) to Pt deposited on the surface of CdS clusters; (2) to carbon atoms on the RGO sheets; and (3) to Pt located on the RGO nanosheets. Eventually, the electrons react with the adsorbed H^+ ions to form H_2. Although the CB edge of CdS is more negative than the reduction potential of H^+/H_2, the H_2-production rate is negligible. This can be explained by the rapid recombination rate of CB electrons and VB holes. Once RGO is introduced to CdS nanoparticles, it can serve as an electron collector and transporter to efficiently separate the photogenerated electron-hole pairs, effectively lengthening the lifetime of the charge carriers. Furthermore, the unique features of RGO allow photocatalytic reactions to take place not only on the surface of semiconductor catalysts but also on the RGO sheet, greatly enlarging the reaction space.

Later, Gao et al. and Peng et al. studied the visible-light driven photocatalytic H_2-production activities of the RGO combined with CdS, and the results showed that RGO improved the H_2-production activities in their systems due to the effective charge separation as an electron acceptor and transporter.[52] Dong et al.[53] further investigated the interfacial electron-hole separation mechanism of the graphene-supported CdS nanomaterials using first-principles calculations. It was found that upon visible light irradiation, the excited electron in CdS quantum dots (QDs) injected into graphene and transported along graphene layer through π^* orbitals to achieve interfacial electron-hole separation due to the strong coupling effect between them, which was in accordance with our experimental results.

8.2.2 RGO-$Zn_xCd_{1-x}S$ Nanocomposites

ZnS is one of the most often applied photocatalysts due to its rapid generation of electron-hole pairs upon photoexcitation and highly negative reduction potentials of excited electrons.[54] Also, it shows the high activity for hydrogen evolution under UV-light irradiation even without noble metal co-catalysts. Unfortunately, the bandgap of ZnS is 3.66 eV, which is too large to respond under visible light.[55]

Therefore, numerous attempts have been made on development of the doped ZnS and preparation of solid solutions (e.g., $Cd_xZn_{1-x}S$, Cu-$Zn_xCd_{1-x}S$, CuS-In_2S_3-ZnS) in order to shift the absorption edge of ZnS into the visible light range.[56–63] These ZnS-based solid solutions have been fully proved to be an efficient, low-toxic, and cost-effective visible-light driven photocatalysts for hydrogen production from water splitting. However, the hydrogen production rate observed for these solid solutions is far from exciting and sometimes expensive Pt co-catalyst is still needed in order to achieve higher quantum efficiency. The cocatalyst is typically a noble metal or metal oxide or a combination of them, loaded on the surface of the photocatalyst to produce reactive sites and reduce the activation energy for gas evolution. Usually, noble metals possess higher work function than those of semiconductors, so noble metal nanoparticles depositing onto the surface of semiconductor always make electrons transfer from the CB of excited semiconductor to metal and result in the formation of schottky barrier, which can efficiently suppress the recombination of electrons and holes. Furthermore, most of metals have much lower overpotentials for H_2O reduction, indicating that they can provide plenty of reaction sites for absorbed protons. Therefore, noble metal cocatalysts have been widely applied for the enhancement of photocatalytic activities. However, noble metals are costly and scarce; so much attention has been focused on seeking novel low-cost cocatalysts recently. Although combining graphene with semiconductor photocatalysts enhanced the photocatalytic hydrogen production performance, in most cases, the hydrogen production efficiencies of the graphene-based composites were still lower than that of semiconductors containing the Pt co-catalyst under the same conditions.[64–70] Therefore, we synthesized the noble metal-free RGO-$Zn_xCd_{1-x}S$ nanocomposite to study the photocatalytic H_2-production activity.[71]

It is well known that the photocatalytic activity of $Zn_xCd_{1-x}S$ can be altered by varying the Cd/Zn ratio. Our preliminary experiment shows that the prepared $Zn_xCd_{1-x}S$ solid solution without RGO exhibits the highest solar H_2-production rate at $x = 0.8$. Thus, $Zn_{0.8}Cd_{0.2}S$ solid solution was used in the following study. The RGO-$Zn_{0.8}Cd_{0.2}S$ nanocomposites were prepared by a simple and environmently-friendly co-precipitation-hydrothermal strategy[72] using $Cd(Ac)_2{\cdot}2H_2O$, $Zn(Ac)_2{\cdot}2H_2O$, and Na_2S as precursors in the presence of GO. During this process, GO was reduced to RGO as proved by XPS, FT-IR, and Raman data in Fig. 8.9 (a–d) and $Zn_{0.8}Cd_{0.2}S$ nanoparticles simultaneously formed on the RGO sheets. The weight ratios of GO to $Zn_{0.8}Cd_{0.2}S$ were 0, 0.1%, 0.25%, 0.5%, 1%, 2%, and 5%, and the corresponding samples were labeled as GS0, GS0.1, GS0.25, GS0.5, GS1, GS2, and GS5, respectively. The energy-dispersive X-ray (EDX) spectrum of sample GS0.25 shown in Fig. 8.9e confirms the presence of S, Zn, and Cd elements. Figure 8.10a shows the typical TEM image of sample GS0.25 with $Zn_{0.8}Cd_{0.2}S$ nanoparticles well dispersed on the RGO nanosheets. The HRTEM image of GS0.25 (Fig. 8.10b) clearly shows the lattice fringes, suggesting the well-defined crystal structure. The lattice spacing is about 0.313 nm, which can be assigned to the (111) plane of $Zn_{0.8}Cd_{0.2}S$. In comparison to the pure cubic ZnS (0.312 nm), the observed value of the (111) plane of $Zn_{0.8}Cd_{0.2}S$ is a little bit bigger due to the larger radius of Cd^{2+} than that of Zn^{2+},[56] which is in agreement with the XRD results (Fig. 8.10c). The XRD peaks of all RGO-$Zn_{0.8}Cd_{0.2}S$ samples exhibit an obvious left-shift compared with the standard sphalerite ZnS [JCPDS No. 05-0566] due to the formation of $Zn_{0.8}Cd_{0.2}S$ solid solution. The peak position and intensity of these RGO-$Zn_{0.8}Cd_{0.2}S$ samples do not change much with increasing the RGO amount, implying that RGO does not have a significant influence on the phase structure and crystallinity of $Zn_{0.8}Cd_{0.2}S$ solid solutions. No characteristic diffraction peaks for RGO can be observed on these patterns because of the low amount and relatively low diffraction intensity.[65]

A comparison of the UV-vis diffuse reflectance spectra of the RGO-$Zn_{0.8}Cd_{0.2}S$ samples is depicted in Fig. 8.11a. For all the samples studied, a significant increase in the absorption at wavelengths shorter than 470 nm can be assigned to the intrinsic bandgap of $Zn_{0.8}Cd_{0.2}S$ solid solution. The bandgap energy (~2.53 eV) of the pure $Zn_{0.8}Cd_{0.2}S$ solid solution was estimated by extrapolating the straight portion of the $(ahv)^2$ versus photon energy (hv) curve to $a = 0$ according to the Kubelka–Munk function[56,73], as illustrated in the inset of Fig. 8.11a. In addition, a broad background absorption in the visible-light region (500–800 nm) for the RGO-$Zn_{0.8}Cd_{0.2}S$

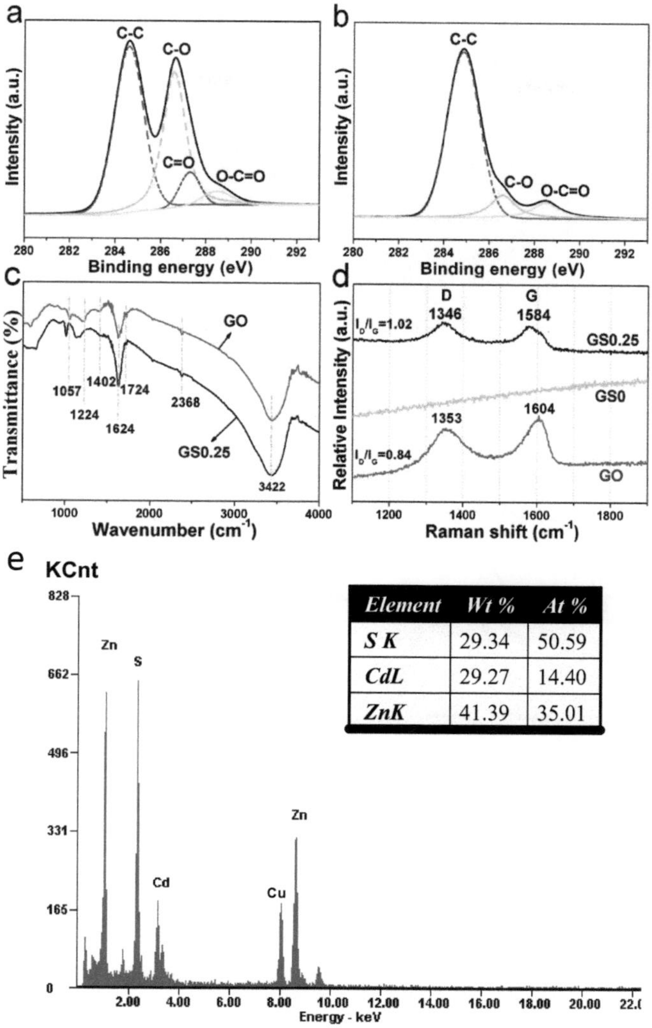

Element	Wt %	At %
S K	29.34	50.59
CdL	29.27	14.40
ZnK	41.39	35.01

Figure 8.9 High-resolution XPS spectra of C 1s from (a) GO and (b) GS0.25; (c) FT-IR spectra of GO and GS0.25; (d) Raman spectra of GO, GS0, and GS0.25 samples; (e) The electron dispersive X-ray (EDX) spectrum for GS0.25 sample; this spectrum was recorded by using an Oxford Instruments X-ray analysis system. The Cu peak originates from the TEM copper grid used as a sample holder. The EDX spectrum confirms the presence of S, Zn, and Cd elements, and the molar ratio of Zn to Cd is found to be ca. 0.72:0.28. Reprinted with permission from Ref. [71]. Copyright 2012. American Chemical Society.

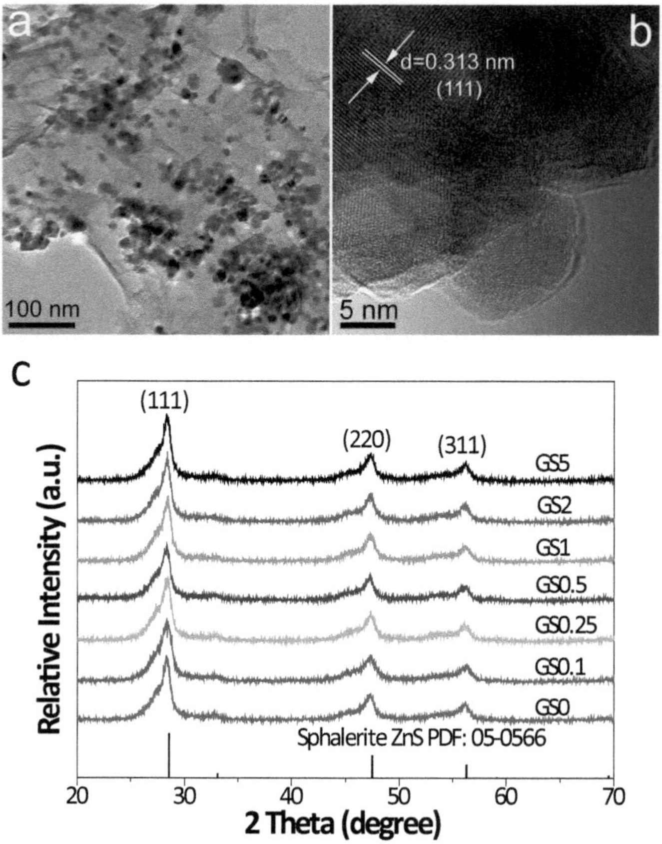

Figure 8.10 TEM (a) and HRTEM (b) images of sample GS0.25. (c) XRD patterns of the RGO-$Zn_{0.8}Cd_{0.2}S$ samples. Reprinted with permission from Ref. [71]. Copyright 2012. American Chemical Society.

nanocomposite samples can be observed, and such absorption gradually increases by increasing RGO content. The corresponding color of the nanocomposite samples becomes darker, that is, from light yellow to dark green. This is attributed to the presence of RGO in RGO-$Zn_{0.8}Cd_{0.2}S$, which reduces reflection of light and thus enhances the absorption.[65,66] Further observation indicates that all the RGO-$Zn_{0.8}Cd_{0.2}S$ samples show almost the same absorption edge as that of the pure $Zn_{0.8}Cd_{0.2}S$ solid solution, implying that RGO is not incorporated into the lattice of $Zn_{0.8}Cd_{0.2}S$.

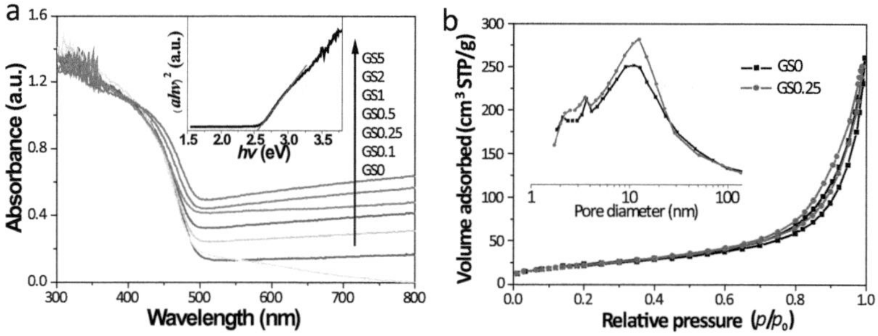

Figure 8.11 (a) UV-vis diffuse reflectance spectra of the RGO-$Zn_{0.8}Cd_{0.2}S$ samples. (b) Nitrogen adsorption-desorption isotherms and the corresponding pore size distribution curves of the GS0 and GS0.25 samples. Reprinted with permission from Ref. [71]. Copyright 2012. American Chemical Society.

Heterogeneous photocatalysis is a surface-based process, and a large surface area provides more surface active sites for the adsorption of reactant molecules, making the photocatalytic process more efficient.[74,75] Figure 8.11b shows the typical nitrogen adsorption-desorption isotherms and the corresponding pore size distribution curves of the GS0 and GS0.25 samples. The nitrogen adsorption-desorption isotherms are of type IV [Brunauer-Deming-Deming-Teller (BDDT) classification], indicating the presence of mesopores (2–50 nm). In addition, the isotherms show high adsorption at high relative pressures (P/P_0) (approaching 1.0), suggesting the formation of large mesopores and macropores.[76] The pore-size distributions (inset in Fig. 8.11b) of the samples indicate a wide pore-size distribution from 2 to 100 nm, further confirming the presence of mesopores and macropores. Table 8.2 displays that all RGO-$Zn_{0.8}Cd_{0.2}S$ nanocomposites have larger BET surface areas than that of pure $Zn_{0.8}Cd_{0.2}S$ due to the presence of RGO with the extremely high surface area.

The photocatalytic H_2-production activity of the pristine $Zn_{0.8}Cd_{0.2}S$ solid solution and RGO-$Zn_{0.8}Cd_{0.2}S$ nanocomposites from an aqueous solution containing 0.35 M Na_2S and 0.25 M Na_2SO_3 under simulated solar irradiation (AM1.5, 100 mW cm^{-2}), together with that of Pt-GS0 and pure RGO for the purpose of comparison is shown in Fig. 8.12. $Zn_{0.8}Cd_{0.2}S$ solid solution without RGO shows some photocatalytic H_2-production activity, but the rate of H_2

Table 8.2 Effect of RGO content on the physical properties of the RGO-$Zn_{0.8}Cd_{0.2}S$ samples

Samples	RGO content (wt%)	S_{BET} (m^2/g)	Pore volume (cm^3/g)	Average pore size (nm)
GS0	0	79	0.27	13.6
GS0.1	0.1	80	0.28	14.0
GS0.25	0.25	81	0.30	15.1
GS0.5	0.5	83	0.30	14.4
GS1	1	89	0.29	14.8
GS2	2	91	0.27	13.3
GS5	5	101	0.28	13.1

evolution is low (403 mol/h/g). The activity of the GS0.1 sample is remarkably enhanced in the presence of a small amount of RGO. The rate of H_2 evolution over RGO-$Zn_{0.8}Cd_{0.2}S$ increases with increasing RGO content, achieving a maximum of 1824 mol/h/g at RGO content of 0.25 wt%. The corresponding QE of 23.4% at 420 nm is higher than those of the graphene-based photocatalysts reported so far,[65,69] showing that the RGO-$Zn_{0.8}Cd_{0.2}S$ sample represents one of the most highly active metal sulfide photocatalyts in the absence of noble metal co-catalysts.[77–79] A further increase in the content of RGO leads to a reduction in the photocatalytic activity. It is clearly visible in the case of GS5, the activity for H_2-produciton decreases dramatically and even below that of the pure $Zn_{0.8}Cd_{0.2}S$ solid solution. This is probably due to the following factors: (i) introduction of a large percentage of black RGO leads to shielding the incident light, thus preventing the generation of electrons from the inside of the $Zn_{0.8}Cd_{0.2}S$ nanoparticles;[80,81] (ii) excessive amount of RGO may cover the active sites on the surface of $Zn_{0.8}Cd_{0.2}S$ and also could hinder the contact of the sacrificial agents with $Zn_{0.8}Cd_{0.2}S$.[65] Therefore, a suitable content of RGO is crucial for optimizing the photocatalytic activity of the composite photocatalysts, which can also be proved by the previous studies.[65,66,69] It should be noted that no hydrogen was detected when RGO alone was used as a photocatalyst, suggesting that bare RGO is unlikely active for photocatalytic H_2 production under our experimental conditions. In addition, control experiments indicate that no appreciable hydrogen production was detected in the absence of either irradiation or photocatalyst, suggesting that hydrogen was produced by photocatalytic reaction. It has been fully demonstrated that noble metals, especially Pt, can function as an efficient H_2-production promoter for many photocatalysts. However, the activity of Pt-$Zn_{0.8}Cd_{0.2}S$ is still lower than that of the RGO-$Zn_{0.8}Cd_{0.2}S$ photocatalyst (GS0.25). In addition, as carbon material, graphene is cheaper and more abundant than noble metals. Thus, it is fascinating to use RGO as a new co-catalyst for replacing noble metals in hydrogen production from water splitting.

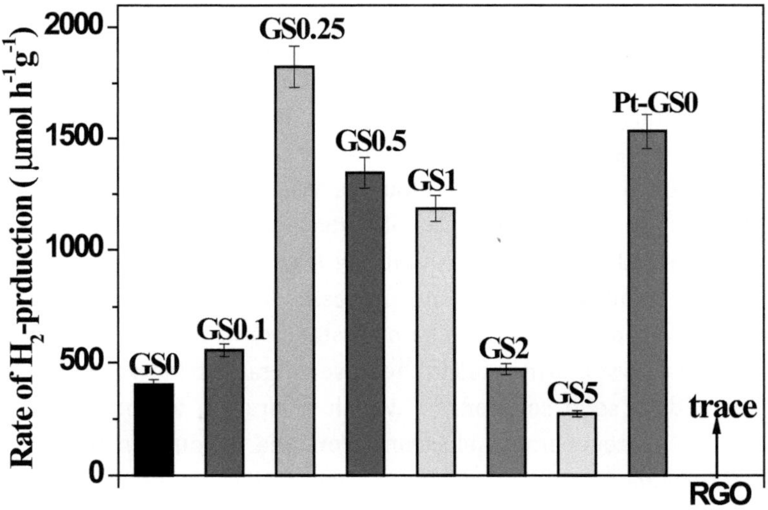

Figure 8.12 Comparison of the photocatalytic H_2-production activity under simulated solar irradiation over GS0, GS0.1, GS0.25, GS0.5, GS1, GS2, GS5, Pt-GS0, and RGO samples. Pt-GS0: GS0 loaded with the optimized 1 wt % of Pt cocatalyst. Reprinted with permission from Ref. [71]. Copyright 2012. American Chemical Society.

Although there are numerous reports about powdered photocatalysts for hydrogen production under visible light irradiation, only a few reports indicate solar energy conversion efficiencies especially for metal sulfide powdered photocatalysts.[59,62,63] Kudo et al. reported the H_2 gas evolved at a rate of 3.3 and 7.9 l/m^2/h for the samples of Pt-loaded $(AgIn)_{0.22}Zn_{1.56}S_2$ and Ru-loaded $Cu_{0.25}Ag_{0.25}In_{0.5}ZnS_2$ under one-sun (AM 1.5G) illumination,[59,63] and the solar energy conversion efficiencies were calculated to be ca. 0.85% and 2.0%, respectively. In addition, Ikeda et al. reported the H_2 evolution rate of more than 13.3 μmol/h over the sample of Ru-loaded $(Cu_xAg_{1-x})_2ZnSnS_4)_{0.9}(ZnS)_{0.4}$ under simulated solar irradiation (AM 1.5G).[62] In our work, the solar energy conversion efficiency of GS0.25 was determined to be 0.36%, which is lower than that of Pt-loaded $(AgIn)_{0.22}Zn_{1.56}S_2$ and Ru-loaded $Cu_{0.25}Ag_{0.25}In_{0.5}ZnS_2$, but higher than that of Ru-loaded $((Cu_xAg_{1-x})_2ZnSnS_4)_{0.9}(ZnS)_{0.4}$. Also, it should be noted that no expensive metals such as Pt, Ru, Ag as well as In were used in our RGO-$Zn_xCd_{1-x}S$ system.

Metal sulfides are usually not stable during the photocatalytic reaction and subjected to photocorrosion.[82,83] In our experiment, the stability of the RGO-$Zn_{0.8}Cd_{0.2}S$ (GS0.25) and pure $Zn_{0.8}Cd_{0.2}S$ (GS0) was evaluated by performing the recycle experiments under the same conditions. After four recycles, the photocatalytic H_2-production rate of sample GS0 decreases gradually; however, sample GS0.25 does not exhibit any loss of activity, indicating the better stability of this nanocomposite for hydrogen generation as displayed in Fig. 8.13. The high stability of the RGO-$Zn_{0.8}Cd_{0.2}S$ photocatalyst is attributed to the close interaction between RGO and $Zn_{0.8}Cd_{0.2}S$ solid solution,[84,85] which favors the vectorial transfer of the photogenerated electrons from the conduction band (CB) of $Zn_{0.8}Cd_{0.2}S$ to RGO. This space separation of the photogenerated electrons and holes is beneficial for preventing the reduction of Cd^{2+} and Zn^{2+}. Furthermore, high Na_2S concentration in the solution can suppress the oxidation of S^{2-} on $Zn_{0.8}Cd_{0.2}S$.

The high H_2-production activity of GS0.25 sample under solar irradiation can be explained in Fig. 8.14a. Under solar irradiation, electrons are excited from the valance band (VB) populated by S 3p to the formed CB by hybridizing Zn 4s4p with Cd 5s5p,[86] which create holes in VB. Normally, these charge carriers recombine rapidly resulting in a low photocatalytic H_2-production rate of $Zn_{0.8}Cd_{0.2}S$ itself. However, when $Zn_{0.8}Cd_{0.2}S$ nanoparticles are immobilized on the surface of RGO, those photogenerated electrons in CB of $Zn_{0.8}Cd_{0.2}S$ tend to transfer to RGO, leading to the hole-electron separation. RGO

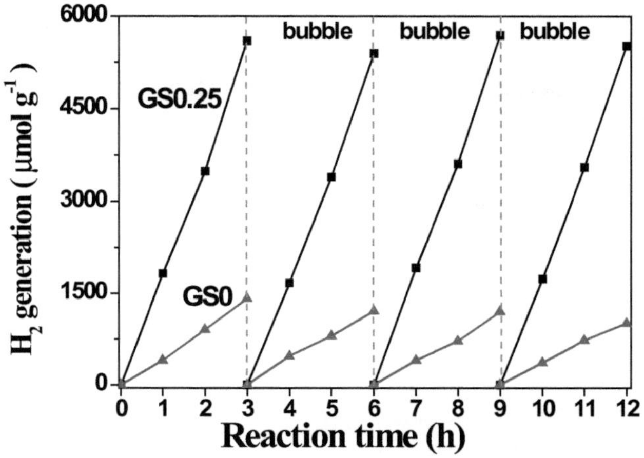

Figure 8.13 Time course of photocatalytic H_2-production over samples GS0.25 and GS0; every 3 h, the reaction system is bubbled with N_2 for 30 min to remove the H_2 inside.

can function as an electron collector and transporter to lengthen the lifetime of the charge carriers, consequently improving the charge separation and photocatalytic activity.[64–69] In addition, the unique features of RGO facilitate the photocatalytic reaction to take place not only on the surface of semiconductor catalysts but also on the RGO sheet, greatly enlarging the reaction space.[65] Furthermore, as the potential of graphene/graphene$^{\cdot-}$ is about −0.08 V (versus SHE, pH = 0) as illustrated in Fig. 8.14b,[87,88] it can act as a co-catalyst to promote the separation and transfer of the photogenerated electrons from CB of $Zn_{0.8}Cd_{0.2}S$ to the RGO, where H^+ is reduced to hydrogen molecule ($E^0_{H+/H2}$ = 0 versus SHE, pH = 0); meanwhile, the holes in VB of $Zn_{0.8}Cd_{0.2}S$ can be consumed by the sacrificial agents (S^{2-}, SO_3^{2-}).

XPS characterization of sample GS0.25 was carried out to examine the change in chemical status of Zn and Cd elements before and after photocatalytic reaction for 2 h. As shown in Fig. 8.15a, the binding energies of 1021.7 and 1044.9 eV match well those of Zn $2p_{3/2}$ and $2p_{1/2}$ XPS peaks of Zn^{2+} in ZnS. The two peaks recorded in the Cd region at 404.9 and 411.6 eV (Fig. 8.15b) are in agreement with the literature data for Cd^{2+} in CdS.[89] In addition, due to the formation

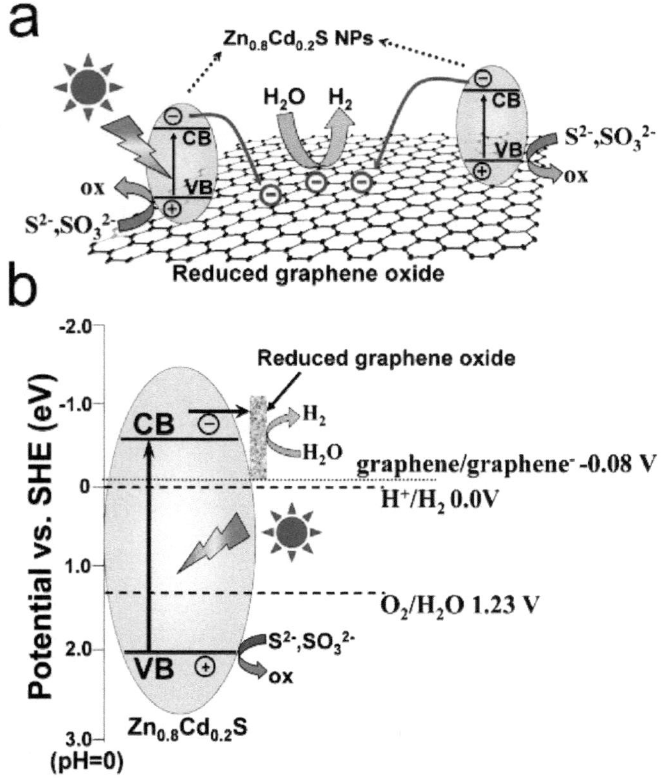

Figure 8.14 (a) Schematic illustration for the charge transfer and separation in the RGO-$Zn_{0.8}Cd_{0.2}S$ system; (b) Proposed mechanism for photocatalytic H_2-production under simulated solar irradiation. Reprinted with permission from Ref. [71]. Copyright 2012. American Chemical Society.

of the solid solution, the XPS peaks of Zn and Cd exhibit a slight shift compared with that of the pure phase.[90] It should be noted after the 2 h photocatalytic reaction, the XPS peak positions of Zn 2p and Cd 3d all exhibit a slight right-shift; especially for Cd 3d, the right-shift is expected. This can be further concluded that under the solar irradiation, the electrons are excited from VB to CB of $Zn_{0.8}Cd_{0.2}S$ and then transfer out from CB of $Zn_{0.8}Cd_{0.2}S$ into RGO, and thus greatly enhance the separation of charge carriers and photocatalytic H_2-production performance.

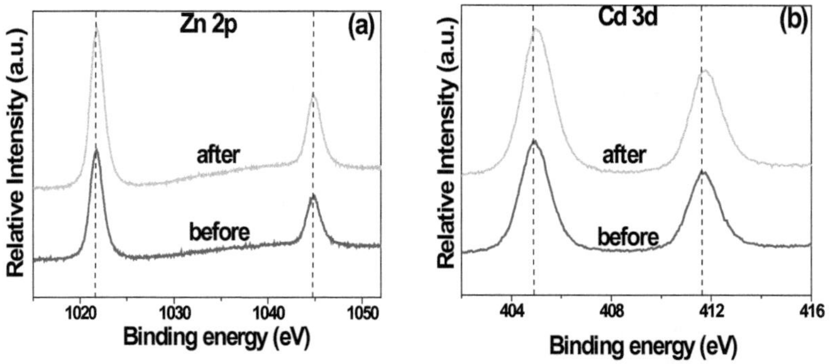

Figure 8.15 High-resolution XPS spectra of Zn 2p (a) and Cd 3d (b) for the GS0.25 sample before and after 2 h photocatalytic H_2-production under simulated solar irradiation.

To provide additional evidence for the above suggested photocatalytic mechanism, the transient photocurrent response of $Zn_{0.8}Cd_{0.2}S$ (GS0), RGO-$Zn_{0.8}Cd_{0.2}S$ (GS0.25), and Pt-GS0 (GS0 loaded with optimized 1 wt % of Pt cocatalyst) electrodes was recorded for several on-off cycles of solar irradiation. A comparison of the photocurrent-time (*I*-*t*) curves for the above three samples with typical on-off cycles of intermittent solar irradiation at a bias potential of 0.4 V is shown Fig. 8.16a. An apparently boosted photocurrent response appears for all three samples under solar illumination, and the on-off cycles of the photocurrent are reproducible. This indicates that most of the photogenerated electrons are transported to the back contact electrode across the sample to produce the photocurrent under solar irradiation.[91] Notably, for the GS0 sample, the photocurrent curve has an obvious anodic photocurrent spike, which is enlarged in Fig. 8.16b at the initial time of irradiation. After the spike current has been attained, a continuous decrease in the photocurrent with time can be observed until reaching a constant current. Such photocurrent decay indicates that the recombination process occurs.[92] During the decay process, the holes accumulated at the $Zn_{0.8}Cd_{0.2}S$ surface competitively recombine with electrons from the CB of $Zn_{0.8}Cd_{0.2}S$ instead of being trapped or captured by reduced species in the electrolyte. In contrast, the anodic photocurrent spikes are not visible and the transient photocurrent shows a relatively slow response when the light is switched on and off for the Pt-GS0 and GS0.25 samples as displayed respectively in panels b and c of Fig.

8.16a. The electrons trapped on the Pt nanoparticles and RGO in the composites are responsible for their slow photocurrent response. It is not difficult to understand because Pt and graphene have lower Fermi level than that of the conduction band electrons and the trapping of conduction band electrons is a very rapid event.[93] Therefore, upon irradiation, the traps in graphene are gradually filled with electrons, and then the trapped electrons further move to the deep traps. During this time, only a part of photoelectrons can transport to the back contact electrode until the deep traps are filled, which causes a slowly increased photocurrent response. Similarly, the slow response of the photocurrent until decaying to zero by switching off irradiation is controlled by charge carriers released from the traps.[93] It suggests that RGO (or Pt nanoparticles) separate and store the photo-generated electrons, consequently reducing the recombination rate of the photogenerated electrons and holes and enhancing the production of photocurrent. Therefore, it is not surprising that the GS0 sample exhibits relative weak photocurrent because of the high recombination rate of electrons and holes.

In addition, the highest transient photocurrent was obtained for GS0.25 sample. It is due to the fact that RGO is an excellent electron-acceptor with superior conductivity due to its two-dimensional conjugation structure;[94] in the RGO-$Zn_{0.8}Cd_{0.2}S$ system, the excited electrons of $Zn_{0.8}Cd_{0.2}S$ can transfer from the CB of $Zn_{0.8}Cd_{0.2}S$ to RGO. Thus, in GS0.25 sample, RGO serves as an electron collector and transporter that effectively suppresses the charge recombination, leaving more charge carriers to form reactive species, which in turn results in the highest photocurrent response and photocatalytic H_2-production rate. The comparison of the *I-V* characteristics for the three samples in Fig. 8.16c also proves the highest photocurrent in GS0.25 sample, which further confirms that introduction of RGO can promote the separation of electrons and holes and thus enhance the photocatalytic activities.

Electrochemical impedance spectra (EIS) analysis has become a powerful tool in studying the charge transfer process occurring in the three-electrode system and the EIS Nynquist plots of the three samples as shown in Fig. 8.16d. The high-frequency response is due to the charge transfer (or electrochemical reaction) at the Pt counter

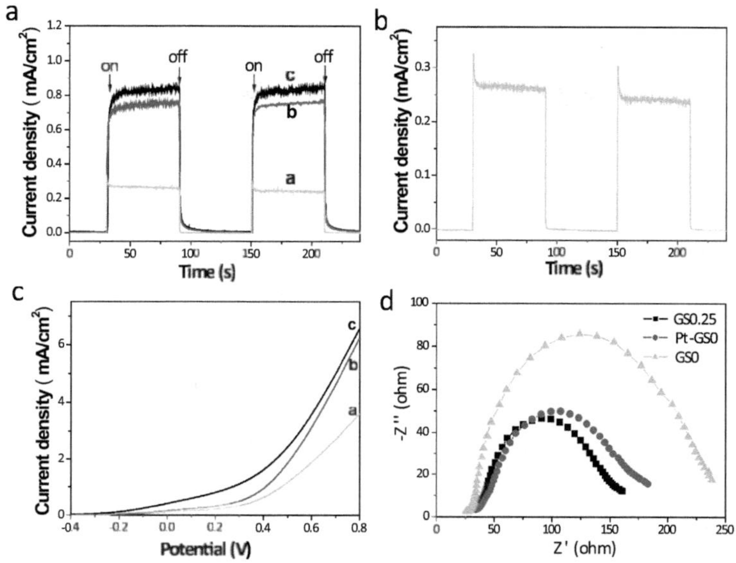

Figure 8.16 (a) Transient photocurrent responses of GS0 (a), GS0 loaded with 1 wt% Pt (b), and GS0.25 (c) samples in 0.1 M Na_2S + 0.02 M Na_2SO_3 mixed aqueous solution under solar irradiation. (b) Individual transient photocurrent responses of sample GS0 in 0.1 M Na_2S + 0.02 M Na_2SO_3 aqueous solution under solar irradiation. (c) *I-V* characteristics of GS0 (a), GS0 loaded with 1 wt% Pt (b), and GS0.25 (c) samples under solar irradiation. (d) Nyquist plots of GS0, GS0 loaded with 1 wt% Pt, and GS0.25 electrodes in 0.1 M Na_2S + 0.02 M Na_2SO_3 aqueous solution under solar irradiation. Reprinted with permission from Ref. [71]. Copyright 2012. American Chemical Society.

electrode, while the intermediate-frequency response is associated with the electron transport and transfer at the $Zn_{0.8}Cd_{0.2}S$/electrode interface.[95,96] The RGO-$Zn_{0.8}Cd_{0.2}S$ sample (GS0.25) shows the smallest semicircle in the middle-frequency region in comparison to the pristine $Zn_{0.8}Cd_{0.2}S$ (GS0) and GS0 loaded with 1 wt % Pt (Pt-GS0) electrodes, which indicates the fastest interfacial electron transfer.[65,95] That is, due to the excellent conductivity, the introduction of RGO can benefit the charge transfer in the RGO-$Zn_{0.8}Cd_{0.2}S$ system and thus lower the charge recombination. Overall, RGO can function as an electron collector and transporter in the composite and inhibit the charge recombination, and thus significantly enhance the photocatalytic H_2-production activity.

8.3 Perspectives and Challenges

From the above two systems, the main roles of RGO in photocatalytic hydrogen production over metal sulphide nanocomposites can be summarized as follows: efficiently separate the photogenerated electrons-hole pairs as an electron collector and transporter, lengthen the lifetime of the charge carriers, enlarge reaction space from the surface of catalysts to the RGO sheet, supersede noble metals, broaden light absorption, prohibit photocorrosion, and improve the stability of the photocatalysts. Other roles of RGO or graphene are out of the scope of this chapter, and please refer to the related comprehensive review.[97]

Graphene has been widely applied in hydrogen generation from solar water splitting as a new type of carbon nanomaterial. The extended honeycomb network of graphene is the basic building block of other important allotropes of carbon. Before the discovery of graphene, fullerene, CNT, and graphite have already been recognized as useful materials in solar water splitting systems.[98] Particularly, CNT is extensively applied to improve the energy conversion efficiency in solar energy conversion field owing to its outstanding structural and electrical properties. Although it is too early to get a decisive answer for graphene's superiority over other carbon materials, graphene do have some advantages over CNT in the application for photocatalytic hydrogen evolution. First, graphene possesses an ultrahigh theoretical specific surface area of 2600 m^2/g and optical transparency of 97.7%, which are both higher than those of CNT (1315 m^2/g for specific surface area[99] and 95.7% for optical transparency[100]). It makes graphene more desirable to provide sufficient reactive sites and other special functions for hydrogen generation than CNT. Second, the contact area or interfacial interaction between semiconductors and graphene is commonly larger than that between semiconductors and CNT even using the same preparation strategy, because CNT is a tubular structure with a curvature. The sheet-structured graphene also has better ability to disperse other nanomaterials for obtaining larger overall specific surface area than CNT with a tube-structure.

Third, graphene has higher conductivity and mobility than CNT, as the degree of curvature presented by the extended π-aromatic structure in graphene is lower than that in CNT. For example, the room-temperature mobility of electrons is 200000 $cm^2\ V^{-1}\ s^{-1}$ for pristine graphene, which is almost three times higher than that for CNT (79000 $cm^2\ V^{-1}\ s^{-1}$)[101]. Thus, we can figure that graphene could be better than CNT for use in solar hydrogen generation regarding to its larger specific surface area, better contact with semiconductor, and higher electrical conductivity than CNT. For example, Ye et al.[102] reported that CdS/RGO composite was more efficient than CdS/CNT composite in photocatalytic H_2 generation under visible light irradiation due to the better contact between CdS and RGO. In addition, graphene is nontoxic and biocompatible,[103] although CNT was reported to be cytotoxic in many papers.[104] So, graphene might be more suitable for future large-scale production without detrimental effect on environment compared with CNT.

Despite of great progress on the graphene-based nanomaterials for hydrogen generation from solar water splitting, this research is still at the early stage, and there are many challenges for significantly improving the photocatalytic performance or possibly leading to breakthrough. Some issues need to be considered:

(1) Synthesis of graphene-based nanomaterials. The photocatalytic activity of graphene-based nanomaterials is highly dependent on their physicochemical properties, such as composition and morphology. Most of graphene-related nanomaterials for water splitting are prepared following a chemical route from GO in the shape of sheet. Substantially enhancing the hydrogen generation efficiency of graphene-based nanomaterials can be expected by tuning the composition and optimizing the morphology. Therefore, extra efforts have to be made for rational design and synthesis of graphene-based photocatalysts so that the properties and functions of graphene can be fully utilized. Take graphene grown by chemical vapor deposition (CVD) as an example, the conductivity is much better than that synthesized by chemical methods and the scalability is more suitable for large-size photocatalytic materials; single and multiple-doping of heteroatom in graphene can tune its many properties. Furthermore, the doped graphene can give

rise to multiple exciton generation (MEG), that is, the formation of several electron-hole pairs after absorption of one photon.[105] This MEG effect will extremely enhance the hydrogen production efficiency if applied to solar water splitting. In addition, graphene with different dimensionality ranging from 0D graphene quantum dots (GQDs), 1D graphene nanoribbons, 2D graphene sheets, to 3D porous frameworks, has special electronic and optical properties.[106] For instance, GQDs have light response ranging from UV through visible into infrared depending on their specific size, geometry, and boundary conditions. More importantly, GQDs have remarkable upconversion ability to convert low-energy photons into high-energy photons and exceptionally long lifetime of hot carriers,[107] which have been successfully applied to improve the photocatalytic activity of semiconductors as the electron-acceptor or photoabsorber.[108] The unique properties of GQDs might be beneficial to separation of photoinduced charges and absorption of light in solar water splitting. Therefore, design and synthesis of graphene with desired composition, size, edge structure, and dimensionality is worthy of exploration for enormously improving the performance of the photocatalytic hydrogen evolution.

(2) Comprehensive roles of graphene. A few roles of RGO in hydrogen production from water splitting over metal sulphide nanocomposites were discussed in this chapter, although it is far from completed and there are also many questions related to it. It is reported that the strong synergetic interaction between graphene and semiconductor improves the charge separation and transfer properties in the photoexcited semiconductor and affects the band structure of graphene. The coupling interaction usually opens up a bandgap in the zero-bandgap graphene and induces doping effect in the carbon monolayer. However, the semiconducting properties of graphene in the composite are usually ignored when graphene is used as the electron acceptor and transporter, even with RGO as the building block. Moreover, the structure defects of graphene might be the active sites for hydrogen evolution, which is necessary for cocatalyst. These defects in

carbon nanomaterials are of importance for their electronic properties and play a critical role in enhancement of the catalytic activity of the supported catalysts,[109] which should be paid much attention to in the future work. So, the functions of graphene in photocatalytic reactions need to be better studied for a thorough and clear picture of the underlying mechanism.

(3) Theoretical calculation of graphene-based nanomaterials. Numerous experimental results have already demonstrated the crucial roles of graphene in enhancement of solar conversion efficiency. However, the inherent mechanism behind enhancement of hydrogen production activity is still not well clarified for graphene-based nanomaterials at the present stage, as the current experimental technologies cannot provide direct evidence for the strengthened interfacial charge transfer property, whereas theoretical calculation can almost perfectly simulate all these specific microstructures, and provide in-depth insight to the electronic properties of nanomaterials, giving us a better understanding of the experimental phenomena.[110] Therefore, theoretical calculation has been considered as one of the most powerful means to investigate the photoinduced charge transfer between graphene and semiconductor at the molecular level, which is pivotal for revealing the underlying photocatalytic mechanism of the composite semiconductors. Furthermore, considering the various morphologies of graphene and the diverse choice of photoactive materials, it is incredible to try out all the possible combinations for hydrogen generation from solar water splitting experimentally. Theoretical calculation has been demonstrated to be capable of simulating catalytic reactions at surfaces with the detail and accuracy required for computational results that compare favourably with experiments. Hence, theoretical calculation holds great promise for screening for catalysts with increased activity, designing new photocatalysts, and engineering the electronic structures of the active surface by changing the composition and structure of the graphene-based nanocomposites, except for describing the photocatalytic reactions.

Study targeting efficient hydrogen generation from solar water splitting by graphene-based nanomaterials is inherently interdisciplinary, involving chemistry, physics, materials, and engineering. It is a very exciting research avenue, and challenges as well as opportunities coexist. The intense and rational efforts and close cooperation among researchers in different fields would make a crucial breakthrough for hydrogen energy development by utilizing graphene-based nanomaterials and be beneficial for human being on both economic and environmental aspects.

Acknowledgments

The authors acknowledge partial financial support for this work from the National Natural Science Foundation of China (Nos. 91123003, 21005023), the National Basic Research Program of China (No. 2011CB933401), and the Knowledge Innovation Program of the Chinese Academy of Sciences.

References

1. N. S. Lewis, *Science* **2007**, *315*, 798.
2. J. A. Turner, *Science* **1999**, *285*, 687.
3. O. Akhavan, *ACS Nano* **2010**, *4*, 4174.
4. P. Millet, F. Andolfatto, R. Durand, *Int. J. Hydrogen Energy* **1996**, *21*, 87.
5. W. Kreuter, H. Hofmann, *Int. J. Hydrogen Energy* **1998**, *23*, 661.
6. T. Pregger, D. Graf, W. Krewitt, C. Sattler, M. Roeb, S. Möller, *Int. J. Hydrogen Energy* **2009**, *34*, 4256.
7. M. Ni, D. Y. C. Leung, M. K. H. Leung, K. Sumathy, *Fuel Process. Technol.* **2006**, *87*, 461.
8. M. G. Walter, E. L. Warren, J. R. McKone, S. W. Boettcher, Q. Mi, E. A. Santori, N. S. Lewis, *Chem. Rev.* **2010**, *110*, 6446.
9. A. Fujishima, K. Honda, *Nature* **1972**, *238*, 37.
10. A. J. Bard, *J. Photochem.* **1979**, *10*, 59.
11. (a) A. Kudo, Y. Miseki, *Chem. Soc. Rev.* **2009**, *38*, 253; (b) X. Chen, S. Shen, L. Guo, S. S. Mao, *Chem. Rev.* **2010**, *110*, 6503; (c) K. Maeda, K. Domen, *J. Phys. Chem. Lett.* **2010**, *1*, 2655; (d) M. A. Fox, M. T. Dulay, *Chem. Rev.* **1993**, *93*, 341; (e) A. Fujishima, X. Zhang, D. A. Tryk, *Int. J. Hydrogen Energy* **2007**, *32*, 2664; (f) P. V. Kamat, *J. Phys. Chem. C* **2007**, *111*, 2834; (g) A. J. Esswein, D. G. Nocera, *Chem. Rev.* **2007**, *107*, 4022;

(h) H. Tong, S. Ouyang, Y. Bi, N. Umezawa, M. Oshikiri, J. Ye, *Adv. Mater.* **2012**, *24*, 229; i) X. Chen, S. S. Mao, *Chem. Rev.* **2007**, *107*, 2891; (j) A. L. Linsebigler, G. Lu, J. T. Yates, *Chem. Rev.* **1995**, *95*, 735; (k) F. E. Osterloh, *Chem. Mater.* **2007**, *20*, 35.

12. (a) C. J. Li, G. R. Xu, B. Zhang, J. R. Gong, *Appl. Catal., B* **2012**, *115–116*, 201; (b) C. J. Li, J. N. Wang, B. Wang, J. R. Gong, Z. Lin, *Mater. Res. Bull.* **2012**, *47*, 333; (c) C. J. Li, J. N. Wang, B. Wang, J. R. Gong, Z. Lin, *J. Nanosci. Nanotechnol.* **2012**, *12*, 2496.
13. (a) K. Domen, A. Kudo, T. Onishi, *J. Catal.* **1986**, *102*, 92; (b) K. Domen, S. Naito, T. Onishi, K. Tamaru, *Chem. Phys. Lett.* **1982**, *92*, 433.
14. (a) H. Kato, A. Kudo, *Catal. Lett.* **1999**, *58*, 153; (b) A. Kudo, H. Kato, *Chem. Phys. Lett.* **2000**, *331*, 373.
15. M. Matsumura, S. Furukawa, Y. Saho, H. Tsubomura, *J. Phys. Chem.* **1985**, *89*, 1327.
16. Reber, J. F.; Rusek, M. *J. Phys. Chem.* **1986**, *90*, 824.]
17. (a) J. S. Jang, H. G. Kim, J. S. Lee, *Catal. Today* **2012**, *185*, 270; (b) A. Kubacka, M. Fernández-García, G. Colón, *Chem. Rev.* **2012**, *112*, 1555; (c) P. D. Tran, L. H. Wong, J. Barber, J. S. C. Loo, *Energy Environ. Sci.* **2012**, *5*, 5902; (d) A. Kudo, *Int. J. Hydrogen Energy* **2007**, *32*, 2673.
18. (a) A. J. Cowan, J. R. Durrant, *Chem. Soc. Rev.* **2013**, *42*, 2281; (b) F. E. Osterloh, *Chem. Soc. Rev.* **2013**, *42*, 2294.
19. K. S. Novoselov, a. K. Geim, S. V. Morozov, D. Jiang, Y. Zhang, S. V. Dubonos, I. V. Grigorieva, A. A. Firsov, *Science* **2004**, *306*, 666.
20. (a) S. Guo, S. Dong, *Chem. Soc. Rev.* **2011**, *40*, 2644; (b) F. Bonaccorso, Z. Sun, T. Hasan, A. C. Ferrari, *Nat. Photonics* **2010**, *4*, 611; (c) J. Li, Y. Liu, D. Chen, h. zhang, *Energy Environ. Sci.* **2013**, 10.1039/C3EE23586F; (d) P. V. Kamat, *J. Phys. Chem. Lett.* **2011**, *2*, 242.
21. A. K. Geim, K. S. Novoselov, *Nat. Mater.* **2007**, *6*, 183.
22. (a) Y. Xu, H. Bai, G. Lu, C. Li, G. Shi, *J. Am. Chem. Soc.* **2008**, *130*, 5856; (b) X. Huang, Z. Yin, S. Wu, X. Qi, Q. He, Q. Zhang, Q. Yan, F. Boey, H. Zhang, *Small* **2011**, *7*, 1876; (c) D. Chen, H. Feng, J. Li, *Chem. Rev.* **2012**, *112*, 6027.
23. D. R. Dreyer, S. Park, C. W. Bielawski, R. S. Ruoff, *Chem. Soc. Rev.* **2010**, *39*, 228.
24. V. Singh, D. Joung, L. Zhai, S. Das, S. I. Khondaker, S. Seal, *Prog. Mater. Sci.* **2011**, *56*, 1178.
25. S. Mao, H. Pu, J. Chen, *RSC Adv.* **2012**, *2*, 2643.
26. (a) C. Gómez-Navarro, J. C. Meyer, R. S. Sundaram, A. Chuvilin, S. Kurasch, M. Burghard, K. Kern, U. Kaiser, *Nano Lett.* **2010**, *10*, 1144;

(b) C. Gómez-Navarro, R. T. Weitz, A. M. Bittner, M. Scolari, A. Mews, M. Burghard, K. Kern, *Nano Lett.* **2007**, *7*, 3499; (c) F. Tuinstra, J. L. Koenig, *J. Chem. Phys.* **1970**, *53*, 1126.

27. (a) L. Han, P. Wang, S. Dong, *Nanoscale* **2012**, *4*, 5814; (b) N. Zhang, Y. Zhang, Y. J. Xu, *Nanoscale* **2012**, *4*, 5792; (c) Q. Xiang, J. Yu, *J. Phys. Chem. Lett.* **2013**, *4*, 753.
28. A. J. Hoffman, H. Yee, G. Mills, H. Yee, M. R. Hoffmann, *J. Phys. Chem.* **1992**, *96*, 5546.
29. Y. X. Li, Y. Z. Xie, S. Q. Peng, G. X. Lu, S. B. Li, *Chemosphere* **2006**, *63*, 1312.
30. H. J. Yan, J. H. Yang, G. J. Ma, G. P. Wu, X. Zong, Z. B. Lei, J. Y. Shi, C. Li, *J. Catal.* **2009**, *266*, 165.
31. W. F. Shangguan, A. Yoshida, *J. Phys. Chem. B* **2002**, *106*, 12227.
32. O. Enea, A. J. Bard, *J. Phys. Chem.* **1986**, *90*, 301.
33. T. Sato, H. Okuyama, T. Endo, M. Shimada, *React. Solids* **1990**, *8*, 63.
34. T. Hirai, H. Okubo, I. Komasawa, *J. Phys. Chem. B* **1999**, *103*, 4228.
35. J. C. Meyer, A. K. Geim, M. I. Katsnelson, K. S. Novoselov, T. J. Booth, S. Roth, *Nature* **2007**, *446*, 60.
36. L. Shen, M. G. Zeng, S. W. Yang, C. Zhang, X. F. Wang, Y. P. Feng, *J. Am. Chem. Soc.* **2010**, *132*, 11481.
37. G. Eda, G. Fanchini, M. Chhowalla, *Nat. Nanotechnol.* **2008**, *3*, 270.
38. Q. Li, B. D. Guo, J. G. Yu, J. Ran, B. H. Zhang, H. Yan, J. R. Gong, *J. Am. Chem. Soc.* **2011**, *133*, 10878.
39. A. N. Cao, Z. Liu, S. S. Chu, M. H. Wu, Z. M. Ye, Z. Cai, Y. L. Chang, S. F. Wang, Q. H. Gong, Y.F. Liu, *Adv. Mater.* **2010**, *22*, 103.
40. Y. X. Xu, H. Bai, G. W. Lu, C. Li, G.Q. Shi, *J. Am. Chem. Soc.* **2008**, *130*, 5856.
41. N. Z. Bao, L. M. Shen, T. Takata, K. Domen, *Chem. Mater.* **2008**, *20*, 110.
42. G. Williams, B. Seger, P. V. Kamat, *ACS nano* **2008**, *2*, 1487.
43. J. G. Yu, H. G. Yu, B. Cheng, M. H. Zhou, X. J. Zhao, *J. Mol. Catal. A* **2006**, *253*, 112.
44. K. S.W. Sing, D. H. Everett, R. Haul, L. Moscou, R. A. Pierotti, J. Rouquerol, T. Siemieniewska, *Pure Appl. Chem.* **1985**, *57*, 603.
45. S. Stankovich, R. D. Piner, X. Q. Chen, N. Q. Wu, S. T. Nguyen, R. S. Ruoff, *J. Mater. Chem.* **2006**, *16*, 155.
46. S. Stankovich, D. A. Dikin, R. D. Piner, K. A. Kohlhaas, A. Kleinhammes, Y. Y. Jia, Y. Wu, S. T. Nguyen, Ruoff, R. S. *Carbon* **2007**, *45*, 1558.

47. T. Szabó, E. Tombácz, E. Illés, I. Dékány, *Carbon* **2006**, *44*, 537.

48. N. Alonso-Vante, H. Colell, U. Stimming, H. Tributsch, *J. Phys. Chem.* **1993**, *97*, 7381.

49. G. R. Bamwenda, S. Tsubota, T. Nakamura, M. Haruta, *J. Photochem. Photobiol. A* **1995**, *89*, 177.

50. S. Sakthivel, M. V. Shankar, M. Palanichamy, B. Arabindoo, D. W. Bahnemann. V. Murugesan, *Water Res.* **2004**, *38*, 3001.

51. J. G Yu, T. T. Ma, S. W. Liu, *Phys. Chem. Chem. Phys.* **2011**, *13*, 3491.

52. (a) T. Peng, K. Li, P. Zeng, Q. Zhang, X. Zhang, *J. Phys. Chem. C* **2012**, *116*, 22720; (b) P. Gao, J. Liu, S. Lee, T. Zhang, D. D. Sun, *J. Mater. Chem.* **2012**, *22*, 2292.

53. C. Dong, X. Li, P. Jin, W. Zhao, J. Chu, J. Qi, *J. Phys. Chem. C* **2012**, *116*, 15833.

54. J. Zhang, J. G. Yu, Y. M. Zhang, Q. Li, J. R. Gong, *Nano Lett.* **2011**, *11*, 4774–4779.

55. J. F. Reber, K. Meier, *J. Phys. Chem.* **1984**, *88*, 5903–5913.

56. L. Wang, W. Z. Wang, M. Shang, W. Z. Yin, S. M. Sun, L. Zhang, *Int. J. Hydrogen Energy* **2010**, *35*, 19–25.

57. J. Zhang, S. W. Liu, J. G. Yu, M. Jaroniec, *J. Mater. Chem.* **2011**, *21*, 14655–14662.

58. J. G. Yu, J. Zhang, M. Jaroniec, *Green Chem.* **2010**, *12*, 1611–1614.

59. I. Tsuji, H. Kato, A. Kudo, *Chem. Mater.* **2006**, *18*, 1969–1975.

60. S. H. Shen, L. Zhao, Z. H. Zhou, L. J. Guo, *J. Phys. Chem. C* **2008**, *112*, 16148–16155.

61. Y. X. Li, G. Chen, Q. Wang, X. Wang, A. K. Zhou, Z. Y. Shen, *Adv. Func. Mater.* **2010**, *20*, 3390–3398.

62. S. Ikeda, T. Nakamura, T. Harada, M. Matsumura, *Phys. Chem. Chem. Phys.* **2010**, *42*, 13943–13949.

63. I. Tsuji, H. Kato, H. Kobayashi, A. Kudo, *J. Am. Chem. Soc.* **2004**, *126*, 13406–13413.

64. X. Y. Zhang, H. P. Li, X. L. Cui, Y. H. Lin, *J. Mater. Chem.* **2010**, *20*, 2801–2806.

65. Q. Li, B. D. Guo, J. G. Yu, J. R. Ran, B. H. Zhang, H. J. Yan, J. R. Gong, *J. Am. Chem. Soc.* **2011**, *133*, 10878–10884.

66. L. Jia, D. H. Wang, Y. X. Huang, A. W. Xu, H. Q. Yu, *J. Phys. Chem. C* **2011**, *115*, 11466–11473.

67. Q. J. Xiang, J. G. Yu, M. Jaroniec, *J. Phys. Chem. C* **2011**, *115*, 7355–7363.

68. Y. H. Ng, A. Iwase, A. Kudo, R. Amal, *J. Phys. Chem. Lett.* **2010**, *1*, 2607–2612.

69. A. Mukherji, B. Seger, G. Q. Lu, L. Z. Wang, *ACS Nano* **2011**, *5*, 3483–3492.

70. X. J. Lv, W. F. Fu, H. X. Chang, H. Zhang, J. S. Cheng, G. J. Zhang, Y. Song, C. Y. Hu, J. H. Li, *J. Mater. Chem.* **2012**, *22*, 1539–1546.

71. J. Zhang, J. Yu, M. Jaroniec, J. R. Gong, *Nano Lett.* **2012**, *12*, 4584.

72. Y. Zhou, Q. L. Bao, L. A. L. Tang, Y. L. Zhong, K. P. Loh, *Chem. Mater.* **2009**, *21*, 2950–2956.

73. P. Kubelka, *J. Opt. Soc. Am.* **1948**, *38*, 448–457.

74. J. G. Yu, L. J. Zhang, B. Cheng, Y. R. Su, *J. Phys. Chem. C* **2007**, *111*, 10582–10589.

75. J. Yang, J. Zhang, L. W. Zhu, S. Y. Chen, Y. M. Zhang, Y. Tang, Y. L. Zhu, Y. W. Li, *J. Hazard. Mater.* **2006**, *137*, 952–958.

76. K. S. W. Sing, D. H. Everett, R. A. W. Haul, L. Moscou, R. A. Pierotti, J. Rouquerol, T. Siemieniewska, *Pure Appl. Chem.* **1985**, *57*, 603–619.

77. A. Kudo, Y. Miseki, *Chem. Soc. Rev.* **2009**, *38*, 253–278.

78. R. M. Navarro, M. C. Alvarez-Galvan, de la Mano, J. A. V.; Al-Zahrani, S. M.; Fierro, J. L. G. *Energy Environ. Sci.* **2010**, *3*, 1865–1882.

79. T. C. Johnson, D. J. Morris, M. Wills, *Chem. Soc. Rev.* **2010**, *39*, 81–88.

80. J. G. Yu, J. R. Ran, *Energy Environ. Sci.* **2011**, *4*, 1364–1371.

81. J. G. Yu, Y. Hai, B. Cheng, *J. Phys. Chem. C* **2011**, *115*, 4953–4958.

82. A. J. Frank, K. Honda, *J. Phys. Chem.* **1982**, *86*, 1933–1935.

83. N. Bao, L. Shen, T. Takata, K. Domen, K. Gupta, A. Yanagisawa, *J. Phys. Chem. C* **2007**, *111*, 17527–17534.

84. F. Zhou, R. Shi, Y. F. Zhu, *J. Mol. Catal. A* **2011**, *340*, 77–82.

85. T. G. Xu, L. W. Zhang, H. Y. Cheng, Y. F. Zhu, *Appl. Catal. B* **2011**, *101*, 382–387.

86. G. J. Liu, L. Zhao, L. J. Ma, L. J. Guo, *Catal. Commun.* **2008**, *9*, 126–130.

87. K. F. Zhou, Y. H. Zhu, X. L. Yang, X. Jiang, C. Z. Li, *New J. Chem.* **2011**, *35*, 353–359.

88. X. Wang, L. J. Zhi, K. Mullen, *Nano Lett.* **2008**, *8*, 323–327.

89. J. F. Moulder, W. F. Stickle, P. E. Sobol, K. D. Bomben, J. Chastain, Handbook of X-ray Photoelectron Spectroscopy; Physical Electronics, Inc.: Eden Prairie, MN, **1992**.

90. A. Deshpande, P. Shah, R. S. Gholap, N. M. Gupta, *J. Collo. Inter. Sci.* **2009**, *333*, 263–268.

91. A. Hagfeldt, H. Lindstrom, S. Sodergren, S. Lindquist, *J. Electroanal. Chem.* **1995**, *381*, 39–46.

92. Z. H. Xu, J. G. Yu, *Nanoscale* **2011**, *3*, 3138–3144.

93. J. G. Yu, G. P. Dai, B. B. Huang, *J. Phys. Chem. C* **2009**, *113*, 16394–16401.

94. K. S. Novoselov, A. K. Geim, S. V. Morozov, D. Jiang, Y. Zhang, S. V. Dubonos, I. V. Grigorieva, A. A. Firsov, *Science* **2004**, *306*, 666–669.

95. J. G. Yu, J. J. Fan, B. Cheng, *J. Power Sour.* **2011**, *196*, 7891–7898.

96. A. Jana, C. Bhattacharya, J. Datta, *Electro. Acta.* **2010**, *55*, 6553–6562.

97. G. Xie, K. Zhang, B. Guo, Q. Liu, L. Fang, J. R. Gong. *Adv. Mater.* **2013**, DOI: 10.1002/adma.201301207.

98. (a) R. Leary, A. Westwood, *Carbon* **2011**, *49*, 741; (b) B. Chai, T. Peng, P. Zeng, X. Zhang, *Dalton Trans.* **2012**, *41*, 1179; (c) Y. K. Kim, H. Park, *Energy Environ. Sci.* **2011**, *4*, 685; (d) L. X. Xu, C. F. Ma, L. X. Sang, Q. W. Li, H. X. Dai, H. Hong, *Chin. J. Catal.* **2007**, *28*, 1083; (e) J. Yu, B. Yang, B. Cheng, *Nanoscale* **2012**, *4*, 2670; (f) Y. Ou, J. Lin, S. Fang, D. Liao, *Chem. Phys. Lett.* **2006**, *429*, 199; (g) Z. Liu, W. Hou, P. Pavaskar, M. Aykol, S. B. Cronin, *Nano Lett.* **2011**, *11*, 1111.

99. A. Peigney, C. Laurent, E. Flahaut, R. R. Bacsa, A. Rousset, *Carbon* **2001**, *39*, 507.

100. C. Biswas, Y. H. Lee, *Adv. Funct. Mater.* **2011**, *21*, 3806.

101. T. Dürkop, S. A. Getty, E. Cobas, M. S. Fuhrer, *Nano Lett.* **2003**, *4*, 35.

102. A. Ye, W. Fan, Q. Zhang, W. Deng, Y. Wang, *Catal. Sci. Technol.* **2012**, *2*, 969.

103. (a) Y. Chang, S. T. Yang, J. H. Liu, E. Dong, Y. Wang, A. Cao, Y. Liu, H. Wang, *Toxicol. Lett.* **2011**, *200*, 201; (b) Z. Liu, J. T. Robinson, X. Sun, H. Dai, *J. Am. Chem. Soc.* **2008**, *130*, 10876.

104. Y. Liu, Y. Zhao, B. Sun, C. Chen, *Acc. Chem. Res.* **2013**, *46*, 702.

105. K. J. Tielrooij, J. C. W. Song, S. A. Jensen, A. Centeno, A. Pesquera, A. Zurutuza Elorza, M. Bonn, L. S. Levitov, F. H. L. Koppens, *Nat. Phys.* **2013**, *9*, 248.

106. (a) K. A. Ritter, J. W. Lyding, *Nat. Mater.* **2009**, *8*, 235; (b) Q. Liu, B. Guo, Z. Rao, B. Zhang, J. R. Gong, *Nano Lett.* 10.1021/nl400368v; (c) X. Huang, K. Qian, J. Yang, J. Zhang, L. Li, C. Yu, D. Zhao, *Adv. Mater.* **2012**, *24*, 4419; (d) Y. H. Chang, C. T. Lin, T. Y. Chen, C. L. Hsu, Y. H. Lee, W. Zhang, K. H. Wei, L. J. Li, *Adv. Mater.* **2013**, *25*, 756; (e) Z. Z. Zhang, K. Chang, *Phys. Rev. B* **2008**, *77*, 1; (f) S. Schumacher, *Phys. Rev. B* **2011**, *83*, 081417.

107. (a) J. Shen, Y. Zhu, C. Chen, X. Yang, C. Li, *Chem. Commun.* **2011**, *47*, 2580; (b) M. L. Mueller, X. Yan, J. A. McGuire, L. S. Li, *Nano Lett.* **2010**, *10*, 2679. 108. (a) D. Pan, J. Zhang, Z. Li, M. Wu, *Adv. Mater.* **2010**, *22*, 734; (b) S. Zhuo, M. Shao, S. T. Lee, *ACS Nano* **2012**, *6*, 1059; (c) W. Chen, S. Li, C. Chen, L. Yan, *Adv. Mater.* **2011**, *23*, 5679; (d) C. Hou, Q. Zhang, Y. Li, H. Wang, *J. Hazard. Mater.* **2012**, *205–206*, 229; (e) S. H. Lee, H. W. Kim, J. O. Hwang, W. J. Lee, J. Kwon, C. W. Bielawski, R. S. Ruoff, S. O. Kim, *Angew. Chem., Int. Ed.* **2010**, *49*, 10084; (f) D. I. Son, B. W. Kwon, D. H. Park, W. S. Seo, Y. Yi, B. Angadi, C. L. Lee, W. K. Choi, *Nat. Nanotechnol.* **2012**, *7*, 465; (g) Y. Li, Y. Hu, Y. Zhao, G. Shi, L. Deng, Y. Hou, L. Qu, *Adv. Mater.* **2011**, *23*, 776.

109. (a) I. Fampiou, A. Ramasubramaniam, *J. Phys. Chem. C* **2012**, *116*, 6543; (b) D. S. Su, S. Perathoner, G. Centi, *Catal. Today* **2012**, *186*, 1.

110. (a) J. Wang, D. N. Tafen, J. P. Lewis, Z. Hong, A. Manivannan, M. Zhi, M. Li, N. Wu, *J. Am. Chem. Soc.* **2009**, *131*, 12290; (b) X. Chen, L. Liu, P. Y. Yu, S. S. Mao, *Science*, **2011**, *331*, 746.

Chapter 9

Micro/Nanostructured Materials and Their Structurally Enhanced Performances for Environment

Xianbiao Wang,[a,b] Weiping Cai,[a] and Guozhong Wang[a]
[a]*Key Laboratory of Materials Physics,*
Anhui Key Laboratory of Nanomaterials and Nanotechnology,
Institute of Solid State Physics, Chinese Academy of Sciences,
Hefei 230031, Anhui, China
[b]*School of Materials and Chemical Engineering,*
Anhui Jianzhu University, Hefei 230601, China
wpcai@issp.ac.cn

9.1 Introduction

Environmental pollution refers to that the hazardous materials released by human activities result in the decrease of environmental quality [1]. The environmental pollution induced by industrial development has attracted more and more attention in recent years, especially in developing countries. The harmful materials, such as heavy metal ions, organic pollutants have been released to air, water, and soil, which directly threaten the people's health.

Nanomaterials in Energy and Environmental Applications
Edited by Junhui He

ISBN 978-981-4463-78-2 (Hardcover), 978-981- 4463-79-9 (eBook)
www.panstanford.com

Till now, many useful methods and techniques have been developed for the pollution remediation. Among them, removal of pollutants by the method of adsorption and/or catalysis using adsorptive and/or catalystic materials is an important and effective way. In this case, the key issue is the high efficiently absorptive and/or catalystic materials. The materials with high surface activity, high specific surface area, strong selective adsorption, and stable structure are expected. Although the nanosized particle powders could be the candidate for such materials for their high surface area and surface activity [2–9], they are very easily aggregated, leading to the unwanted reduction in the active surface area and surface activity, and hence the bad reusable property. In addition, the used nanosized adsorbents or photocatalysts are suspended in the solution and difficult to separate from bulk solution.

Micro/nanostructured materials, which are composed of the microsized objects with nanostructures [10,11], could overcome such disadvantages. The microsized objects could resist aggregation and the nanostructure could supply high surface area and surface activity during usage. Therefore, these materials with micro/nano-architectures not only possess large surface to-volume ratio, high structural stability against aggregation, and are very easily separated from solution during application in the pollution remediation, but also exhibit strong structurally enhanced adsorption and catalysis performances, and hence can be used for high efficient removal of contaminants in water or solutions. In this chapter, we will introduce some recent progresses in this area at our lab, mainly including mass production of new micro/nanostructured materials for adsorbents; structurally enhanced catalysis properties; and structurally enhanced and selective adsorption performances.

9.2 Mass Production of Novel Micro/Nanostructured Materials

Mass production of the micro/nanostructured materials plays an important role in their applications in environmental remediation. There have been many methods developed for such micro/nanostructured materials. Here, we will only introduce solvothermal/hydrothermal (soft template), template-etching (hard template), and electrospinning methods.

9.2.1 Solvothermal/Hydrothermal Method

Solvothermal/hydrothermal method is known for its low cost, easy operation, and easy control for preparation of micro/nanostructured materials. The general strategy is illustrated in Fig. 9.1. Briefly, structural directors (surfactants or organic solvents) are added in the precursor solution and then maintained at 150°C–250°C for reaction. Metal oxides or sulfides with novel micro/nanostructures are then obtained after reaction for a certain time (several hours) or subsequent heat treatment.

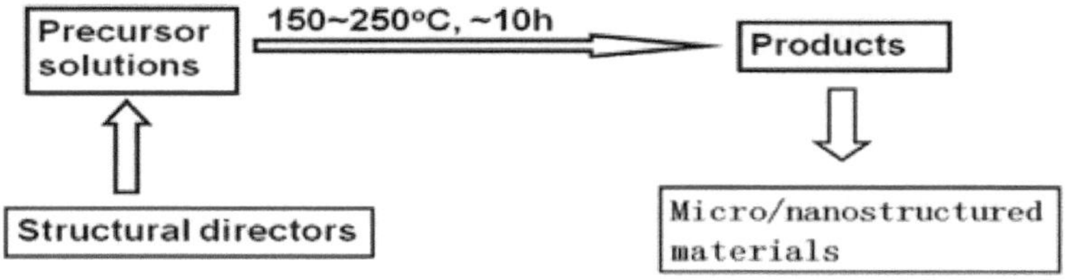

Figure 9.1 General synthetic route for solvothermal/hydrothermal preparation of micro/nanostructured materials.

9.2.1.1 Novel micro/nanostructured ZnO

Lu et al [12] prepared micro/nanostructured ZnO with core-shell structure that contains pyramid-like core and nanosheet-built network shell, showing honeycomb-like structure, by solvothermal method using ethylenediamine (EDA) as the structural director. Importantly, this new ZnO micro/nanostructure is of single crystal nature (Fig. 9.2). Further experiments show that the formation of micro/nanostructure depends on the concentration of EDA solution and the type of zinc source. The formation of ZnO could be described by the following reactions [12]:

$$\mathrm{Zn} + m\mathrm{EDA} + 2\mathrm{H_2O} \rightarrow [\mathrm{Zn(EDA)}_m](\mathrm{OH})_2 + \mathrm{H_2} \quad (9.1)$$

$$[\mathrm{Zn(EDA)}_m]^{2+} \leftrightarrow \mathrm{Zn}^{2+} + m\mathrm{EDA} \quad (9.2)$$

$$\mathrm{Zn}^{2+} + 2\mathrm{OH}^- \rightarrow \mathrm{Zn(OH)_2} \rightarrow \mathrm{ZnO} + \mathrm{H_2O} \quad (9.3)$$

Figure 9.2 Morphology and structure of the honeycomb-like micro/nanostructured ZnO [12]. Copyright (2008) John Wiley and Sons.

Zn atoms are firstly oxidized to form $[Zn(EDA)_m]^{2+}$ with existence of EDA (reaction (9.1)). Then, $[Zn(EDA)_m]^{2+}$ can release Zn^{2+} to form $Zn(OH)_2$ (see reactions (9.2) and (9.3). Finally, ZnO is produced by decomposition of $Zn(OH)_2$. The formation of the honeycomb-like micro/nanostructured ZnO was described by a two-step growth model, as illustrated in Fig. 9.3. In the first step, the pyramid-like core is formed owing to growth habit of ZnO and controllable release of Zn^{2+}. In the second step, the crosslinked nanosheets are formed by growth of outshoots on the surface of pyramid-like microcrystal. In this work, EDA acted as the stabilizer of Zn^{2+} and then controlled the release of Zn^{2+}.

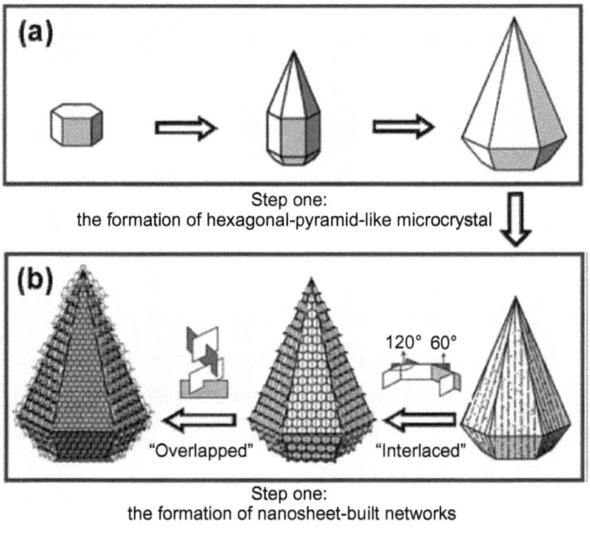

Figure 9.3 Two-step formation mechanism of the honeycomb-like micro/nanostructured ZnO [12]. Copyright (2008) John Wiley and Sons.

Later, Wang et al. [13] reported alternative micro/nanostructured ZnO, which consists of the standing porous ZnO nanoplate-built hollow microspheres through hydrothermal process using citrate as structure director. In this case, flower-like hydrozincite $Zn_5(OH)_6(CO_3)_2$ (or ZnHC for short) hollow spheres with standing nanoplates on the surface were first obtained after hydrothermal reaction (Fig. 9.4a, b). Then, the ZnHC hollow spheres were annealed at 400°C for 2 h to produce porous ZnO with almost the same morphology (Fig. 9.4c, d). For the unique morphology, the key issue is the formation of flower-like nanosheet-built ZnHC hollow spheres by hydrothermal process with presence of citrate.

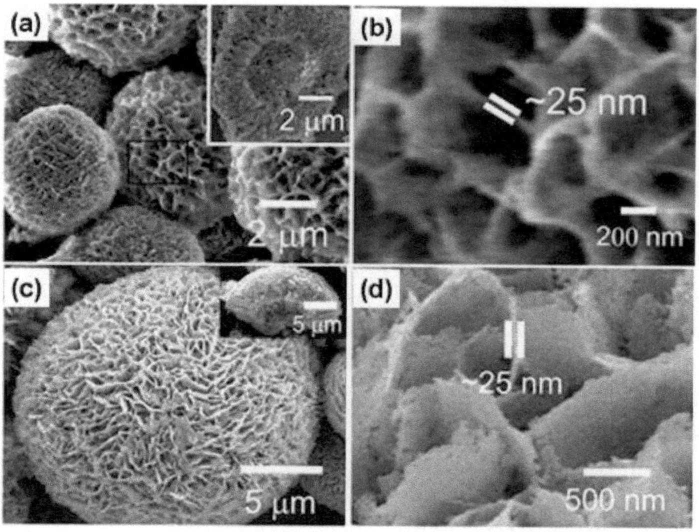

Figure 9.4 SEM images of the as-prepared hydrozincite hollow spheres [13]. SEM images of (a, b) the as-prepared hydrozincite hollow spheres and (c, d) the corresponding ZnO hollow spheres after annealling [13]. Copyright (2012) Cambridge University Press.

The formation process contains three steps. Firstly, the citrate adsorption-induced formation of amorphous hydrozincite spheres. At the beginning, ZnHC molecules were produced on the basis of reactions (9.4–9.7) [13].

$$CO(NH_2)_2 + 3H_2O \rightarrow 2NH_3 \bullet H_2O + CO_2 \quad (9.4)$$

$$NH_3 \bullet H_2O \leftrightarrow NH_4^+ + OH^- \quad (9.5)$$

$$CO_2 + 2OH^- \rightarrow CO_2 + H_2O \quad (9.6)$$

$$5Zn^{2+} + 2CO_3^{-2} + 6OH^- \rightarrow Zn_5(CO_3)_2(OH)_6 \quad (9.7)$$

When the concentration reached supersaturation, ZnHC colloids would be precipitated. The Gibbs free energy (G) of a single spherical ZnHC colloid was determined by Equation (9.8) [14]

$$G = f(R,\sigma) = \frac{4}{3}\pi R^3 G_v + 4\pi R^2 \sigma \quad (9.8)$$

where R is the radius of the colloid, G_v and σ are the free energy in unit volume and the specific surface energy for the colloid, respectively. When the colloid is small in size, the surface energy accounts for a decent proportion of the total free energy. Generally, the G value of crystal state is lower than that of amorphous state because the crystal is stable than amorphous, as shown in curves 1 and 2 of Fig. 9.5. In this case, the adsorption of citrate ions on the colloids would lead to a decrease of surface energy. Therefore, the G value of amorphous colloids is lower than that of crystal at the first stage of colloids' growth (diameter smaller than 5 μm), as shown in curves 1'and 2 of Fig. 9.5. When the size of amorphous colloids is bigger (>3–5 μm), the crystallized phase will be more stable than that of amorphous ones. Crystallization will be formed from the surface of the colloids. Then, further inward crystallization will be difficult owing to the crystallization-induced structural contraction [14]. Thus, the amorphous core will go through dissolution process, leading to formation of hollow structure. Meanwhile, the ZnHC nanosheets were formed due to the layered structure of ZnHC and grew on the surface of colloids, forming ZnHC microspheres with standing nanoplates on the surface. Finally, the standing porous nanoplate-built ZnO hollow microspheres were produced by subsequent annealing process due to the decomposition of ZnHC. Figure 9.6 shows formation of such standing porous ZnO nanoplate-built hollow microspheres.

Recently, the separate ZnHC nanoplates were fabricated by the solvothermal method using ethylene glycol as the morphology director [15]. Subsequently, the micro/nanostructured porous ZnO nanoplates were produced by annealing the product ZnHC. In the formation process, ZnHC molecules were first produced according to reactions (9.4–9.7). Then, the ZnHC will precipitate and grow depending on the layered structure of ZnHC. Meanwhile, EG molecules could adsorb on the (100) plane of ZnHC crystal, leading to formation

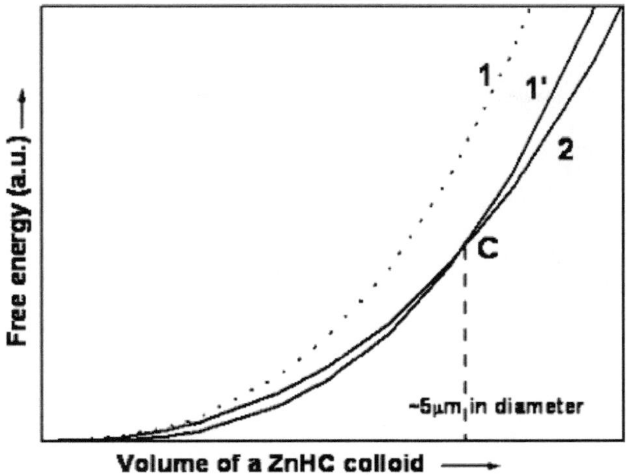

Figure 9.5 Relationship between the free energy (G) of a solid ZnHC sphere and the volume of a ZnHC colloid. Curves 1 refers to an amorphous ZnHC sphere without $C_6H_5O_7{}^{3-}$ adsorption. Curve 1' refers to that with $C_6H_5O_7{}^{3-}$ adsorption. Curve 2 refers to a crystal ZnHC sphere [13]. Copyright (2012) Cambridge University Press.

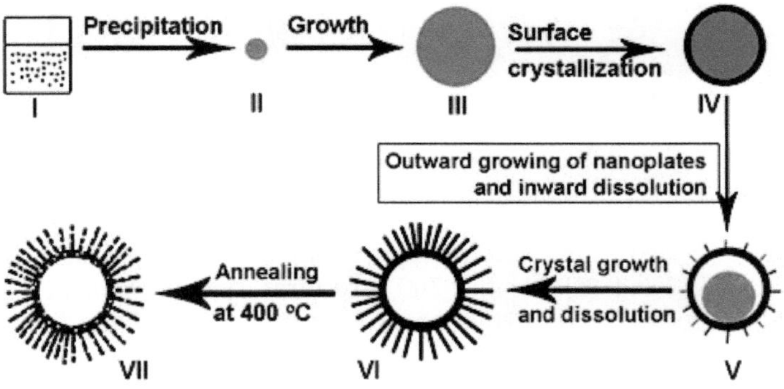

Figure 9.6 The schematic illustration of the formation mechanism of the micro/nanosturctured ZnO hollow microsphere [13]. Copyright (2012) Cambridge University Press.

of ZnHC nanoplates with exposed surface of (100), as shown in Fig. 9.7. Finally, the micro/nanostructured porous ZnO nanoplates were formed by decomposition and structural transformation of ZnHC during annealing process, as shown in Fig. 9.8.

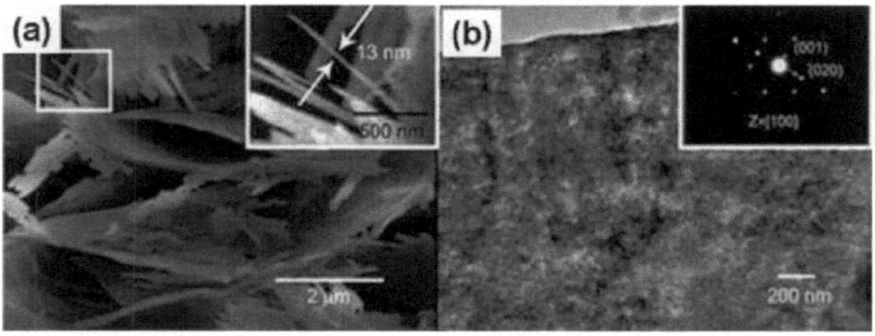

Figure 9.7 SEM (a) and TEM (b) images of the ZnHC produced by solvothermal method [15]. Copyright (2010) Royal Society of Chemistry.

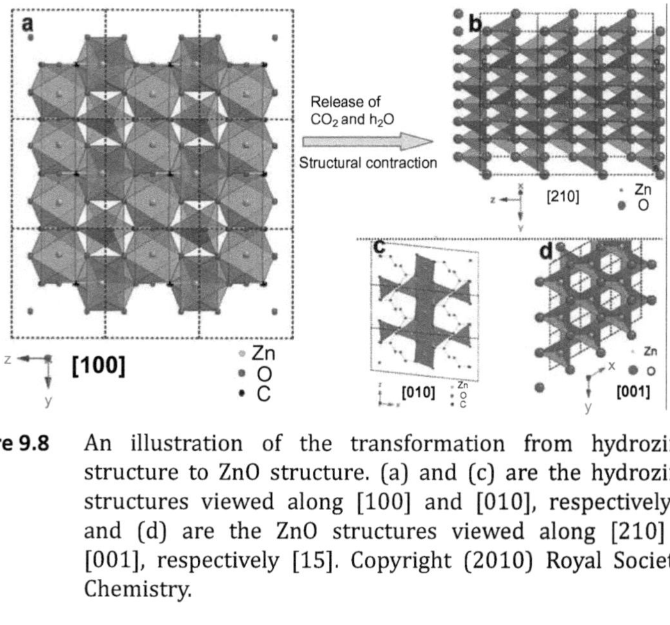

Figure 9.8 An illustration of the transformation from hydrozincite structure to ZnO structure. (a) and (c) are the hydrozincite structures viewed along [100] and [010], respectively. (b) and (d) are the ZnO structures viewed along [210] and [001], respectively [15]. Copyright (2010) Royal Society of Chemistry.

9.2.1.2 Micro/nanostructured porous Fe_3O_4 nanofibers

Also, the micro/nanostructured porous Fe_3O_4 nanofibers were produced by a protein-assisted hydrothermal treatment at 200°C, which exhibited structurally enhanced adsorption property for persistent organic pollutants (POPs) and heavy metal ions [16]. The fibers showed cylindrical shape with length of several micrometers and pore size below 5 nm (Fig. 9.9a, b, d) [16].

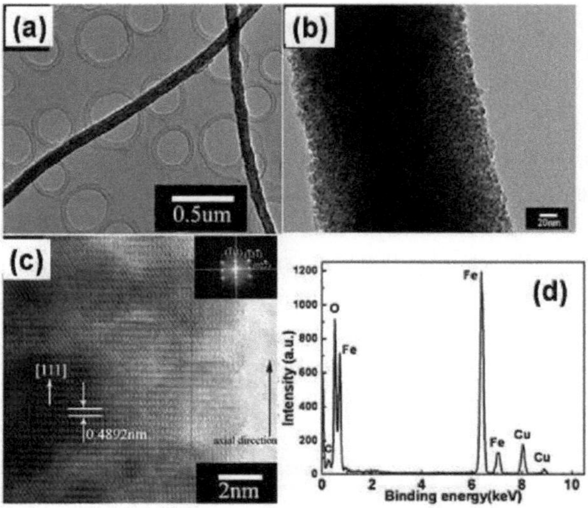

Figure 9.9 (a) TEM image of two Fe_3O_4 nanofibers at low magnification. (b) TEM image of a single fiber at high magnification. (c) HRTEM pattern of a partial fiber ([211] zone). (d) EDX pattern of a single Fe_3O_4 nanofiber (the signals of C and Cu are from the carbon-coated Cu grids) [16]. Copyright (2011) Royal Society of Chemistry.

The precursor solution contains $FeCl_3.6H_2O$, citric acid, and glutin. Then, the pH in the solution was adjusted to 7–8 using ammonia. Formation mechanism of such Fe_3O_4 fibers was deduced on the basis of a series of experiments and influence factors, combining the chemical reactions during the hydrothermal treatment of the precursor solution at 200°C as follows [16]:

$$Fe^{3+} + 2C_6H_5O_7^{3-} + 2OH^- \xrightarrow{25^\circ C} [Fe(C_6H_4O_7)_2]^{5-} + 2H_2O \quad (9.)$$

$$2[Fe(C_6H_4O_7)_2]^{5-} + 2OH^- \xrightarrow{200^\circ C}$$

$$2Fe(OH)_2 + 4C_6H_3O_7^{3-} + 2H_2O \quad (9.10)$$

$$6Fe(OH)_2 + O_2 \xrightarrow{200^\circ C} 2Fe_2O_4 + 6H_2O \quad (9.11)$$

Firstly, the Fe^{3+} ions were coordinated with citrate ions to form $[Fe(C_6H_4O_7)_2]^{5-}$ and further produced $Fe(OH)_2$ in alkaline condition (see reactions (9.9) and (9.10)). Finally, Fe_3O_4 was obtained by reaction of $Fe(OH)_2$ and oxygen (see reaction (9.11)). As the reactions proceeded, Fe_3O_4 began to precipitate after the solution was supersaturated. Then, Fe_3O_4 nanoparticles were formed after nucleation and crystal growth. The added glutin plays an important role in formation of the final fiber morphology. When the glutin was dissolved in precursor solution, protein colloids were formed. The protein molecules exhibited positive charge at pH 7–8. At the same time, Fe_3O_4 nanoparticles were adsorbed by citrate ions, leading to negatively charged surface. Therefore, the negatively charged Fe_3O_4 nanoparticles would adsorb surrounding the surface of positively charged protein chains owing to the electrostatic interaction, as shown in Fig. 9.10a,b [16]. The magnetic Fe_3O_4 nanoparticles would assemble along the [111] direction due to their magnetic dipole, as shown in Fig. 9.9c and Fig. 9.10c,d [16].

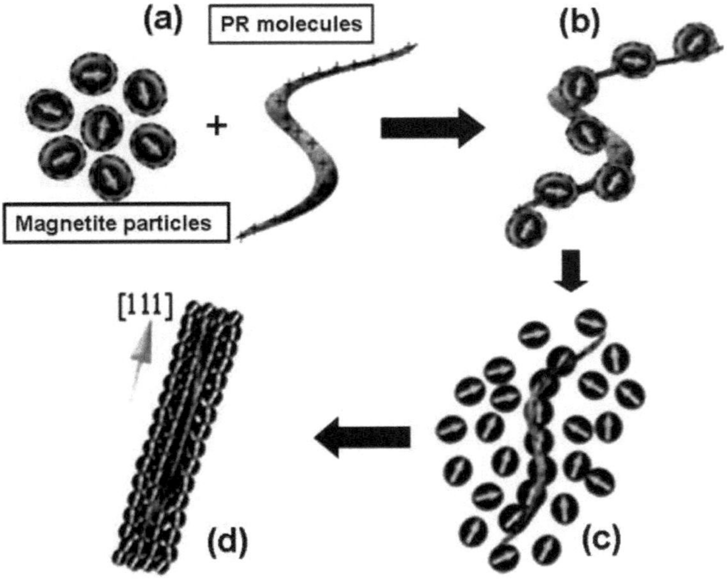

Figure 9.10 Schematic illustration for formation process of porous oriented Fe_3O_4 fibers [16]. Copyright (2010) Royal Society of Chemistry.

The template effect of such protein was demonstrated by the structure of the Fe_3O_4 fibers, which were annealed at 600°C under a vacuum. The tube-like morphology, from observation of Fig. 9.11a, indicates that the protein chains were removed during annealing process. Figure 9.11b also verified the [111] orientation of magnetic Fe_3O_4.

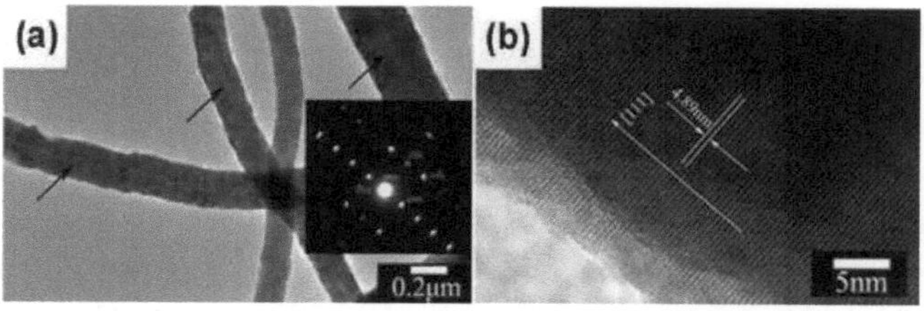

Figure 9.11 The structure of the as-prepared magnetite fibers after annealing. (a) TEM image. Inset: the SAED pattern. (b) HRTEM image [16]. Copyright (2010) Royal Society of Chemistry.

9.2.1.3 Tremella-like micro/nanostructured Fe_3S_4/C composites

Among all the adsorbents for environmental remediation, magnetic carbon adsorbents are known for their chemical stability, high surface area, high adsorption amount, and easy separation properties. Generally, the carbon adsorbents are produced by thermal treatment of precursors, such as polymers [17,18]. The surface functional groups will be decreased during annealing process. Hydrothermal treatment of sugar at low temperature can obtain carbonaceous materials with many functional groups, which can act as adsorption sites. Hence, hydrothermally prepared carbonaceous materials have been an ideal adsorbent and attracted much more attention in recent years [19]. As mentioned in the above section, the adsorbents with micro/nanostructure have advantages of high surface area and high adsorption efficiency during adsorption process. Further, the adsorbents will be easily separated from water after use by introduction of magnetic materials in them.

Figure 9.12 SEM images of Fe_3S_4/C composites. (a) Image at low magnification (inset: EDS results), (b) image at high magnification (inset: enlarged local image) [20]. Copyright (2013) Royal Society of Chemistry.

Recently, we have obtained tremella-like magnetic micro/nanostructured Fe_3S_4/C composites by solvothermal method using EG as solvent and glucose as carbon source. The tremella-like micro/nanostructure is built of curved Fe_3S_4/C nanoplates with thickness of nearly 50 nm as shown in Fig. 9.12. For a single nanoplate, it shows single crystal nature of face-centered cubic Fe_3S_4 phase (JCPDS card No. 89-1998, space group Fd^3m) with exposed (111) plane (Fig. 9.13a,b). The carbonaceous materials are dispersed uniformly on the surface of nanoplates from observation of Fig. 9.13c. The curved fringes with the spacing of 0.34 nm are attributed to the graphite phase of surface-distributed carbon (Fig. 9.13d).

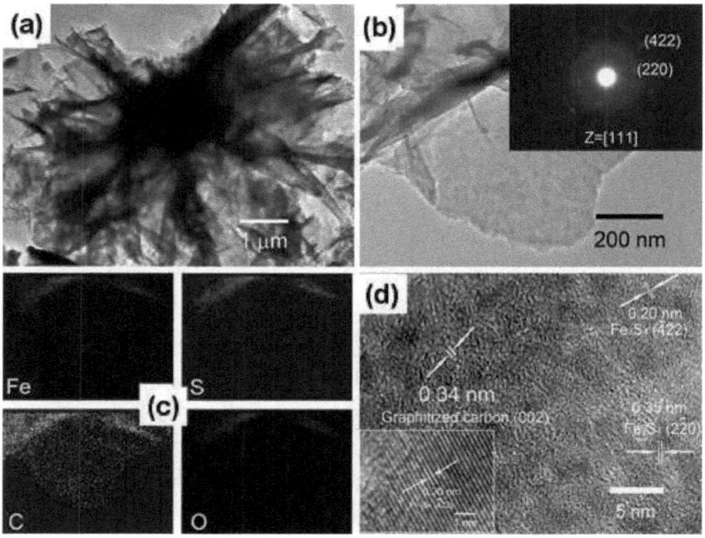

Figure 9.13 TEM results and composition mapping of Fe_3S_4/C. (a) TEM morphology of a single object. (b) Enlarged pattern of a fringe. (c) EDS mapping (Fe, S, C, O) of the single nanoplate. (d) HRTEM result of the nanoplate [20]. Copyright (2010) Royal Society of Chemistry.

Further investigation reveals that glucose plays an important role in the formation of such micro/nanostructure in addition to acting as a carbon source. Without glucose, only aggregated Fe_3S_4 naoparticles were obtained Fig. 9.14a). By addition of glucose, small nanosheets were formed gradually (Fig. 9.14b,c). The tremella-like morphology contains curved nanoplates which were produced with addition of appropriate amount of glucose (Fig. 9.12).

Figure 9.14 SEM results of the as-obtained products prepared with (a) 0 g/l, (b) 10 g/l, (c) 50 g/l and (d) 100 g/l of glucose in the precursor solution [20]. Copyright (2010) Royal Society of Chemistry.

The formation of such tremella-like Fe_3S_4/C composites can be summarized in Fig. 9.15. Firstly, formation of Fe_3S_4 was based on the following reactions [20].

$$HOCH_2CH_2OH \rightarrow CH_3CHO + H_2O \tag{9.12}$$

$$2Fe^{3+} + 2CH_3CHO \rightarrow CH_3COCOCH_3 + 2Fe^{2+} + 2H^+ \tag{9.13}$$

$$CH_3CSNH_2 + H_2O \rightarrow CH_3CONH_2 + H_2S \tag{9.14}$$

$$H_2S \rightarrow 2H^+ + S^{2-} \tag{9.15}$$

$$2Fe^{3+} + Fe^{2+} + 4S^{2-} \rightarrow Fe_2S_4 \tag{9.16}$$

Part of Fe^{3+} ions is reduced to Fe^{2+} by EG molecules (see reactions (9.12) and (9.13)). Then, Fe^{2+} will react with Fe^{3+} and S^{2-}, which were produced by hydrolyzation of thioacetamide (TAA) to produce Fe_3S_4 (see reactions (9.14–9.16)). As the reaction goes on, the concentration of Fe_3S_4 in the solution would increase and be saturated to produce Fe_3S_4 crystal nuclei in the solution. At the same time, glucose molecules would be oxidized by Fe^{3+} to gluconic acid or

$$CH_2OH(CHOH)_4CHO \xrightarrow{[O]} CH_2OH(CHOH)_4COOH \tag{9.17}$$

and then release $CH_2OH(CHOH)_4COO^-$ anions. Such anions would adsorb on the positively charged (111) plane of Fe_3S_4 by electrostatic interaction. The adsorption would have great influence on anisotropic growth process or results in termination of the growth in direction [111] [21], forming Fe_3S_4 nanoplates. Meanwhile, the adsorbed glucose will further go through carbonization process during solvothermal treatment, leading to surface tension on the Fe_3S_4/C nanoplates. Thus, the curved Fe_3S_4/C nanoplates were formed. Then, the solvent EG molecules would adsorb on Fe_3S_4/C nanoplates, resulting in charge difference between the surfaces and the edges of nanoplates [22], which would create self-assembly of the curved nanoplates. Thus, tremella-like morphology was formed.

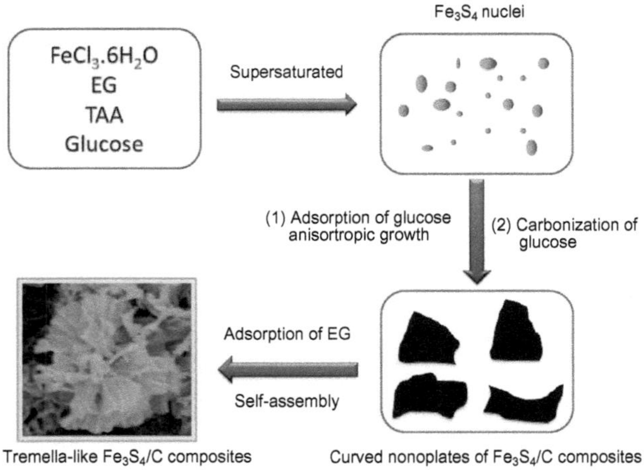

Figure 9.15 The schematic illustration for formation mechanism of the tremella-like Fe_3S_4/C composites [20]. Copyright (2010) Royal Society of Chemistry.

9.2.2 Template-Etching Strategy

Micro/nanostructured hollow spheres are widely applied in environmental remediation as adsorbent [13,23], photocatalyst [24,25], and so on, due to their large surface area and high working efficiency. Template methods are commonly used in fabrication of them. The template method contains soft template and hard template [26]. Among them, soft template strategy generally produces the hollow spheres with micro/nanostructure using organic molecules, such as EG [27], PEG [28], and so on. In addition, surfactants are also acted as soft template to produce the micro/nanostructured hollow spheres, such as hexadecyltrimethylammonium bromide (CTAB) [29]. The formation mechanism in such process is generally associated with the Ostwald ripening effect [30], Kirkendall effect [31] and kinetically controlled amorphous spheres' dissolution process [13], and so on. However, the diameter of the hollow structure is not uniform but easily controlled by using soft template. The other strategy to produce the micro/nanostructured materials with hollow structure is hard template. Generally, polymer (e.g., polystyrene) or silica spheres are used as hard template to produce the hollow spheres. In a typical layer-by-layer (LbL) process, precursors of the hollow spheres are adsorbed on the surface of the template (e.g., colloidal particles) layer by layer, leading to multilayer-coated particles [32]. Then, the colloidal particles are removed by calcination or exposure to solvent to obtain the hollow structures [32]. The hard template method to produce hollow structures has the merits of low cost and easy control of hollow spheres in size, which depends on the diameter of template. Here, a chemical-template-synthesis strategy is introduced to demonstrate fabrication of the micro/nanostructured hollow spheres. In this strategy, the colloidal particles not only act as template to determine the diameter of hollow structure but are also used as reactant for production of hollow spheres. Recently, our group has successfully prepared micro/nanostructured metal silicate hollow spheres by chemical-template-synthesis strategy using silica spheres as template.

9.2.2.1 Magnesium silicate hollow microspheres

Using magnesium chloride as metal ions source, silica colloidal spheres as template, and also with the existence of ammonia and ammonia chloride, micro/nanostructured magnesium silicate hollow spheres were obtained on the basis of a hydrothermal process, as shown in Fig. 9.16. Further observation shows that the surface of the hollow spheres is rough and consists of lamellae. The shell thickness was uniformly nearly 50 nm.

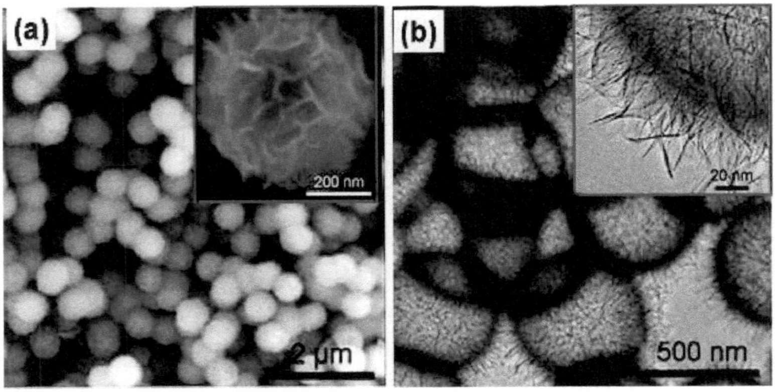

Figure 9.16 The morphology of the magnesium silicate hollow microspheres. (a) SEM image. (b) TEM results. The insets are the corresponding local magnification [33]. Copyright (2010) John Wiley and Sons.

In order to investigate the formation mechanism of such micro/nanostructured hollow spheres, the products at different synthesis stage were collected and shown in Fig. 9.17. Before reaction, the template silica colloidal particles are uniform smooth spheres with diameter nearly 500 nm (Fig. 9.17a). After hydrothermal treatment for 0.5 h, there are some rough particles around the silica spheres (Fig. 9.17b), indicating that the silica spheres were dissolved and new materials were produced in alkaline condition. As we know, silica spheres could be dissolved in alkaline condition and silicate anions would be obtained. The magnesium silicate would be produced by reaction of magnesium cations with silicate anions, forming the rough particles shown in Fig. 9.17b. As the concentration of silicate anions is large around the silica colloidal particles, the magnesium silicate particles are aggregated to form shell-like morphology. As the process goes on (3 h), the silica colloidal spheres were further etched, showing irregular shape and the rough shell became thicker, leading to yolk morphology (Fig. 9.17c). When the hydrothermal treatment time was prolonged to 12 h, the silica colloidal particles were completely dissolved and the hollow structure was thus obtained (Fig. 9.17d).

Further observation shows that the shells of the hollow spheres are composed of nanosized magnesium silicate lamellae. The lamella structure is formed according to the crystal growth habits of talc phase of magnesium silicate [33]. Thus, magnesium silicate lamellae-built hollow spheres are formed by the chemical-template-synthesis strategy, which is illustrated in Fig. 9.18.

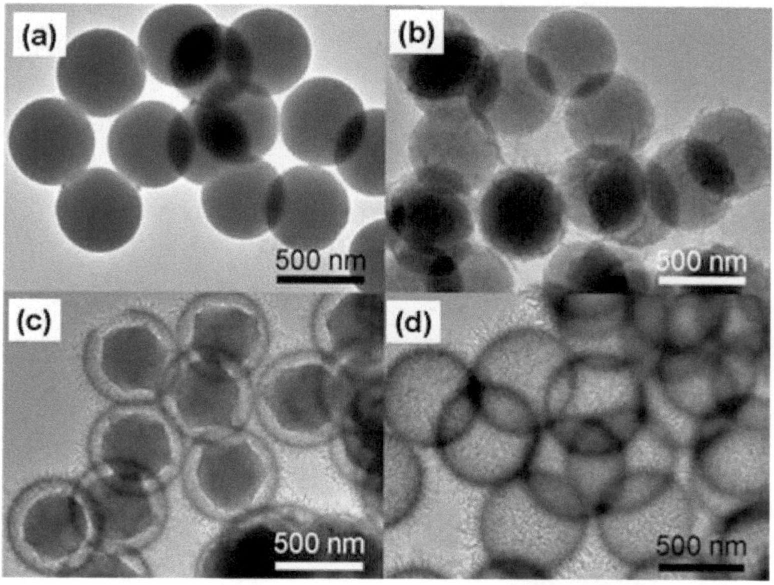

Figure 9.17 TEM records for formation of magnesium silicate hollow spheres after reaction for a different time: (a) 0 h; (b) 0.5 h; (c) 3 h and (d) 12 h [33]. Copyright (2010) John Wiley and Sons.

It is worthy to note that the addition of ammonia chloride plays an important role in formation of such micro/nanostructured hollow structure. Without addition of ammonia chloride, the magnesium cations will be precipitated in alkaline condition to produce magnesium hydroxide, or the reaction [33]

$$Mg^{2+}\ 2NH_3 \bullet H_2O \rightleftarrows Mg(OH)_2 + 2NH_4^+ \quad (9.18)$$

takes place. In this case, there are few free magnesium cations in the precursor solution that are not enough to react with silicate anions around silica colloidal particles. Therefore, the silicate anions would diffuse into the solution, leading to formation of irregular magnesium silicate particles. Adding ammonia chloride would increase the concentration of free magnesium cations in the solution (reaction (9.18)), which reacts with silicate anions to produce magnesium silicate shell. Appropriate amount of ammonia chloride would lead to formation of micro/nanostructured hollow spheres. Thus, the existence of ammonia chloride in precursor solution could prevent magnesium cations from precipitation and would supply enough magnesium cations to produce magnesium silicate.

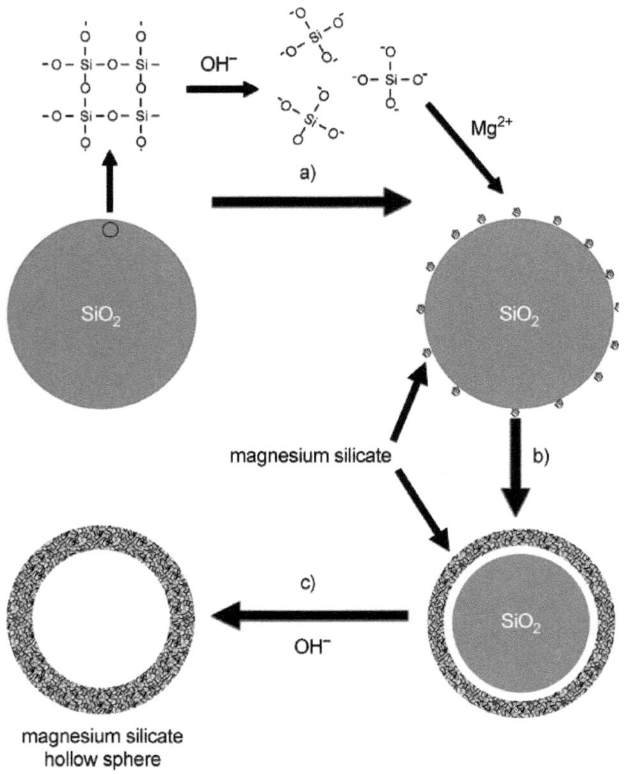

Figure 9.18 Schematic illustration for the formation mechanism of magnesium silicate hollow spheres [33]. Copyright (2010) John Wiley and Sons.

9.2.2.2 Copper silicate hollow microspheres

Similarly, micro/nanostructured copper silicate hollow spheres were also obtained by the above chemical-template-synthesis strategy, using $Cu(NO_3)_2{\cdot}3.5H_2O$ as the copper source [34]. The obtained hollow spheres have rough surface of the shell layer composed of copper silicate nanotubes, as shown in Fig. 9.19a,b. The nanotubes are standing on the surface of the hollow spheres with outer and inner diameters about 8 nm and 3.5 nm, respectively.

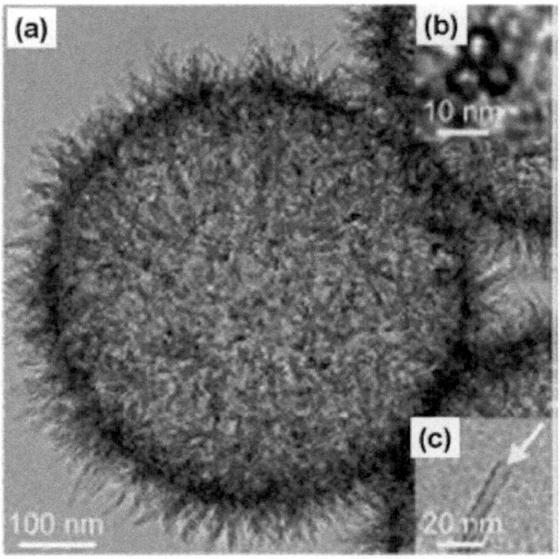

Figure 9.19 TEM results of (a) a single copper silicate hollow spheres; (b) three nanotubes on the shell; (c) a single nanotube [34]. Copyright (2008) Royal Society of Chemistry.

The formation process was investigated by time-dependent experiments. Figure 9.20 shows morphologies of the as-prepared products after reaction for a different time. Similar to that of the magnesium silicate hollow spheres [33], the copper silicate was formed at the surface of silica colloidal particles. After reaction for 10 h, complete copper silicate hollow spheres were obtained.

It is worthy to note that ammonia chloride is not needed in the precursor solution because copper–ammonia complex ions (not precipitate) would be formed when copper cations were mixed with ammonia solution based on the reaction [34]

$$Cu^{2+} + 4NH_3 \rightleftarrows Cu(NH_3)_4^{2+} \quad (9.19)$$

Therefore, adequate copper cations could react with silicate anions to form copper silicate.

According to crystal growth habits of copper silicate, clay-type structure composed of asymmetric alternating sheets of silica tetrahedra and copper oxide octahedra will be formed [34]. In this case, copper silicate nanotubes were formed according to the theory of L. Pauling, which says that layer crystal tends to curve in the circumstance of the two crystal faces of a constituent layer being not equivalent [34,35].

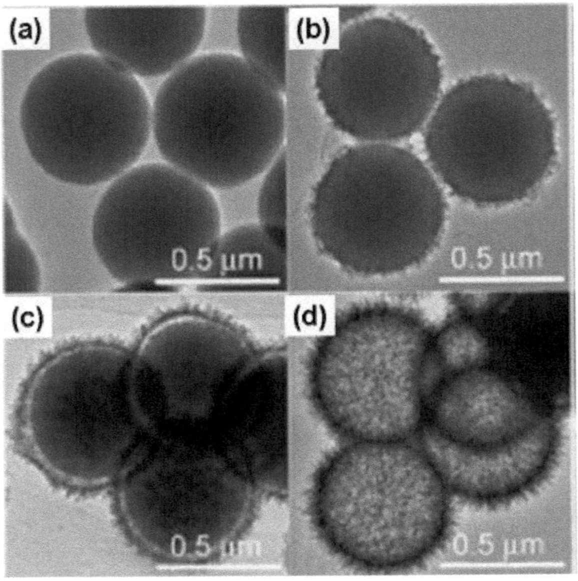

Figure 9.20 TEM results of the as-obtained products collected at different reaction times: (a) 0 h; (b) 1 h; (c) 2 h; and (d) 10 h [34]. Copyright (2010) Royal Society of Chemistry.

9.2.2.3 Hierarchical SiO_2@γ-AlOOH microspheres

We can also obtain hierarchical SiO_2@γ-AlOOH spheres by template-induced -synthesis method using silica colloidal spheres as template [36]. In the preparation procedure, sodium aluminate was used as aluminum source, urea was used to produce hydroxide ions, and deionized water was solvent. The above solutions were mixed with silica colloidal spheres and subsequently sealed and maintained at 160°C for a certain time. Finally, SiO_2@γ-AlOOH spheres were obtained, as shown in Fig. 9.21. Further observation shows that the product has a core–shell structure with the shell thickness about 100 nm. The shell consists of the lamellar structures standing on the surface of silica core. The average thickness is nearly 5 nm by observation of Fig. 9.21d.

To further investigate the template-induced-synthesis method, products at different reaction time were collected and analyzed by TEM examination, as shown in Fig. 9.22. As the process goes on, there are small lamellar structures produced around silica colloidal spheres at initial stage. When the reaction time was prolonged to 5 h and 12 h, the lamellar structures became thicker and a complete core-shell structure was formed. Its formation can be described on the basis of the following reactions [36].

$$AlO_2^- + 2H_2O \rightleftarrows Al(OH)_3 + OH^- \tag{9.20}$$

$$Al(OH)_3 \rightarrow \gamma\text{-}AlOOH + H_2O \tag{9.21}$$

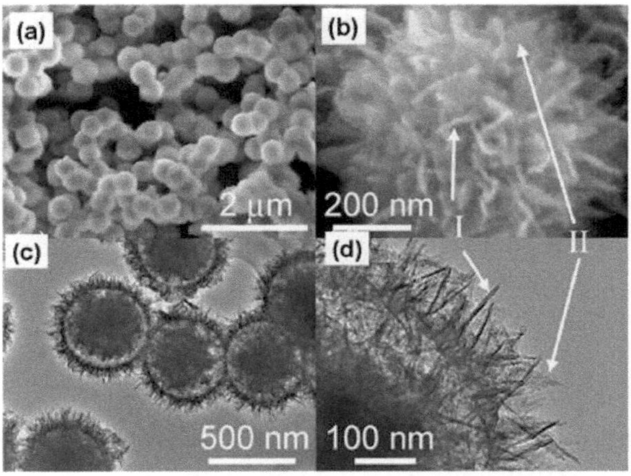

Figure 9.21 The morphology of hierarchical $SiO_2@\gamma$-AlOOH spheres. (a) and (b) SEM images of the general view and single sphere; (c) and (d) TEM images [36]. Copyright (2009) Institute of Physics.

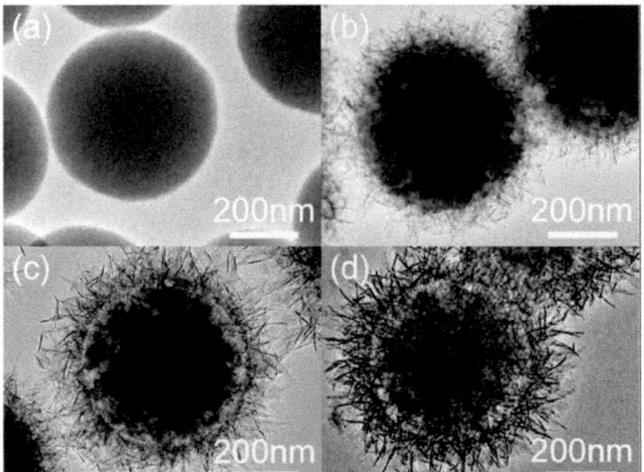

Figure 9.22 TEM results of the as-obtained products at different reaction times: (a) 0 h, (b) 1 h, (c) 5 h, and (d) 12 h [36]. Copyright (2009) Institute of Physics.

Firstly, AlO_2^- anions released from sodium aluminate would hydrolyze to produce $Al(OH)_3$ and hydroxide ions (reaction (9.20)). Then, unstable $Al(OH)_3$ could further pyrolyse to form γ-AlOOH precipitate (reaction (9.21)). It is well known that silica colloidal spheres could be etched in alkaline environment. Thus, the hydroxide ions produced from hydrolysis of AlO_2^- could further react with silica colloidal spheres, leading to the chemical equilibrium of reaction (9.20) moving toward producing more $Al(OH)_3$ around the silica colloidal spheres. Meanwhile, the unstable $Al(OH)_3$ was converted to γ-AlOOH. So, the produced γ-AlOOH would cover on the surface of silica colloidal spheres and grow along with template etching. The core-shell structures were thus formed. The lamellar morphology of γ-AlOOH was constructed on the basis of its layered crystal structure [36,37]. Finally, hierarchical $SiO_2@\gamma$-AlOOH spheres were produced.

9.2.2.4 Nobel-metal/ZnO hollow nanospheres

In addition, we also obtained ZnO or nobel-metal/ZnO hollow nanoparticles (HNPs) by etching Zn@ZnO structure using weak acid based on different etching rate between metal and metal oxide [38]. In this work, Zn/ZnO core/shell structure was first obtained by laser ablation in liquid (LAL) method, as shown in Fig. 9.23 [38]. There are many uniform particles with average size nearly 20 nm (see Fig. 9.23a). Enlarged image indicates that a single particle is of an obvious core/shell structure. Zinc is the core part and zinc oxide is the shell part as shown in Fig. 9.23c. Subsequently, different sorts of weak acids were used as etching reagents to produce HNPs, as illustrated in Fig. 9.24. When the weak acid was dropped into the Zn/ZnO colloidal solution, both components (Zn, ZnO) could react with weak acid, or the reactions [38].

$$Zn + 2H^+ \rightarrow Zn^{2+} + H_2\uparrow \tag{9.22}$$

$$ZnO + 2H^+ \rightleftarrows Zn^{2+} + H_2O \tag{9.23}$$

took place. The reaction rate in reaction (9.22) is faster than that in reaction (9.23) because zinc is more active than ZnO when it reacts with weak acid. Therefore, the Zn core (template) could be etched out and the ZnO HNPs were left.

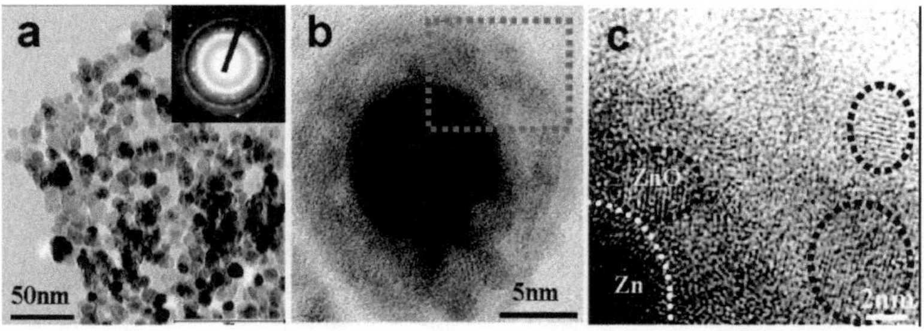

Figure 9.23 TEM results of Zn/ZnO core/shell nanoparticles. (a) General view, insert: SAED pattern, (b) the view of a single nanoparticles, and (c) HRTEM image of nanoparticles [38]. Reprinted with permission from H. B. Zeng et al., *ACS Nano*, 2008, 2 (8), 1661–1670 Copyright (2008) American Chemical Society.

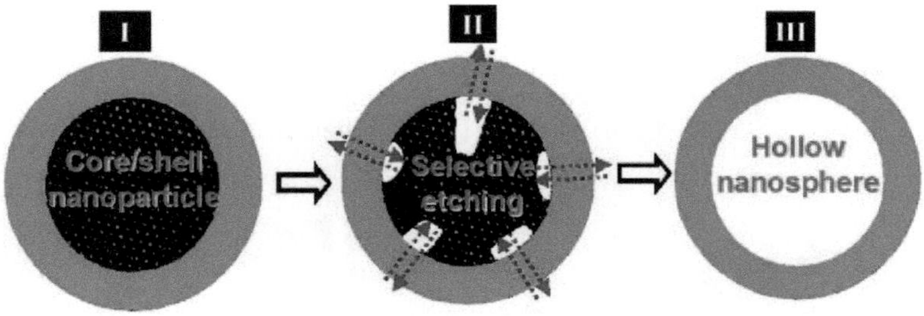

Figure 9.24 Schematic illustration of the selective etching strategy for preparation of ZnO HNPs using a sacrificial template [38]. Reprinted with permission from H. B. Zeng et al., *ACS Nano*, 2008, 2 (8), 1661–1670 Copyright (2008) American Chemical Society.

On the basis of above observations, pure ZnO HNPs would be obtained if we used tartaric acid (TA, $C_4H_6O_6$) to etch the core/shell structure. Importantly, nobel metal/ZnO HNPs would be achieved by using weak acid with nobel metal ions as the etching reagent. Figure 9.25 shows the results after using chloroaurie acid (CA, $HAuCl_4$) to etch Zn/ZnO nanoparticles. Au/ZnO HNPs are produced. Au ultrafine nanoparticles are embedded into the shell layer.

The noble metal ions would diffuse through the shell during the etching. Then, nobel metal nanoparticles would deposite into the ZnO shell because of redox reaction [38], leading to Au/ZnO HNPs. The introduction of nobel metal into ZnO HNPs could improve their photocatalystic performance for environmental remediation [38], which will be described in the following section. Similarly,

Pt/ZnO HNPs and Au/Pt/ZnO HNPs would be produced by using chloroplatinic acid (PA, H_2PtCl_6) etching or CA-PA co-etching, as shown in Fig. 9.26.

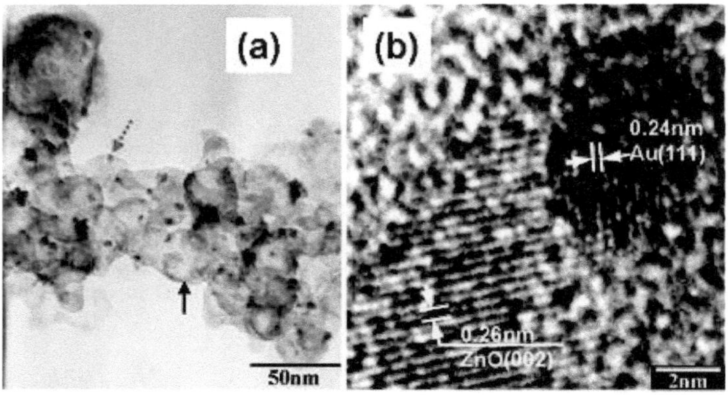

Figure 9.25 TEM images of Au/ZnO HNPs by CA etching [38]. Reprinted with permission from H. B. Zeng et al., *ACS Nano*, 2008, 2 (8), 1661–1670 Copyright (2008) American Chemical Society.

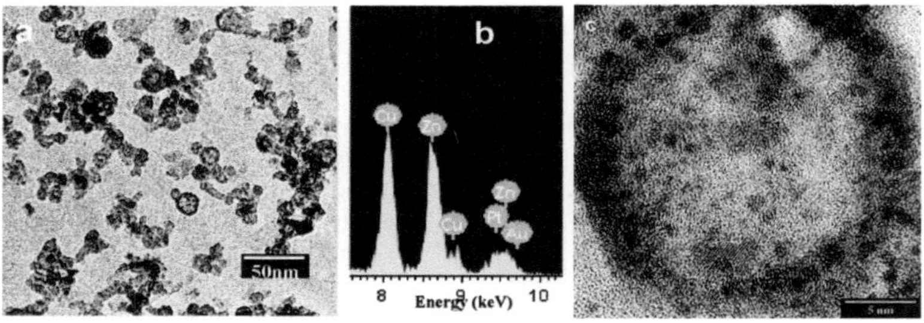

Figure 9.26 TEM and EDX (b) results of Au/Pt/ZnO HNPs by CA-PA etching [38]. (a) Low magnification; (c) TEM image of a single ZnO HNP embedded with Pt and Au ultrafine clusters. Reprinted with permission from H. B. Zeng et al., *ACS Nano*, 2008, 2 (8), 1661–1670 Copyright (2008) American Chemical Society.

9.2.3 Electrospinning Method

Electrospinning is a method to draw fine fibers from a liquid using an electrical charge [39]. The electrospinning apparatus is shown in Fig. 9.27. This method has advantages of easy operation, low cost, and does not need the use of coagulation chemistry or high temperature to obtain solid micro/nanostructured fibers from solution. Figure 9.28 presents a typical result corresponding to the polyacrylonitrile (PAN)/ferrous chloride composite porous nanofibers by electrospinning technology [40]. The obtained composite nanofibers are noodle-like and 100–300 nm in diameter. The specific surface area was measured to be 10 m^2/g (Fig. 9.28c), which indicates the porous structure for the fibers. The precursor solution contains dimethylformamide (DMF), polyacrylonitrile (PAN, 10% wt, MW ~60000), and $FeCl_2 \cdot 4H_2O$ [40]. The obtained composites exhibit high Cr-removal performance [40] in environmental remediation, which will be introduced in following section.

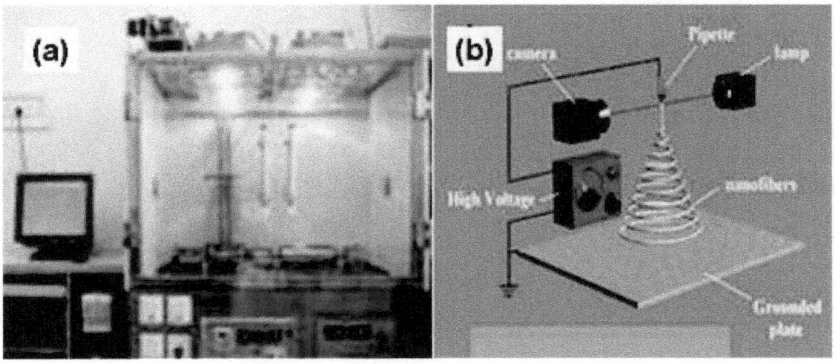

Figure 9.27 Electrospinning apparatus. (a) General view and (b) schematic illustration.

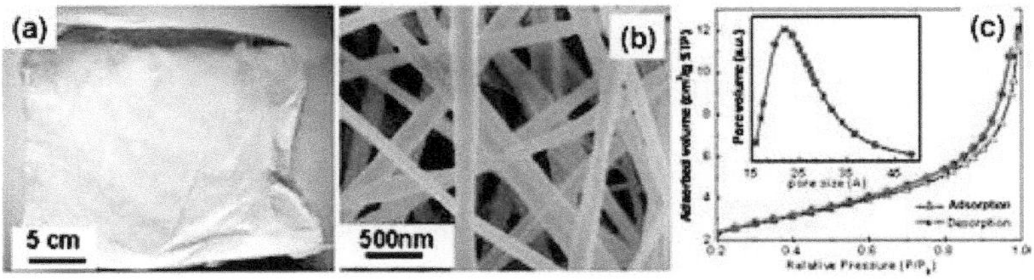

Figure 9.28 Structure and morphology of the obtained polyacrylonitrile/ferrous chloride composite porous nanofibers by electrospun. (a) Photograph image of the fiber film, (b) SEM image, (c) N_2 sorption isotherm results [40]. Copyright (2011) Royal Society of Chemistry.

Further, the PAN@γ-AlOOH fibers were also produced using PAN fibers as template [41]. Typically, the as-prepared PAN fiber film was mixed with aluminum powder and hexamethylenetetramine ($C_6H_{12}N_4$, or HMTA for short) aqueous solution (1% wt) [41]. Then, the above solution was sealed and went through hydrothermal treatment at 110°C for 10 h [41]. The morphology and structure of PAN template and the obtained PAN@γ-AlOOH fibers are shown in Fig. 9.29. The PAN@γ-AlOOH fibers exhibit micro/nanostructure, with several tens of microns in length and nearly 1.2 um in width. Importantly, there are crosslinked γ-AlOOH nanoplates standing on the PAN fibers with thickness of 20–40 nm (Fig. 9.29c,d) [41].

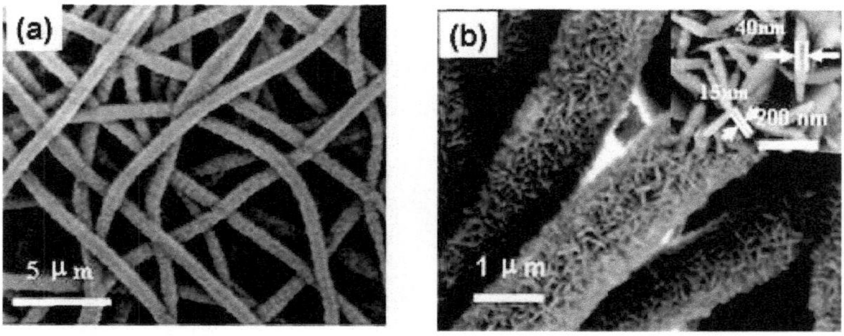

Figure 9.29 SEM images for the PAN@γ-AlOOH fibers. (a) Low magnification; (b) high magnification; the inset is a local magnified image [41]. Copyright (2012) Royal Society of Chemistry.

The formation of such PAN@γ-AlOOH fiber is easy to be understood. When the hydrothermal treatment starts, the following reactions occur [41].

$$C_6H_{12}N_4 + 6H_2O \rightarrow 6HCHO + 4NH_3 \tag{9.24}$$

$$NH_3 + H_2O \rightarrow NH_4^+ + OH^- \tag{9.25}$$

$$Al + 2H_2O + OH^- \rightarrow Al(OH)_4^- + H_2 \tag{9.26}$$

$$Al(OH)_4^- \rightarrow AlOOH + H_2O + OH^- \tag{9.27}$$

When the concentration of AlOOH reaches supersaturation, γ-AlOOH nanoplates are formed on the basis of their layered structure [41,42]. Therefore, theγ-AlOOH nanoplates would grow on the surface of PAN fibers to decrease their surface energy, leading to formation of micro/nanostructured PAN@γ-AlOOH fibers. Such fibers show strong and recyclable Cr(VI)-removal performance in environmental remediation [41]. Interestingly, PAN template can be removed by calcination at 550°C and tubular Al_2O_3 hollow fibers can be produced [41]. Figure 9.30 schematically shows such formation process.

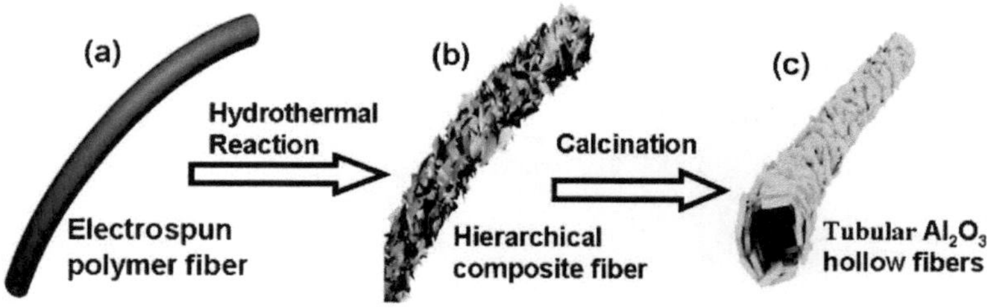

Figure 9.30 Schematic illustration for preparation of polymer@γ–AlOOH fibers [41]. Copyright (2012) Royal Society of Chemistry.

All in all, the methods mentioned above are simple, flexible, controllable, and hence suitable for mass production of micro/nanostructured materials with benign condition, good reproducibility, and friendly environment. These materials possess high specific surface area, structural stability and anti-aggregation, and hence good contaminant-removal capacity.

9.3 Structurally Enhanced Photocatalystic Properties

The semiconducting oxide with micro/nanostructures has exhibited significantly enhanced photocatalystic properties, due to their special structure, compared with the same material but with normally structures. This is so-called "structurally enhanced photocatalystic properties," or enhancement of photocatalystic performance by structural tuning. Here, we take the micro/nanostructured ZnO as an example to demonstrate such enhancement.

9.3.1 Honeycomb-Like Micro/Nanostructured ZnO

As we know, ZnO is an excellent photocatalyst in environmental remediation [43–47]. For the honeycomb-like micro/nanostructured ZnO shown in Fig. 9.2, they exhibit strong photocatalystic property for decomposition of methyl orange (MO), as shown in Fig. 9.31. Under UV irradiation, the micro/nanostructured ZnO could decolor MO solution with high efficiency. The absorbance of MO decreased along with irradiation time and the color of MO solution became lighter (inset of Fig. 9.31). After irradiation for 80 min, the MO was decomposed completely.

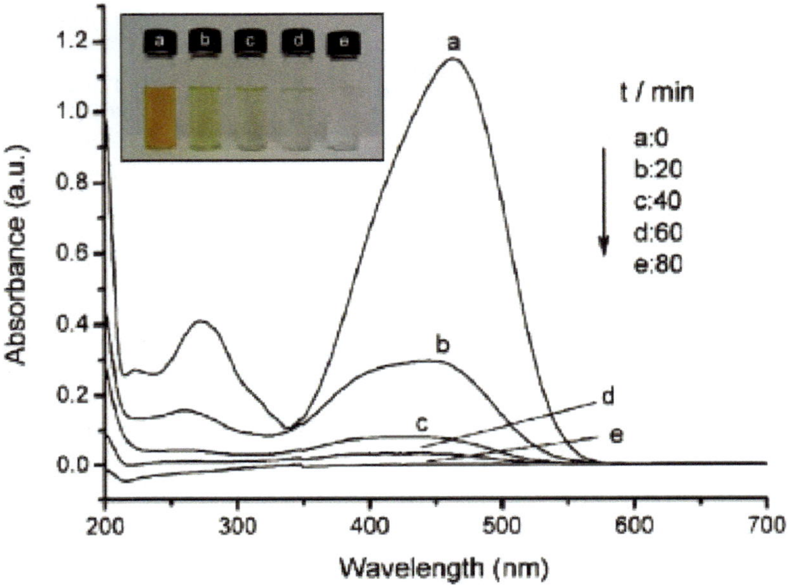

Figure 9.31 Time-dependent color change (inset) and corresponding optical absorbance spectra for the MO solutions by photodecomposition using the micro/nanoarchitectured ZnO as catalyst [12]. Copyright (2008) John Wiley and Sons.

Further, to evaluate the advantage of such micro/nanostructured ZnO, the photocatalysis experiments were carried out compared with the ZnO with different nanostructures (nanosheets, nanoparticles, and nanoneedles) and Degussa P25 titania. The photocatalystic effects of the ZnO with different architectures and Degussa P25 are shown in Fig. 9.32. Obviously, the ZnO with micro/nanostructure is much better than that with other nanostructures in photocatalystic performance. The former can decompose MO within 80 min in our experimental conditions, and be even better than Degussa P25 with complete decomposition time 100 min. The good performance of the micro/nanostructured ZnO lies in their unique structure [12]. Firstly, this micro/nanostructured ZnO has high surface area (185 m^2/g), which is larger than that of Degussa P25 (~50 m^2/g), ZnO nanoparticles (~30.6 m^2/g), and so on [12]. Secondly, the micro/

nanostructure could effectively resist aggregation to maintain larger active surface area than other nanostructures [12]. Thirdly, the unique micro/nanostructure could reduce the photogenerated electron/hole pair recombination and thus enhance their photocatalystic performance.

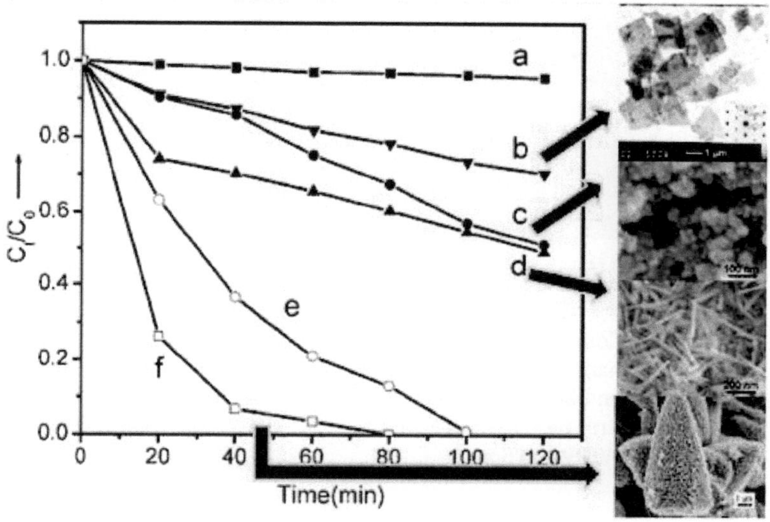

Figure 9.32 The time-dependent concentration change of MO in the solution under photodecomposition by different catalysts (a) without catalyst, (b) ZnO nanosheets, (c) ZnO nanoparticles, (d) ZnO nanoneedles, (e) Degussa P25 titania, and (f) the micro/nanoarchitectured ZnO [12]. Copyright (2008) John Wiley and Sons.

Importantly, the durability of the micro/nanostructured ZnO is very good, as shown in Fig. 9.33. After three cycles, the decoloration percentage of ZnO nanoneedles decreased obviously. Instead, the micro/nanostructured ZnO could maintain the decolouration percentage above 90% after three cycles. The good durability of the micro/nanostructured ZnO is due to their single crystalline nature [12]. But for Degussa P25, the photocatalystic effect keeps constant during each cycle, as titania would not be photo-corroded [12,48]. The decrease of photocatalystic performance of ZnO is induced by the photocorrosion. Obviously, the honeycomb-like micro/nanostructured ZnO shows relatively low photocorrosion effect.

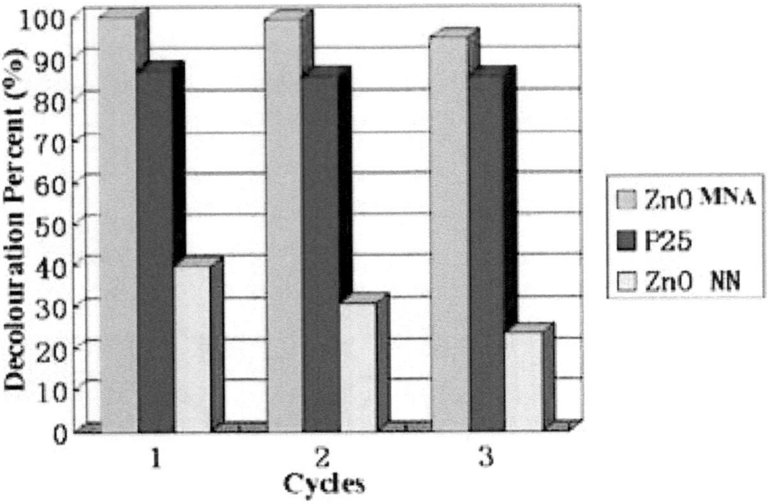

Figure 9.33 Comparison of photodecomposition performance of micro/nanoarchitectured ZnO (ZnO MNA), Degussa P25 titania (P25), and ZnO nanoneedles (ZnO NN) by three cycles (1: original, 2: first recycled, 3: second recycled) [12]. Copyright (2008) John Wiley and Sons.

9.3.2 Nobel Metal/ZnO Hollow Nanospheres

Further, the photocatalystic properties of the prepared ZnO HNPs and nobel metal clusters embedded ZnO HNPs, shown in Figs. 9.25 and 9.26, were evaluated and presented in Fig. 9.34. The ZnO HNPs exhibit limited photocatalystic performance, whereas Pt/ZnO HNPs show faster decomposition rate than P25. The reason lies in the introduction of nobel metal clusters, which can facilitate the charge transfer and effective electron-hole separation [38,49]. Also, the high dispersion of the Pt clusters into the ZnO shell contributes to the excellent photocatalystic performance.

As a whole, the micro/nanostructured materials show the stronger photocatalytic activity and the higher catalytic durability than the nanostructured ones. Such structurally enhanced photocatalysis properties can be attributed to the high specific surface area and high anti-aggregation (or structural stability), which keep high active surface area during use.

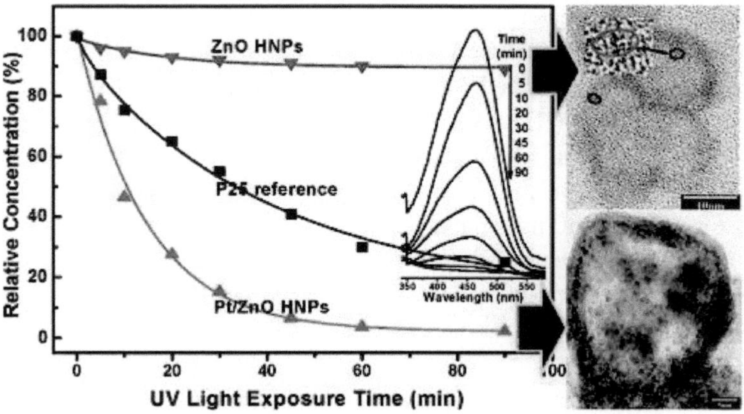

Figure 9.34 MO photodecomposition rate by using pure ZnO and Pt/ZnO HNPs as photocatalyst, compared with reference P25 powders [38]. Reprinted with permission from H. B. Zeng et al., *ACS Nano*, 2008, 2 (8), 1661–1670 Copyright (2008) American Chemical Society.

9.4 Structurally Enhanced and Selective Adsorption Performances

Similarly, micro/nanostructured materials also exhibit the enhanced adsorption performance compared with the normally structured ones due to their high specific surface area and structural stability. Therefore, such materials could be the candidates of high efficient adsorbents for removal of some pollutants, such as heavy metal ions, organic contamination in water (or solutions). Here, we give some examples to show the possibility and validity of the oxide micro/nanostructured materials as the promising adsorbents for contaminant-removal.

9.4.1 Micro/Nanostructured Porous ZnO Plates

ZnO is an environmental-friendly material and also a promising candidate for catalyst, gas sensor, solar cell, and so on [15]. On the contrary, there usually exist many hydroxyl groups on its surface, which could be the active adsorption sites for heavy metal ions removal [15,50]. However, the study about ZnO as an adsorbent is very limited. Recently, we have investigated the adsorption property of micro/nanostructured porous ZnO nanoplates, which has shown structurally enhanced and selective adsorption performance for heavy metal Cu(II) ions in environmental remediation [15].

As shown in Fig. 9.35, the micro/nanostructured porous ZnO nanoplates have much stronger adsorption capacity than that of commercial ZnO nanoparticles. In addition, the adsorption behavior fits well with Freundlich model for the porous ZnO nanoplates and is subject to Langmuir model for the commercial ZnO nanoparticles.

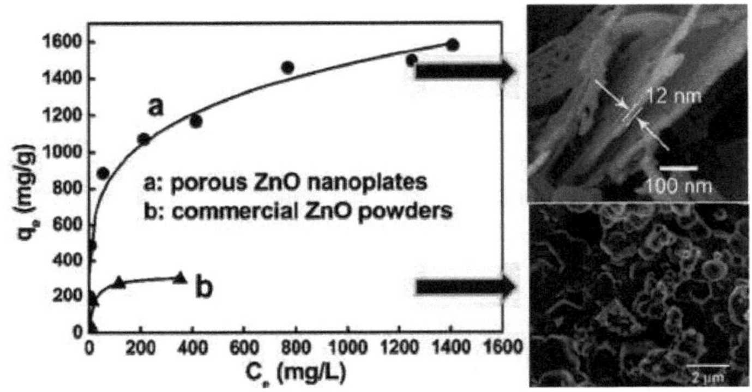

Figure 9.35 The adsorptive isotherms of Cu(II) ions on the porous ZnO nanoplates (a) and the commercial ZnO nanopowders (b) [15]. Copyright (2010) Royal Society of Chemistry.

The strong adsorption capacity of porous ZnO nanoplates is related to their micro/nanostructure. Firstly, the micro/nanostructured porous ZnO nanoplates have high surface area of 147 m^2/g, whereas the commercial ZnO nanoparticles only have nearly 5m^2/g. The high specific surface area could result in high adsorption capacity. Secondly, the micro/nanostructures have the advantage of aggregation resistance, which will be helpful for high adsorption performance. Thirdly, the polar plane in such porous ZnO nanoplates could interact with Cu(II) ions strongly through hydroxyl groups, which results in multilayer adsorption behavior. The XPS analysis has confirmed the existence of the Cu-O bonds between surface of porous ZnO nanoplates and Cu(II) ions [15]. Thus, the plausible adsorption mechanism is summarized and illustrated in Fig. 9.36. Also, the destructive adsorption or surface precipitation of Cu(II) on the surface of porous ZnO nanoplates is also acceptable [51].

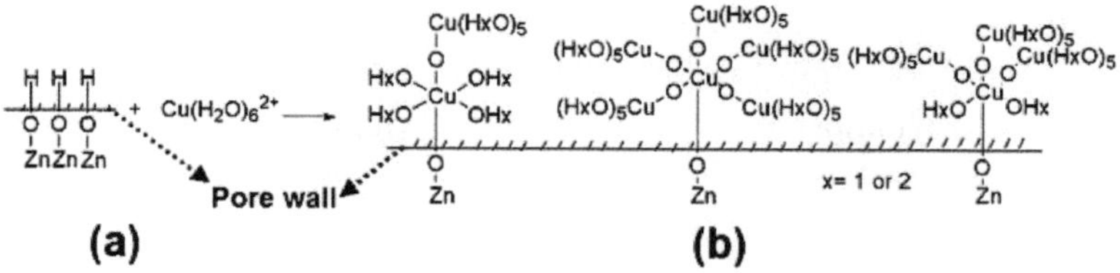

Figure 9.36 Schematic illustration of Cu(II) ion adsorption on the polar sites of the porous ZnO nanoplates [15]. Copyright (2010) Royal Society of Chemistry.

Importantly, such micro/nanostructured porous ZnO nanoplates show selective adsorption performance for cationic pollutants. Methyl orange (MO) was designed as a target pollutant to demonstrate the selective adsorption performance, as shown in Fig. 9.37. There is almost no change in MO's concentration in the solution after addition of the porous ZnO nanoplates, indicating that the nanoplates have strong selective adsorption performance for cationic contaminants. The reason could be attributed to the surface negative charge offered by hydroxyl groups.

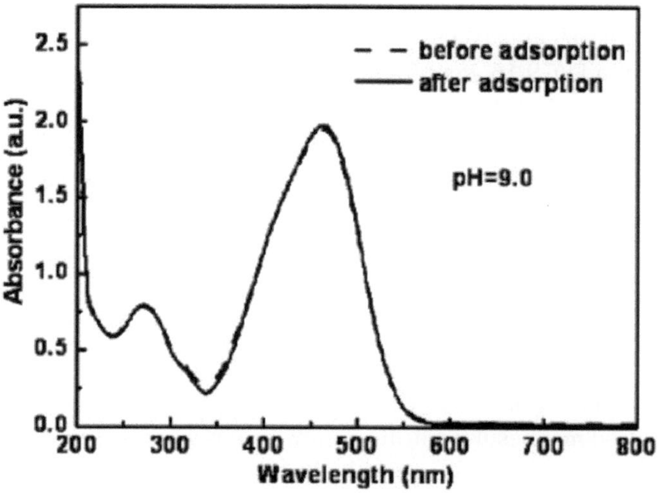

Figure 9.37 Optical absorption spectra of the anionic MO solutions, before and after presence of porous ZnO nanoplates at a pH value of 9 [15]. Copyright (2010) Royal Society of Chemistry.

Further investigation [11] has indicated that the micro/nanostructured ZnO hollow spheres mentioned above, shown in

Fig. 9.4 [13], exhibit stronger structurally enhanced adsorption efficiency for removal of heavy metals Cu(II) ions, compared with the micro/nanostructured porous ZnO nanoplates. The reason lies in the unique micro/nanostructure for the former, which can avoid the overlapping of nanoplates [11]. The adsorption to the heavy metals Pb(II) and Cd(II) demonstrates that the adsorption performances (adsorption amount, adsorption model, and so on) depend on the electronegativity of heavy metals [11]. Among the heavy metals Cu(II), Pb, and Cd(II) ions, the ZnO as an adsorbent shows selective adsorption performance for Cu(II) ions owing to the high electronegativity of Cu [11].

Although ZnO is a promising adsorbent for selective removal of heavy metal ions, its chemical stability is relatively low. There is chemical corrosion when the ZnO is used in water (solution) with pH less than 4 [15]. Therefore, as an adsorbent, it is conditional and should be used in the neutral or weak acidic (basic) solution.

9.4.2 Removal of Organic Cationic Pollutants

The micro/nanostructured magnesium silicate hollow spheres shown in Fig. 9.16 were found to be the good adsorbent for pollutants' removal [33]. About 20 mg micro/nanostructured magnesium silicate adsorbents could almost remove methylene blue (MB) from 40 ml MB solution with concentration of 75 mg/l completely. They can remove more than 90% of MB in 75mg/l MB solution within 10 min [33], showing removal of the dye pollutant MB with high capacity and high rate, as demonstrated in Fig. 9.38a. The high adsorption capacity and efficiency are attributed to their unique micro/nanostructure, which has a high specific surface area (521 m^2/g) and the aggregation-resistance [33]. Importantly, the used micro/nanostructured magnesium silicate hollow spheres can be regenerated by calcination in air. After regeneration for three times, the adsorbent still has more than 85% adsorption capacity tahn that of the fresh adsorbent, as shown in Fig. 9.38b.

Similarly, the copper silicate hollow microspheres with nanotubes on the shell, shown in Fig. 9.19, also show high efficiency in MB removal [34]. Their adsorption amount could reach 160 mg/g. Importantly, the MB-adsorbed copper silicate can be renewed by calcination at 300°C for 4 h in air. The renewed adsorbent still had 140 mg/g MB removal capacity [34].

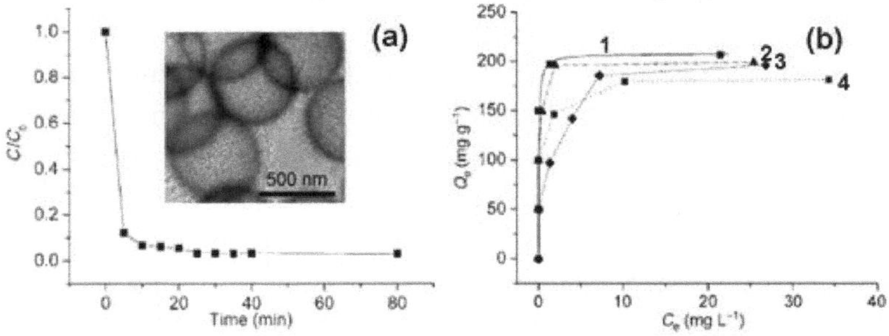

Figure 9.38 (a) Time-dependent concentration change of MB solutions after addition of the as-prepared magnesium silicate hollow spheres, and (b) Isotherms of MB adsorption on (1) the newly prepared magnesium silicate hollow spheres, (2–4) the hollow spheres that have been regenerated once, twice, and three times, respectively [33]. Copyright (2010) John Wiley and Sons.

The organic pollutants such as MB could be decomposed by combustion in air after adsorption treatment. But the heavy metals cannot be decomposed by combustion process. In addition, the structure of adsorbents could change during subsequent thermal treatment, which will diminish their adsorption performance. Thus, using the combustion to regenerate adsorbents is very limited. In addition, the calcination would release toxic gases for some organic pollutants, leading to air pollution. Therefore, a general and environmental-friendly method to renew adsorbents is expected. Recently, we have developed a tremella-like micro/nanostructured Fe_3S_4/C composite adsorbent, shown in Fig. 9.12a, which has structurally enhanced adsorption performance for environmental remediation and can be regenerated by alcohol washing [20]. Figure 9.39a shows the adsorption isotherms of MB on the tremella-like Fe_3S_4/C and Fe_3S_4. It can be seen that the Fe_3S_4/C composites have more adsorption capacity to MB than that of Fe_3S_4 nanoparticles. Such enhanced adsorption capacity lies in two aspects. One is that the micro/nanostructured tremella-like Fe_3S_4/C composites have larger specific surface area than Fe_3S_4 nanopowders. In addition, the effective specific surface area of the micro/nanostructures is also larger than the nanopowders due to the aggregation-resistance for the former. The other is the carbonaceous materials on surface of the composites, which have many functional groups, such as C-O, C=O, -OH, C=C, and so on, and would contribute the enhanced adsorption performance [20]. More importantly, the Fe_3S_4/C composites can be regenerated by an alcohol-washing process, which is easier to operate. During desorption by the washing, the hydroxyl groups of alcohol molecules would replace the O-H...N, O-H...S bonding between MB and hydroxyl groups of the adsorbent and form much stronger hydrogen bonding (O-H...O), leading to desorption of MB and regeneration of the adsorbents [20]. After recycling for two times, the composites still have 92% adsorption efficiency, as shown in Fig. 9.39b.

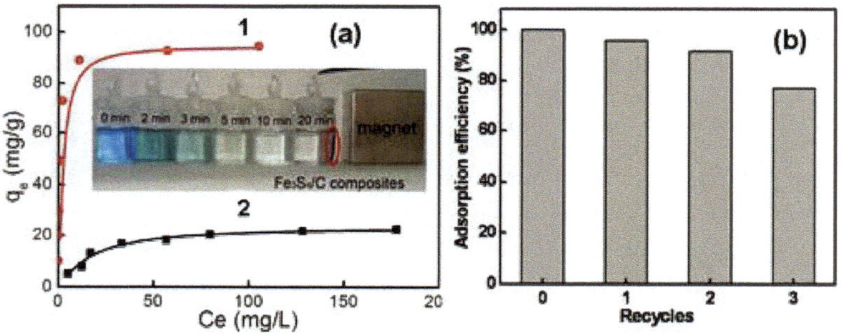

Figure 9.39 (a) Adsorption isotherms of MB on (1) Fe_3S_4/C composite and (2) Fe_3S_4 micro/nanostructured materials. The solid lines are the Langmuir fitting. The inset is photos of the MB solutions after adsorption for different time, and separation of the composites from the solution by a magnet. (b) The adsorption efficiency of the Fe_3S_4/C composite as a function of recycling times [20]. Copyright (2013) Royal Society of Chemistry.

We also used water as a solvent to regenerate the Fe_3S_4/C composites. But the regeneration efficiency is not as high as the alcohol-washing. There are many hydrophobic groups on surface of the adsorbent, such as C-C, C=C, and so on, which would resist water molecules approaching surface of the Fe_3S_4/C composites. Whereas alcohol molecules have both organic groups and hydroxyl groups, which could approach the adsorbent's surface more easily and replace the MB molecules.

9.4.3 Removal of Anionic Pollutants

Hereto, we have illustrated the removal of cationic pollutants from wastewater. But some sort of anionic pollutants are also dangerous to the health of human beings, such as, typically, Cr(VI) species that often exist in the forms of $Cr_2O_7^{2-}$ or $HCrO_4^-$ at pH less than 6.5 [19]. In terms of removal of Cr(VI) from wastewater, we have developed some efficient adsorbents with hierarchical micro/nanostructures, such as hierarchically structutred SiO_2@γ-AlOOH microspheres shown in Fig. 9.21, PAN/ferrous chloride composite porous nanofibers shown in Fig. 9.28, and magnetic Fe_3O_4 fibers shown in Fig. 9.9, and so on [16,36,40]. For instance, The SiO_2@γ-AlOOH microspheres can remove Cr(VI) effectively from wastewater through ion-exchange mechanism. Importantly, they show structurally enhanced adsorption performance compared withγ-AlOOH nanopowers [36]. In addition, the PAN/ferrous chloride composite porous nanofiber films are excellent adsorbent for Cr(VI) removal [40]. Further, the films after use can be easily separated from solution, as shown in Fig. 9.40. The ferrous chloride is very important for effective removal of Cr(VI), owing to the coordination bonds (PAN...Fe(II)), which can adsorb the Cr(VI) and reduce it to Cr(III) [40].

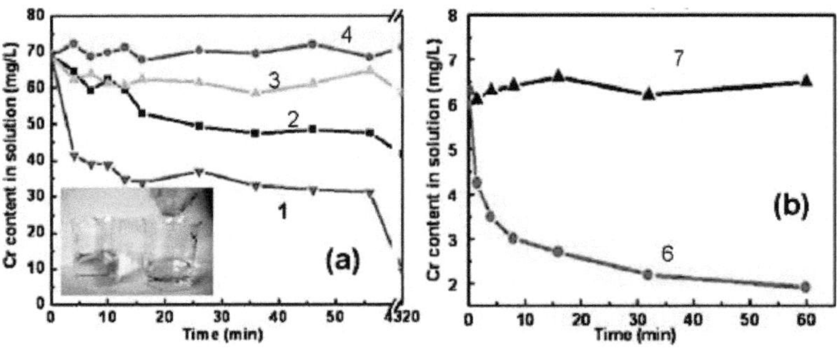

Figure 9.40 Time-dependent Cr-removal with initial pH = 5, Cr concentrations 69.1 ppm (a) and 6.3 ppm (b) by adsorption of (1, 6) PAN/$FeCl_2$ nanofibers, (2) PAN/ $FeCl_2$, (3) the mixture of $FeCl_2$ powders with PAN NFs and (4, 7) pure PAN NFs. Inset: photo of $K_2Cr_2O_7$ solution before and after adsorption of PAN/$FeCl_2$ nanofibers [40]. Copyright (2011) Royal Society of Chemistry.

9.4.4 Removal of Hydrophobic Organic Pollutants

Some organic pollutants, especially hydrophobic organic pollutants or non-polar contaminant molecules in solution, are usually difficult to remove. The POPs are typical examples that are very deleterious to human, even in a trace amount. It has been demonstrated that the micro/nanostructured porous magnetite fibers, shown in Fig. 9.9, are really good adsorbents to effectively remove some non-polar contaminant molecules in solution, typically the polychlorinated biphenyl (PCB-77, or $C_{12}H_6Cl_4$) (a kind of POPs) in n-hexane solution, even at a very low level (<0.01 μmol/l), owing to their unique porous morphology at a high specific surface area (123 m^2/g).

There exist PCBs in some soils or grounds, which can be washed out and dissolved in organic solutions such as acetone and n-hexane (C_6H_{14}). How to effectively enrich the trace of PCBs in the organic solutions has been expected and in challenge. Here, we show the removal behavior of PCB-77 in the n-hexane solution by the as-prepared magnetite porous fibers shown in Fig. 9.9 as adsorbent at room temperature. Figure 9.41a shows the PCB-77 content in the solution as a function of amount of the fibers, added in the 100 ml solution with initial PCB-77 concentration of 0.1 μmol/l (a very low concentration) in 5 min [16]. After adsorption for 5 min, the content of PCB-77 in the solution decreases with increasing magnetite amount, exhibiting nearly linear relation. After the addition of 15 mg porous nanofibers in the solution, more than 90% PCB-77 can be removed in 5 min. Figure 9.41b shows the time evolution of PCB-77 content in 1000 ml solution with much higher initial PCB-77 concentration (4.1 μmol/l) after addition of 1.0 g magnetite porous fibers. Around 82% PCB-77 in the solution can be removed within 2 h. The removal capacity of the porous fibers in this case can be estimated to be more than 1 mg PCB-77 per 1 g fibers. So, the micro/nanostructured porous magnetite fibers can effectively and expediently remove a trace of PCBs-77 in the solution.

The high PCB-77 removal capacity could be attributed to the porous structure and high specific surface area of the fibers. Compared with the normal adsorbents, such as activated carbon, an obvious advantage of the porous fibers as adsorbent is easy magnetic separation from the solution. Similarly, for the other non-polar contaminant molecules, such as methyl orange and methyl blue, the porous fibers also show high removal capacity [16].

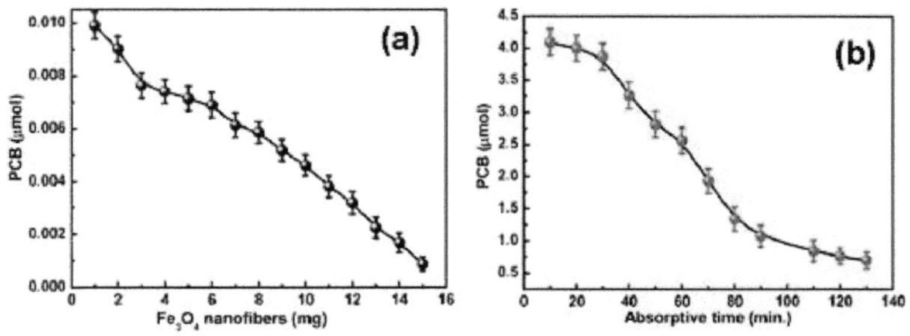

Figure 9.41 Adsorption of PCB-77 on the porous magnetite fibers, shown in Fig. 9.9, in n-hexane solution. (a) PCB-77 content in the 100 ml n-hexane solution, with initial PCB-77 concentration 0.1 μmol/l, as a function of addition amount of the fibers (soaking in the solution for 5 min). (b) The evolution of PCB-77 content in the solution with adsorption time, for 1.00 mg magnetite fibers in the 1000 ml solution with initial PCB-77 concentration 4.1 μmol/l [16]. Copyright (2011) Royal Society of Chemistry.

9.5 Conclusion and Outlook

In the past decade, considerable progress has been made in synthesis of micro/nanostructured materials. In this chapter, we have introduced three main synthetic strategies: (1) solvothermal/hydrothermal methods; (2) template-etching strategies; and (3) electrospinning method. These methods are simple, controllable, and hence suitable for mass production of various micro/nanostructired materials. The micro/nanostructured materials by these methods are of high specific surface area and anti-aggregation (structural stability), and keep high active surface area during use. They have exhibted much stronger photocatalytic activity and adsorption capacity, and hence much higher contaminant-removal ability than the normal nanopowder materials, showing significantly structurally enhanced performances. So, they are really the promising materials for contaminant-removal and environmental remediation.

It should be mentioned that surface functionalization of the micro/nanostructured materials is still a challenge. This problem could be partialy overcome by surface modification. More importantly, their structurally enhanced performances for environmental remediation are attracting more and more attention. The advantage of micro/nanostructured materials lies in their high surface area, aggregation-resistance, and easy collection. The application potentials of these aspects, though still in infancy, offer exciting opportunities and challenges for the newcomers.

Acknowledgments

We acknowledge the National Key Basic Research Program of China (Grant No. 2013CB934303), the National Natural Science Foundation of China (Grant Nos. 21371005, 21001002, 21177132), and the China Postdoctoral Science Foundation (Grant No. 2013T60632).

References

1. W. X. Zhang, Environmental technologies at the nanoscale, *Environ. Sci. Technol.*, 2003, 37, 102A–108A.
2. S. M. Ponder, J. G. Darab, T. E. Mallouk, Remediation of Cr(VI) and Pb(II) aqueous solutions using supported nanoscale zero-valent iron, *Environ. Sci. Technol.*, 2000, 34, 2564–2569.
3. P. He, D. Y. Zhao, Preparation and characterization of a new class of starch–stabilized bimetallic nanoparticles for degradation of chlorinated hydrocarbons in water, *Environ. Sci. Technol.*, 2005, 39, 3314–3320.
4. C. B. Wang, W. X. Zhang, Synthesizing nanoscale iron particles for rapid and complete dechlorination of TCE and PCBs, *Environ. Sci. Technol.*, 1997, 31, 2154–2156.
5. P. Varanasi, A. Fullana, S. Sidhu, Remediation of PCB contaminated soils using iron nano-particles, *Chemosphere*, 2007, 66, 1031–1038.
6. Y. S. Keum, Q. X. Li, Reductive debromination of polybrominated diphenyl ethers by zerovalent iron, *Environ. Sci. Technol.*, 2005, 39, 2280–2286.
7. J. Theron, J. A. Walker, T. E. Cloete, Nanotechnology and water treatment: applications and emerging opportunities, *Crit. Rev. Microbiol.*, 2008, 34, 43–69.

8. V. R. Shinde, T. P. Gujar, T. Noda, D. Fujita, A. Vinu, M. Grandcolas, J. Ye, Growth of shape-and size-selective zinc oxide nanorods by a microwave-assisted chemical bath deposition method: effect on photocatalysis properties, *Chem. Eur. J.* 2010, 16, 10569–10575.
9. W. Yantasee, Y. H. Lin, G. E. Fryxell, K. L. Alford, B. J. Busche, C. D. Johnso, Selective removal of copper(II) from aqueous solutions using fine-grained activated carbon functionalized with amine, *Ind. Eng. Chem. Res.*, 2004, 43, 2759–2764.
10. L. S. Zhong, J. S. Hu, H. P. Liang, A. M. Cao, W. G. Song, L. J. Wan, Self-assembled 3D flowerlike iron oxide nanostructures and their application in water treatment, *Adv. Mater.*, 2006, 18, 2426–2431.
11. X. B. Wang, W. P. Cai, S. W. Liu, G. Z. Wang, Z. K. Wu, H. J. Zhao, ZnO hollow microspheres with exposed porous nanosheets surface: structurally enhanced adsorption towards heavy metal ions, *Colloids Surf. A: Physicochem. Eng. Aspects*, 2013, 422, 199–205.
12. F. Lu, W. P. Cai, Y. G. Zhang, ZnO hierarchical micro/nanoarchitectures: solvothermal synthesis and structurally enhanced photocatalytic performance, *Adv. Funct. Mater.*, 2008, 18, 1047–1056.
13. X. B. Wang. W. P. Cai, G. Z. Wang, C. H. Liang, Standing porous ZnO nanoplate-built hollow microspheres and kinetically controlled dissolution/crystal growth mechanism, *J. Mater. Res.*, 2012, 27, 951–958.
14. J. D. Verhoeven, *Fundamentals of Physical Metallurgy*, Ch. 8 (John Wiley & Sons, NY, 1975).
15. X. B. Wang, W. P. Cai, Y. X. Lin, G. Z. Wang, C. H. Liang, Mass production of micro/nanostructured porous ZnO plates and their strong structurally enhanced and selective adsorption performance for environmental remediation, *J. Mater. Chem.*, 2010, 20, 8582–8590.
16. C. L. Han, W. P. Cai, W. Tang, G. Z. Wang, C. H. Liang, Protein assisted hydrothermal synthesis of ultrafine magnetite nanoparticle built-porous oriented fibers and their structurally enhanced adsorption to toxic chemicals in solution, *J. Mater. Chem.*, 2011, 21, 11188–11196.
17. X. B. Wang, J. Liu, Z. Li, The graphite phase derived from polyimide at low temperature, *J. Non-Cryst. Solids*, 2009, 355, 72–75.
18. X. B. Wang, J. Liu, W. Z. Xu, T. Cao, X. J. Song, C. L. Cheng, Preparation of carbon microstructures by thermal treatment of thermosetting/thermoplastic polymers and their application in water purification, *Micro Nano Lett.*, 2012, 7(9), 918–922.
19. X. B. Wang, J. Liu, W. Z. Xu, One-step hydrothermal preparation of amino-functionalized carbon spheres at low temperature and their enhanced

adsorption performance towards Cr(VI) for water purification, *Colloids Surf. A: Physicochem. Eng. Aspects,* 2012, 415, 288–294.

20. X. B. Wang, W. P. Cai, G. Z. Wang, Z. K. Wu, H. J. Zhao, One-step fabrication of high performance tremella-like Fe_3S_4/C magnetic adsorbent with easy recovery and regeneration properties, *CrystEngComm*, 2013, 15, 2956–2965.
21. J. M. Ma, L. Chang, J. B. Lian, Z. Huang, X. C. Duan, X. D. Liu, P. Peng, T. Kim, Z. F. Liu, W. J. Zheng, Ionic liquid-modulated synthesis of ferrimagnetic Fe_3S_4 hierarchical superstructures, *Chem. Commun.*, 2010, 46, 5006–5008.
22. J. Yang, C. X. Li, Z. W. Quan, C. M. Zhang, P. P. Yang, Y. Y. Li, C. C. Yu, J. Lin, Self-assembled 3D flowerlike Lu_2O_3 and Lu_2O_3:Ln^{3+} (Ln = Eu, Tb, Dy, Pr, Sm, Er, Ho, Tm) microarchitectures: ethylene glycol-mediated hydrothermal synthesis and luminescent properties, *J. Phys. Chem. C*, 2008, 112, 12777–12785.
23. X. W. Zhao, L. M. Qi, Rapid microwave-assisted synthesis of hierarchical ZnO hollow spheres andtheir application in Cr(VI) removal, *Nanotechnology*, 2012, 23, 235604 (7 pp.).
24. Y. Zeng, X. Wang, H. Wang, Y. Dong, Y. Ma, J. N. Yao, Multi-shelled titania hollow spheres fabricated by a hard template strategy: enhanced photocatalytic activity, *Chem. Commun.*, 2010, 46, 4312–4314.
25. G. C. Xi, Y. Yan, Q. Ma, J. F. Li, H. F. Yang, X. J. Lu, C. Wang, Synthesis of multiple-shell WO_3 hollow spheres by a binary carbonaceous template route and their applications in visible light photocatalysis, *Chem. Eur. J.*, 2012, 18, 13949–13953.
26. X. W. (David) Lou, L. A. Archer, Z. C. Yang, Hollow micro-/nanostructures: synthesis and applications, *Adv. Mater.*, 2008, 20, 1–33.
27. Z. B. Huang, F. Q. Tang, Preparation, structure, and magnetic properties of mesoporous magnetite hollow spheres, *J. Colloid Interface Sci.*, 2005, 281, 432–436.
28. Y. Y. Xu, D. R. Chen, X. L. Jiao, K. Y. Xue, Nanosized Cu_2O/PEG400 composite hollow spheres with mesoporous shells, *J. Phys. Chem. C*, 2007, 111, 16284–16289.
29. H. L. Xu, W. Z. Wang, Template synthesis of multishelled Cu_2O hollow spheres with a single-crystalline shell wall, *Angew. Chem. Int. Ed.*, 2007, 46, 1489–1492.
30. H. C. Zeng, Ostwald ripening: a synthetic approach for hollow nanomaterials *Curr. Nanosci.*, 2007, 3, 177–181.

31. Y. D. Yin, R. M. Rioux, C. K. Erdonmez, S. Hughes, G. A. Somorjai, A. P. Al ivisatos, Formation of hollow nanocrystals through the nanoscale Kirkendall effect, *Science*, 2004, 304, 711–714.

32. F. Caruso, R. A. Caruso, H. Möhwald, Nanoengineering of inorganic and hybrid hollow spheres by colloidal templating, *Science*, 1998, 282, 1111–1114.

33. Y. Q. Wang, G. Z. Wang, H. Q. Wang, C. H. Liang, W. P. Cai, L. D. Zhang, Chemical-template synthesis of micro/nanoscale magnesium silicate hollow spheres for waste-water treatment, *Chem. Eur. J.*, 2010, 16, 3497–3503.

34. Y. Q. Wang, G. Z. Wang, H. Q. Wang, W. P. Cai, L. D. Zhang, One-pot synthesis of nanotube-based hierarchical copper silicate hollow spheres, *Chem. Commun.*, 2008, 6555–6557.

35. L. Pauling, The structure of the chlorites, *Proc. Natl. Acad. Sci. U. S. A.*, 1930, 16, 578–582.

36. Y. Q. Wang, G. Z. Wang, H. Q. Wang, W. P. Cai, C. H. Liang, L. D. Zhang, Template-induced synthesis of hierarchical SiO_2@γ -AlOOH spheres and their application in Cr(VI) removal, *Nanotechnology*, 2009, 20, 155606 (6 pp.).

37. H. W. Hou, Y. Xie, Q. Yang, Q. X. Guo, C. R. Tan, Preparation and characterization of γ-AlOOH nanotubes and nanorods, *Nanotechnology*, 2005, 16, 741–745.

38. H. B. Zeng, W. P. Cai, P. S. Liu, X. X. Xu, H. J. Zhou, C. Klingshirn, H. Kalt, ZnO-based HNPs by selective etching: elimination and reconstruction of metal-semiconductor interface, improvement of blue emission and photocatalysis, *ACS Nano*, 2008, 2 (8), 1661–1670.

39. Ziabicki, A. *Fundamentals of Fiber Formation*, John Wiley and Sons, London, 1976, ISBN 0-471-98220-2.

40. Y. X. Lin, W. P. Cai, X. Y. Tian, X. L. Liu, G. Z. Wang, C. H. Liang, Polyacrylonitrile/ferrous chloride composite porous nanofibers and their strong Cr-removal performance, *J. Mater. Chem.*, 2011, 21, 991–997.

41. Y. X. Lin, W. P. Cai, H. He, X. B. Wang, G. Z. Wang, Three-dimensional hierarchically structured PAN@γ–AlOOH fiber films based on a fiber templated hydrothermal route and their recyclable strong Cr(VI)-removal performance, *RSC Adv.*, 2012, 2, 1769–1773.

42. W. Q. Cai, J. G. Yu, S. Mann, Template-free hydrothermal fabrication of hierarchically organized γ-AlOOH hollow microspheres, *Micro. Meso. Mater.*, 2009, 122, 42–47.

43. B. Gong, Q. Peng, J. S. Na, G. N. Parsons, Highly active photocatalytic ZnO nanocrystalline rods supported on polymer fiber mats: synthesis using atomic layer deposition and hydrothermal crystal growth, *Appl. Catal. A: Gen.*, 2011, 407(1–2), 211–216.

44. T. G. Xu, L. W. Zhang, H. Y. Cheng, Y. F. Zhu, Significantly enhanced photocatalytic performance of ZnO via graphene hybridization and the mechanism study, *Appl. Catal. B: Environ.*, 2011, 101, (3-4), 382–387.

45. T. Ohno, L. Bai, T. Hisatomi, K. Maeda, K. Domen, Photocatalytic water splitting using modified GaN:ZnO solid solution under visible light: long-time operation and regeneration of activity, *J. Am. Chem. Soc.*, 2012, 134 (19), 8254–8259.

46. T. Kavitha, A. I. Gopalan, K.-P. Lee, S.-Y. Park, Glucose sensing, photocatalytic and antibacterial properties of graphene–ZnO nanoparticle hybrids, *Carbon*, 2012, 50 (8), 2994–3000.

47. X. J. Wang, Q. L. Zhang, Q. Wan, G. Z. Dai, C. J. Zhou, B. S. Zou, Controllable ZnO architectures by ethanolamine-assisted hydrothermal reaction for enhanced photocatalytic activity, *J. Phys. Chem. C*, 2011, 115 (6), 2769–2775.

48. V. van Dijken, A. H. Janssen, M. H. P. Smitsmans, D. Vanmaekelbergh, K. Meijerink, Size-selective photoetching of nanocrystalline semiconductor particles, *Chem. Mater.*, 1998, 10, 3513–3522.

49. M. J. Height, S. E. Pratsinis, O. Mekasuwandumrong, P. Praserthdam, Ag-ZnO catalysts for UV photodegradation of methylene blue, *Appl. Catal. B: Environ.*, 2006, 63, 305–312.

50. S. Mahdavi, M. Jalali, A. Afkhami, Removal of heavy metals from aqueous solutions using Fe_3O_4, ZnO, and CuO nanoparticles, *J. Nanopart. Res.*, 2012, 14, 846 (18 pp.).

51. S. Rajagopalan, O. Koper, S. Decker, K. J. Klabunde, Nanocrystalline metal oxides as destructive adsorbents for organophosphorus compounds at ambient temperatures, *Chem. Eur. J.*, 2002, 8, 2602–2607.

Chapter 10

Utilization of Biological Polysaccharides as Eco-Friendly Structural Materials

Mineo Hashizume and Kazutoshi Iijima
Department of Industrial Chemistry, Faculty of Engineering, Tokyo University of Science, 12-1 Ichigayafunagawara-machi, Shinjuku-ku, Tokyo 162-0826, Japan
mhashizu@ci.kagu.tus.ac.jp

10.1 Introduction

Polysaccharides, in which saccharide (sugar) molecules are polymerized and linked each other using ether bonds, are produced by various living organisms. Some of them are referred as biomass and used in various applications.[1] In this chapter, after brief explanation about classification of polysaccharides, applications of polysaccharides as functional materials, non-structural materials, and structural materials are described. Development of processes for obtaining structural materials from biological polysaccharides is important from the viewpoint not only of preparing biocompatible or biodegradable materials but also of effective usage of abundant biomass.

Nanomaterials in Energy and Environmental Applications
Edited by Junhui He

ISBN 978-981-4463-78-2 (Hardcover), 978-981- 4463-79-9 (eBook)
www.panstanford.com

10.2 Classification of Polysaccharides

Polysaccharides are generally classified according to their biological functions or their chemical structures.[2] Chemical structures of some representative polysaccharides are shown in Fig. 10.1.

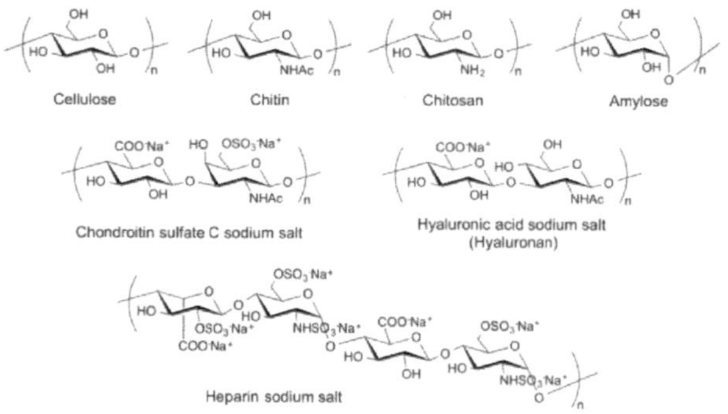

Figure 10.1 Chemical structures of some representative polysaccharides.

10.2.1 Structural Polysaccharides

Structural polysaccharides such as cellulose and chitin are used to construct and maintain external structures of living organisms. Cellulose is a homopolymer of D-glucose that linked via $\beta(1 \rightarrow 4)$ linkage. Cellulose is the most abundant carbon source in biosphere and the major component of cell wall. Sea squirts and some bacteria also synthesize cellulose.[2,3] Chitin is a homopolymer of *N*-acetyl-D-glucosamine that linked via $\beta(1 \rightarrow 4)$ linkage. Chitin is the second abundant biomass following cellulose and the major components of exoskeletons of crustaceans and insects.[1,2] Both bulk cellulose and chitin are not soluble in water because of their high crystallinities (see 10.4.1). Hydrolysis of acetyl groups of chitin resulted in formation of chitosan that can be dissolved in acidic aqueous solutions such as aqueous acetic acid and formic acid.[1,4] Degree of deacetylation of chitin affects the solubility.

10.2.2 Reserve Polysaccharides

Reserve (or storage) polysaccharides, starch and glycogen, are also homopolymer of D-glucose, but their linkages are different from that of cellulose. Starch is produced by plants and usually consisted of amylose and amylopectin.[2] Amylose is a linear homopolymer of D-glucose that linked via $\alpha(1 \rightarrow 4)$ linkage. Amylopectin is mainly contained in polymers of $\alpha(1 \rightarrow 4)$-linked D-glucose and there are branches, in every 24–30 glucose residues, created using $\alpha(1 \rightarrow 6)$ linkages. Dextrin is the collective term for hydrolyzed product of starch. Glycogen is one of animal reserve polysaccharides. Molecular structures of glycogen are similar to those of amylopectin but contain more branches (in every 8–14 glucose resides) than amylopectin. Dextran, synthesized from sucrose by lactic acid bacteria, is also a homopolymer of D-glucose that mainly linked via $\alpha(1 \rightarrow 6)$ linkage and has some branched structures.

10.2.3 Other Polysaccharides

Regarding other kinds of polysaccharides, glycosaminoglycans are one of the well-known categories.[2,5,6] They are acidic alternate copolymers of amino sugars (usually *N*-acetylated) and uronic acids. Some of them are bioactive: in other words, they play important roles in living bodies. Hyaluronic acid (or hyaluronan) has repeated structures of D-gluclonic acid $\beta(1 \rightarrow 3)$ *N*-acetyl-D-glucosamine unit via $\beta(1 \rightarrow 4)$ linkages. It is the major glycosaminoglycan component in vitreous body and spinal fluid. Hyaluronan molecules strongly bind water molecules or metal ions, which produce characteristic viscous properties. In living bodies, such properties play the role as the shock absorber or lubricant. Many kinds of glycosaminoglyans are sulfated and usually conjugated with proteins to form proteoglycans that exist as the major component of extracellular matrices in various animals. Chondroitin sulfate A (or chondroitin 4-sulfate) is consisted of D-gluclonic acids and *N*-acetyl-D-glucosamine whose 4-position of hydroxy group is sulfated, whereas 6-position of hydroxy group of *N*-acetyl-D-glucosamine is sulfated in chondroitin sulfate C (or chondroitin 6-sulfate). They are found in cartilage as the parts of proteoglycans. Heparin contains multiple sulfate groups in its repeating structure and is known to exhibit anticoagulant activity.

There are polysaccharides obtainable from seaweed such as agar, alginic acid, and carraggenan.[1] They are used in various practical applications such as food and biochemical industries.

10.3 Polysaccharides as Non-Structural Materials

In addition to use of polysaccharides as food and bioactive agents based on their biological functions, application of polysaccharides as functional materials has been receiving attention in various fields. Firstly, application of polysaccharides as non-structural materials, in other words application of polysaccharides dispersed or suspended in solutions, is concisely reviewed. Concentrated aqueous solution of polysaccharides might be referred to as hydrogels. Polysaccharide hydrogels intended to use as structural materials are described in the later sections.

10.3.1 Industrial Applications

Because of their high molecular weights and containing a large number of hydroxy or ionic functional groups, aqueous solutions of polysaccharides are usually viscous. Therefore, some of them, for example, those made of gellan (gellan gum), carrageenan, and carboxymethylcellulose, are used as thickening agents in food or other industries.[1] Starch and their hydrolyzed products are also used for production of various edible chemicals. High water-adsorbent ability of hyaluronan makes it a moisturizing ingredient in cosmetic applications.

10.3.2 Biomedical and Biochemical Applications

Natural polysaccharides are generally referred to as biocompatible substances: therefore, their uses in biomedical applications are reasonable. Nowadays, some glycosaminoglycans such as hyaluronan and chondroitin sulfates are sold to customers as health foods. Regarding non-structural materials, the use of polysaccharide particles, mainly chitosan particles, as the carriers for drug delivery system (DDS) has been widely investigated.[4,7,8] Chemical modifications of chitosan molecules are sometimes conducted to add structural stability, affinity to drug or genes, or molecular recognizability to the resulting particles. Chitosan nano and microparitcles can be used for colon-targeted drug delivery or mucosal delivery. Polysaccharide carriers can also be used as the gene or drug carriers having ligand-mediated endocytosis abilities by utilizing carbohydrate–lectin interactions.

Recent development of characterization techniques for nanomaterials has focused renewed attention on polymer hydrogels having nanometer sizes as "nanogels." Akiyoshi *et al.* have been extensively investigating preparation and applications of nanogels prepared from polysaccharide derivatives (Fig. 10.2).[9] These nanogels are formed by self-association of polysaccharides molecules in which small amounts of their hydroxy groups are modified with hydrophobic groups such as cholesterol. Some of these nanogels act as drug delivery carriers, vaccines, and exhibit chaperon-like functions.

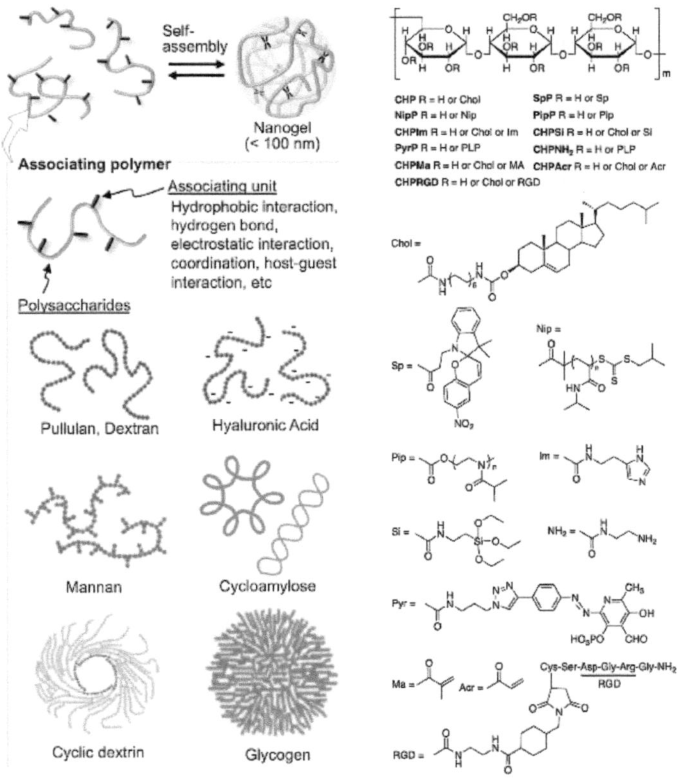

Figure 10.2 Concept of nanogels (a) and examples of chemical structures of polysaccharide derivatives used for preparation of nanogels (b). Reprint with permission from Ref. [9]. Copyright 2012 Chemical Society of Japan.

Exploring novel chemical functions of polysaccharides is an attractive research field. Very recently, Serizawa *et al.* demonstrated that water-dispersible crystalline cellulose nanofibers obtained from bulk cellulose have the capability to hydrolyze the ester, monophosphate, and amide bonds of organic molecules.[10] Similar to this study, other unknown chemical functions of polysaccharides may still exist.

10.4 Structural Materials Made of Structural Polysaccharides

Although molecular structures of cellulose and chitin contain a large number of hydroxy groups, they are water-insoluble. The $\beta(1 \rightarrow 4)$ linkages among the glucose monomer units in these polysaccharides make their conformation linearly-extended, which make them self-assemble to form crystalline fiber structures. In these crystal structures, intramolecular and intermolecular hydrogen-bonding networks can form to increase the stability of the structures (Fig. 10.3 for the case of cellulose).[3,11-14] Chitosan, can be dissolved in water under acidic conditions, also forms crystalline fibers in the solid state (Fig. 10.4).[15]

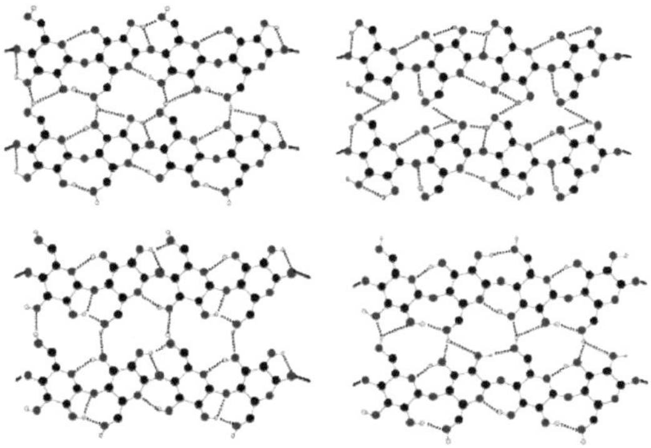

Figure 10.3 Hydrogen-bonding patterns in cellulose Iα and Iβ. Top left and right: the two alternative hydrogen-bond networks in cellulose Iα. Bottom: the dominant hydrogen-bond network in cellulose Iβ. (left) chains at the origin of the unit cell; (right) chains at the center of the unit cell. Some of the hydrogen atoms have been omitted, and hydrogen bonds are shown as dashed lines. Reprint with permission from Ref. [11] (the original figure is illustrated in color). Copyright 2004 American Chemical Society.

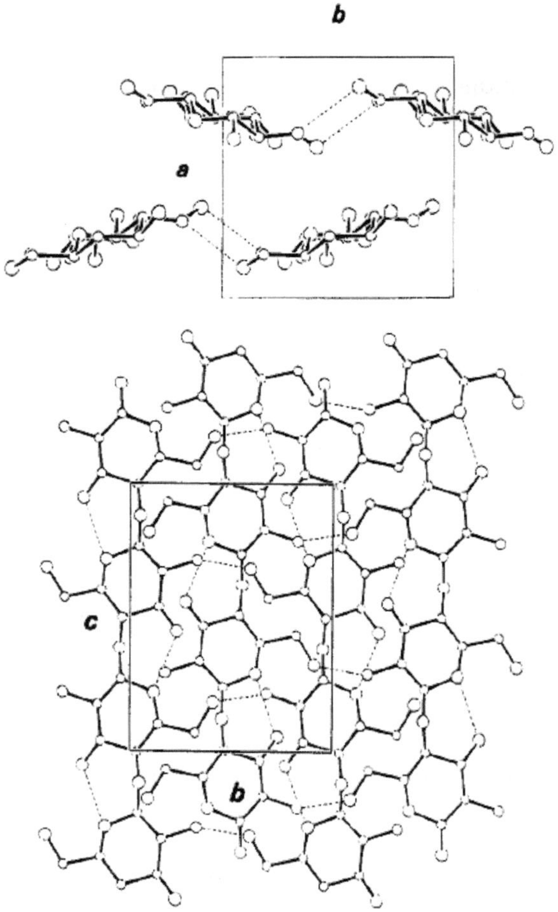

Figure 10.4 Projection of the crystal structure on the *ab* (top) and *bc* (bottom) base planes of chitosan. All hydrogen atoms have been omitted, and hydrogen bonds are shown as dashed lines. Reprint with permission from Ref. [15]. Copyright 1994 American Chemical Society.

10.4.1 Cellulose and Its Derivatives

In addition to conventional use of cellulose as structural (or solid) materials such as woods, pulps, and papers, the use of cellulose derivatives has also been widely developed.[3] Nitrocellulose has been used as film materials. Regenerative cellulose fibers are known as rayon. Some cellulose esters form linearly extended molecular structures and their intermolecular interactions can be controlled by designing the substituent groups. Cellulose triacetates are good liquid crystal materials and used as the coating layers for liquid crystal displays (LCDs).[16] Molecular chains of other types of cellulose esters produce chiral environments and exhibit good performance as the components of solid phase of chiral separation columns.[17] Cellulose esters also have potential application in biomedical and cosmetic fields. Some of them can be used as the matrices having controlled release properties and as semi-permeable membranes.[16]

Sheet materials prepared from microbial cellulose exhibit good wound healing ability due to their unique microstructures that can keep the wound moist during the healing process.[18]

Cellulose nanofibers can be obtained by acid treatments of cellulose fibers.[12,19] For example, cellulose nanofibers from wood are 3–5 nm in width and 100–200 nm in length, whereas those from tunicate, a sea aminal, are *ca*. 10–20 nm in width and 500–2000 nm in length. Sometimes, proper surface chemical modifications such as oxidation, cationization, esterification, or polymer grafting are conducted. Cellulose nanofibers are attractive materials not only because of their crystalline, rod-like structures but also because of their chemical and physical properties. They can self-organize into liquid crystalline arrangements in their concentrated suspensions.[12] They are also used as reinforcing fillers in polymer-based nanocomposit materials.[12,19] Cellulose nanofibers treated with 2,2,6,6-Tetramethylpiperidine-1-oxyl (TEMPO)-mediated oxidation can be formed into transparent gas barrier films.[20]

One interesting use of cellulose sheet or fiber materials is that their use as the removable templates for fabrication of inorganic or hybrid materials having nanostructures.[21–23]

10.4.2 Chitin and Chitosan

Regarding the use in structural material applications, chitosan is basically used rather than chitin because of its easier handling. Chemically functionalized, cross-linked chitosan beads are used as insoluble absorbents in wastewater treatments.[24] For such purposes, the advantages of chitosan are its low cost and biocompatibility.

Chitosan can form hydrogels with the combination of acidic polymers (see 10.5.1). Cross-linked chitosan hydrogels can be converted into chitin hydrogels by their *N*-acetylation. Chitosan films are readily prepared using solution casting processes. Porous chitosan materials are prepared by freezing and lyophilization of chitosan solutions. Wet spinning and electrospinning are useful to prepare chitosan fibers and their woven and non-woven mats. These materials of various shapes are used in various biomedical applications such as biodegradable materials, drug release carriers, wound dressing, skin substitutions, scaffolds for tissue engineering, and regenerative medicine.[4,25,26] Some special solvents such as DMF containing lithium chloride are used for dissolving chitin and such solutions are used for preparation of chitin fibers.[4] Chitosan can also be formed into tablets usable for drug administrations. In addition, chitosan is used as the coating layer to add biocompatibility or antibacterial property to the base materials.

10.5 Structural Materials Made of Non-Structural Polysaccharides

Not limited to structural polysaccharides, other polysaccharides that are usually water-soluble can also be used as the components of structural (solid) materials as they are or after proper treatments.

10.5.1 Bulk Materials and Hydrogels

Starch is the raw material of starch glues. Cushioning materials made of cornstarch are commercially available. Cross-linked starch and cyclodextrins that are further modified with proper functional groups are used as insoluble absorbents for heavy metal ions in wastewater, similar to those made of chitosan.[24]

Some polysaccharides can form hydrogels. Agarose is frequently used in gel electrophoresis. Agarose and carboxymethyl dextran are used as the component of commercial gels for gel permeation chromatography (Sepharose® and Sephadex®) by conducting cross-linking and/or other chemical modifications. Alginate can gelate by complexing with calcium ions. Alginate gels are well used in food industry and biomedical applications, in addition to their use as a shear-thinning agent in textile/paper-printing industry.[27]

Acidic polysaccharides such as glycosaminoglycans can form insoluble polyion complexes with chitosan.[28,29] Such polyion complexes generally have porous structures when they are isolated from the aqueous solution used for their preparation and dried by lyophilization (Fig. 10.5).[29] These dried gels can re-swell in water and are expected as cell culturing scaffolds or matrices for sustained drug release.

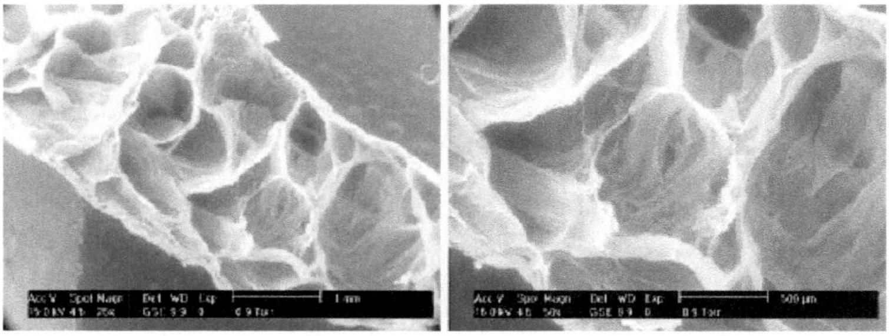

Figure 10.5 Scanning electron microscopic images of dried chitosan/hyaluronan polyion complexes. Scale bar: 1 mm for the left image and 500 μm for the right image. Reprinted with permission from Ref. [29]. Copyright 1994 Springer.

10.5.2 Films, Coatings, and Fibers

In addition to cellulose ethers and chitosan, various polysaccharides such as starch, *k*-carrageenan, and alginate are used for edible films and coatings having antimicrobial activity.[30] Preparation of these films and coating is generally conducted using conventional casting techniques.

Layer-by-layer (LbL) assembly technique is useful for preparation of ultrathin films of oppositely charged polyelectrolytes at molecular precisions.[31–33] For example, LbL films of poly(sodium styrenesulfonate) and chitosan,[34] dextran suflate and chitosan,[35] hyaluronan and chitosan,[36] are readily obtained. When ultrathin films are prepared on sacrificial layers formed on solid substrates, free-standing ultrathin films are obtainable by dissolution of the sacrifical layers.[37–39] Using such an approach, free-standing LbL films consisted of (alginate/chitosan) alternate layers (thickness: *ca.* 30 nm) are obtained. Such films have potential application as biocompatible, nano-adhesive plasters.[40]

On the contrary, we, the authors of this chapter, have been investigating fabrication of film materials, made of polysaccharides, having thicknesses of several tenths of micrometers. Such thicker films should have sufficient mechanical strength and much easier handling than those having nanometer thicknesses. Recently, we have demonstrated that free-standing polysaccharide composite films were obtained successfully by utilizing hot press technique.[41] Briefly, chondroitin sulfate/chitosan polyion complexes are prepared at first by mixing an aqueous solution of chondroitin sulfate with acetic acid solution of chitosan. The polysaccharide polyion complexes are dried by lyophilization and then formed into thin films by hot pressing. Thickness of the film is controllable by conducting hot pressing with plastic sheets, having (rectangular) holes corresponding to film areas, as the spacer. Free-standing heparin/chitosan and hyaluronan/chitosan films are also obtainable using similar procedures. The resulting films have homogeneous, smooth surfaces, and dense structures at microscopic level (Fig. 10.6). Polyion complexes also can be formed into thin films by casting of their suspension: however, it is hard to obtain films having homogeneous thickness and surface smoothness. The polysaccharide composite films obtained by our process swell but are not dissolved in water, that is, water-insoluble films are obtained from water-soluble polysaccharides using neither cross-linking agents nor chemical modifications. The degrees of swelling are suppressed compared with similar polyion complex gels without hot press treatments (see 10.5.1). It was also found that these polysaccharide composite films induced hydroxyapatite deposition on their surfaces under body fluid conditions. Such a property supports potential use of these films in regenerative medicine. In addition, our resent findings support that these films can be useful as sustained drug release carriers because they can incorporate small water-soluble organic molecules and release them when they are immersed in water.

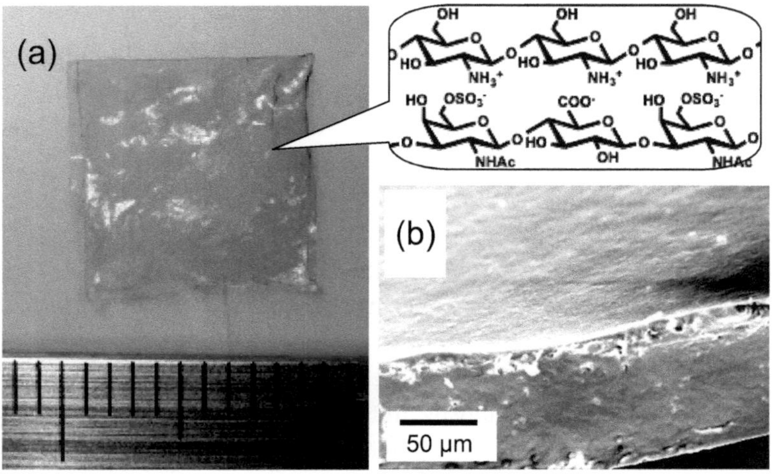

Figure 10.6 Macroscopic view with illustration (a) and scanning electron microscopic image (b) of free-standing chondrotin sulfate/chitosan composite films.

Yamamoto *et al.* have reported that some polysaccharide polyion complexes such as those made of chitosan and gellan can be formed into fibers and capsules.[42] The process for obtaining those fibers and capsules is based on polyion complex formation at the interface between the chitosan solution and the gellan solution. Using a similar approach, we can also prepare chondroitin sulfate/chitosan composite fibers.[43] These fibers have potential application as cell culture scaffolds that can align cells along the axis of the fibers. Using a different approach, Takeuchi *et al.* recently reported the preparation of meter-long, core-shell fibers based on alginate gels that can incorporate cells inside the fibers using microfluidic technique.[44]

10.6 Conclusion

Polysaccharide-based structural materials may have some disadvantages such as weaker mechanical strength, lower stability against various chemical or thermal stimuli, and lower mass productivity than conventional industrial materials made of commonly used synthetic polymers. However, their advantages such as high biocompatibility or biodegradability and relatively low-cost fabrication processes make them useful materials in some special applications. For example, they should be useful as cell culture scaffolds for tissue engineering, implantable materials for regenerative medicine, and sustained drug release carriers. In addition, further development of utilization of biological polysaccharides as eco-friendly materials including structural materials leads to expansion of biomass consumptions. Material chemists can contribute to solve energy and environmental problems from such a point of view.

References

1. M. Rinaudo, *Polym. Int.* 2008, 57, 397.
2. D. Voet, J. G. Voet, C. W. Pratt, Fundamental of Biochemistry: Life at the Molecular Level, Fourth Edition, Wiley, 2012.
3. D. Klemm, B. Heublein, H.-P. Fink, A. Bohn, *Angew. Chem. Int. Ed.* 2005, 44, 3358.
4. M. N. V. R. Kumar, R. A. A. Muzzarelli, C. Muzzarelli, H. Sashiwa, A. J. Domb, *Chem. Rev.* 2004, 104, 6017.
5. L. Kjellén, U. Lindahl, *Annu. Rev. Biochem.* 1991, 60, 443.
6. K. R. Taylor, R. L. Gallo, *FASEB J.* 2006, 20, 9.
7. S. A. Agnihotri, N. N. Mallikarjuna, T. M. Aminabhavi, *J. Control. Release* 2004, 100, 5.
8. T. Sato, T. Ishii, Y. Okahata, *Biomaterials* 2001, 22, 2075.
9. Y. Sasaki, K. Akiyoshi, *Chem. Lett.* 2012, 41, 202. (Highlight review).
10. T. Serizawa, T. Sawada, H. Okura, M. Wada, *Biomacromolecules* 2013, 14, 613.
11. A. Šturcová, I. His, D. C. Apperley, J. Sugiyama, M. C. Jarvis, *Biomacromolecules* 2004, 5, 1333.
12. Y. Habibi, L. A. Lucia, O. J. Rojas, *Chem. Rev.* 2010, 110, 3479.
13. R. Minke, J. Blackwell, *J. Mol. Biol.* 1978, 120, 167.
14. P. Sikorski, R. Hori, M. Wada, *Biomacromolecules* 2009, 10, 1100.
15. T. Yui, K. Imada, K. Okuyama, Y. Obata, K. Suzuki, K. Ogawa, *Macromolecules* 1994, 27, 7601.
16. K. J. Edgar, C. M. Buchanan, J. S. Debenham, P. A. Rundquist, B. D. Seiler, M. C Shelton, D. Tindall, *Prog. Polym. Sci.* 2001, 26, 1605.
17. T. Ikai, Y. Okamoto, *Chem. Rev.* 2009, 109, 6077.

18. W. K. Czaja, D. J. Young, M. Kawecki, R. M. Brown, Jr., *Biomacromolecules* 2007, 8, 1.
19. M. A. S. Azizi Samir, F. Alloin, A. Dufresne, *Biomacromolecules* 2005, 6, 612.
20. H. Fukuzumi, T. Sato, T. Iwata, Y. Kumamoto, A. Isogai, *Biomacromolecules* 2009, 10, 162.
21. J. He, T. Kunitake, T. Watanabe, *Chem. Commun.* 2005, 795.
22. J. Huang, N. Matsunaga, K. Shimanoe, N. Yamazoe, T. Kunitake, *Chem. Mater.* 2005, 17, 3513.
23. Y. Gu, J. Huang, *J. Mater. Chem.* 2009, 19, 3764.
24. G. Crini, Prog. *Polym. Sci.* 2005, 30, 38.
25. J.-K. F. Suh, H. W. T. Matthew, *Biomaterials* 2000, 21, 2589.
26. A. D. Martino, M. Sittinger, M. V. Risbud, *Biomaterials* 2005, 26, 5983.
27. A. D. Augst, H. J. Kong, D. J. Mooney, *Macromol. Biosci.* 2006, 6, 623.
28. W.-B. Chen, L.-F. Wang, J.-S. Chen, S.-Y. Fan, *J. Biomed. Mater. Res.* 2005, 75A, 128.
29. C. Peniche, M. Fernández, G. Rodrígez, J. Parra, J. Jimenez, A. L. Bravo, D. Gómez, J. S. Román, *J. Mater. Sci., Mater. Med.* 2007, 18, 1719.
30. C. A. Campos, L. N. Gerschenson, S. K. Flores, *Food Bioprocess Technol.* 2011, 4, 849.
31. G. Decher, *Science* 1997, 277, 1232.
32. V. Gribova, R. Auzely-Velty, C. Picart, *Chem. Mater.* 2011, 24, 854.
33. K. Ariga, Q. Ji, J. P. Hill, Y. Bando, M. Aono, *NPG Asia Mater.* 2012, 4, e17.
34. Y. Lvov, M. Onda, K. Ariga, T. Kunitake, *J. Biomater. Sci., Polym. Ed.* 1998, 9, 345.
35. T. Serizawa, M. Yamaguchi, T. Matsuyama, M. Akashi, *Biomacromolecules* 2000, 1, 306.
36. M. Thierry, F. M. Winnik, Y. Merhi, J. Silver, M. Tabrizian, *Biomacromolecules* 2003, 4, 1564.
37. Arif A. Mamedov, N. A. Kotov, *Langmuir* 2000, 16, 5530.
38. M. Hashizume, T. Kunitake, *Langmuir* 2003, 19, 10172.
39. H. Watanabe, R. Vendamme, T. Kunitake, *Bull. Chem. Soc. Jpn.* 2007, 80, 433. (Accounts)
40. T. Fujie, Y. Okamura, S. Takeoka, *Adv. Mater.* 2007, 19, 3549.
41. M. Hashizume, H. Kobayashi, M. Ohashi, *Colloids Surf. B* 2011, 88, 534.

42. M. Amaike, Y. Senoo, H. Yamamoto, *Macromol. Rapid. Commun.* 1998, 19, 287.

43. K. Iijima, K. Yuyama, K. Asaine, K. Irie, M. Hashizume, *Kobunshi Ronbunshu* 2014, 71, 11.

44. H. Onoue, T. Okitsu, A. Itou, M. Kato-Negishi, R. Gojo, D. Kiriya, K. Sato, S. Miura, S. Iwanaga, K. Kuribayashi-Shigetomi, Y. T. Matsunaga, Y. Shimaoka, S. Takeuchi, *Nat. Mater.* 2013, 12, 584.

Chapter 11

Catalysts for Indoor Formaldehyde Control

Hua Tian and Junhui He
*Functional Nanomaterials Laboratory
and Key Laboratory of Photochemical
Conversion and Optoelectronic Materials,
Technical Institute of Physics and Chemistry,
Chinese Academy of Sciences,
Zhongguancundonglu 29,
Haidianqu, Beijing 100190, China*
jhhe@mail.ipc.ac.cn

11.1 Introduction

People often spend most of their time indoors, such as workplace, public buildings, and residential dwellings, while indoor air pollution is becoming one of the top risks to human health [1]. Volatile organic compounds (VOCs) are among the most abundant indoor air pollutants [2], and formaldehyde (HCHO) is one of the most common and most noxious indoor air pollutants.

Formaldehyde is an important chemical for the global economy, widely used in construction, wood processing, furniture, textiles, carpeting, and in the chemical industry. In China, about 1 million tons of formaldehyde could be consumed every year for construction and decoration materials [3]. It can be found in particle board, household

Nanomaterials in Energy and Environmental Applications
Edited by Junhui He

ISBN 978-981-4463-78-2 (Hardcover), 978-981- 4463-79-9 (eBook)
www.panstanford.com

products, fibreboard, and plywood. Already being classified as a known human carcinogen, formaldehyde is a sensory and respiratory irritant, and prolonged formaldehyde exposure has been associated with reduced pulmonary function and asthma. Also, formaldehyde causes nasopharyngeal cancer and probably leukemia [4]. Because of its harmful properties and the progressive increase of formaldehyde emissions, the release of formaldehyde into the indoor environment is being controlled by increasingly stringent regulations. For instance, an exposure limit of 0.25 mg/m^3 (8 h) has been proposed by the EU Scientific Committee on Occupational Exposure limits [5]. In China, the residential indoor formaldehyde maximum allowable concentration established by the Chinese Ministry of Health is 0.1 mg/m^3 (or ca. 0.08 ppm) [6].

As a result of continuing awareness of and concern about the health impacts of formaldehyde, the development of new effective methods for its abatement is urgent and mandatory. The most efficient method for formaldehyde destruction is catalytic degradation, which is becoming most popular because in many cases, it is more versatile and economical, especially for low concentrations. However, the selection of catalytic materials and the optimization of catalyst formulation do not appear an easy task. To date, three classes of catalysts, that is, photocatalysts, transition metal oxides, and noble metals have been widely explored for the formaldehyde abatement.

In this review, an overview of the three classes of catalysts will be presented. The effects of preparation method, metal loading and particle size, nature of transition metal oxides will be discussed on the catalytic efficiency. Some representative examples of the latest progress in the development of new catalysts will be given as well.

11.2 Photocatalysts

Since the first report of Fujishima et al. on the UV-light induced redox chemistry over titanium dioxide (TiO_2), photocatalysis has received remarkable interest as a green and sustainable technology for energy and environmental issues [7–10]. The photocatalytic oxidation process combines oxygen, water, and VOCs on the catalyst surface to produce CO_2 and H_2O under UV irradiation (Fig. 11.1). Activation of TiO_2 is achieved through the absorption of a photon (hv) from UV irradiation source. This results in the promotion of an electron (e^-) from the valence band to the conduction band, with the generation of highly reactive positive holes (h^+) in the valence band. This can cause aggressive oxidation of the surface-adsorbed VOC molecules and converts them into CO_2 and water. In the photocatalytic process of formaldehyde, formic acid was found on the TiO_2 surface as the partial oxidation byproduct [11].

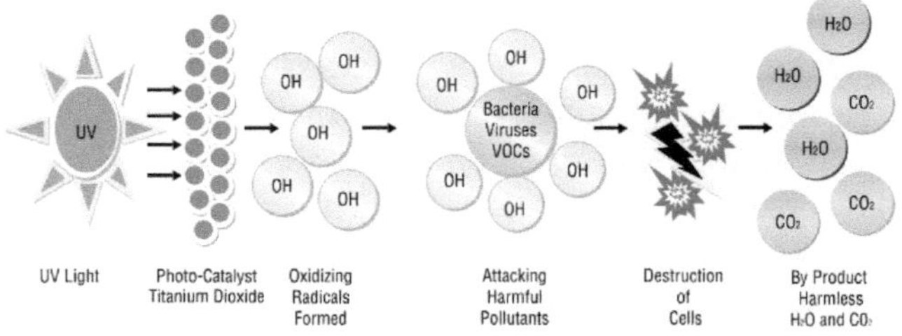

Figure 11.1 Photcatalytic oxidation process on TiO_2 under UV irradiation [12]. Copyright © 2013 Elsevier Ltd.

It is known that there are many factors that can exert significant influence on photocatalytic performance, including the size, specific surface area, pore structure, crystalline phase, and exposed surface facets of photocatalysts. Structural dimensionality is also a factor that has a significant impact on the properties of catalyst materials and eventually affects the photocatalytic performance (Fig. 11.2) [13].

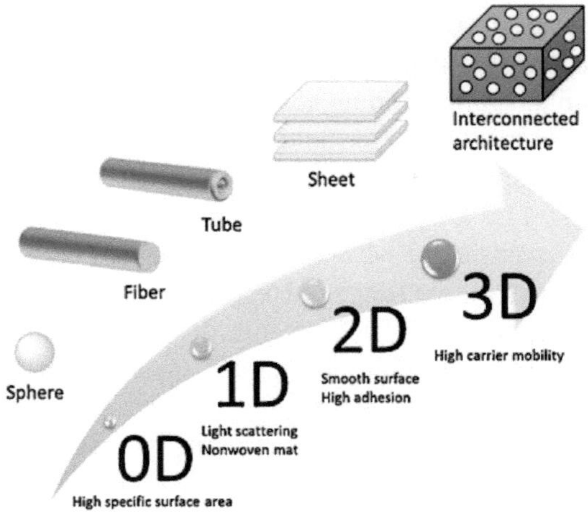

Figure 11.2 Schematic illustration of structural dimensionality of materials with expected properties [13]. Copyright © 2012 Elsevier B.V.

For example, nanoparticles with zero dimensionality have high specific surface area, while one-dimensional fibers or tubes possess the advantage of light-scattering properties. Designing photocatalysts with appropriate properties enables us to gain unique performance of the photocatalysts. Thus, performance improvements by adjusting these factors remain the focus of photocatalysis research.

To date, many experimental and theoretical studies have been carried out to study the photochemistry of formaldehyde on photocatalysts, and most photocatalysis studies with valuable results were based on TiO_2 [14–17]. TiO_2, a nontoxic, inexpensive, and highly stable white powder, has been considered the most promising photocatalyst for the formaldehyde degradation. For example, Liu et al. [18] studied the kinetic process of photocatalytic oxidation of gaseous formaldehyde over TiO_2. It is reported that the photocatalytic rates of formaldehyde can be described by the simplified Langmuir-Hinshelwood model. The apparent first-order reaction coefficient is lower and the half-life of photocatalysis is longer for low concentration than for high concentration.

Though TiO_2 has been extensively studied and widely used, the application of pure TiO_2 is limited because it requires UV activation. To expand the optical response of TiO_2 from the UV to the visible light region, considerable efforts have been directed, including doping with metals or nonmetals, or compounding with transition metal oxides. The doped metal atoms can suppress the recombination of photo-induced electron-hole pairs, and consequently increase the quantum efficiency. In addition, some doped transition metal oxides have a low band gap and could absorb more visible light for the photocatalytic degradation of formaldehyde.

The UV-vis spectra of TiO_2 doped with varied content of ZrO_2 are shown in Fig. 11.3 for example [14]. Due to the new energy level or lattice defects observed with Zr doping, the TiO_2 catalysts with Zr doping demonstrate a stronger adsorption in the UV light and visible light ranges, than pure TiO_2. The results of formaldehyde degradation by different $Zr_xTi_{1-x}O_2$ samples under UV-vis light indicate that the $Zr_xTi_{1-x}O_2$ samples show visible light-induced degradation of formaldehyde, and the catalytic activities of these samples are all superior to the P25 and TiO_2 samples. The formaldehyde removal with only 45 W UV-vis light is 92.4%. Among these Zr doped samples, $Zr_{0.08}Ti_{0.92}O_2$ shows the best activity, and the concentration of formaldehyde can be reduced to below 0.08 mg/m^3 (WHO guideline) in 48 h.

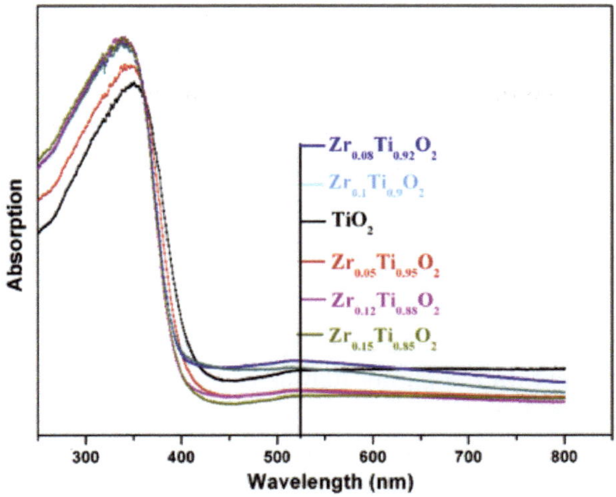

Figure 11.3 Diffuse reflectance spectra of $Zr_xTi_{1-x}O_2$ samples prepared by the sol–gel method with varied Zr contents [14]. Copyright © 2012 Elsevier B.V.

Noble metals (Pt, Ag) and rare earth oxides (e.g., Ce) are also used to improve the photocatalytic activity of TiO_2 for the formaldehyde oxidation. Figure 11.4 illustrates the photocatalytic performance of silver and cerium doped over TiO_2 for the formaldehyde degradation [19]. It is noted that the doped silver or cerium ion can enhance the photoactivity of TiO_2, and cerium performs better than silver.

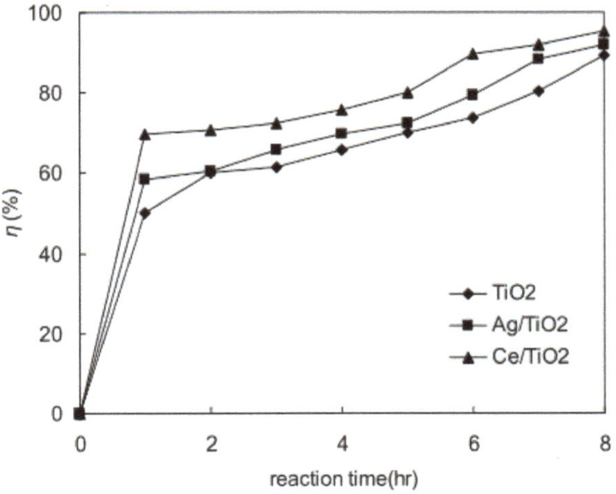

Figure 11.4 Effects of Ag or Ce-doped TiO_2 on decomposition of formaldehyde [19]. Copyright © 2011 Elsevier Ltd.

Recently, a new type of nitrogen-doped TiO_2 nanostructure with binary and/or ternary polymorphs has been achieved by a hydrazine-mediated solvothermal approach [20]. These prepared materials not only modify the band structure of TiO_2 to make it more responsive to visible light but also suppress charge recombination and render TiO_2 to have an enhanced photocatalytic activity. The diffuse absorbance spectra of the samples (Fig. 11.5) show that the absorption in the range of 400–550 nm for the nitrogen -doped samples is higher in intensity than that for the undoped TiO_2 sample. The band gap decreases with nitrogen doping and with the gradual increase in rutile content. In the photocatalysis experiments of formaldehyde, the nitrogen-doped TiO_2 sample with anatase, brookite, and rutile structures exhibits enhanced photocatalytic activity, and its catalytic efficiency could be 3.14 times that of P25 under visible light irradiation.

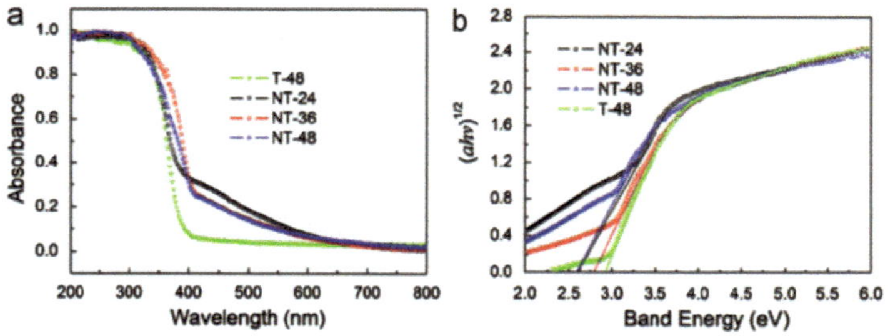

Figure 11.5 (a) Diffuse reflection spectra of the samples, and (b) plots of $(\alpha h\nu)^{1/2}$ versus $h\nu$. [20]. Copyright © 2013 Elsevier Inc.

11.3 Transition Metal Oxide Catalysts

For indoor air cleaning, low energy consumption and low formaldehyde concentration strongly call for a catalyst that would exhibit the complete oxidation of formaldehyde, preferably at a relatively low temperature or even at room temperature. Conversion of formaldehyde into H_2O and CO_2 using heterogeneous catalysts has proven to be an efficient and practical technology for controlling formaldehyde emission and indoor air purification. Unlike TiO_2, the great benefit of these catalysts is that they do not require any extraneous equipment to function (e.g., UV lights).

Great efforts have recently been made to develop efficient catalysts for complete oxidation of formaldehyde. Transition metal oxides and their combinations, such as Co_3O_4, MnO_x, SnO_2, and CeO_2, were found to show high performance in formaldehyde oxidation reactions [21–25].

11.3.1 Ceria Oxide

Attributing to the well-known oxygen storage capacity of ceria and the strong interactions between metal and ceria support, ceria-based catalysts generally show high performance in formaldehyde oxidation. Zhang et al. [26] synthesized a three-dimensionally ordered macroporous (3DOM) CeO_2 catalysts, and investigated their catalytic performance for formaldehyde oxidation. It is reported that a 100% formaldehyde conversion at temperatures around 355°C and 370°C could be achieved over 3DOM CeO_2 with 80 and 130 nm pore sizes, respectively. Ma et al. [27] reported the catalytic oxidation of formaldehyde using Co_3O_4, Co_3O_4-CeO_2, and other supported catalysts. It is found that the catalytic performance is markedly influenced by the morphology and composition of the catalysts. For instance, the Co_3O_4 nanosheet catalyst prepared by the precipitation method oxidizes 2.5% of the initial formaldehyde to CO_2 at 25°C, whereas the mesoporous Co_3O_4 catalyst prepared by the "nanocasting" method exhibits nearly 10-fold increase in CO_2 yield (20.3%). However, the CeO_2 catalyst prepared by precipitation method shows no formaldehyde absolute oxidation activity at 25°C. The CO_2 yield over mesoporous Co_3O_4–CeO_2 (7:3) is 13.2% higher than Co_3O_4 or CeO_2 catalysts and similar to the nanocast Co_3O_4 catalyst.

11.3.2 Manganese Oxide

11.3.2.1 Brief introduction to manganese oxide

Manganese oxide (MnO_x), a common mineral abundant in nature, is an efficient catalyst for the removal of formaldehyde and other VOCs at low temperatures [28–31]. Studies have already indicated that manganese oxides are more promising for formaldehyde emission control than other transition metal oxides and show catalytic activity as high as or slightly higher than supported noble metals [24, 32].

Manganese oxides occur in many forms, and are roughly classified into one-dimensional tunnel structures, two-dimensional layer phases, and three-dimensional spinels [33,34]. Among the various manganese oxides studied, microporous manganese oxides with tunnel structure continue to attract considerable scientific interest as inexpensive and nontoxic materials. The combination of porous structure, mixed valence framework, and high mobility of oxygen results in excellent performance, with potential use in heterogeneous catalysis, hazardous waste remediation, and rechargeable battery technology [35–39]. This large class of compounds with MnO_2 or near-MnO_2 stoichiometry displays a variety of structures based on edge-shared MnO_6 octahedra. Birnessite and cryptomelane are the most common crystalline manganese oxides, and play an important role in catalytic reactions. Their microporous structures are shown in Fig. 11.6.

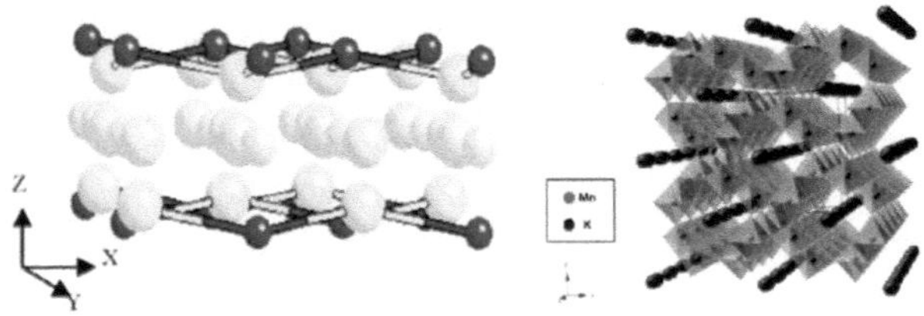

Figure 11.6 Structures of microporous manganese oxides: layered birnessite (left, the dark and light blue balls represent Mn and O atoms and the green and yellow balls represent H_2O, and Na^+ atoms) [40]. Copyright © 2004 Elsevier Inc., and tunneled cryptomelane [right, this image was generated using CrystalMaker®: CrystalMaker Software Ltd, Oxford, England (www.crystalmaker.com)].

Birnessite is a common name for layered manganese oxides, formed by two-dimensional layers of edge-sharing MnO_6 octahedra, with metal cations (usually alkaline or alkaline-earth cations such as Na^+, K^+, and Ca^{2+}) in the interlayer, together with water molecules. These cations balance the negative electric charge of the layers. The structural formula can be written as $A_xMnO_{2\pm y}\cdot z(H_2O)$, where A stands for the interlayer cation. The oxidation state of manganese ranges between 3.6 and 3.8. The common interlayer spacing is close to 7 Å for samples containing interlayer alkaline or alkaline-earth cations, but upon hydration it can increase up to 10 Å [41, 42]. Cryptomelane is characterized by a 2 × 2 tunnel structure (4.6 Å side). Its name is currently replaced by α-MnO_2 or OMS-2 (octahedral molecular sieves). The tunnel has a chemical composition of $KMn_8O_{16}\cdot nH_2O$ with the K^+ cation residing in the center of the tunnel [43]. The average manganese oxidation state is 3.8 in the presence of a mixture of Mn^{4+}, Mn^{3+}, and Mn^{2+} ions [44], and the Lewis acidity is 0.7 [45]. Birnessite and cryptomelane have many attractive properties such as abundant porous and tunnel structure, ion-exchange ability, facile evolution of lattice oxygen, and reasonable thermal stability [46,47]. These properties make them promising materials for the total oxidation of formaldehyde and other VOCs.

11.3.2.2 Effects of nanostructure, surface area, and morphology

Chen et al. [24] prepared three manganese oxides with different square tunnel sizes, including pyrolusite of 0.23 nm × 0.23 nm, cryptomelane of 0.46 nm × 0.46 nm, and todorokite of 0.69 nm × 0.69 nm. The cryptomelane shows extremely high catalytic activity in the reaction of formaldehyde oxidation, and achieves 100% formaldehyde conversion into H_2O and CO_2 at 140°C (Fig. 11.7). Under identical conditions, however, pyrolusite and todorokite only give a formaldehyde conversion of ca. 20% and 40%, respectively. The temperatures required for complete oxidation of formaldehyde are 180°C and 160°C. The effective tunnel diameter of cryptomelane is similar to the dynamic diameter of the formaldehyde molecule, which is proposed to be a major factor that gave rise to its higher catalytic activity than the other two manganese oxides.

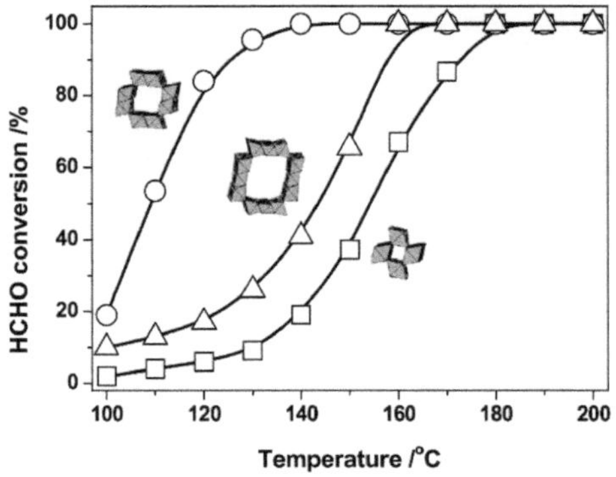

Figure 11.7 Tunnel effect of the 1×1, 2×2, and 3×3 tunnel manganese oxide catalysts in complete oxidation of formaldehyde. Reaction conditions: catalyst = 0.2 g, HCHO = 400 ppm, O_2 = 10.0 vol%, and N_2 balance, gas rate = 100 ml/min [24]. Copyright © 2009 Elsevier Inc.

By a simple soft-chemistry route, we prepared honeycomb and hollow manganese oxide (K_xMnO_2) nanospheres using oleic acid and $KMnO_4$ as starting materials [48]. Figure 11.8 shows typical SEM and TEM images of the obtained products. The honeycomb and hollow K_xMnO_2 nanospheres exhibit excellent robustness and high catalytic activities for oxidative decomposition of formaldehyde at low temperatures, even higher than those of previously reported OMS-2 nanorods, MnO_x powders, and Mn-Pd/Al_2O_3 catalysts. Complete conversion of formaldehyde to CO_2 and H_2O can be achieved at very low temperature without harmful byproducts detected in effluent gases. The catalytic activity of manganese oxide hollow nanospheres is much higher than that of honeycomb nanospheres.

Figure 11.8 SEM (a,c) and TEM (b,d) images of honeycomb (a,b) and hollow (c,d) K_xMnO_2 nanospheres [48]. Copyright © 2007, American Chemical Society.

We further developed the research of hollow manganese oxide nanostructures by a hydrothermal treatment, and obtained a series of flower-like manganese oxide nanospheres, as shown in Fig. 11.9

Figure 11.9 SEM images of MnO_2 nanostructures obtained by hydrothermal reaction at 120°C for varied periods of time: 6 h (a), 12 h (b), 18 h (c), and 24 h (d) [49]. Copyright © 2011 Elsevier Ltd.

[49]. It is found that control over the morphology and structure of product can be achieved by tuning reaction conditions, such as hydrothermal temperature, reaction time, pH value, and kinds of anion. The increase of reaction temperature leads to the formation of MnOOH nanorods, and cryptomelane and ramsdellite MnO_2 nanorods can be obtained by adding sulfuric acid and nitric acid, respectively, in the reaction system. The results of catalytic experiments demonstrate that the manganese oxide nanostructures have high catalytic activities for oxidative decomposition of formaldehyde. The crystallographic structure of the products has great influence on the catalytic performance. Among the MnO_2 products, cryptomelane-type MnO_2 shows the highest catalytic activity. Around 75.9% of

formaldehyde can be oxidized into CO_2 at 100°C over cryptomelane MnO_2 nanorods, while the birnessite MnO_2, ramsdellite MnO_2, and monoclinic MnOOH achieve a formaldehyde conversion of about 61.2%, 52.1%, and 34.6%, respectively.

An earlier study on VOCs oxidation revealed that high surface area is much crucial for high catalytic activity of catalyst. Recently, our group [50, 51] reported a novel microemulsion method for synthesis of birnessite manganese oxides. The surface area of the obtained crystallized birnessite can reach 154 m^2/g by tailoring the synthesis conditions, higher than those reported for crystallized K-type birnessite to date. The results of formaldehyde oxidation demonstrate that BSW-80 with lower surface area has a lower activity. BSW-100 and BSW-120 have comparable surface areas, but BSW-120 is the most active; 100% of 460 ppm formaldehyde can be oxidized into CO_2 and H_2O at 100°C. The reason might be that compared with BSW-100, BSW-120 has smaller pore size and consists of some nanospheres, which can prolong the stay time of formaldehyde molecules in the catalyst and enhance formaldehyde oxidation. The similar results were also reported by our previous work [48]. It is noted that hollow K_xMnO_2 nanospheres exhibit larger catalytic activity than honeycomb K_xMnO_2 nanospheres because the former can adsorb and retain formaldehyde for a longer period of time than the latter, and eventually enhance the formaldehyde oxidation. All of these results indicate that the morphology, pore size, and surface area of catalysts are among the major factors governing their catalytic activity.

It is known that the preparation method may affect the morphology of catalysts and eventually affect the catalyst properties. Methods reported for preparing K-OMS-2 (cryptomelane) include reflux, solid-state chemical reactions, sol–gel, and hydrothermal treatment to produce nanorods, nanoneedles, and paper-like materials. An acidic condition is usually needed to form K-OMS-2 materials. Recently, K-OMS-2 nanoparticles were prepared in an initially neutral media with cetyltrimethylammonium bromide (CTAB) and benzyl alcohol [52]. The obtained products are shown in Fig. 11.10.

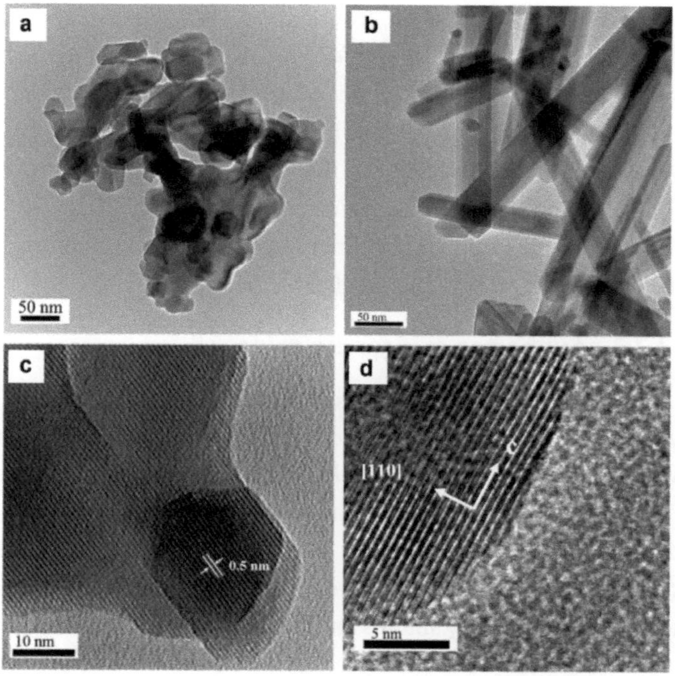

Figure 11.10 TEM and HRTEM images of K-OMS-2 prepared at 25°C (a and c) and 100°C (b and d), respectively. Panel c shows a two-dimensional Fourier-filtered lattice image [52]. Copyright © 2010 Elsevier Inc.

K-OMS-2 crystallization usually proceeds through two steps. First, a disordered or layered manganese oxide precursor is formed, which subsequently transforms into fibrous K-OMS-2 via a dissolution–recrystallization process. Then, the dissolution/collapse of the precursor in acidic solution results in the formation of primary MnO_2 crystallites. During this process, several experimental parameters, such as temperature, pH value, and reactant concentrations, can influence the crystal structure and morphology of the final product. As shown in Fig. 11.10, the c-direction is that parallel to the long side of nanoparticles or nanorods, and is unambiguously the growth direction of the K-OMS-2 material. This might be the growth mechanism of K-OMS-2 nanoparticles into nanorods in aqueous solution. With the templating effect of CTAB, the primary manganese oxide crystallites directly transform into nanoparticle-shaped K-OMS-2 at the reaction temperature of 25°C. At a higher reaction temperature such as 100°C, a longitudinal growth proceeds, and nanorod-shaped K-OMS-2 forms. The reaction of formaldehyde oxidation reveals that the morphology greatly influences the catalytic performance. Over the K-OMS-2 nanorods, there is little CO_2 produced at temperatures lower than 80°C. In contrast, the K-OMS-2 nanoparticles show much higher catalytic activity for formaldehyde oxidation. About 12% CO_2 yield is observed at 20°C. After increasing the reaction temperature to 100°C, the CO_2 yield can rise sharply to 64%.

11.3.2.3 Supported manganese oxide

Small particle sizes of catalysts including manganese oxides are advantageous for catalysis reactions. However, it is known that these powdered or granular catalysts have many engineering limitations for their field applications, such as giving rise to dust contamination and difficulty in the separation process from the treated exhaust gas. Furthermore, these powdered or granular catalysts cannot be readily formed into any shapes and cannot be directly used in indoor equipment such as air-conditioners or air-cleaners. Hence, supported catalysts are favorable because of their forming properties.

Recently, manganese oxide nanosheets supported on cellulose fibers have been synthesized successfully in our group [53]. Porous cellulose fibers that are composed of microfibrils of 10–30 nm width were used as template and support. Manganese oxide catalysts were grown in situ onto cellulose fibers by an impregnating method. Figure 11.11 displays the TEM images of the obtained composite. It can be seen that manganese oxide nanosheets are arranged roughly perpendicular to the surface of the cellulose fiber. This MnO_2/cellulose composite shows excellent catalytic activity. The formaldehyde conversion over the MnO_2/cellulose composite reaches 16.7% at 60°C. This value is about 9–17 times as high as that of unsupported MnO_2 powder when the formaldehyde conversion per milligram of MnO_2 is calculated. These results demonstrate that the introduction of cellulose fibers not only facilitates the applicability of manganese oxide catalysts but also increases the catalyst activity of MnO_2/cellulose.

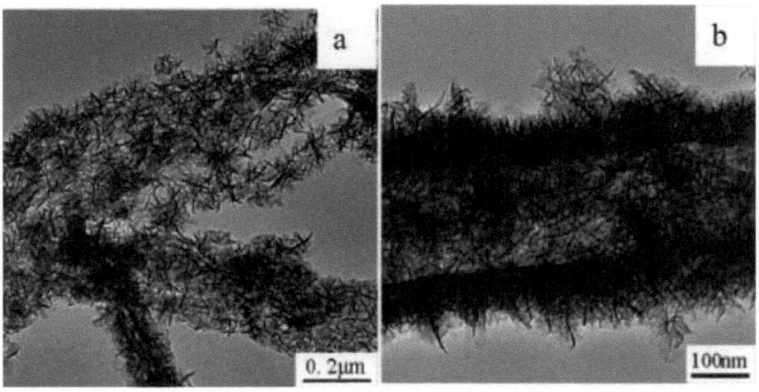

Figure 11.11 TEM (a) and magnified TEM (b) images of MnO_2 nanosheets coated cellulose fibers prepared from aqueous $KMnO_4$ of 12.6 mM [53]. Copyright © 2011, American Chemical Society.

Wang et al. [54] chose an alumite plate as support and synthesized Pt/TiO_2 alumite catalyst. They firstly deposited TiO_2 onto an anodic alumite plate with electro-deposition technology, and then supported platinum on the alumite plate using a wet impregnation method. Because aluminum can be formed into any shapes, this developed catalyst can be formed into various shapes such as mesh, plate, fin, serrate, and so on, as shown in Fig. 11.12. Moreover, the Pt/TiO_2 alumite catalyst shows fine catalytic activity toward the formaldehyde oxidation and can give 94% of formaldehyde conversion at 30°C.

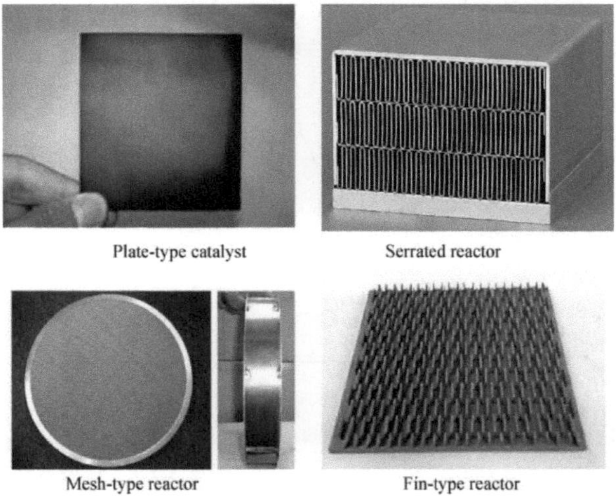

Figure 11.12 Structured catalysts developed using anodic alumite as support [54]. Copyright © 2009 Elsevier B.V.

11.4 Supported Noble Metal Catalysts

11.4.1 Noble Metal Catalysts

As catalytic oxidation at even lower temperature is desirable for indoor air purification, the development of a catalyst for total formaldehyde oxidation at room temperature is of great interest. The noble metal based catalysts, in spite of their high costs, are the preferred ones because of their high specific activity, resistance to deactivation, and ability of regeneration. On account of their low temperature activity in formaldehyde oxidation, oxide-supported noble metallic (Pt, Pd, Au, Ag) catalysts have been widely studied [55–59].

Zhang et al. [60] studied the catalytic performance of TiO_2-supported noble metal (Au, Pd, Rh, and Pt) catalysts in the formaldehyde oxidation. Figure 11.13 shows the formaldehyde conversions over these catalysts. The order of activity is Pt/TiO_2 >> Rh/TiO_2 > Pd/TiO_2 > Au/TiO_2 >> TiO_2. The Pt/TiO_2 shows the highest activity, and formaldehyde can be completely oxidized into CO_2 and H_2O over Pt/TiO_2 even at room temperature. In contrast, the other catalysts are much less effective for formaldehyde oxidation under identical reaction conditions. Formaldehyde conversion to CO_2 is only 20% over the Rh/TiO_2 at 20°C. The Pd/TiO_2 and Au/TiO_2 show no activity for formaldehyde oxidation at 20°C.

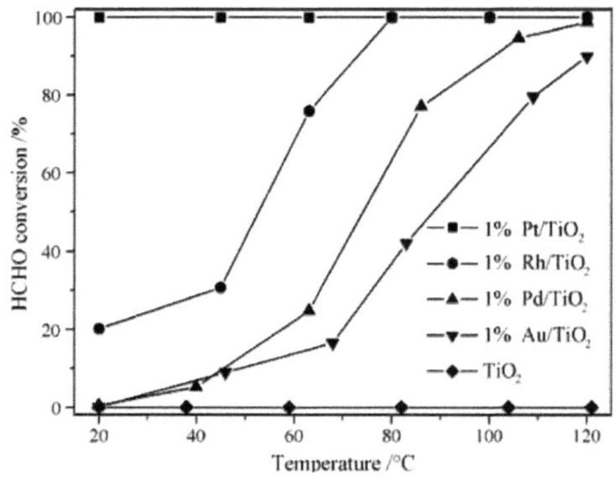

Figure 11.13 Formaldehyde conversions over TiO_2 (♦); Pt/TiO_2 (■); Rh/TiO_2 (●); Pd/TiO_2 (▲); Au/TiO_2 (▼) catalysts at various temperatures. Reaction conditions: Formaldehyde 100 ppm, O_2 20 vol%, He balance, total flow rate: 50 cm^3/min, GHSVs: 50000 h^{-1} [60]. Copyright © 2007 Elsevier B.V.

Recently, Zhang et al. [61] developed Pt/TiO_2 catalysts with alkali metal ions (such as Li^+, Na^+, and K^+). It is found that the addition of alkali-metal ions to Pt/TiO_2 catalyst stabilizes an atomically dispersed Pt-O(OH)x–alkali-metal species on the catalyst surface, activates H_2O, and catalyzes the reaction between surface hydroxyl groups and formate species to total oxidation products, significantly promoting the catalytic activity for the formaldehyde oxidation. As shown in Fig. 11.14, the pure Pt/TiO_2 catalyst has a very low activity with formaldehyde conversion being only about 19% at 15°C. With 1 wt% Na addition, the formaldehyde conversion reaches 96% at 15°C and 100% at 40°C. With 2 wt% Na addition, 100% formaldehyde conversion to CO_2 and H_2O can be obtained at 15°C. Moreover, these Na-promoted samples have excellent stability, and 80% of formaldehyde conversion can be maintained over a 72 h long test.

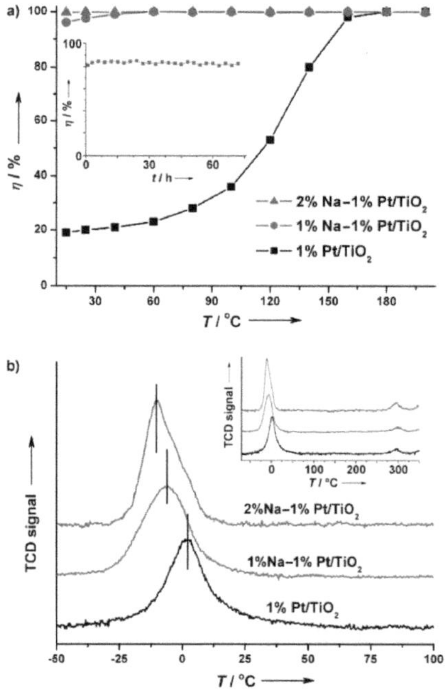

Figure 11.14 (a) Formaldehyde conversion η over x wt% Na-1 wt% Pt/TiO_2 (x = 0, 1, and 2) catalysts as a function of temperature T. Reaction conditions: Formaldehyde 600 ppm, O_2 20 vol%, relative humidity: about 50%, He balance, total flow rate of 50 cm^3/min, and GHSV 120000 h^{-1} (inset: stability test of 2 wt% Na-1% Pt/TiO_2 at 25 °C, GHSV 300000 h^{-1}). (b) H_2 TPR profiles of x wt% Na-1 % Pt/TiO_2 (x = 0, 1, and 2) [61]. Copyright © 2012 WILEY-VCH Verlag GmbH %26 Co. KGaA, Weinheim.

Formaldehyde oxidation is often easily enhanced by the presence of noble metal species at low temperature, and the noble metal loading greatly affects the catalytic activity. Figure 11.15 [62] shows the effect of Pt loading on the activity of Pt/TiO_2 for formaldehyde oxidation at 20°C. It can be seen that 1 % Pt/TiO_2 is the most active, giving 100% of formaldehyde conversion at 20°C [62, 63].

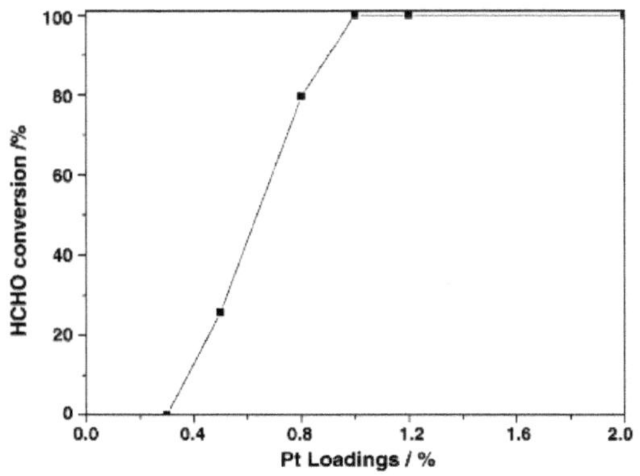

Figure 11.15 Formaldehyde conversions on Pt/TiO_2 catalysts with various Pt loadings at 20°C. Reaction conditions: Formaldehyde 100 ppm, O_2 20 vol%, He balance, total flow rate 50 cm^3/min, GHSV 50000 h^{-1} [62]. Copyright © 2006 Elsevier B.V.

In addition, reduction treatments of noble metals greatly affect catalytic activity, as active species and support oxides can strongly interact with each other. Recently, our group prepared birnessite-supported Pt catalysts and investigated the effect of Pt reduction methods on the catalytic activity for the complete oxidation of formaldehyde [64]. We found that birnessite-supported Pt reduced by KBH_4 has higher specific surface area, pore volume, and better dispersion of Pt nanoparticles and exhibits much higher catalytic activity than that reduced by H_2 -plasma. Over this catalyst, 460 ppm formaldehyde could be rapidly and completely converted into CO_2 and H_2O at 50°C.

11.4.2 Effect of support nature

For supported noble metal catalysts, the nature of supports is one of the key factors affecting their catalytic performance, although the concreted role of supports is still under debate. So far, a few oxides, such as TiO_2, CeO_2, Al_2O_3, Fe_2O_3, SiO_2, have been reported to be suitable supports to prepare highly active noble metal based catalysts for low-temperature oxidation of formaldehyde [65–70]. Tang et al. [67] found that complete oxidization of formaldehyde into harmless CO_2 and H_2O was achieved over MnO_x-CeO_2 catalysts at 423 K, and the introduction of silver could decrease this temperature to 373 K.

In addition to the support composition, the catalytic performance of noble metal-supported samples is also strongly influenced by the morphology, porosity, and acid-base properties of the supports. Liu et al. [71] prepared three-dimensionally ordered macroporous (3DOM) Au/CeO_2-Co_3O_4 catalysts via a precursor thermal decomposition-assisted colloidal crystal templating method. As shown in Fig. 11.16, the 3DOM Au/CeO_2-Co_3O_4 catalysts possess well-defined 3DOM structures with adjustable pore sizes. The 3DOM Au/CeO_2-Co_3O_4 catalysts display superior catalytic activity for formaldehyde oxidation with a 100% conversion at temperatures as low as ~39°C. By comparing the catalytic activity of 3DOM Au/CeO_2-Co_3O_4 catalysts with 80, 130, and 240 nm pore sizes, the catalyst with 80 nm pores has the highest catalytic activity, which may be due to its high surface area, small pore size, and more uniform distribution of small Au nanoparticles.

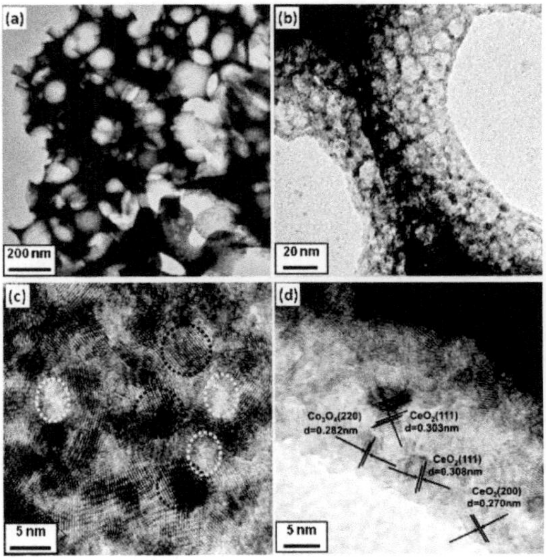

Figure 11.16 TEM (a and b) and HRTEM (c and d) images of 3DOM ~3 wt% Au/(2.5:1)CeO_2 -Co_3O_4 catalyst [71]. Copyright © 2012 Elsevier B.V.

Using ascorbic acid, poly(vinyl pyrrolidone) (PVP), sodium citrate as reducing agent, dispersant, and protective agent, Pt nanoparticles were successfully deposited on three different MnO_2 nanostructures (cocoon-, urchin-, and nest-like MnO_2), as shown in Fig. 11.17 [72]. It is noted that the activities of the Pt/MnO_2 catalysts are significantly higher than those of the corresponding MnO_2

Figure 11.17 SEM (a,c,e) and TEM (b,d,f) images of three MnO_2 nanostructures: (a,b) cocoon-like MnO_2, (c,d) urchin-like MnO_2, (e,f) nest-like MnO_2 ; insets in d and f are corresponding HRTEM images [72]. Copyright © 2012, American Chemical Society.

nanostructures, and the most effective catalyst is 2 wt% Pt/nest-like MnO_2. The catalytic performance at 20°C and the temperature for complete oxidation of formaldehyde over catalysts are listed in Table 11.1. The morphologies and crystal forms of MnO_2 nanostructures have significant influence on the catalytic performance. The catalytic activity nearly follows the order: 2 wt% Pt/nest-like MnO_2 > 2 wt% Pt/urchin-like MnO_2 > 2 wt% Pt/cocoon-like MnO_2.

Table 11.1 Catalytic performance of Pt/MnO_2 catalysts at 20°C and temperature for complete conversion of formaldehyde [72]. Copyright © 2012, American Chemical Society

MnO_2 nanostructures	Pt loading (wt%)	Conversion at 20°C[a]	$T_{100\%}$[b]
cocoon-like MnO_2	0	0	>200°C
	2	24.3%	90°C
urchin-like MnO_2	0	0	>200°C
	2	30.9%	80°
nest-like MnO_2	0	0	200°C
	2	41.6%	70°C

[a]Conversion of formaldehyde at 20°C. [b]Temperature for complete decomposition of formaldehyde.

It is possible to modify the monometallic catalyst by adding a second noble metal to improve the activity toward VOCs oxidation. The synergistic effect between bimetal nanoparticles and supports, as well as the synergistic effect between noble metals can enhance the catalytic performance of catalysts. In recent years, supported bimetallic catalysts, such as Pt/Pd, Au/Pt, and Pd/Au, were investigated in heterogeneous catalysis, which showed super catalytic activities [74–77]. For instance, Au-Pt bimetallic catalysts supported on nest-like MnO_2 exhibit high activity in formaldehyde combustion at low temperature [77]. The effects of noble metal and Au/Pt molar ratio on the catalytic activity are displayed in Fig. 11.18. It can be seen that $Au_{0.5}Pt_{0.5}/MnO_2$ has the highest catalytic activity among all the as-prepared metal nanoparticles supported MnO_2 catalysts, and the temperature for complete oxidation of formaldehyde reaches as low as 313 K.

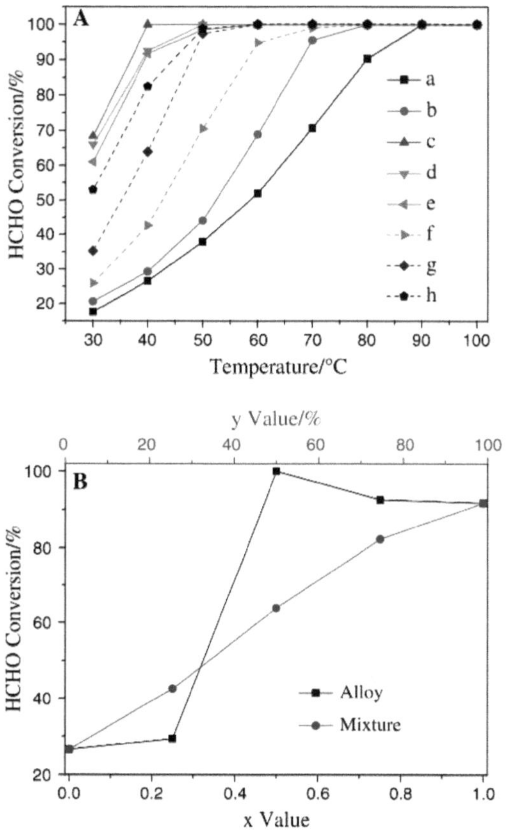

Figure 11.18 Catalytic performance of noble metal NPs-supported MnO_2 catalysts: A (a) Au NPs/MnO_2, (b) $Au_{0.75}Pt_{0.25}/MnO_2$, (c) $Au_{0.5}Pt_{0.5}/MnO_2$, (d) $Au_{0.25}Pt_{0.75}/MnO_2$, (e) Pt/MnO_2, (f) (75 %Au/25 %Pt)/MnO_2, (g) (50 %Au/50 %Pt)/ MnO_2, and (h) (25 %Au/75 %Pt)/ MnO_2. B Dependence of the catalytic performance of noble metal NPs-supported MnO_2 catalysts at 313 K on varied x and y value, x is the content of Au in the $Au_{1-x}Pt_x/MnO_2$ catalysts, y is the content of Au in the [(100 %-y)Au/yPt]/ MnO_2 catalysts [77]. Copyright © 2012, Springer Science+Business Media Dordrecht.

11.4.3 Effect of Operating Parameters of Formaldehyde Oxidation

For heterogeneous catalytic reactions, it is known that reactant concentrations, space velocitiy, and the amount of catalyst greatly influence the performance of catalyst and reactant conversion. This is why Pt/TiO_2 catalyst can give 100% of formaldehyde conversion at room temperature, but the formaldehyde conversion just reaches only about 20% using the same Pt/TiO_2 as catalyst, as seen in Figs. 11.13 and 11.14.

Tang et al. [78] investigated the effect of formaldehyde concentration over a Pt/MnO_x-CeO_2 catalyst and found that the performance of formaldehyde oxidation decreases rapidly with increasing formaldehyde concentration. Around 100% of formaldehyde conversion can be achieved when formaldehyde concentration is below 100 ppm, but decreases to 54% when formaldehyde concentration increases to 580 ppm. However, this linear relation of formaldehyde concentration and its conversion is not unaltered, and may be changed with a different reaction system. Peng et al. [63] measured and analyzed the reaction rate variation with the molar ratio of oxygen to formaldehyde (O_2/HCHO). As shown in Fig. 11.19, when the molar ratio of O_2/HCHO is below 3, the reaction rate increases with O_2/HCHO, but when the ratio is above 3, the reaction rate levels off. This means that with relatively high formaldehyde concentration, the catalytic oxidation of formaldehyde is strongly dependent on formaldehyde concentration, but if much low, the formaldehyde concentration will have less influence on the reaction rate of formaldehyde oxidation.

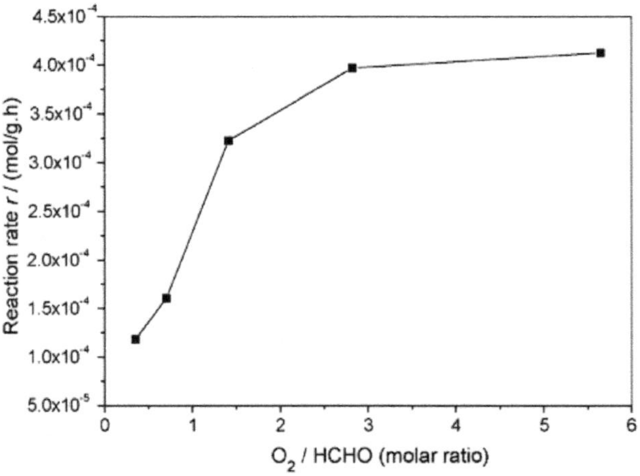

Figure 11.19 The reaction rate varied with the molar ratio O_2/HCHO, reaction conditions: HCHO 100 mg/m^3, O_2 22 vol.%, N_2 balance, total flow rate 1200 cm^3 and GHSV 1,500,000 cm^3/g h [63]. Copyright © 2006 Elsevier B.V.

Huang et al. [79] investigated the effect of formaldehyde concentration on formaldehyde conversion at room temperature in a range of 5–30 ppm, which is relatively lower than those experimental concentrations reported previously. As shown in Fig. 11.20, formaldehyde conversion decreases with increasing initial formaldehyde concentration. However, it is noted that the change of formaldehyde removal is generally small. Formaldehyde conversion is 100% at 5 ppm, 99.1% at 10 ppm, and 95.8% at 30 ppm.

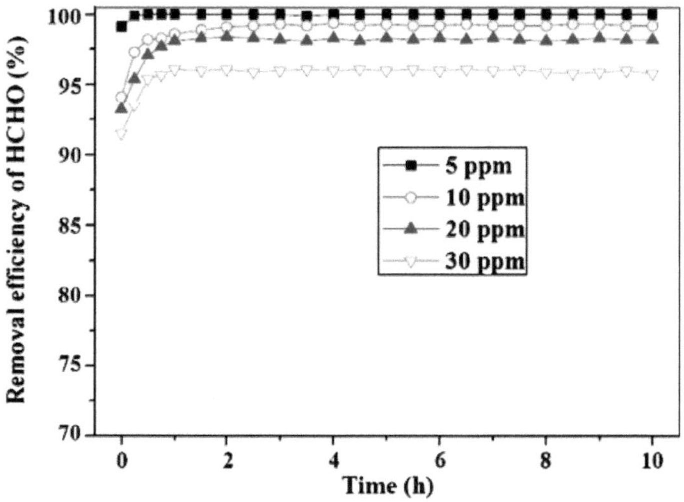

Figure 11.20 Effect of initial formaldehyde concentration on formaldehyde conversion over the 0.1% Pt–TiO_2 catalyst (Relative humidity = 50 %, GHSV = 80000 h^{-1}) [79]. Copyright © 2011 Elsevier Inc.

Huang and co-workers also studied other influencing factors of the catalytic reaction of formaldehyde, including relative humidity and air hourly space velocity (GHSV). They found that under different relative humidity, the formaldehyde conversion follows the order: 25 % ≈ 50 % > 75 % > 97.5 % >> 0 %. For Pt-TiO_2 used as a catalyst, the presence of moisture does not inhibit but enhances formaldehyde removal at room temperature. Nearly 100% formaldehyde conversion can be achieved at 25% and 50% humidity while it is only 45.2% with 0% of humidity. The results of formaldehyde oxidation with different GHSV (40000–240000 h^{-1}) indicated that formaldehyde conversion decreases with increasing GHSV value, due to less contact time during catalytic reaction.

11.4.4 Reaction Mechanism

The validity of each mechanism strongly depends on the catalyst (noble metal and nature of the support) as well as on the character of the formaldehyde molecule, and it is much difficult to generalize [76]. Many researchers directly investigate the reaction mechanism through kinetic experiments, or specific techniques (DRIFT (diffuse reflectance infrared Fourier transform, GC, and so on). Tang et al. [67] proposed a consecutive oxygen transfer mechanism starting from the oxygen reservoir of CeO_2 to active Ag_2O sites through MnO_x, as shown below.

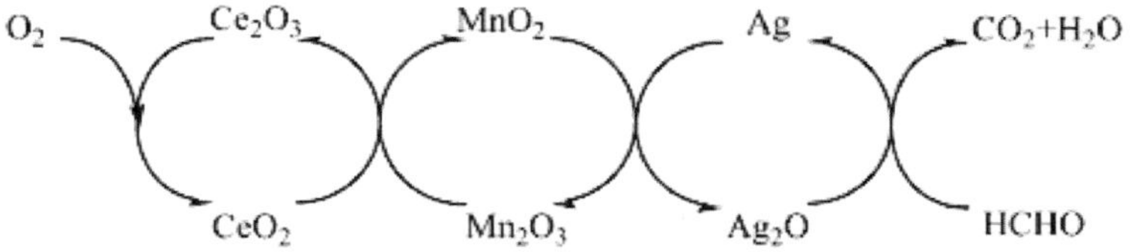

The oxygen species released from the decomposition of Ag_2O participate in the oxidation of formaldehyde, and the reoxidation of Ag to Ag_2O is achieved through the oxygen species from MnO_2. Meanwhile, the transformation of the produced Mn_2O_3 to MnO_2 is realized by the oxygen species from the oxygen reservoir of CeO_2, and the Ce_2O_3 formed can be reoxidized into CeO_2 by the oxygen in the feed stream.

Ma et al. [27] carried out a DRIFT study suggesting a reaction mechanism of formaldehyde on meseporous Co_3O_4, Au/ Co_3O_4, and Au/ Co_3O_4-CeO_2 catalysts. They proposed that gold nanoparticles of less than 5 nm diameter are embedded or partly entered within the Co_3O_4 or CeO_2 pore walls. The Co_3O_4 (110) facets composed mainly of Co^{3+} cations are the main active facets for formaldehyde oxidation, and these facets adsorb and activate formaldehyde species on Co_3O_4 surface. Over these active facets, adsorbed formaldehyde can be oxidized into formate, which can be further oxidized into carbonate, and then dissociate to CO_2 (Fig. 11.21). The reaction mechanism that they proposed centers on a pathway where only C—H bonds are broken (C—O σ and C—O π bonds kept).

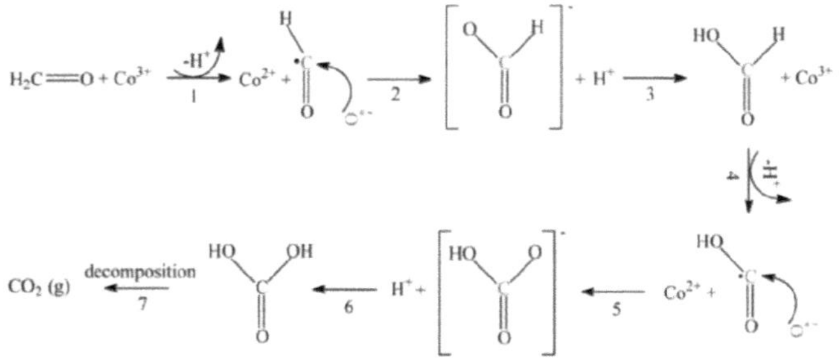

Figure 11.21 Microreaction process for formaldehyde oxidation over mesoporous Co_3O_4, Au/ Co_3O_4, and Au/ Co_3O_4-CeO_2 catalysts at 25°C [27]. Copyright © 2011, American Chemical Society.

Recently, on the basis of FT-IR, XPS, GC, TPD, and TPR measurements, Liu and coworkers [55] proposed a mechanism for the formaldehyde oxidation over Au/ CeO_2 catalysts, involving the participation of both Au^{3+} and Au^0 in the formaldehyde combustion (Fig. 11.22). They suggested that two reaction processes occur on ionic Au^{3+} and metallic Au^0, involving HCOOH and formate as intermediates, carbonate and hydrocarbonate formed being responsible for the catalyst deactivation. They found that the process on Au^{3+} shows higher activity than that one on Au^0.

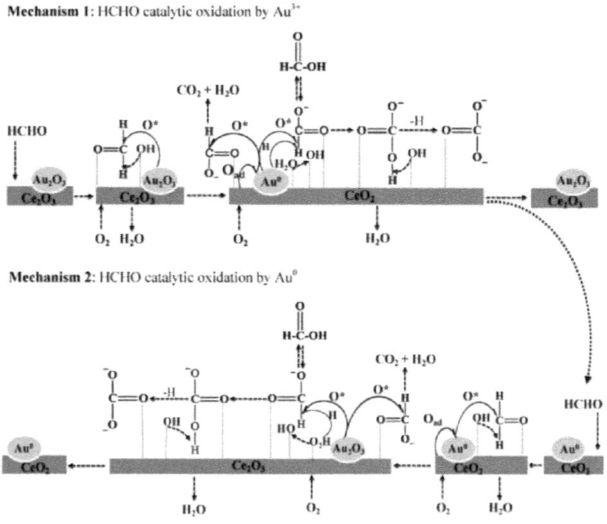

Figure 11.22 The proposed catalytic mechanism of 3DOM Au~0.56 wt%/ CeO_2 catalyst for the catalytic oxidation of formaldehyde [55]. Copyright © 2011 Elsevier B.V.

11.5 Conclusion and Outlook

The present review clearly points out that the catalytic degradation has become the most popular method for formaldehyde abatement, being more versatile and economical; transition metal oxides are receiving growing attention as catalysts for the complete oxidation of formaldehyde because of their high activity; noble metals supported on metal oxides exhibit excellent catalytic performance for formaldehyde oxidation at room or low temperatures, and many significant progresses on improving the low-temperature activity of formaldehyde oxidation have been made recently over a few kinds of supported noble metal catalysts.

Studies up to now carried out by different research groups allow to establish that formaldehyde oxidation over heterogeneous catalysts is governed by both the catalyst species and their structures and properties (nanostructure, surface area, porosity, reduction property, and so on). Therefore, the current researches mainly focus on the design and preparation of the most suitable metal oxide and/or noble metal catalysts, aiming to understand how different operative variables, such as preparation methods, reaction conditions, and presence of promoters, influence the catalyst properties, and consequently the catalytic activity. The mechanism of each influencing parameter toward the formaldehyde oxidation is still always ambiguous up to know, but we believe that the research in this field can reach a satisfactory level of knowledge in future.

Formaldehyde concentration and space velocity are proved to inhibit the formaldehyde oxidation. Due to their hydrophilic surface, the catalysts may be cloaked and the activity is reduced. Moreover, the catalytic performance in the formaldehyde oxidation is always influenced by the presence of a mixture of other VOC compounds and usually an inhibitor effect is found.

In the last years, significant progresses have been made in the development of effective catalysts for complete oxidation of formaldehyde even at room temperature. But we also see some questions deserving further research studies. One aspect is the possibility to drive the formaldehyde oxidation in air at room temperature with non-noble metal or low-cost catalysts. Moreover, the development of simple, effective, and economical advantageous methods for the formaldehyde abatement is expected for the effective removal of large-scale formaldehyde.

Acknowledgments

This work was supported by the National Natural Science Foundation of China (Grant 21007076), Technical Institute of Physics and Chemistry, the Knowledge Innovation Program of the Chinese Academy of Sciences (CAS) (Grant KSCX2-YW-G-059), and the National Basic Research Program of China (Grant 2010CB934103).

References

1. EPA, U. S. **2009**.
2. Carter, E. M., Katz, L. E., Speitel, G. E., Jr., Ramirez, D. *Environmental Science & Technology* **2011,** 45, (15), 6498–6503.
3. Weng, R. *Material Journal (Chinese)* **2000,** 14, 19–21.
4. Xing, Y., Chen, B., Guo, T., Ying, G. *Environmental Progress & Sustainable Energy* **2013,** 32, (1), 115–119.
5. Wolkoff, P., Nielsen, G. D. *Environment International* **2010,** 36, (7), 788–799.
6. Tang, X., Bai, Y., Duong, A., Smith, M. T., Li, L., Zhang, L. *Environment International* **2009,** 35, (8), 1210–1224.
7. Yamada, N., Suzumura, M., Koiwa, F., Negishi, N. *Water Research* **2013,** 47, (8), 2770–2776.
8. Guo, Q., Xu, C., Yang, W., Ren, Z., Ma, Z., Dai, D., Minton, T. K., Yang, X. *The Journal of Physical Chemistry C* **2013,** 117, (10), 5293–5300.
9. Fisher, M. B., Keane, D. A., Fernández-Ibáñez, P., Colreavy, J., Hinder, S. J., McGuigan, K. G., Pillai, S. C. *Applied Catalysis B: Environmental* **2013,** 130–131, 8–13.
10. Autin, O., Hart, J., Jarvis, P., MacAdam, J., Parsons, S. A., Jefferson, B. *Water Research* **2013,** 47, (6), 2041–2049.
11. Tetsuro Noguchi, Fujishima, A. *Environmental Science & Technology* **1998,** 32, (23), 3831–3833.
12. Uraz, A. *Filtration + Separation* **2013,** 50, (1), 38–40.
13. Nakata, K., Fujishima, A. *Journal of Photochemistry and Photobiology C: Photochemistry Reviews* **2012,** 13, (3), 169–189.
14. Huang, Q., Ma, W., Yan, X., Chen, Y., Zhu, S., Shen, S. *Journal of Molecular Catalysis A: Chemical* **2013,** 366, 261–265.
15. Liao, Y., Xie, C., Liu, Y., Huang, Q. *Journal of Alloys and Compounds* **2013,** 550, 190–197.

16. Martono, E., Vohs, J. M. *The Journal of Physical Chemistry C* **2013,** 117, (13), 6692–6701.

17. Muir, J. M. R., Idriss, H. *Surface Science* **2013,** 607, 187–196.

18. Liu, H., Lian, Z., Ye, X., Shangguan, W. *Chemosphere* **2005,** 60, (5), 630–635.

19. Liang, W., Li, J., Jin, Y. *Building and Environment* **2012,** 51, 345–350.

20. Gai, L., Mei, Q., Duan, X., Jiang, H., Zhou, G., Tian, Y., Lu, X. *Journal of Solid State Chemistry* **2013,** 199, 271–279.

21. Tang, X., Huang, X., Shao, J., Liu, J., Li, Y., Xu, Y., Shen, W. *Chinese Journal of Catalysis* **2006,** 27, (2), 97–99.

22. Tang, X., Li, Y., Huang, X., Xu, Y., Zhu, H., Wang, J., Shen, W. *Applied Catalysis B: Environmental* **2006,** 62, (3–4), 265–273.

23. Xu, J., Jia, X., Lou, X., Xi, G., Han, J., Gao, Q. *Sensors and Actuators B: Chemical* **2007,** 120, (2), 694–699.

24. Chen, T., Dou, H., Li, X., Tang, X., Li, J., Hao, J. *Microporous and Mesoporous Materials* **2009,** 122, (1–3), 270–274.

25. Wen, Y., Tang, X., Li, J., Hao, J., Wei, L., Tang, X. *Catalysis Communications* **2009,** 10, (8), 1157–1160.

26. Zhang, J., Jin, Y., Li, C., Shen, Y., Han, L., Hu, Z., Di, X., Liu, Z. *Applied Catalysis B: Environmental* **2009,** 91, (1–2), 11–20.

27. Ma, C., Wang, D., Xue, W., Dou, B., Wang, H., Hao, Z. *Environmental Science & Technology* **2011,** 45, (8), 3628–3634.

28. Sidheswaran, M. A., Destaillats, H., Sullivan, D. P., Larsen, J., Fisk, W. J. *Applied Catalysis B: Environmental* **2011,** 107, (1–2), 34–41.

29. Kim, S. C., Shim, W. G. *Applied Catalysis B: Environmental* **2010,** 98, (3–4), 180–185.

30. Wu, Y., Lu, Y., Song, C., Ma, Z., Xing, S., Gao, Y. *Catalysis Today* **2013,** 201, 32–39.

31. Genuino, H. C., Dharmarathna, S., Njagi, E. C., Mei, M. C., Suib, S. L. *The Journal of Physical Chemistry C* **2012,** 116, (22), 12066–12078.

32. Sekine, Y. *Atmospheric Environment* **2002,** 36, 5543–5547.

33. Portehault, D., Cassaignon, S., Baudrin, E., Jolivet, J.-P. *Journal of Materials Chemistry* **2009,** 19, (16), 2407.

34. Stephanie L. Brock, N. D., Zheng Rong Tian, Oscar Gialdo, Hua Zhou, and Steven L. Suib. *Chemical Materials* **1998,** 10, 2619–2628.

35. Zhang, T., Zhang, X., Ng, J., Yang, H., Liu, J., Sun, D. D. *Chemical Communications* **2011,** 47, (6), 1890–1892.

36. Wang, C., Sun, L., Cao, Q., Hu, B., Huang, Z., Tang, X. *Applied Catalysis B: Environmental* **2011,** 101, (3–4), 598–605.

37. Liu, J., Makwana, V., Cai, J., Suib, S. L., Aindow, M. *The Journal of Physical Chemistry B* **2003,** 107, (35), 9185–9194.

38. Pizzigallo, M. D., Leo, P. D., Ancona, V., Spagnuolo, M., Schingaro, E. *Chemosphere* **2011,** 82, (4), 627–634.

39. Jiang, H., Zhao, T., Ma, J., Yan, C., Li, C. *Chemical Communications* **2011,** 47, (4), 1264–1266.

40. Boonfueng, T., Axe, L., Xu, Y. *Journal of Colloid and Interface Science* **2005,** 281, (1), 80–92.

41. Prieto, O., del Arco, M., Rives, V. *Thermochimica Acta* **2003,** 401, (2), 95–109.

42. Yeager, M., Du, W., Si, R., Su, D., Marinković, N., Teng, X. *The Journal of Physical Chemistry C* **2012,** 116, (38), 20173–20181.

43. Iyer, A., Galindo, H., Sithambaram, S., King'ondu, C., Chen, C.-H., Suib, S. L. *Applied Catalysis A: General* **2010,** 375, (2), 295–302.

44. Hernández, W. Y., Centeno, M. A., Ivanova, S., Eloy, P., Gaigneaux, E. M., Odriozola, J. A. *Applied Catalysis B: Environmental* **2012,** 123–124, 27–35.

45. Thamayanthy Sriskandakumar, N. O., Chun-Hu Chen, Aimee Morey, Cecil King'ondu, and Steven L. Suib. *Journal of Physical Chemistry A* **2009,** 113, 1523–1530.

46. Santos, V. P., Soares, O. S. G. P., Bakker, J. J. W., Pereira, M. F. R., Órfão, J. J. M., Gascon, J., Kapteijn, F., Figueiredo, J. L. *Journal of Catalysis* **2012,** 293, 165–174.

47. Sun, M., Yu, L., Ye, F., Diao, G., Yu, Q., Hao, Z., Zheng, Y., Yuan, L. *Chemical Engineering Journal* **2013,** 220, 320–327.

48. Hongmin Chen, J. H., Changbin Zhang, Hong He. *Journal of Physical Chemistry C* **2007,** 111, 18033–18038.

49. Zhou, L., Zhang, J., He, J., Hu, Y., Tian, H. *Materials Research Bulletin* **2011,** 46, (10), 1714–1722.

50. Tian, H., He, J. H., Liu, L. L. *Advanced Materials Research* **2011,** 181–182, 485–488.

51. Tian, H., He, J., Liu, L., Wang, D., Hao, Z., Ma, C. *Microporous and Mesoporous Materials* **2012,** 151, 397–402.

52. Tian, H., He, J., Zhang, X., Zhou, L., Wang, D. *Microporous and Mesoporous Materials* **2011,** 138, (1–3), 118–122.

53. Zhou, L., He, J., Zhang, J., He, Z., Hu, Y., Zhang, C., He, H. *The Journal of Physical Chemistry C* **2011,** 115, (34), 16873–16878.

54. Wang, L., Sakurai, M., Kameyama, H. *Journal of Hazardous Materials* **2009,** 167, (1–3), 399–405.

55. Liu, B., Li, C., Zhang, Y., Liu, Y., Hu, W., Wang, Q., Han, L., Zhang, J. *Applied Catalysis B: Environmental* **2012,** 111–112, 467–475.

56. Nie, L., Yu, J., Li, X., Cheng, B., Liu, G., Jaroniec, M. *Environmental Science & Technology* **2013,** 47, (6), 2777–2783.

57. Chen, B.-B., Shi, C., Crocker, M., Wang, Y., Zhu, A.-M. *Applied Catalysis B: Environmental* **2013,** 132-133, 245–255.

58. An, N., Zhang, W., Yuan, X., Pan, B., Liu, G., Jia, M., Yan, W., Zhang, W. *Chemical Engineering Journal* **2013,** 215–216, 1–6.

59. Qu, Z., Shen, S., Chen, D., Wang, Y. *Journal of Molecular Catalysis A: Chemical* **2012,** 356, 171–177.

60. Zhang, C., He, H. *Catalysis Today* **2007,** 126, (3–4), 345–350.

61. Zhang, C., Liu, F., Zhai, Y., Ariga, H., Yi, N., Liu, Y., Asakura, K., Flytzani-Stephanopoulos, M., He, H. *Angewandte Chemie* **2012,** 51, (38), 9628–9632.

62. Zhang, C., He, H., Tanaka, K.-i. *Applied Catalysis B: Environmental* **2006,** 65, (1–2), 37–43.

63. Peng, J., Wang, S. *Applied Catalysis B: Environmental* **2007,** 73, (3–4), 282–291.

64. Liu, L., Tian, H., He, J., Wang, D., Yang, Q. *Journal of Environmental Sciences* **2012,** 24, (6), 1117–1124.

65. Álvarez-Galván, M. C., de la Peña O'Shea, V. A., Fierro, J. L. G., Arias, P. L. *Catalysis Communications* **2003,** 4, (5), 223–228.

66. O'Shea, V. A. d. l. P., Álvarez-Galván, M. C., Fierro, J. L. G., Arias, P. L. *Applied Catalysis B: Environmental* **2005,** 57, (3), 191–199.

67. Tang, X., Chen, J., Li, Y., Li, Y., Xu, Y., Shen, W. *Chemical Engineering Journal* **2006,** 118, (1–2), 119–125.

68. Jia, M., Bai, H., Zhaorigetu, Shen, Y., Li, Y. *Journal of Rare Earths* **2008,** 26, (4), 528–531.

69. Li, C., Shen, Y., Jia, M., Sheng, S., Adebajo, M. O., Zhu, H. *Catalysis Communications* **2008,** 9, (3), 355–361.

70. Shi, C., Chen, B.-B., Li, X.-S., Crocker, M., Wang, Y., Zhu, A.-M. *Chemical Engineering Journal* **2012,** 200-202, 729–737.

71. Liu, B., Liu, Y., Li, C., Hu, W., Jing, P., Wang, Q., Zhang, J. *Applied Catalysis B: Environmental* **2012,** 127, 47–58.

72. Yu, X., He, J., Wang, D., Hu, Y., Tian, H., He, Z. *The Journal of Physical Chemistry C* **2012,** 116, (1), 851–860.

73. Hosseini, M., Barakat, T., Cousin, R., Aboukaïs, A., Su, B. L., De Weireld, G., Siffert, S. *Applied Catalysis B: Environmental* **2012,** 111–112, 218–224.

74. Scirè, S., Liotta, L. F. *Applied Catalysis B: Environmental* **2012,** 125, 222–246.

75. Liotta, L. F. *Applied Catalysis B: Environmental* **2010,** 100, (3-4), 403–412.

76. Yu, X., He, J., Wang, D., Hu, Y., Tian, H., Dong, T., He, Z. *Journal of Nanoparticle Research* **2012,** 14, (11), 1260.

77. Tang, X., Chen, J., Huang, X., Xu, Y., Shen, W. *Applied Catalysis B: Environmental* **2008,** 81, (1–2), 115–121.

78. Huang, H., Leung, D. Y. C. *Journal of Catalysis* **2011,** 280, (1), 60–67.

Chapter 12

Nanostructured Copper Oxide for Sensing Hydrogen Cyanide Gas

Mingqing Yang and Junhui He
Functional Nanomaterials Laboratory and Key Laboratory of Photochemical Conversion and Optoelectronic Materials, Technical Institute of Physics and Chemistry, Chinese Academy of Sciences, Zhongguancundonglu 29, Haidianqu, Beijing 100190, China
jhhe@mail.ipc.ac.cn

12.1 Introduction

Hydrogen cyanide (HCN) is a colorless or pale blue liquid or a gas. It has the following physicochemical properties: bitter almond odor, mp: −13.4°C, bp: 25.6°C, liquid density at 20°C: 0.688 g/cm^3, gas density (air = 1 g/cm^3) at 31°C: 0.947 g/cm^3, very weakly acid, soluble in water and alcohol, and slightly soluble in ether. HCN is extremely dangerous to human beings and animals, as it inhibits the consumption of oxygen by body tissues via breath, ingestion, skin absorption, and ocular routes [1–3]. Exposure levels of 100 ppm

Nanomaterials in Energy and Environmental Applications
Edited by Junhui He

ISBN 978-981-4463-78-2 (Hardcover), 978-981- 4463-79-9 (eBook)
www.panstanford.com

HCN will result in death in about 1 h or less in some cases, while exposure levels of 500 ppm HCN will result in death within 15 min; higher concentration levels will result in faster onset of symptoms or death. Although HCN is extremely toxic, it is a very useful raw material. It is used in a variety of syntheses, including the production of adiponitrile (for nylon), methyl methacrylate, sodium cyanide, cyanuric chloride, chelating agents (e.g., ethylenediaminetetraacetic acid, EDTA), pharmaceuticals, and other specialty chemicals. In addition, HCN has some insecticide and fungicide applications. For human safety, it is extremely important and highly desired to detect any possible presence or leakage of HCN during its application. Approaches to analyze HCN include spectrophotometric, atomic absorption spectrophotometric, and electrochemical methods [4–8]. The spectrophotometric analysis of HCN has been widely known, which uses pyridine-pyrazolone, isonicotinic acid-pyrazolone, and pyridine-barbituric acid as ligand. Although it is sensitive, the analysis procedure is complicated and time-consuming. The atomic absorption spectrophotometric method is sensitive, stable, and easy in operation. However, it uses large apparatuses that are not available for sensing application. Thus, the spectrophotometric and atomic absorption spectrophotometric methods are not suited for fast in situ and real-time detection. The electrochemical method uses a cyanide ion-selective membrane electrode and is the most applicable technique among the current three methods for in situ and real-time detection of HCN. It is sensitive and easy in operation, and needs little operator attention. Unfortunately, the electrochemical analysis is frequently interfered by other ions, resulting in low selectivity. Therefore, development of novel sensors with fast response, high sensitivity, and high selectivity is attractive for fast in situ and real-time detection of HCN.

As Sauerbrey found experimental verification of the mass-frequency shift relation for quartz crystal resonators, quartz crystal microbalance (QCM) has been widely exploited in the field of chemical and biological sensors because of its many advantages, such as intrinsic high sensitivity, low cost, easy installation, and inherent ability to monitor analytes in real time [9–22]. Thus far, how to modify resonator surface properties becomes an essential issue for enhancing the range of application of QCM sensors. A variety of materials, such as ceramics, polymers, dendrimers, oils, and waxes, have been employed as coatings on the surfaces of QCM devices in efforts to attain fast response, high sensitivity, and high selectivity for analytes [23,24]. Until now, nanostructured materials, especially nanostructured metal oxides, have attached great attention as sensing materials due to their superior features, such as lower cost, very high surface to volume ratios, and the ease with which they can be fabricated as compared with their bulk or thin film counterparts [25–52]. Copper oxide (CuO), as an important p-type metal oxide with a narrow band gap (E_g = 1.2 eV), has been extensively studied owing to its important properties and practical applications (e.g., electrode materials, catalysts, optical switches, fungicides, pigments, field emission emitters, magnetic storage media, and gas sensors) [53–76]. Nanostructured CuO may exhibit unique properties that can be significantly different from those of their bulk counterparts. For example, CuO nanostructures were tested as Li-ion b attery anodes and showed high Li-ion storage capacities, which suggests that CuO can offer high-charge storage capacity through a redox reaction [58,77]. CuO nanorods were synthesized, and their diffuse reflectance properties were investigated in the near infrared region. It was found that the as-prepared CuO nanorods are better diffuse reflectors than other CuO and Cu nanomaterials [78]. CuO microflower/nanorod array structures were directly fabricated on the surface of copper plates by a simple one-step solution-immersion process. After being modified with fluoroalkylsilane, the CuO films showed an analogous extreme nonwetting property but high contrast water adhesion that ranges from extremely low to very high [79]. Nanosized copper oxides were successfully prepared by microwave-assisted hydrothermal reaction, and the synthesized copper oxide exhibited high catalytic activity for olefin epoxidation and the oxidation of CO to CO_2 [80]. Novel cotton-candy-like CuO 3D hierarchical microspheres were successfully synthesized by a facile surfactant-free solvothermal route and subsequent calcination process. Arsenic (|||) adsorption experiments indicated that the as-synthesized CuO hierarchical nanostructures exhibited fast adsorption rates and higher removal capacity than other reported nanostructured CuO adsorbents [61]. Hierarchically porous CuO hollow spheres were fabricated via a one-pot template-free method. The unique CuO spheres exhibited excellent sensing performance, such as low detection limit, high sensitivity, and quick response and recovery, due to the special structure with open macropores for gas molecule buffering, mesopores, and micropores for fast H_2S molecule diffusion and effective sensing, and good conductivity of the quasi-single crystal nonosheets for fast cattier transportation [81]. Therefore, the synthesis and study of CuO nanostructures are of both practical and fundamental importance.

In our recent works, using QCM as detection platform, we synthesized CuO nanostructures in a facile way, from which CuO functionalized QCM sensors were fabricated and explored for HCN sensing [82–86]. The sensors response to HCN was found to be in an opposite direction as compared with other common volatile substances, offering excellent selectivity for HCN detection. The sensitivity was very high, and the response and recovery were very fast. The CuO functionalized QCM sensors are considered to be a promising candidate for HCN gas detection in practical application.

12.2 Nanostructures of CuO

The optical, electrical, and gas-sensing properties and catalytic performance of materials are strongly influenced by their structure and crystallographic form. In our works, CuO products were prepared and studied in detail. X-ray diffraction (XRD), X-ray photoelectron spectroscopy (XPS), thermogravimetry (TG), scanning electron microscopy (SEM), transmission electron microscopy (TEM), selected area electron diffraction, nitrogen adsorption-desorption measurements, and UV-visible diffuse reflectance spectra were employed to characterize the obtained CuO nanostructures. $Cu(OH)_2$ nanowires were prepared firstly, after calcination of the $Cu(OH)_2$ nanowires at 400°C for 3 h, a black CuO powder was obtained [82]. From SEM and TEM images of the CuO powder (Fig. 12.1), both elongated particles and nanoparticles of CuO were produced. Apparently, the $Cu(OH)_2$ nanowires fragmented into the elongated particles and nanoparticles of CuO after thermolysis.

Figure 12.1 SEM (a,c) and TEM (b,d) images of as-synthesized $Cu(OH)_2$ nanowires (a,b) and CuO nanomaterial (c,d) [82], Copyright 2011 Elsevier.

Figure 12.2 Typical SEM images of $Cu(OH)_2$ precursor (a and b) and CuO products obtained in presence of PEG and $NH_3{\cdot}H_2O$ by hydrothermal treatment at 150°C for varied periods of time: 0.5 h (c and d), 1 h (e and f), 2 h (g and h) [83], Copyright 2011 Elsevier.

Flower-like, boat-like, plate-like and ellipsoid-like CuO nanostructures were obtained by simple modulation of reactants and the time and temperature of hydrothermal treatment (Fig. 12.2) [83]. Figures 12.2a,b are SEM images of the $Cu(OH)_2$ precursor. It consists of uniform well-defined sheaves that consist of nanowires nearly 3 μm in length and several tens of nanometers in diameter. When the time of hydrothermal treatment was 0.5 h, the $Cu(OH)_2$ precursor was converted to CuO, which was confirmed through XRD. As shown in Figs. 12.2c,d, flower-like CuO nanostructures formed. An individual flower-like structure was assembled by CuO nanoplates of ca. 80 nm in thickness. When the time of hydrothermal treatment was 1 h, boat-like CuO nanostructures already formed (Figs. 12.2e,f). The boat-like structure was composed of nanoplates of ca. 100 nm in thickness. When the time of hydrothermal treatment was 2 h, well-defined boat-like structures formed on a large scale (Fig. 12.2g). The average length and width of the boat-like structure were estimated to be 10 μm and 5 μm, respectively. From the enlarged image (Fig. 12.2h), it can be seen that the boat-like structure is composed of nanoplates. The thicknesses of the nanoplates are in the range of 100–150 nm. Thus, the flower-like and boat-like CuO have the same crystal structure and were in fact all assembled from the CuO nanoplates. Two comparative experiments were carried out in order to investigate the effects of polyethylene glycol (PEG) and $NH_3{\cdot}H_2O$ on the morphology of CuO nanostructures. PEG and $NH_3{\cdot}H_2O$ were added alternatively under otherwise identical conditions. When the reagents were hydrothermally treated without addition of PEG, 2D platelets of polygonal shape were obtained on a large scale. The thicknesses of the platelets are ca. 100 nm. Without addition of $NH_3{\cdot}H_2O$, ellipsoids of uniform shape and size were obtained. A magnified image of the ellipsoid shows that the ellipsoids consist of stacked CuO layers. In another work, nanostructured CuO of even higher surface area was synthesized by calcinating copper oxalate precursors [86]. Figures 12.3a,b show SEM images of CuO from cushion-like copper oxalate. The overall morphology of CuO products remained as cushion-like geometry, but the surface had changed significantly. As shown in Fig. 12.3b, the a-faces and b-faces were composed of nanoparticles of ca. 93 nm in diameter. This interesting morphology resembles a famous Chinese cake, Sacima (Fig. 12.3a, inset). Figure 12.3d shows SEM image of CuO from sphere-like copper oxalate. The morphology of CuO products remained as sphere-like geometry. From the enlarged image in Fig. 12.3e, it can be clearly seen that the surface of sphere-like CuO is composed of nanoparticles of ca. 127 nm in diameter. There are some broken sphere-like CuO in the as-prepared CuO product. The inside of sphere-like CuO is also composed of nanoparticles, the diameter of which is ca. 63 nm.

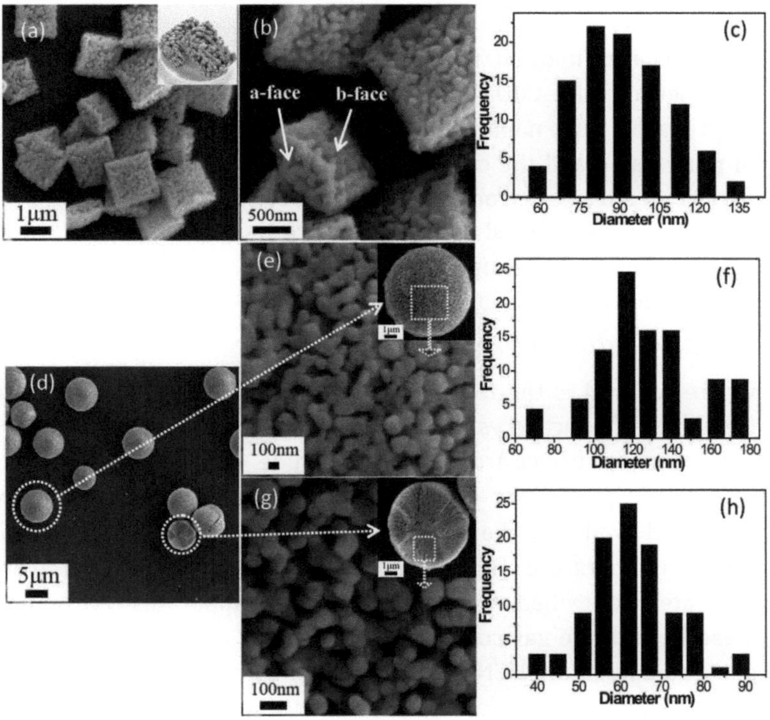

Figure 12.3 SEM images of CuO (a and b) from cushion-like copper oxalate ($H_2C_2O_4$ and $CuCl_2$ concentrations: 0.01 mol/l), (d, e, and g) from sphere-like copper oxalate ($H_2C_2O_4$ and $CuCl_2$ concentrations: 0.05 mol/l) and size histograms of CuO nanoparticles in the cushion-like CuO (c) and on the surface (f) and inside (h) of the sphere-like CuO [86], Copyright 2013 Royal Society of Chemistry.

Nitrogen physisorption represents the most widely used technique to measure materials surface area. By nitrogen adsorption–desorption measurements, the specific surface areas of flower-like, boat-like, plate-like, ellipsoid-like, and sphere-like CuO were measured to be 5.2, 5.1, 9.3, 1.9, and 73 m^2/g, respectively, following the order of sphere-like CuO > plate-like CuO > flower-like CuO ≈ boat-like CuO > ellipsoid-like CuO.

12.3 Sensing Properties of CuO-Functionalized QCM Sensors

A drop-coating method was used to coat both sides of a silver-coated QCM sensor with CuO nanomaterials. In a typical procedure, the as-prepared CuO nanomaterials were dispersed homogeneously in pure water, and the obtained suspension was dropped onto the surface of silver electrode using a microsyringe. The QCM sensors with CuO nanomaterials thin films were quickly dried at 30°C in vacuum. The CuO coating amount on QCM was calculated according to the Sauerbrey equation [9]:

$$\Delta F = -2.26 \times 10^{-6} F_0^2(\Delta m/A)$$

where ΔF (Hz) is the frequency shift, F_0 (Hz) is the fundamental frequency of quartz crystal, Δm (g) is the mass change, and A (cm^2) is the sensing surface area of QCM electrode.

Sensing properties of CuO-functionalized QCM sensors were examined by placing them in a glass chamber inside of an incubator where temperature (25°C) and relative humidity (10%) were accurately controlled. A dynamic gas-mixing apparatus was used to steadily generate gas containing HCN of low concentration. Vapor generated by liquid HCN was taken away by a mass flow controller manipulated gas flow and was further diluted in proportion by another steady gas flow. In order to simulate real environments, all gas flows were provided air by an air compressor. A four-way valve was applied to switch between air flow and diluted HCN flow. Frequency shifts were recorded by an Agilent 53131A universal counter linked to a computer [82,84,86].

The concentration of HCN in the carrier gas (air) was controlled at 5–60 ppm. A naked silver-coated QCM sensor was also tested as control for comparison. From the test, the naked silver-coated QCM sensor had no significant response to HCN. Compared with the naked silver-coated QCM sensor, all the CuO-functionalized QCM sensors have fast, significant response to HCN. Analysis of the profiles of the frequency shift of CuO-functionalized QCM sensors upon exposure to HCN shows that they can be broken into two parts: the "rapid" (steep slope) and "slow" (shallow slope) response. When the HCN was introduced into the testing chamber, the QCM frequency began to rise. The frequency shift initially increased rapidly. Then, the frequency shift increased less rapidly. When the testing chamber was flushed by air, the frequency shift decreased rapidly. Finally, the frequency shift recovered basically. The sensitivities of CuO-functionalized QCM sensors were assessed by the (frequency shift)/(coating amount) ratio, in order to compare the performance of different CuO nanomaterials. The results show that the sensitivity of the sphere-like CuO-functionalized QCM sensor is significantly higher (6.53 Hz/μg) than those of plate-like (2.26 Hz/μg), flower-like (0.83 Hz/μg), boat-like (0.80 Hz/μg), ellipsoid-like (0.49 Hz/μg), and commercial (0.31 Hz/μg) CuO-functionalized QCM sensors. A detailed comparison of the sensitivities of these QCM sensors was made in Table 12.1. As shown in the second entry of Table 12.1, the specific surface areas of as-prepared CuO and commercial CuO follow the order of sphere-like CuO > plate-like CuO > flower-like CuO ≈ boat-like CuO > ellipsoid-like CuO > commercial CuO.

Table 12.1 BET specific surface areas of CuO nanomaterials, frequency shifts of CuO-functionalized QCM resonators 10 s after introduction of 50 ppm HCN, and the corresponding sensitivities

CuO nanostructures	BET specific surface area (m^2/g)	Frequency shift at 10 s (Hz)	Sensitivity (Hz at 10s/μg)
Commercial CuO	1.5	11±2	0.31
Ellipsoid-like CuO	1.9	17±2	0.49
Boat-like CuO	5.1	28±3	0.80
Flower-like CuO	5.2	29±3	0.83
Plate-like CuO	9.3	79±5	2.26
Sphere-like CuO	73	85±2	6.53

The third entry of Table 12.1 shows the frequency shifts recorded 10 s after HCN exposure. Clearly, the sensitivities of as-prepared CuO and commercial CuO-functionalized QCM sensors follow the order of sphere-like CuO > plate-like CuO > flower-like CuO ≈ boat-like CuO > ellipsoid-like CuO > commercial CuO, in striking agreement with that of their specific surface areas. The fourth entry of Table 12.1 shows the sensitivities of CuO-functionalized QCM sensors. Similarly, the sensitivities of as-prepared CuO and commercial CuO-functionalized QCM sensors also follow the order of their specific surface areas. From these experimental results, the specific surface area of CuO plays a crucial role in the sensor response and thus the sensitivity of these CuO-functionalized QCM sensors: the sensor response and sensitivity increase with increase of the specific surface area of the CuO used. Future efforts would be directed to the development of nanomaterials with even larger surface areas and to full utilization of varied surface reactions.

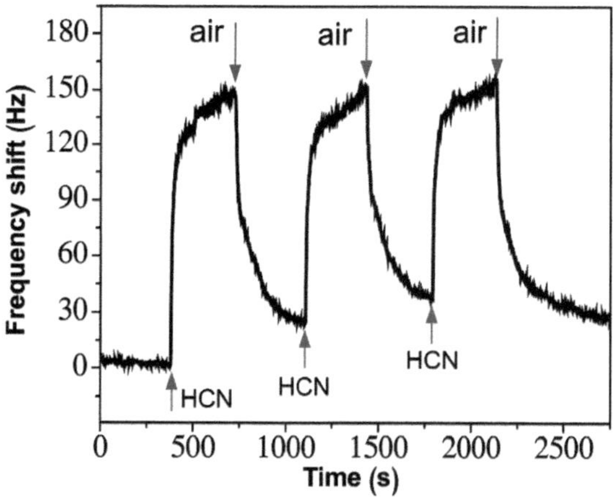

Figure 12.4 Profiles of frequency shift of plate-like CuO-functionalized QCM resonator upon exposure to air containing 50 ppm HCN at 25°C in three continuous cycles [84], Copyright 2011 American Chemical Society.

Reproducibility and stability are regarded as important aspects for sensors. As shown in Fig. 12.4, the response curves are similar for three continuous cycles with nearly no changes in response, response time, and recovery time, indicating good reproducibility and stability. The high reproducibility and stability also indicate the

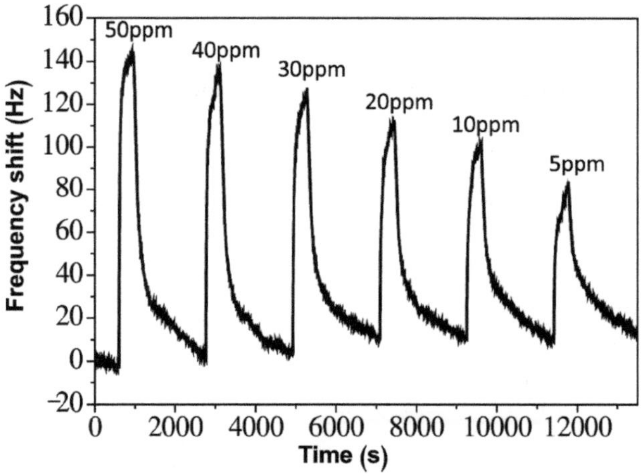

Figure 12.5 Profiles of frequency shift of plate-like CuO-functionalized QCM resonator upon exposure to HCN gases of varied concentrations at 25°C [84], Copyright 2011 American Chemical Society.

robustness of the sensing layers, though they were applied to QCM electrodes by simple drop-coating method. Doubtlessly, such high reproducibility and stability are critical for applicable sensing devices. Figure 12.5 displays the response curves of a plate-like CuO-functionalized QCM sensor to HCN of varied concentrations. When a HCN gas of given concentration was introduced into the testing chamber, a frequency shift was recorded and all responses were fast within the initial few seconds toward HCN of varied concentrations. The response profiles are similar in shape, and the response of the sensor decreases proportionally with the decrease of HCN concentration. Clearly, the frequency shift shows a linear relationship with HCN concentration in the HCN concentration range of 0–50 ppm. A linear equation was derived as follows:

$$\Delta F = 1.52c + 0.43, \mathrm{r} = 0.9935,$$

where ΔF is frequency shift (Hz), c is the HCN concentration (ppm), and r is regression coefficient. Selectivity is one of the most important factors that determine the eventual applicability of any chemical sensors. The selectivity of the plate-like CuO-functionalized QCM resonator was investigated by exposing the resonator to saturated vapors of a variety of common volatile chemicals (25°C), including acetone, ethyl ether, water, n-hexane, benzene, acrylic acid, acetic acid, benzyl alcohol, and ethanol. As shown in Fig. 12.6, the plate-like CuO-functionalized QCM resonator not only showed large and rapid response to all the examined chemicals but also demonstrated fast recovery when the chemical vapors were removed. However, surprisingly and interestingly, reversed signals were observed for the examined chemicals as compared with HCN. These characteristics would allow unambiguous report on HCN among these volatile substances.

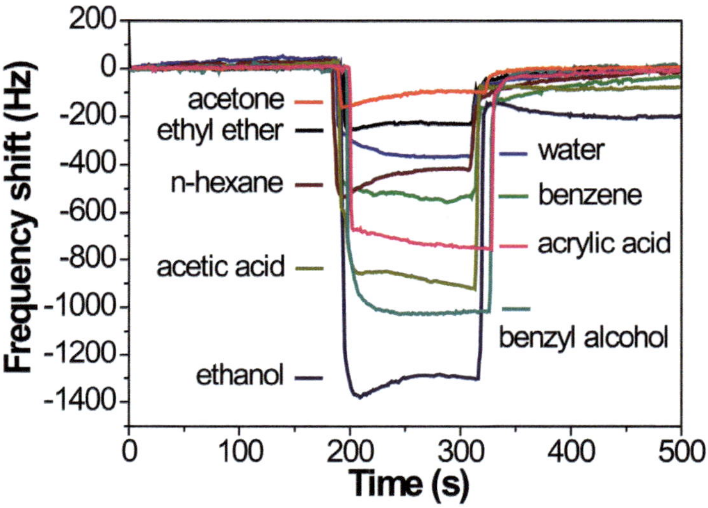

Figure 12.6 Profiles of frequency shift of plate-like CuO-functionalized QCM resonator upon exposure to saturated vapors of acetone, ethyl ether, water, n-hexane, benzene, acrylic acid, acetic acid, benzyl alcohol, and ethanol [84], Copyright 2011 American Chemical Society.

12.4 Sensing Mechanism of CuO-Functionalized QCM Sensors

In order to understand the sensing mechanism of the CuO-functionalized QCM sensor, the effluent air flow passing the testing chamber was analyzed in real time by a mass spectrometer for clarification of possible sensing mechanisms. Interestingly, after the

CuO-functionalized QCM sensor was exposed to HCN, a new species of m/z = 52 was detected by the mass spectrometer, which is attributed to $(CN)_2$. We also monitored the variation of $(CN)_2$ ion intensity with time. The variation profile of $(CN)_2$ ion intensity is similar to that of QCM frequency shift. When air containing HCN was introduced into the testing chamber, the $(CN)_2$ ion was detected. When the testing chamber was flushed by air, the $(CN)_2$ ion intensity began to decrease and reached the initial base line eventually. The $(CN)_2$ ion intensity variations are also similar for the three continuous cycles. Such striking agreements between the QCM frequency shift and the $(CN)_2$ ion intensity doubtlessly point to the production of $(CN)_2$ as a result of the contact of HCN with the CuO. It is important to unveil the underlying sensing mechanism. Frequency increase usually means loss of mass or change of viscoelastic property of deposited materials. In case of the current work, CuO nanomaterials are rigid solid, and the change of its viscoelastic property upon gas adsorption is hardly believed. Thus, some mass loss must have occurred when CuO nanomaterials came into contact with HCN. On the basis of comprehensive analyses of experimental results, an applausible mechanism was proposed. As shown in Scheme 12.1, a surface redox reaction might have occurred reversibly between CuO and Cu_2O. When CuO nanomaterials were exposed to HCN, the surface CuO was reduced to Cu_2O by HCN, forming $(CN)_2$ species (Equation (12.1)):

$$2CuO + 2HCN \rightarrow Cu_2O + (CN)_2 + H_2O \qquad 12.1$$

In this process (from CuO to Cu_2O), one oxygen atom was lost. Thus, the total mass on QCM decreased, and a positive frequency shift was recorded according to the Sauerbrey equation. Initially, the speed of the redox reaction was high because of the high HCN concentration, which resulted in the "rapid" response in the detection. Then, the production of Cu_2O on the surface of CuO reduced the speed of redox reaction, which resulted in the "slow" response. When the testing chamber was flushed by air, the surface Cu_2O was re-oxidized to CuO by O_2 (Equation (12.2)), and the frequency shift recovered:

$$Cu_2O + 1/2O_2 \rightarrow 2CuO \qquad 12.2$$

These processes are reversible, which brings the excellent reproducibility and stability of the CuO-functionalized QCM sensor.

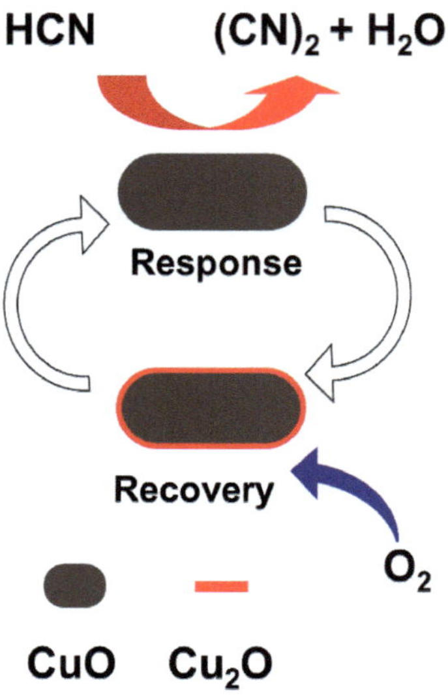

Scheme 12.1 Schematic illustration of the processes on the CuO-functionalized QCM sensors [82], Copyright 2011 Elsevier.

12.5 Conclusion and Outlook

In summary, CuO nanomaterials, including nanoparticles and flower-like, boat-like, plate-like, ellipsoid-like, and sphere-like CuO nanostructures, were prepared and applied to the sensing of HCN gas. The HCN selectivity of CuO-functionalized QCM sensor is extremely high because of its reversed response to HCN as compared with a variety of other common volatile substances. In addition, the sensitivity is very high, and the response and recovery are very fast. The specific surface area of CuO plays a crucial role in the sensor response and thus the sensitivity of the CuO-functionalized QCM sensors: the sensor response and sensitivity increase with an increase of the specific surface area of the CuO used. On the basis of our experimental results, a sensing mechanism was proposed in which a surface redox reaction occurs between CuO and Cu_2O on the CuO nanostructures reversibly upon contact with HCN and air, respectively. The high sensitivity, rapid response and fast recovery, and unprecedented high HCN selectivity would allow fast, sensitive, and unambiguous detection of HCN gas, which would be of particular benefits in the area of public security and environmental applications. In order to improve the sensing performances (especially the sensitivity), nanostructured CuO with even larger surface areas should be designed and prepared. The enhancement of surface area may be realized by designing and building novel CuO frameworks and by fine control over their porosity, pore structure, and pore size.

Acknowledgements

The work described here was supported by the National Natural Science Foundation of China (Grant No. 21271177), the National Basic Research Program of China (Grant No. 2010CB934103), the Knowledge Innovation Program of the Chinese Academy of Sciences (CAS) (Grant No. KGCX2-YW-111-5), "Hundred Talents Program" of CAS. It was also partially supported by Key Laboratory of Photochemical Conversion and Optoelectronic Materials, Technical Institute of Physics and Chemistry (TIPC), CAS.

References

1. Patnaik, P. *Cyanides, Inorganic*, John Wiley & Sons, Inc., **2006**, 317–335.
2. Ballantyne, B., Bismuth, C., Hall, A. H. *Cyanides: Chemical Warfare Agents and Potential Terrorist Threats*, John Wiley & Sons, Ltd, **2007**, 495–542.
3. Porter, T. L., Vail, T. L., Eastman, M. P., Stewart, R., Reed, J., Venedam, R., Delinger, W. *Sens. Actuators B: Chem.* **2007**, *123*, 313–317.
4. Toth, K., Pungor, E. *Anal. Chim. Acta* **1970**, *51*, 221–230.
5. Meyer, R. E., Lantz, P. M. *J. Electroanal. Chem.* **1975**, *61*, 155–163.
6. Sweileh, J. A. *Anal. Chim. Acta* **1996**, *336*, 131–140.
7. Nakano, N., Yamamoto, A., Kobayashi, Y., Nagashima, K. *Anal. Chim. Acta* **1999**, *398*, 305–310.
8. Ma, J., Dasgupta, P. K. *Anal. Chim. Acta* **2010**, *673*, 117–125.
9. Sauerbrey, G. *Zeitschrift Für Physik* **1959**, *155*, 206–222.
10. O'Sullivan, C. K., Guilbault, G. G. *Biosens. Bioelectron.* **1999**, *14*, 663–670.

11. Gomes, M. T. S. R., Nogueira, P. S. T., Oliveira, J. A. B. P. *Sens. Actuators B: Chem.* **2000**, *68*, 218–222.
12. Marx, K. A. *Biomacromolecules* **2003**, *4*, 1099–1120.
13. Xu, X., Cang, H., Li, C., Zhao, Z. K., Li, H. *Talanta* **2009**, *78*, 711–716.
14. Wang, J., Zhu, Z., Ma, H. *Anal. Chem.* **2013**, *85*, 2096–2101.
15. Nalam, P. C., Daikhin, L., Espinosa-Marzal, R. M., Clasohm, J., Urbakh, M., Spencer, N. D. *J. Phys. Chem. C* **2013**, *117*, 4533–4543.
16. Zhang, Z., Fan, J., Yu, J., Zheng, S., Chen, W., Li, H., Wang, Z., Zhang, W. *ACS Appl. Mater. Interfaces* **2012**, *4*, 944–949.
17. Rehman, A., Hamilton, A., Chung, A., Baker, G. A., Wang, Z., Zeng, X. *Anal. Chem.* **2011**, *83*, 7823–7833.
18. Hou, K., Rehman, A., Zeng, X. *Langmuir* **2011**, *27*, 5136–5146.
19. Buttry, D. A., Ward, M. D. *Chem. Rev.* **1992**, *92*, 1355–1379.
20. Shinar, R., Liu, G., Porter, M. D. *Anal. Chem.* **2000**, *72*, 5981–5987.
21. Hagleitner, C., Hierlemann, A., Lange, D., Kummer, A., Kerness, N., Brand, O., Baltes, H. *Nature* **2001**, *414*, 293–296.
22. Vaiyapuri, R., Greenland, B. W., Elliott, J. M., Hayes, W., Bennett, R. A., Cardin, C. J., Colquhoun, H. M., Etman, H., Murray, C. A. *Anal. Chem.* **2011**, *83*, 6208–6214.
23. Liang, C., Yuan, C., Warmack, R. J., Barnes, C. E., Dai, S. *Anal. Chem.* **2002**, *74*, 2172–2176.
24. Lee, S., Takahara, N., Korposh, S., Yang, D., Toko, K., Kunitake, T. *Anal. Chem.* **2010**, *82*, 2228–2236.
25. Nguyen, H., El-Safty, S. A. *J. Phys. Chem. C* **2011**, *115*, 8466–8474.
26. Li, X., Xiong, Y., Li, Z., Xie, Y. *Inorg. Chem.* **2006**, *45*, 3493–3495.
27. Erol, A., Okur, S., Yagmurcukardes, N., Arlkan, M. Ç. *Sens. Actuators B: Chem.* **2011**, *152*, 115–120.
28. Kimura, M., Sugawara, M., Sato, S., Fukawa, T., Mihara, T. *Chem. Asian J.* **2010**, *5*, 869–876.
29. Della Gaspera, E., Guglielmi, M., Agnoli, S., Granozzi, G., Post, M. L., Bello, V., Mattei, G., Martucci, A. *Chem. Mater.* **2010**, *22*, 3407–3417.
30. Zhao, Y., He, J. H., Yang, M., Gao, S., Zuo, G., Yan, C., Cheng, Z. *Anal. Chim. Acta* **2009**, *654*, 120–126.
31. Zhao, Y., Chen, H., Wang, X., He, J., Yu, Y., He, H. *Anal. Chim. Acta* **2010**, *675*, 36–41.
32. Zhao, Y., Du, X., Wang, X., He, J., Yu, Y., He, H. *Sens. Actuators B: Chem.* **2010**, *151*, 205–211.

33. Yu, H., Yu, J., Liu, S., Mann, S. *Chem. Mater.* **2007**, *19*, 4327–4334.
34. Cao, S. W., Zhu, Y. J. *J. Phys. Chem. C* **2008**, *112*, 6253–6257.
35. Liu, J., Xue, D. *Adv. Mater.* **2008**, *20*, 2622–2627.
36. Zhang, T., Zhang, Q., Ge, J., Goebl, J., Sun, M., Yan, Y., Liu, Y., Chang, C., Guo, J., Yin, Y. *J. Phys. Chem. C* **2009**, *113*, 3168–3175.
37. Zheng, Z., Huang, B., Wang, Z., Guo, M., Qin, X., Zhang, X., Wang, P., Dai, Y. *J. Phys. Chem. C* **2009**, *113*, 14448–14453.
38. Zheng, X., Wu, R., Chen, Y. *Environ. Sci. Technol.* **2011**, *45*, 2826–2832.
39. Lei, A., Qu, B., Zhou, W., Wang, Y., Zhang, Q., Zou, B. *Mater. Lett.* **2012**, *66*, 72–75.
40. Sun, G., Dong, B., Cao, M., Wei, B., Hu, C. *Chem. Mater.* **2011**, *23*, 1587–1593.
41. Toberer, E. S., Seshadri, R. *Adv. Mater.* **2005**, *17*, 2244–2246.
42. Waller, M. R., Townsend, T. K., Zhao, J., Sabio, E. M., Chamousis, R. L., Browning, N. D., Osterloh, F. E. *Chem. Mater.* **2012**, *24*, 698–704.
43. Xi, G., Yue, B., Cao, J., Ye, J. *Chem. Eur. J.* **2011**, *17*, 5145–5154.
44. D'Arienzo, M., Armelao, L., Mari, C. M., Polizzi, S., Ruffo, R., Scotti, R., Morazzoni, F. *J. Am. Chem. Soc.* **2011**, *133*, 5296–5304.
45. Kalantar-zadeh, K., Vijayaraghavan, A., Ham, M., Zheng, H., Breedon, M., Strano, M. S. *Chem. Mater.* **2010**, *22*, 5660–5666.
46. Shibuya, M., Miyauchi, M. *Adv. Mater.* **2009**, *21*, 1373–1376.
47. Du, X., He, J. *Nanoscale* **2011**, *3*, 3984–4002.
48. Yu, X., He, J., Wang, D., Hu, Y., Tian, H., He, Z. *J. Phys. Chem. C* **2011**, *116*, 851–860.
49. Chen, H., Zhao, Y., Yang, M., He, J., Chu, P. K., Zhang, J., Wu, S. *Anal. Chim. Acta* **2010**, *659*, 266–273.
50. Zhou, L., He, J., Zhang, J., He, Z., Hu, Y., Zhang, C., He, H. *J. Phys. Chem. C* **2011**, *115*, 16873–16878.
51. Tian, H., He, J., Liu, L., Wang, D., Hao, Z., Ma, C. *Micropor. Mesopor. Mater.* **2012**, *151*, 397–402.
52. Yu, X., He, J., Wang, D., Hu, Y., Tian, H., Dong, T., He, Z. *J. Nanopart. Res.* **2012**, *14*, 1–14.
53. Xu, Y., Chen, D., Jiao, X. *J. Phys. Chem. B* **2005**, *109*, 13561–13566.
54. Xu, L., Sithambaram, S., Zhang, Y., Chen, C. H., Jin, L., Joesten, R., Suib, S. L. *Chem. Mater.* **2009**, *21*, 1253–1259.
55. Park, J. C. K., J., Kwon, H., Song, H, *Adv. Mater.* **2009**, *21*, 803–807.

56. Umar, A., Rahman, M. M., Al-Hajry, A., Hahn, Y. B. *Electrochem. Commun.* **2009**, *11*, 278–281.

57. Zhang, F., Zhu, A., Luo, Y., Tian, Y., Yang, J., Qin, Y. *J. Phys. Chem. C* **2010**, *114*, 19214–19219.

58. Wan, M., Jin, D., Feng, R., Si, L., Gao, M., Yue, L. *Inorg. Chem. Commun.* **2011**, *14*, 38–41.

59. Zhu, M., Diao, G. *Catal. Sci. Technol.* **2012**, *2*, 82–84.

60. Zhang, Z., Che, H., Wang, Y., Gao, J., She, X., Sun, J., Zhong, Z., Su, F. *RSC Advances* **2012**, *2*, 2254–2256.

61. Yu, X. Y., Xu, R. X., Gao, C., Luo, T., Jia, Y., Liu, J. H., Huang, X. J. *ACS Appl. Mater. Interfaces* **2012**, *4*, 1954–1962.

62. Yu, Q., Huang, H., Chen, R., Wang, P., Yang, H., Gao, M., Peng, X., Ye, Z. *Nanoscale* **2012**, *4*, 2613–2620.

63. Volanti, D. P., Felix, A. A., Orlandi, M. O., Whitfield, G., Yang, D.-J., Longo, E., Tuller, H. L., Varela, J. A. *Adv. Funct. Mater.* **2013**, *23*, 1759–1766.

64. Shao, Q., Wang, L., Wang, X., Yang, M., Ge, S., Yang, X., Wang, J. *Solid State Sci.* **2013**, *20*, 29–35.

65. Zhu, G., Xu, H., Xiao, Y., Liu, Y., Yuan, A., Shen, X. *ACS Appl. Mater. Interfaces* **2012**, *4*, 744–751.

66. Yao, W., Yu, S., Zhou, Y., Jiang, J., Wu, Q., Zhang, L., Jiang, J. *J. Phys. Chem. B* **2005**, *109*, 14011–14016.

67. Liu, J., Huang, X., Li, Y., Sulieman, K. M., He, X., Sun, F. *Cryst. Growth Des.* **2006**, *6*, 1690–1696.

68. Lai, X., Li, X., Geng, W., Tu, J., Li, J., Qiu, S. *Angew. Chem. Int. Ed.* **2007**, *46*, 738–741.

69. Zhang, X., Wang, G., Liu, X., Wu, J., Li, M., Gu, J., Liu, H., Fang, B. *J. Phys. Chem. C* **2008**, *112*, 16845–16849.

70. Taubert, A., Uhlmann, A., Hedderich, A., Kirchhoff, K. *Inorg. Chem.* **2008**, *47*, 10758–10764.

71. Li, Y., Liang, J., Tao, Z., Chen, J. *Mater. Res. Bull.* **2008**, *43*, 2380–2385.

72. Jia, Z., Yue, L., Zheng, Y., Xu, Z. *Mater. Res. Bull.* **2008**, *43*, 2434–2440.

73. Gou, X., Wang, G., Yang, J., Park, J., Wexler, D. *J. Mater. Chem.* **2008**, *18*, 965–969.

74. Gao, S., Yang, S., Shu, J., Zhang, S., Li, Z., Jiang, K. *J. Phys. Chem. C* **2008**, *112*, 19324–19328.

75. Chen, J., Wang, K., Hartman, L., Zhou, W. *J. Phys. Chem. C* **2008**, *112*, 16017–16021.

76. Mortimer, M., Kasemets, K., Vodovnik, M., Marinšek-Logar, R., Kahru, A. *Environ. Sci. Technol.* **2011**, *45*, 6617–6624.

77. Zhang, X., Shi, W., Zhu, J., Kharistal, D. J., Zhao, W., Lalia, B. S., Hng, H. H., Yan, Q. *ACS Nano* **2011**, *5*, 2013–2019.

78. Shrestha, K. M., Sorensen, C. M., Klabunde, K. J. *J. Phys. Chem. C* **2010**, *114*, 14368–14376.

79. Li, J., Liu, X., Ye, Y., Zhou, H., Chen, J. *J. Phys. Chem. C* **2011**, *115*, 4726–4729.

80. Qiu, G., Dharmarathna, S., Zhang, Y., Opembe, N., Huang, H., Suib, S. L. *J. Phys. Chem. C* **2012**, *116*, 468–477.

81. Qin, Y., Zhang, F., Chen, Y., Zhou, Y., Li, J., Zhu, A., Luo, Y., Tian, Y., Yang, J. *J. Phys. Chem. C* **2012**, *116*, 11994–12000.

82. Yang, M., He, J., Hu, X., Yan, C., Cheng, Z., Zhao, Y., Zuo, G. *Sens. Actuators B: Chem.* **2011**, *155*, 692–698.

83. Yang, M., He, J. *J. Colloid Interface Sci.* **2011**, *355*, 15–22.

84. Yang, M., He, J., Hu, X., Yan, C., Cheng, Z. *Environ. Sci. Technol.* **2011**, *45*, 6088–6094.

85. Yang, M., He, J. *J. Colloid Interface Sci.* **2012**, *368*, 41–48.

86. Yang, M., He, J., Hu, X., Yan, C., Cheng, Z. *Analyst* **2013**, *138*, 1758–1763.

Chapter 13

Nanomaterials and Health

Wenjun Ding and Jieting Wang
Laboratory of Environment and Health, College of Life Sciences, University of Chinese Academy of Sciences, No. 19A Yuquan Road, Beijing 100049, China
wangjieting08@mails.ucas.ac.cn, wangjt@ucas.ac.cn

13.1 Introduction

Nanotechnologies are emerging as highly promising technologies in many sectors in the society. However, the prospective widespread use of engineered nanomaterials in energy and environment also raises concerns about inadvertent exposure to these materials and the potential for adverse effects on human health and the environment [1]. Nanomaterials released in the production, transportation, storage, and other aspects are likely to enter the environment, then through uptake, accumulation, transformation, and degradation in organisms [2]. Due to their relatively small size (at least one dimension in the range of 1–100 nm) and large surface area per mass

Nanomaterials in Energy and Environmental Applications
Edited by Junhui He

ISBN 978-981-4463-78-2 (Hardcover), 978-981- 4463-79-9 (eBook)
www.panstanford.com

unit, nanomaterials could gain access to biological sites. Several studies suggested that nanomaterials invade the human body via the dermal, inhalation, ingestion, and ocular pathway [2,3]. Shape, size, density, and other factors can affect the deposition of nanoparticles in different areas of the human body [4]. Deposited nanoparticles may be cleared through different mechanisms. Studies have shown that nanoparticles which deposited in the alveolar region can efficiently phagocytosis and clearance by macrophages, causing lung inflammation, granulomas, and fibrosis and lead to dysfunction of these critical targets, being associated even with diseases [3,5–9].

It is critical to evaluate hazards associated with the use of engineered nanomaterials. Despite several years of intensive investigations, a common paradigm for the understanding of nanomaterial-induced toxicity remains to be firmly established. Thus, the toxic potential of each type of nanomaterials has to be evaluated individually. This review then aims to provide an overview of the impact of nanomaterial on organisms. Meanwhile, information about safety and potential hazards is urgently needed.

13.2 Nanomaterials

13.2.1 Carbon-Based Nanomaterials

13.2.1.1 Fullerenes (C_{60})

Fullerenes are a series of nanomaterials that have a caged structure, consisting purely of carbon. They are being used particularly in the fields of catalyst supports, battery anodes, and proton transport membranes and the like [10].

13.2.1.1.1 *In vivo studies*

The toxicity studies of fullerenes focus on whether fullerenes could induce redox-active, lipophilic molecule could cause oxidative damage. Among toxicity of several nanoparticles, the most controversial is the fullerene.

The first study of fullerenes-caused oxidative damage is the one by Oberdorster et al., which investigated oxyradical-induced lipid and protein damage, and depletion of total glutathione (GSH) in juvenile largemouth bass [11]. Further research needs to be done to evaluate the potential toxicity of manufactured nanomaterials, especially with respect to translocation into the brain [11].

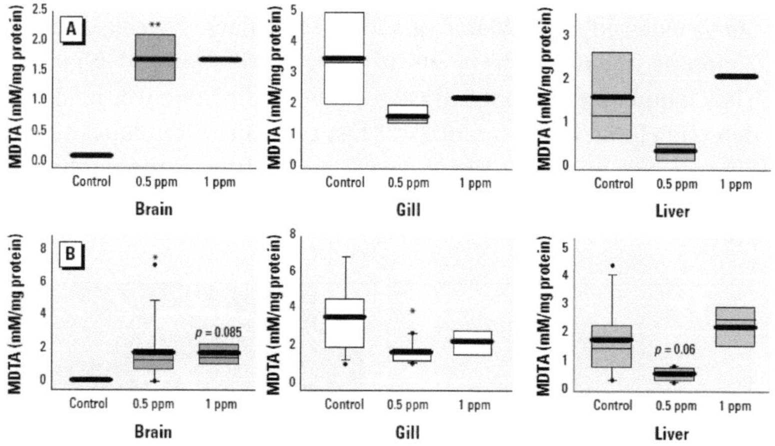

Figure 13.1 Lipid peroxidation of brain, gill, and liver of largemouth bass after 48 h of exposure to 0.5 or 1 ppm nC60. MDTA, 1,1,3,3-tetraethoxy propane. (A) Aquarium averages. (B) Data using individual fish; in the brain, 0.5 ppm nC60 caused a significant 17-fold increase in lipid peroxidation, whereas in gill and liver there is a trend for reduction of lipid peroxidation at 0.5 ppm nC60. Heavy black bands represent the means; thinner lines indicate medians; boxes represent 25th and 75th percentiles; error bars indicate minimum and maximum; and circles represent outliers. *$P < 0.05$. **$P < 0.01$ [11]. Reproduced from Environmental Health Perspectives, http://www.ncbi.nlm.nih.gov/pmc/articles/PMC1247377/

On the contrary, a recent study by Takahashi et al. involved sub-acute oral toxicity study with fullerene C60 in rats. SD rats were exposed to fullerene C60 by oral gavage once daily at 0, 1, 10, 100, or 1000 mg/kg/day for 29 days, followed by a 14-day recovery period. However, they did not found treatment-related histopathological changes in any organs examined at the end of the administration period and at the end of the recovery period. Fullerene was not detected in the liver, spleen, or kidney at the end of the treatment period and recovery period by using liquid chromatography-tandem mass spectrometry. However, liver and spleen weights were increased after the 14-day recovery period [12].

Therefore, it has been suggested that inhaled nanoparticles were cleared by the alveolar macrophages and/or transferred to the pulmonary interstitium [13]. In a recent study, Shinohara et al. determined the deposition, clearance, and translocation of C60 nanoparticles (18–29 nm, 100, 200, and 1000 micrograms per body) exposed male Wistar rats for 18 h, 3 days, 1 week, 1 month, 3 months, and 6 months by intratracheal instillation and inhalation. They found that the concentration of C60 in the brain was below the detection limit. C60 clearance was fast clearance after deposition on lung surface and slow clearance after retention in the epithelium. The calculated biological half clearance time of C60 was 15–28 days [14].

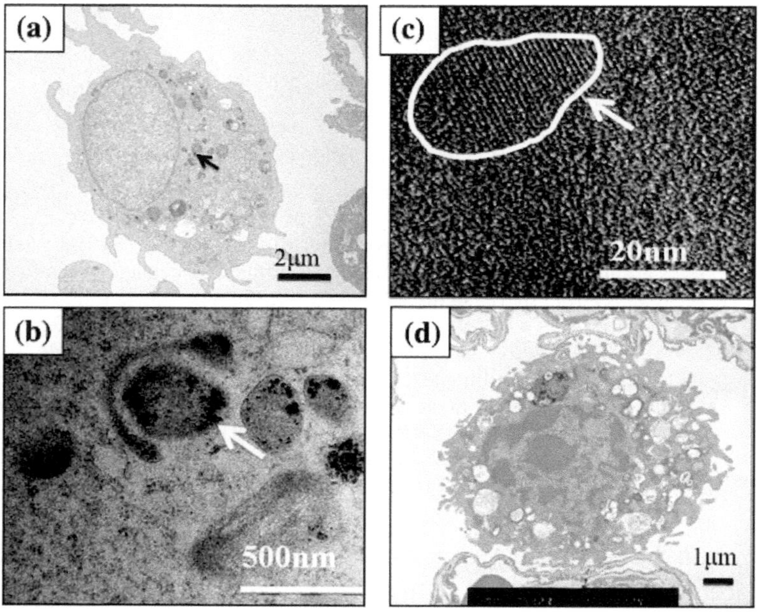

Figure 13.2 TEM images of alveolar macrophage instilled with C60 (0.2 milligrams per rat). (a) Low-magnification image of alveolar macrophage after 1-week instillation exposure. (b) Magnified image of phagosome in alveolar macrophage after 1-week instillation exposure. Many black particles resembling fullerene particles are seen in the phagosome. (c) High-resolution image of the black particles in (b). A crystal lattice order can be observed. (d) Low-magnification image of alveolar macrophage after 6 months instillation. Few fullerene particles can be seen [14]. Copyright © 2010, Oxford University Press.

13.2.1.1.2 *In vitro studies*

Several studies have demonstrated that fullerenes have the ability to quench various free radicals, with evidence demonstrating that fullerenes are able to damage DNA possibly as a result of oxidative stress-based mechanisms [15]. Thus, to define and constrain the potential biomedical applications of C60, it is of great interest to identify the genotoxicity of C60 in mammalian cells.

Although studies have reported that exposure to fullerenes induced cytotoxicity, there are conflicting reports in the previous studies. Jacobsen et al. examined the cell viability, ROS generation, cell cycle, genotoxicity, and mutagenicity induced by C60 fullerenes in the mouse lung epithelial cell line. Fullerenes did not induced cell death and induce DNA strand breakages within 24 h at doses between 0 and 200 mg/ml or during long-term subculture exposure at 100 mg/ml [16]. The findings of the current study are not consistent with those of Dhawan et al. who demonstrated human lymphocytes using single-cell gel electrophoresis assay (Comet assay). They observed genotoxicity of nC60 at concentrations as low as 2.2 μg/l for aqu/nC60 and 4.2 μg/l for EtOH/nC60. The Olive tail moments (OTMs) for these two concentrations were increased statistically in comparison to the negative control [17].

The toxicity of C60 thought to be mediated through oxidative stress induced lipid peroxidation of cell membrane [18]. In a recent study, Isakovic et al. confirmed that C60 caused ROS-associated necrosis characterized by cell membrane damage without DNA fragmentation. In contrast, C60(OH)n caused delayed, ROS-independent cell death with characteristics of apoptosis, including DNA fragmentation and loss of cell membrane asymmetry in the absence of increased permeability in mouse L929 fibrosarcoma, rat C6 glioma, and U251 human glioma cell lines [19].

C60 dramatically increased the mutation yield, which could be abrogated by concurrent treatment with the endocytosis inhibitor, Nystatin. Under confocal scanning microscopy together with the radical probe dihydrorhodamine 123 (DHR 123), Xu et al. demonstrated that C60 induced increase of ONOO- in a dose-dependent manner in live MEF cells, and the protective effects of antioxidants were demonstrated by the nitric oxide synthase

(NOS) inhibitor, NG-methyl-L-arginine (L-NMMA). Furthermore, suppression of cyclooxygenase-2 (COX-2) activity significantly reduced C60 treatment which reduced the mutant fractions. The induction of ONOO- may be a critical signaling event for nanoparticle genotoxicity [20].

13.2.1.2 Carbon nanotubes (SWNTs and MWNTs)

Carbon nanotubes are a promising breakthrough in the field of material science for consumer, industrial, and medical purposes. With their anticipated production and application at a large scale, bioeffects and safety of CNTs to human health need to be addressed. However, unlike chemical agents with well-defined structures and purity, CNTs are different in scale, structure, purity, and morphology, depending upon the preparation and purification methods. Briefly, the toxicity can be attributed to various factors: metal impurities, length and size distribution, surface area, dispersion and aggregation status, coating, type of functionalization, concentration, duration of exposure, or method of exposure [21,22].

13.2.1.2.1 *In vivo study*

SWCNTs were reported to evade phagocytosis and rapidly penetrate the epithelial barrier and enter the interstitium of alveolar walls. Huczko et al. first showed that SWCNT did not show any signs of toxicity skin irritation and allergic risks and did not induce any abnormalities of pulmonary function or measurable inflammation in guinea pigs [23,24]. However, Lam et al. observed that SWCNT induced dose-dependent epithelioid granulomas and peribronchial inflammation in the mice, which were exposed intratracheally instilled with 0, 0.1, or 0.5 mg of carbon nanotubes for 7 days and 90 days [5]. This is particularly evident in studies of single-walled carbon nanotubes (SWCNT), in which granulomatous lesions encased by epithelioid macrophages were produced by large agglomerates.

There was also evidence that SWCNT can reach the alveolar interstitium in prior intratracheal instillation or aspiration [25]. The distribution of MWCNT was determined by using morphometric methods, and they found that MWCNT penetrated into the alveolar macrophages, the alveolar wall, and visceral pleura in C57BL/6J mice [26].

In general, the nanotoxicological effect is related to the specific surface area, and SWCNTs have greater specific surface areas than MWCNTs [27]. Moreover, the aggregation ability of MWCNTs is lower than that of SWCNTs. It is somewhat difficult to compare the cytotoxicity of MWCNTs and SWCNTs, as it is not clear whether nanotoxicity should be related to the same mass concentration of CNTs or the same total surface area [21]. MWCNT might have the potential to translocate or penetrate to the pleura. Studies dealing with oral exposure of SWCNT NPs in mice have demonstrated that dispersed SWCNTs rapidly penetrated the epithelial barrier and entered the alveolar interstitium, inducing an early, transient neutrophilic and inflammatory phase, and granulomatous lesions encased by epithelioid macrophages at 1 h, 1 day, 7 day, and 1 month. This is particularly evident in prior intratracheal instillation or aspiration studies of SWCNT, in which granulomatous lesions were encased by epithelioid macrophages [25].

Although the important question of whether inhaled MWCNT will translocate to these sensitive pleural sites has not been studied yet, results of new studies using dosing by oro-pharyngeal bolus aspiration of MWCNT in mice show dose-dependent penetration into the alveolar and subpleural interstitial space and into the pleural space anulomas observed around the materials, infiltration of inflammatory cells, and increase of IL-8 concentration in BAL-fluid. In a long-term assay in vivo, the authors observed 4 weeks after surgery that an increased inflammatory response was established only by the 825 nm MWCNTs, as indicated by the formation of granulation tissue [28].

Ma-Hock et al. investigated the impact of MWCNT (Nanocyl NC 7000) on rats by head-nose expose for 6 h/day, 5 days/week, and 13 weeks. Their results clearly show that inhaled MWCNT induced increased lung weights, multifocal granulomatous inflammation, diffuse histiocytic and neutrophilic inflammation, and intra-alveolar lipoproteinosis in lung and lung-associated lymph nodes at 0.5 and 2.5 mg/m^3. Moreover, the rats that were exposed to 2.5 mg/m^3 MWCNT had goblet cells hyperplasia and increased incidence of eosinophilic intraepithelial inclusions in the respiratory epithelium [29].

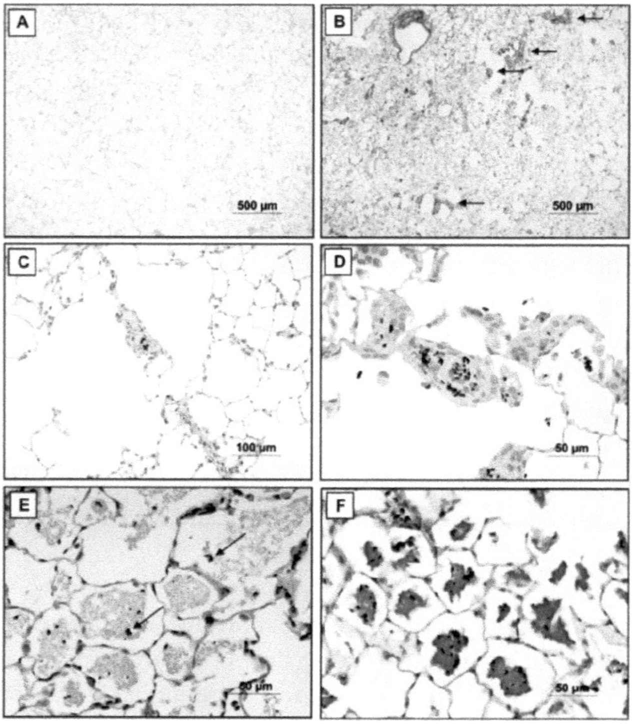

Figure 13.3 Overview of the lung section of (A) control rat and (B) rat exposed to 2.5 mg/m^3 MWCNT; arrows indicate multifocal granulomatous inflammation. (C) is a detailed view of focal granulomatous inflammation observed in a rat exposed to 0.1 mg/m^3 MWCNT, and (D) in a rat exposed to 2.5 mg/m^3 MWCNT. The dark particles appear to be intralesional-located MWCNT. (E) A lung section of a rat exposed to 2.5 mg/m^3 showing intra-alveolar eosinophilic granular material, which was diagnosed as lipoproteinosis. The intermingled black staining clumps (arrows) appear to be CNT. A few alveolar macrophages and neutrophils are present. (F) This intra-alveolar eosinophilic granular material was PAS-positive (A–E: H&E stain, F: PAS) [29]. Copyright © 2009, Oxford University Press.

13.2.1.2.2 *In vitro studies*

The cytotoxic effects of CNT have been controversial, with some investigations reporting toxicity following the exposure of several cell types to both single-walled carbon nanotubes (SWCNT) and multi-walled carbon nanotubes (MWCNTs) [16,28,30–34], whereas others demonstrate very low or insignificant cellular responses [35,36]. The effects of nanoparticle exposure seen in in vitro studies indicate that CNTs can enter human cells and accumulate in the cytoplasm, induce harmful effects such as inflammatory and cell death [37,38]. Previous studies showed that SWCNT did not cause cell death and DNA strand breakages or increases over background mutation frequencies at doses up to 100 mg/ml in mouse lung epithelial cell line [16, 39]. However, there are another two studies that have observed the opposite using the same technique (the comet assay). Kisin and colleagues found that high purity, 1.3 mm length SWCNT induced significant DNA strand breakages at 96 mg/ cm^2 following 3 h and 24 h exposure using the Comet assay [40]. Pacurari and colleages also found significant DNA damage but at lower concentrations of 25 mg/cm^2 using SWCNT with lengths of 2–5 mm and approximately 72% carbon purity [34].

Physical properties of the nanotube such as length and shape affect toxicity. Sato et al. compared the cytotoxicity of 220 nm and 825 nm CNTs in human acute monocytic leukemia cell line (THP-1). They found that both CNTs (220 nm and 825 nm) induced human monocytes in vitro, and no activity appeared following variation in the length of CNTs [28]. However, the degree of inflammation around 825 nm CNTs was stronger than that around 220 nm CNTs. Results also demonstrate that smaller size of MWCNT seems to be more toxic than that of larger one [31]. The previous study evaluated the potential cytotoxicity of two kinds of MWCNT (2 μm and 10 μm in average length) in C6 rat glioma cell line. The results show that both kinds of carbon nanotubes at concentrations between 25 and 400 μg/ml decreased the cell viability in a concentration and time-dependent manner. Interestingly, the smaller size of MWCNT seems to be more toxic than that of larger one [31]. The influence of CNTs length on toxicity should be considered carefully.

The cytotoxicity of CNTs with or without purification and functionalization has been studied in a variety of cell types in vitro. CNTs containing metal catalytic decomposition can induce cytotoxicity. Recent studies suggest that SWCNT stimulated ·OH generation in HaCaT cells. SWCNT-associated ferrous iron can catalyze the decomposition of hydrogen peroxide and organic (lipid) peroxides, yielding hydroxyl radical from hydrogen peroxide or alkoxyl radicals from lipid peroxide. In both cases, reactive intermediates are formed that propagate free radical oxidation and enhance oxidative stress [41]. Furthermore, unpurified CNTs contain metal impurities. The increased DNA damage observed by Pacurari and colleagues could have been due to the high content of metal impurities in the CNT assessed [34]. Furthermore, the exposure of C6 rat glioma cells to MWCNT (200–400 μg/ml) resulted in a concentration-dependent cell apoptosis and G1 cell cycle arrest, and increased the level of oxidative stress [31]. Exposure to Fe-high MWCNTs can reduce cell viability and increase cytoskeletal disruption of undifferentiated PC12 cells, diminish the ability to form mature neurites, and then adversely influence the neuronal dopaminergic phenotype in NGF-treated PC-12 cells. The present results demonstrated the critical role of iron residue in the adverse response to MWCNTs exposure in neural cells [32].

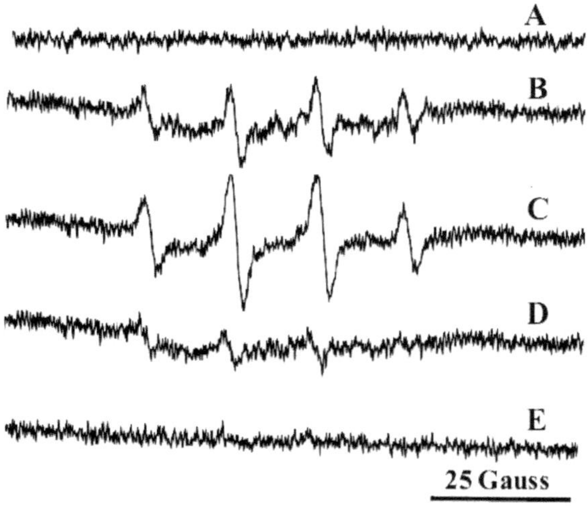

Figure 13.4 Generation of free radicals from SWCNT-stimulated HaCaT cells. (A) HaCaT (10^6 cells/ml) in a phosphate-buffered solution (pH 7.4) containing 40 mM 5,5-dimethyl-1-pyrroline-1-oxide. (B) Electron spin resonance (ESR) spectrum recorded 5 min after the addition of SWCNT (0.24 mg/ml) to HaCaT (10^6 cells/ml) in a phosphate-buffered solution (pH 7.4) containing 40 mM 5,5-dimethyl-1-pyrroline-1-oxide. (C) Same as (B) plus 1 mM H_2O_2 added. (D) Same as B plus 20 U/ml catalase added. (E) Same as (B) plus 0.2 mM deferoxamine added. Instrumental conditions: microwave power, 50 mW; modulation amplitude, 1.0 G; time constant, 40.96 ms; center field, 3435 G; sweep width, 100 G [41]. Copyright © 2003 Taylor & Francis.

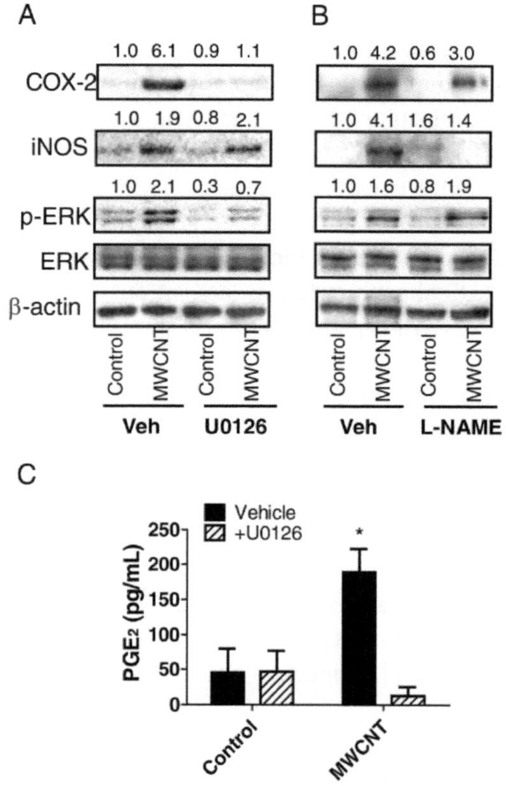

Figure 13.5 Effect of pharmacologic inhibitors on MEK and iNOS on the induction of COX-2 in RAW264.7 cells exposed to MWCNTs. (A) Western blot analysis showing inhibition of COX-2 expression and phosphorylated ERK1, 2 (p-ERK1, 2) by the MEK inhibitor U0126 relative to vehicle (Veh) in cells treated with 50 μg/ml MWCNTs. Controls (C) were treated with serum-free medium alone. Levels of iNOS were not affected by U0126 (10 μM) at 24 h post-treatment with MWCNTs, nor were constitutive levels of total ERK1, 2, or β-actin. (B) Western blot showing inhibition of MWCNT-induced iNOS at 24 h by L-NAME (250 μM) with no significant effect on MWCNT-induced COX-2 or p-ERK. Numeric values above the western blot represent the fold-increase in relative densitometric intensity of each band relative to control (zero) nanoparticle treatment, which was normalized to 1.0. Data for all three experiments were from duplicate studies, which yielded similar results. (C) Results of PGE2 ELISA showing that production of PGE2 and release into cell supernatants after 24 h treatment with MWCNTs (50 μg/ml) is blocked by 10 μM U0126 co-treatment [32]. Copyright 2012 Lee et al.; licensee BioMed Central Ltd.

Recent studies of CNT-induced damage in vivo and in vitro are discussed along with proteomics and transcriptomics approaches to achieve a comprehensive understanding of the complex and interrelated molecular changes in cells and tissues exposed to nanoparticles [42]. It has been demonstrated that CNTs can inhibit mammalian cells growth, decrease cell adhesive ability, and even induce cell apoptosis [21,30,31,33,34,41]. Exposure to SWCNTs induced ROS generation, increased cell death enhanced DNA damage, H2AX phosphorylation, activation of poly (ADP-ribose) polymerase 1 (PARP-1), stimulation of nuclear factor kappaB (NF-kappaB), protein p38, and activation of activator protein-1 (AP-1), and protein serine-threonine kinase (Akt) in a dose-dependent manner in mesothelial cells [34]. Another study also suggests that SWCNTs induced downregulation of G1-associated cdks and cyclins and upregulation of apoptosis-associated genes, which contribute to G1 phase arrest and cell apoptosis [30].

These findings provide useful information for understanding the toxicity and safe application of CNTs. It is becoming increasingly evident that the toxicity of nanomaterials significantly varied in the physico-chemical properties and cell type-specific differences. Surface chemistry is also another key parameter that is often overlooked, but has been reported to have a substantial effect upon toxic potential. Thus, emphasis on the need for complete physico-chemical characterisation is critical in aiding the interpretation and comparison of results in the literature [43].

13.2.2 Metal Oxide Nanoparticles

13.2.2.1 Zinc oxide nanoparticles (ZnO NPs)

ZnO NPs belongs to a group of metallic oxide nanoparticles (MONP), which is widely used in the manufacture of paints, coatings, and environmental cleanup [7,44–46]. Accumulating evidence from in vivo and in vitro studies has revealed that ZnO NPs are toxic to mammals and cells [47–51]. Exposure to ZnO NPs induces a series of toxic effects, such as oxidative damage and apoptosis [49,51,52]. Therefore, it is important to understand the biological effects and risks of ZnO NPs.

13.2.2.1.1 *In vivo studies*

For ZnO NPs, only a few in vivo studies are available and most of them are focused on respiratory tract exposure [53]. Our previous studies also demonstrated that ZnO NPs could decrease serum alanine aminotransferase (ALT), aspartate aminotransferase (AST), alkaline phosphatase (ALP), total protein (TP), creatine kinase (CK), and lactate dehydrogenase (LDH) significantly compared with the unexposed group by treatment of ZnO NPs at 2.5 mg/kg body weight, twice daily for 3 days. Meanwhile, the histopathological examination showed that ZnO NPs caused inflammation, interstitial hyperemia, emphysema, and interstitial substance hyperplasia in liver and lung tissues [54].

Exposure to ZnO NPs in mice leads to an accumulation of nanoparticles in the liver causing oxidative stress mediated DNA damage and apoptosis. Sharma et al. investigated the oral toxicity of ZnO NPs. Their results showed that histopathological changes occurred in liver and kidney induced by ZnO NPs (300 mg/kg) for 14 consecutive days. The liver revealed hepatocellular necrosis and accumulation of mixed inflammatory cells around the necrotic area. The kidney depicted cystic dilation of tubules along with hypertrophied tubules. Moreover, rats exposed to ZnO NPs produced inflammation in the lungs and bronchoalveolar lavage [BAL] fluid lactate dehydrogenase [LDH] values via the oral route measured at 24 h, 1 week, 1 and 3 months postexposure, potent but reversible inflammation that was resolved by 1 month postinstillation exposure (Fig. 13.6) [47].

Several studies suggested that ZnO NPs could be taken up by the olfactory bulb and cross the blood–brain barrier (BBB) to disrupt the brain in rats, and that endocytosis is required for interneuron translocation of these particles [55,56]. Han et al. focused on the effects of ZnO NPs on spatial learning and memory and synaptic plasticity in the hippocampus of young rats, and tried to interpret the underlying mechanism. ZnO NPs were administered in 4-week-old rats for 8 weeks. The data showed that the escape latency was prolonged in the ZnO NPs treated group in Morris water maze (MWM). Moreover, long-term potentiation (LTP) was significantly enhanced in the ZnO NPs treated group, while depotentiation was barely influenced in the dentate gyrus (DG) of the ZnO NPs group.

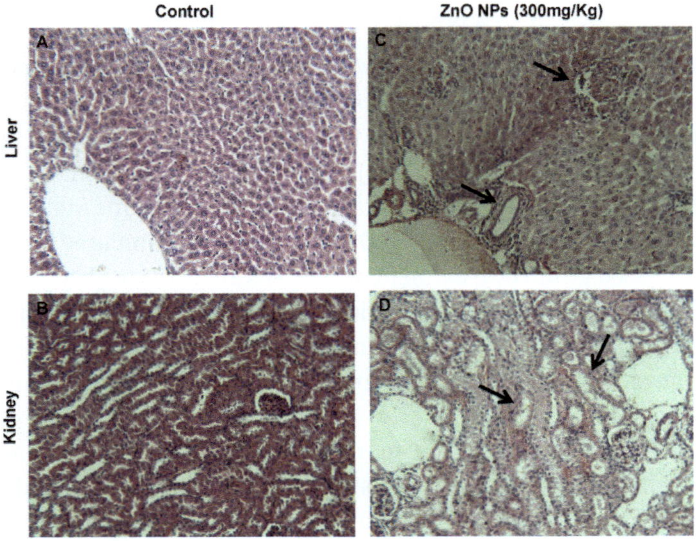

Figure 13.6 Histopathology of liver and kidney tissues in mice treated with ZnO nanoparticle (NPs) for 14 consecutive days. (A) and (B) Control group showing normal liver and kidney. (C) and (D) Pathological alterations in the liver and kidney of ZnO NPs (300 mg/kg) treated group (indicated by arrow); magnification (200×) [47].

This bidirectional effect on long-term synaptic plasticity broke the balance between stability and flexibility of cognition. The spatial learning and memory ability was attenuated by the alteration of synaptic plasticity in ZnO NPs treated rats [57].

The previous study showed that ZnO NPs may induce disturbance of the ionic homeostasis and the physiological functions of neurons. Their study suggests that ZnO NPs affect the properties of voltage-dependent sodium, potassium currents, and evoked action potentials in rat hippocampal CA3 pyramidal neurons at postnatal ages of 10–14 days old rats. The results showed that exposure to 100 μg/ml ZnO NPs decreased the current-voltage curve of sodium current and increased the amplitudes of transient outward potassium current (I_A). Moreover, peak amplitude and overshoot of the evoked single action potential were increased and half-width was diminished in the presence of 100 μg/ml ZnO NPs solution. A prolonged depolarizing current injection enhanced ($P < 0.05$) repetitive firing evoked firing rate at the same time (Fig. 13.7) [52].

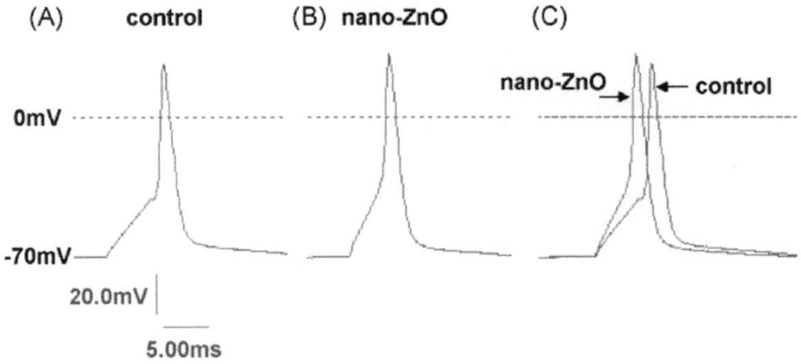

Figure 13.7 Effect of ZnO NPs on evoked single action potential. The single action potentials elicited by brief depolarizing current pulses (5 ms, 100 pA) from a holding potential of 70 mV before (A) and after application of ZnO NPs (B); (C) is the overlapped traces of (A) and (B) [52]. Copyright © 2008 Elsevier Inc. All rights reserved.

13.2.2.1.2 *In vitro studies*

Studies had indicated that ZnO NPs could exert cytotoxicity on both broncho-alveolar lavage cells and white blood cells in rats via interfering with zinc ion homeostasis, the liver, and kidney cells in mice via induction of oxidative stress, DNA damage and apoptosis, human bronchial epithelial cells, human keratinocyte HaCaT cells [58–62]. In addition, ZnO NPs could induce apoptosis in human dermal fibroblasts via p53 and p38 pathways [63], exert cytotoxicity on rat retinal ganglion cells via overexpression of caspase-12, and trigger endoplasmic reticulum stress [64], caspase-7, and PARP cleavage [61] and finally induce apoptosis/necrosis.

A number of studies suggest that the effects of nanoparticles are likely to depend on their shape. A significant example is the cytotoxic effects of ZnO nanoparticles and microspheres on RSC96 Schwann cells in dose and time-dependent manners, whereas no or low cytotoxic effect was observed when the cells were treated with the prism-like and flower-like ZnO [65].

It has been suggested that adverse effects of nanoparticles on cell depend on individual factors such as size of nanoparticle. There is ongoing concern regarding the toxicity of nanoparticles with sizes less than 100 nm as compared with larger particles of the same nominal substance. Moos et al. evaluated effect of two commercial

ZnO types (8–10 nm and <44 μm) powder in human colon derived RKO cells. They found that the ZnO NPs were more cytotoxic than the micrometer-sized ZnO with LC50 values of 15 ± 1 and 29 ± 4 μg/cm^2, respectively. Both ZnO samples induced similar mechanisms of toxicity, but there was a statistically significant increase in potency per unit mass with the smaller particles [66]. However, the study of the relationship between toxicological effects and physico-chemical characteristics of particles suggests that surface area does not play a primary role in the cytotoxicity [67]. Bae et al. investigated the potential of cytotoxicity induced by ZnO NPs in human keratinocyte HaCaT cells. ZnO NPs of 100 nm induced a greater increase of cytosolic ROS levels in 30 min treatment, compared with treatment with 20 nm ZnO NPs at the same concentration [61].

The key factors of ZnO NPs induced toxicity include nanoparticle dose and their solubility. ZnO NPs lead to cellular mitochondrial dysfunction, morphological modifications, and apoptosis at the concentration range of 0.25–1.50 mg/ml and the toxic effects are obviously displayed in a dose-dependent manner [68]. ZnO NPs induced a remarkable cell apoptosis and G2/M cell cycle arrest in RSC96 Schwann cells at a dose of 80 μg/ml for 12 h [65]. The potential cytotoxicity of ZnO NPs with three different sizes has been investigated using human embryonic lung fibroblasts (HELF) cells. The results indicated that ZnO NPs aqueous suspensions with concentrations varying from 2.5 to 150 mg/l were harmful to the HELF cells and the cytotoxicity to HELF showed obviously concentration dependence. With concentrations higher than 20 mg/l, ZnO NPs led to the livability of the tested cells lower than 10%. These results highlighted the need for caution during the use of ZnO NPs above 1.2–1.6 mg/l for safety. Finally, a UV–vis spectroscopic approach was developed for the determination of ZnO NPs concentration in the linear range of 2–50 mg/l, and the detection limit was 1.0 mg/l [69].

Several studies have revealed that it is the CO_2 atmosphere under cell culture condition that promoted the solubilization of ZnO NPs. Transfer of Zn from the solid phase to the cell culture media in the presence of cells was time and concentration-dependent, and the toxicity of particles was independent of the amount of soluble Zn in the cell culture media [66]. They suggests that the cytotoxic effects were associated with the subsequent release of Zn^{2+} [70]. They point out that disruption of cellular Zn homeostasis in in vitro systems has been linked to loss of viability, oxidative stress, and mitochondrial dysfunction [58,71]. This was confirmed by the time-dependent increase of Zn^{2+} concentration in the culture media (Fig. 13.8) [65]. Deng et al. have reported that ZnO NPs and $ZnCl_2$ had the similar toxic effect on mouse NSCs [51]. Similarly, Song et al. found that reduced cell viability and LDH leakage induced by supernatants was similar to ZnO NPs suspensions at dosage under 10 μg/ml by comparing the toxicity of supernatants with ZnO NPs suspensions, which indicated that the toxicity was mainly due to dissolved zinc ions at the low dosage [72]. Furthermore, Müller et al showed that pH-triggered intracellular release of ionic Zn^{2+} is responsible for the toxicity of ZnO nanowires [73]. Thus, zinc ions played a major role in the toxicity of ZnO NPs. These results provide new insights into the toxicity mechanism of ZnO NPs attributed to zinc ions. Other studies showed that free Zn^{2+} ions released from the ZnO NPs were not responsible for the viability decrease [74]. A recent study demonstrated that overexpression of MGST1 could not reverse the cytotoxicity of the ZnO NPs because rapid dissolution of ZnO NPs and the subsequent interaction of $Zn2^{+}$ with cellular targets is likely to contribute to the cytotoxic effects. A direct inhibition of MGST1 by Zn^{2+} could provide a possible explanation for the lack of protection against ZnO NPs [75].

For the in vitro component of the study, Sayes et al. compared the effect of ZnO NPs on three different culture conditions: (1) rat L2 lung epithelial cells, (2) primary alveolar macrophages (AMs), and (3) AM-L2 lung epithelial cell cocultures. The results demonstrated that L2 cells were the most sensitive and exposures to nano or fine-sized ZnO for 4 or 24 h were more cytotoxic than exposures to CS or AS particles. Macrophages essentially were resistant and epithelial macrophage cocultures generally reflected the epithelial results at 4 and 24 h incubation, but not at 48 h incubation. MTT results were also interesting but, except for nano and fine-sized ZnO, did not correlate well with LDH results. Results of in vitro pulmonary inflammation studies demonstrated that L2 cells did not produce MIP-2 cytokines. Measurements of TNF-a in the culture media by particle-exposed cells demonstrated little activity. In addition, IL-6 secretion was measured in CS, AS, and ZnO NPs-exposed cocultures [76].

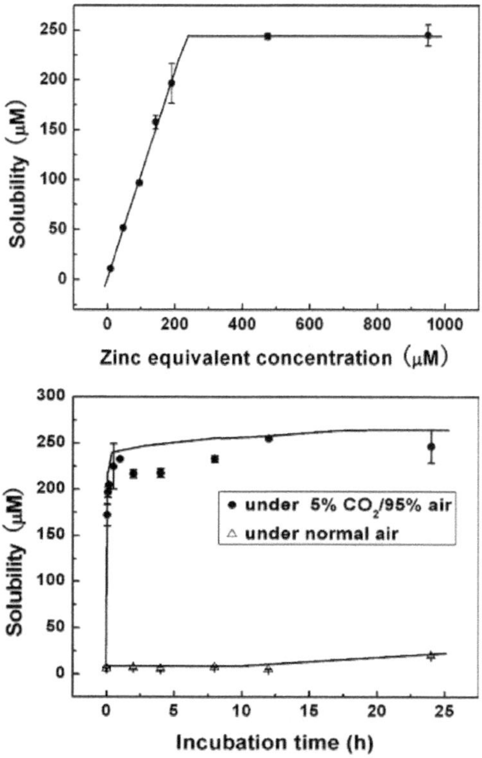

Figure 13.8 Solubilization of ZnO NPs in cell culture medium.(a) A concentration course of the solubilization of ZnO NPs. (b) A time course of the solubilization of ZnO NPs [70]. Copyright © 2010 American Scientific Publishers.

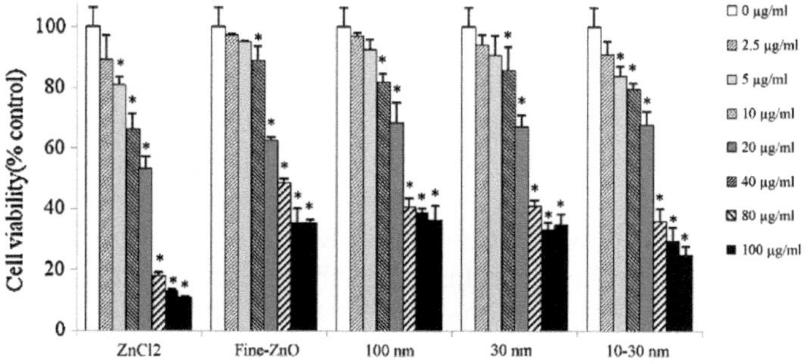

Figure 13.9 Cell viability of Ana-1 cell exposed to different sized ZnO particles. Control cells treated without ZnO particles are considered to have 100% activity. *$P < 0.05$ vs. control cells ($n = 5$) [72]. Copyright © 2011 Elsevier B.V. All rights reserved.

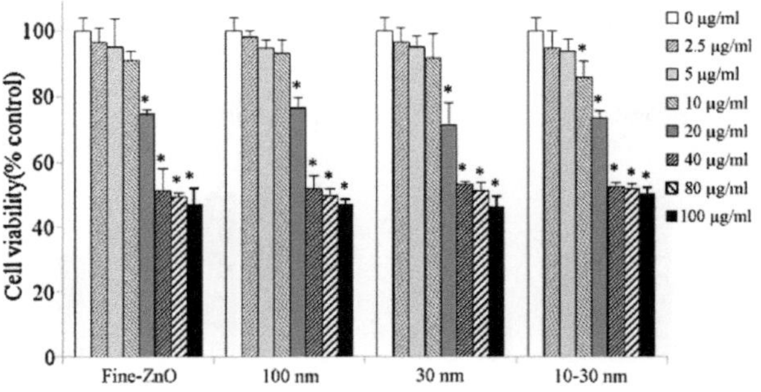

Figure 13.10 Cytotoxicity of supernatant of ZnO particles suspensions incubated for 24 h. *$P < 0.05$ vs control cells ($n = 5$) [72].

Another study showed that ZnO NPs exhibited a preferential ability to kill cancerous HL60 cells as compared with normal PBMCs. The nanoparticles enhanced ultrasound-induced lipid peroxidation in the liposomal membrane. The work also revealed potential utility of ZnO NPs in the treatment of cancer, for their selective toxicity to cancer cells [77].

Ostrovsky et al. examined the cytotoxic effect of ZnO NPs on various human cancer and normal cells. They found that the ZnO NPs exerted a cytotoxic effect on human glioma cell lines, whereas no cytotoxic effect was observed on normal human astrocytes. Similarly, the ZnO NPs induced cell death in breast and prostate cancer cell lines, although no major effect was observed in the respective normal breast and prostate cell lines [78].

ZnO NPs induced genotoxicity in human epidermal cell line (A431). ZnO NPs demonstrated a DNA damaging potential as evident from an increased OTM of 2.13 ± 0.12 (0.8 g/ml) compared with control 1.37 ± 0.12 in the Comet assay after an exposure of 6 h. ZnO NPs were also found to induce oxidative stress in cells indicated by depletion of glutathione (59% and 51%); catalase (64% and 55%) and superoxide dismutase (72% and 75%) at 0.8 and 0.08 g/ml, respectively. ZnO NPs even at low concentrations possess a genotoxic potential in human epidermal cells, which may be mediated through lipid peroxidation and oxidative stress. Hence, caution should be taken in their use in dermatological preparations as well as while handling [45].

Valdiglesias et al. examined the genotoxic effects of ZnO NPs on human SHSY5Y neuronal cells. Results showed that ZnO NPs not only induced cytotoxicity but also genotoxicity, including micronuclei production, H2AX phosphorylation, and DNA damage, both primary and oxidative, on human neuronal cells in a dose and time-dependent manner [74].

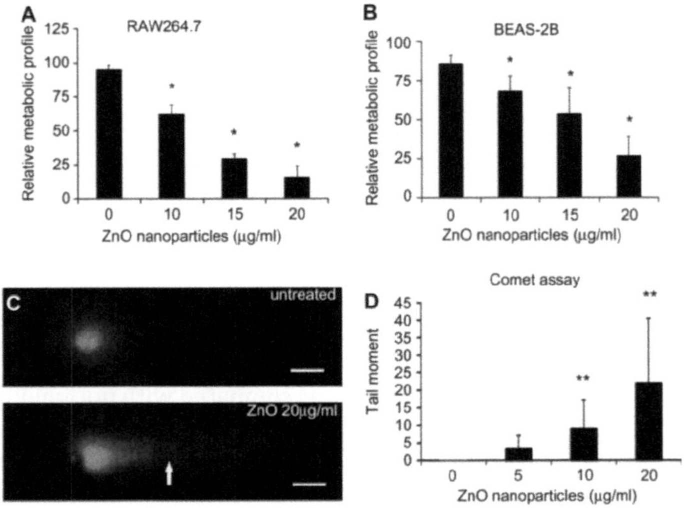

Figure 13.11 Cytotoxic and genotoxic effects of ZnO NPs. WST-8 assay performed on (A) RAW264.7 and (B) BEAS-2B cells showed cytotoxic effects of ZnO NPs in a dose dependent manner after 24 h exposure. (C) DNA damage in BEAS-2B cells was observed using the comet assay. Fluorescent images of encapsulated DNA showed absence of DNA migration in BEAS-2B cultured without nanoparticles (top panel) while cells cultured with 20 mg/ml ZnO NPs for 4 h showed clear migrating DNA tails (bottom panel, arrowhead). Scale bars: 30 mm. (D) Plot of tail moments indicated that there was increasing DNA damage in BEAS-2B cells with increasing ZnO nanoparticle concentration [85]. Copyright © 2011 Elsevier Ltd.

They investigated that the influence of 20 nm ZnO NPs on cytotoxicity, oxidative stress, intracellular calcium homeostasis, and gene expression was studied in human bronchial epithelial cells (BEAS-2B). ZnO caused a concentration and time-dependent cytotoxicity while elevating oxidative stress and causing membrane damage. There was a remarkably steep relationship between concentration and toxicity at concentrations from 5 to 10 μg/ml. Cytotoxicity was completely abolished by the antioxidant N-acetylcysteine (NAC). Exposure to ZnO also increased intracellular calcium levels ([Ca^{2+}] in) in a concentration and time-dependent manner that was partially attenuated by NAC. Nifedipine, a calcium channel blocker, partially attenuated the elevated [Ca^{2+}]in, indicating that some of the excess [Ca^{2+}]in is a result of influx from outside the cell.

Exposure to a sublethal concentration of ZnO (5 μg/ml) increased the expression of four genes that are involved in apoptosis and oxidative stress responses BNIP, PRDX3, PRNP, and TXRND1, by at least 2.5-fold. Therefore, zinc oxide nanoparticles induced gene expression involved in oxidative stress and apoptosis. Thus, ZnO alters transcriptional regulation in BEAS-2B cells [44].

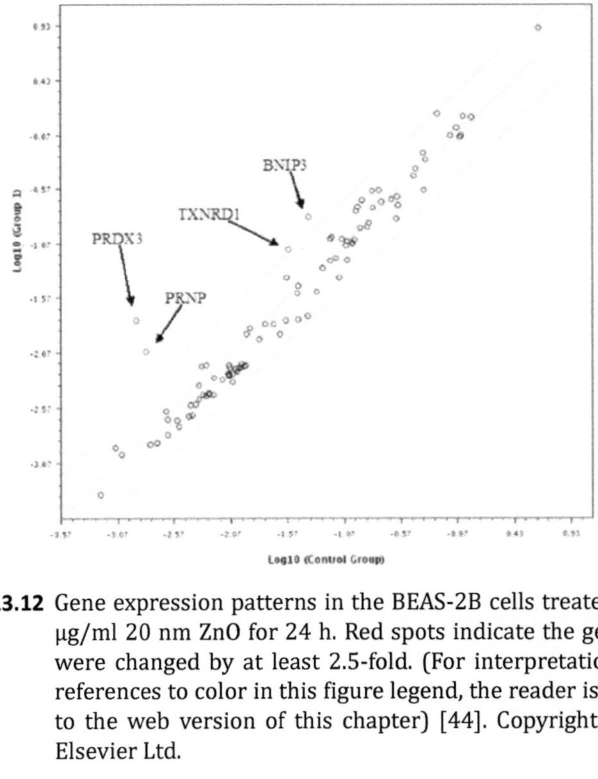

Figure 13.12 Gene expression patterns in the BEAS-2B cells treated with 5 μg/ml 20 nm ZnO for 24 h. Red spots indicate the genes that were changed by at least 2.5-fold. (For interpretation of the references to color in this figure legend, the reader is referred to the web version of this chapter) [44]. Copyright © 2009 Elsevier Ltd.

ZnO NPs have been reported to elicit various adverse cellular effects, including cytotoxicity. However, the underlying mechanisms of these adverse effects have not been fully characterized. The general mechanism of cell death includes generation of ROS, the disruption of mitochondrial function, and apoptosis [77]. Yu et al. determined the potential toxicity of ZnO NPs to assess the toxicity mechanism in normal skin cells. ZnO NPs induced ROS in a concentration and time-dependent manner. To demonstrate ZnO NPs toxicity mechanism related to ROS, we detected abnormal autophagic vacuoles accumulation and mitochondria dysfunction after ZnO NPs treatment. Furthermore, mitochondria membrane potential and ATP production of cell are decreased for culture with ZnO NPs. We conclude that ZnO NPs lead to cell death through autophagic vacuole accumulation and mitochondria damage in normal skin cells via ROS induction [79]. Robust markers of apoptosis, Annexin V staining, loss of mitochondrial potential, and increased generation of superoxide were observed when human colon cancer cells were treated with ZnO NPs [66]. ZnO NPs are cytotoxic with strong indications for apoptotic effects (caspase 3/7 activation, hypodiploid DNA, and phosphatidylserine exposure) in RAW 264.7 macrophages [80].

For instance, a recent study suggests that acute exposure of ZnO NPs on human aortic endothelial cells (HAECs) significantly upregulates mRNA levels of inflammatory markers—Interleukin-8 (IL-8), intracellular cell adhesion molecule-1 (ICAM-1), and monocyte chemotactic protein-1 (MCP) [81]. Another in vitro study has revealed the clastogenic effects of ZnO NPs on Chinese hamster ovary (CHO) cell line (Clastogenicity, photo-clatogenicity, or pseudo-photo-clastogenicity: genotoxic effect of zinc oxide in the dark, in pre-irradiated, or simultaneously irradiated Chinese hamster ovary cells). Some studies have further described the cytotoxicity of these particles in human CD4+ T cells and lung mesothelioma cell line [67,82].

De Berardis assessed the cytotoxicity, oxidative stress, apoptosis, and proinflammatory mediator release induced by ZnO NPs on human colon carcinoma LoVo cells. Treatment with ZnO (5 μg/cm^2 corresponding to 11.5 μg/ml) for 24 h induced a significant decrease of cell viability on LoVo cells, H_2O_2/OH• increase, O^{2-}• and GSH decrease, depolarization of inner mitochondrial membranes, apoptosis, and IL-8 release. Higher doses induced about 98% of cytotoxicity already after 24 h of treatment. The experimental data show that exposure to ZnO-NPs induced evident cellular alterations, as demonstrated both by apoptotic nucleus with condensed and marginalized chromatin (Fig. 13.13B) and by mitochondria with shrunken condensed matrix and swollen cristae (Fig. 13.13C,D) [83].

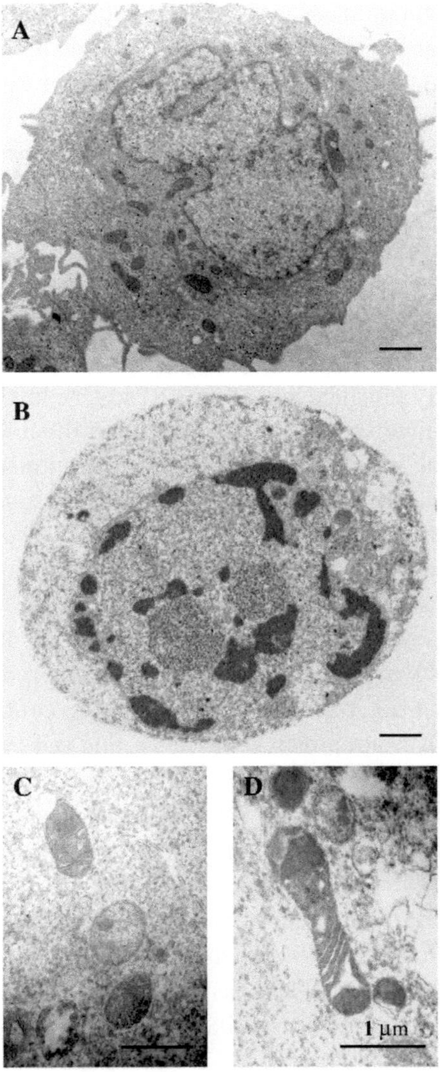

Figure 13.13 Ultrastructural analysis of control (A) and 10 μg/cm^2 ZnO NPs treated cells (B–D) performed by TEM. Treatment induced evident nuclear alterations with condensed chromatin (B) and alterations of mitochondria that appeared swollen and shrunken (C and D) [83]. Copyright © 2010 Elsevier Inc.

Previous studies have reported the molecular mechanisms of apoptosis induced by ZnO NPs via mitogen-activated protein kinase p38 and upregulation cell cycle checkpoint protein p53 pathways in human dermal fibroblasts. MTT-based cell viability assay showed a significant decrease in cell survivorship after ZnO NP exposure, and phase contrast images revealed that ZnO NP treated cells had lower density and a rounded morphology. Apoptosis induction was confirmed by the annexin V assay and western blot analysis showed the upregulation of p53 and phospho-p38 proteins. Furthermore, in ZnO NP exposed cells, p53 protein was phosphorylated at Ser33 and Ser46 sites known to be phosphorylated by p38. Their results suggest that ZnO NPs have the potential to induce apoptosis in human dermal fibroblasts via p53–p38 pathways [63].

The toxicity mechanisms of ZnO NPs were further investigated using human liver cancer HepG2 cells. Both the mRNA and protein levels of tumor suppressor gene p53 and apoptotic gene bax were upregulated whereas the antiapoptotic gene bcl-2 was downregulated in ZnO NP treated HepG2 cells. ZnO NPs were also found to induce activity of caspase-3 enzyme, DNA fragmentation, ROS generation, and oxidative stress in HepG2 cells [84].

ZnO NPs cross-interact with p53, which is one of the most important known tumor suppressors that protects cells from developing cancer phenotypes through its control over major pathways such as apoptosis, senescence, and cell cycle progression. The results showed that the p53 pathway was activated in BJ cells (skin fibroblasts) after ZnO NPs treatment with a concomitant decrease in cell numbers. The p53 knocked down BJ cells became more resistant to ZnO NPs induced cell death and increased cell progression [85]. Ahamed et al. demonstrate that ZnO nanorod induced apoptosis in A549 cells through ROS and oxidative stress via p53, survivin, bax/bcl-2, and caspase pathways. ZnO nanorod was found to induce cytotoxicity, ROS generation, oxidative stress, and activities of caspase-3 and caspase-9 in a dose and time-dependent manner in human alveolar adenocarcinoma (A549) cells. Western blot results showed that ZnO nanorods induced the expression of heat shock protein 70, a first-tier marker of cell damage and a cell-cycle checkpoint protein p53. Moreover, pro-apoptotic protein bax was upregulated and the antiapoptotic proteins, survivin and bcl-2, were downregulated in ZnO nanorod exposed cells [86]. Sharma et al. investigated that the mechanism of toxicity induced by ZnO NPs in HepG2 cells. ZnO NPs induced apoptosis and DNA damage, which was mediated by oxidative stress as evidenced by an increase in Fpg-sensitive sites. ROS triggered a decrease in mitochondria membrane potential and an increase in the ratio of Bax/Bcl2 leading to mitochondria-mediated pathway involved in apoptosis. In addition, ZnO NPs activated JNK, p38, and induced p53Ser15 phosphorylation. However, apoptosis was found to be independent of JNK and p38 pathways [59].

Guo et al. found that ZnO NPs could apparently decrease the mitochondrial membrane potential, increase the production of ROS, and lead to the overexpression of caspase-12 in RGC-5 cells, suggesting that ZnO nanoparticle induced toxicity via ROS overproduction will trigger endoplasmic reticulum stress, lead to the RGC-5 cell damage, and finally induce apoptosis/necrosis; the overexpression of caspase-12 may be involved in cell death in RGC-5 cells (Fig. 13.14) [64].

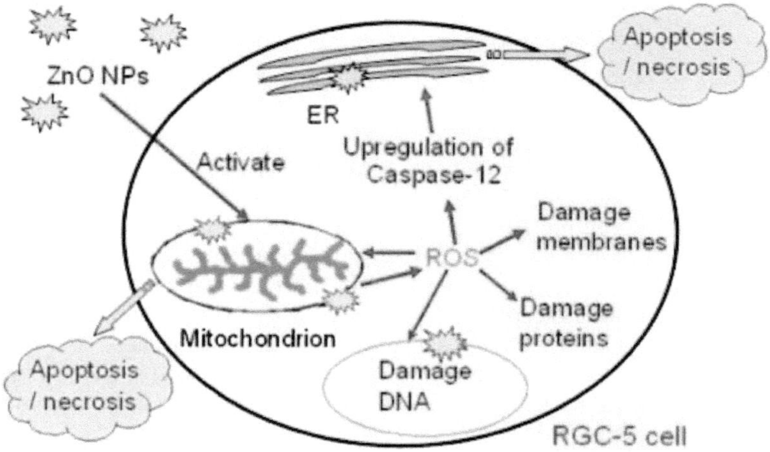

Figure 13.14 ROS-induced cytotoxic effects of zinc oxide nanoparticles in rat retinal ganglion cells [64]. Copyright © 2012 Elsevier Ltd.

Bae et al. investigated that ZnO NPs could induce ROS-mediated apoptosis in human keratinocyte HaCaT cells. Their results demonstrated that intracellular ROS was increased in cells treated with 20 μg/ml ZnO NPs in 24 h. No significant difference according to differences in surface charge was observed. Total levels of caspase-7 and PARP were decreased by ZnO NPs [61].

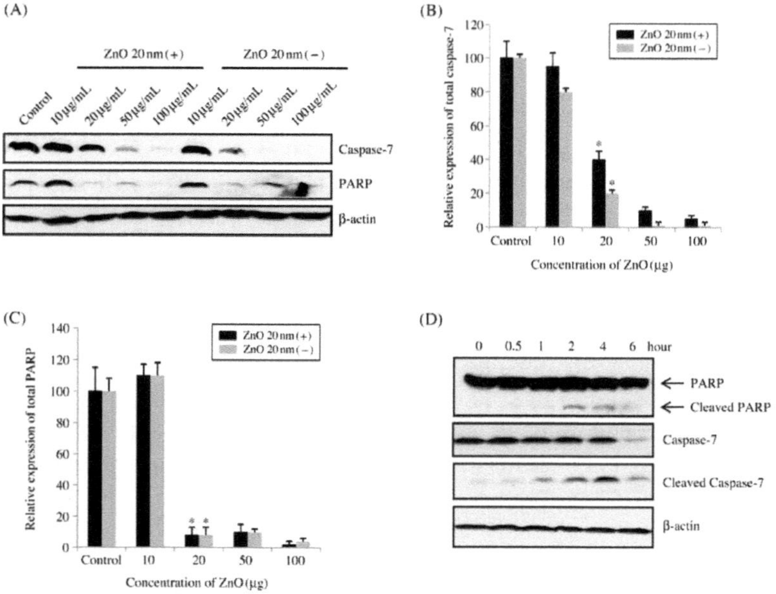

Figure 13.15 Characterization of caspase-7 and PARP cleavage induced by ZnO NPs. HaCaT cells were treated with various concentrations of tested compounds. (A) Western blot analysis for characterization of expression of total Caspase-7 and PARP in HaCaT cells treated with 1–100 μg/ml 20 and 100 nm (+/-) charged ZnO. (B) Bar graph depicted relative expression of caspase-7 and (C) PARP protein after treatment with 1–100 μg/ml 20 and 100 nm (+/-) charged ZnO NPs. Results represent the mean ± SD. (D) Western blot analysis for characterization of expression of cleaved Caspase-7 and PARP in HaCaT cells treated with 20 μg/ml 20 nm (+) charged ZnO [61]. Copyright © 2011, The Korean Society of Toxicogenomics and Toxicoproteomics and Springer Netherlands.

Nanoparticles could potentially cause autophagy dysfunction by overloading or directly damaging the lysosomal compartment, or altering the cell cytoskeleton, resulting in blockade of autophagosome-lysosome fusion [87]. ZnO NPs lead to cell death through autophagic vacuole accumulation. Yu et al. determined the toxicity mechanism of ZnO NPs in normal skin cells. After uptake into cells, ZnO NPs induced ROS in a concentration and time-dependent manner. Furthermore, mitochondria membrane potential and ATP production are decreased for culture with ZnO NPs [79].

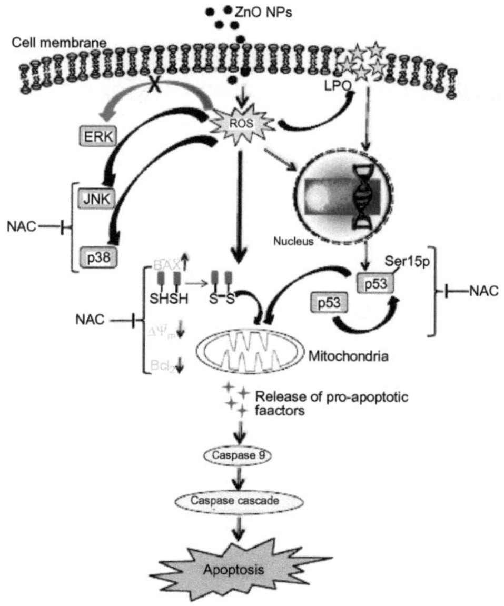

Figure 13.16 Possible mechanisms involved in ZnO NP induced toxicity in human liver cells [59]. Copyright © 2012, Springer Science+Business Media, LLC.

13.2.2.2 Titanium nanoparticles (TiO_2 NPs)

Titanium nanoparticles (TiO_2 NPs) have important applications in the fields of renewable energy and environmental remediation [88]. It also has photocatalytic activity that was used as purification of polluted air or wastewater [89]. Hence, there are increasing opportunities for both human and environmental exposure to this particular nanoparticle. The potential entry routes of TiO_2 NPs into the human body are dermal, ingestion, and inhalation. TiO_2 NPs may translocate and reach other tissues by various transfer routes and mechanisms and pose an exposure risk to humans [90]. When exposed to UV light, TiO_2 catalyzes the generation of ROS, such as superoxide anions, hydrogen peroxide, free hydroxyl radicals, and singlet oxygen in aqueous media [91–93].

13.2.2.2.1 *In vivo studies*

Inhalation is one of the major routes for TiO_2 NPs to gain entry into the human body especially in occupational settings [94]. Recent evidence has pointed out that TiO_2 nanoparticles can cause inflammatory response in airways of mice and rats by subchronic inhalation [95]. The animals were exposed to aerosol concentrations of 0.5, 2.0, or 10 mg/m^3 TiO_2 nanoparticles for 6 h/day, 5 days/week, for 13 weeks. The responses studied were chosen to assess a variety of pulmonary parameters, including inflammation, cytotoxicity, lung cell proliferation, and histopathological alterations. They found that neutrophilic response of rats declined in a time-dependent manner correlating with declining lung burdens but not mice [96–98].

Sager et al. reported that rats were exposed by intratracheal instillation to two particle type of TiO_2 (TiO_2 NPs (21 nm, P-25) and TiO_2 FPs (100% rutile; 1μm)) in rats for 1, 7, or 42 days. The results show that TiO_2 NPs caused more significant inflammation and cytotoxic potential. However, when doses were equalized by surface area, the TiO_2 NPs were only slightly more inflammogenic and cytotoxic than the fine-sized particles [99].

It is reported that TiO_2 NPs translocate from the site of absorption to other parts of the body (Fig. 13.17). NPs including TiO_2 NPs in the size range of 5–100 nm could be translocated across the air–blood barrier [100]. When TiO_2 NPs are translocated into the blood, generally they may be retained in the liver and lymphatic system; the systemic circulation can distribute the particles to all organs and tissues in the body [94]. Li et al. demonstrated the distribution of TiO_2 NPs (3 nm; 13.2 mg/kg; once a week for 4 weeks by intratracheal instillation) in mice. They suggested that TiO_2 NPs might pass through the blood–brain barrier [101]. Other in vivo studies have provided evidence that the instilled TiO_2 (80 nm, rutile and 155 nm, anatase) directly entered the brain through olfactory bulb, especially deposited in the mice hippocampus region. At a 30 days intranasal instillation study, the pathological changes were observed in the hippocampus and olfactory bulb using Nissl staining and transmission electron microscope. The oxidative damage expressed as lipid peroxidation increased significantly, in particular in the exposed group of anatase TiO_2 particles at 30 days postexposure. Exposure to anatase TiO_2 particles also produced higher inflammation responses, in association with the significantly increased tumor necrosis factor alpha (TNF-α) and interleukin (IL-1) levels. We conclude that subtle differences in responses to anatase TiO_2 particles versus the rutile ones could be related to crystal structure [102,103].

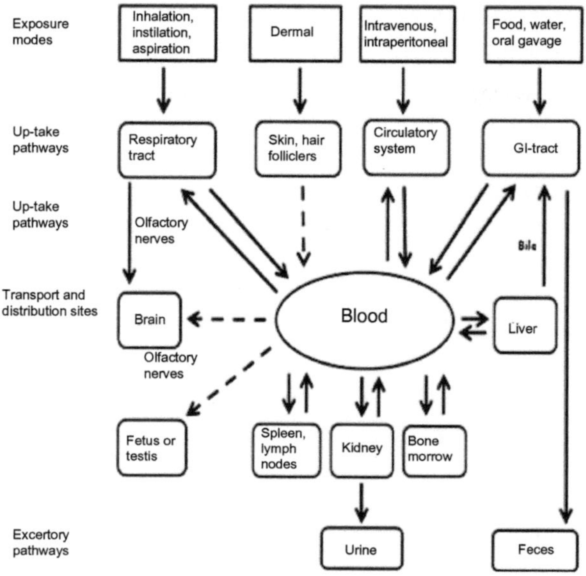

Figure 13.17 Toxicokinetics and accumulation sites of TiO_2 NPs. The arrows in dotted lines represent uncertainties [94]. Copyright © Shi et al.; licensee BioMed Central Ltd. 2013.

13.2.2.2.2 *In vitro studies*

Recent studies indicated that exposed to TiO_2 NPs led to the increased ROS, induction of oxidative stress, and DNA damage in several mammalian cells [49,76,97,98,104–107], accompanied by reduced glutathione (GSH) decrease, and oxidative stress-related genes such as heme oxygenase-1, thioredoxin reductase, glutathione-S-transferase, catalase, and a hypoxia-inducible gene [108], even resulted in cell death through lysosomal membrane destabilization and lipid peroxidation. Cai and colleagues found that cytotoxicity associated with TiO_2 NPs to HeLa cells was enhanced in the presence of UV light and suggested that this was due to the photo-excited TiO_2 promoting oxidative stress [109]. Furthermore, ultrafine titanium dioxide micronuclei and apoptosis, the activities of caspase 3 and PARP, were also increased in parallel to the morphological apoptosis [106,110].

A recent study showed that TiO_2 NPs induced cytotoxicity in peripheral blood lymphocytes. Lymphocytes treated with TiO_2 NPs showed the accumulation of p53, which respond to DNA damage

[111]. These results were confirmed by Wu et al. They found TiO_2 NPs displayed activating oxidative stress, upregulation of JNK, and p53 phosphorylation in PC12 cells [112].

In addition to oxidative stress, inflammatory responses are associated with TiO_2 NPs-inducing cytotoxicity via generation of ROS; it seems likely that exposure to TiO_2 NPs may cause genotoxicity in cultured cells. TiO_2 NPs have been shown to result in increased proportion of DNA breakage in peripheral blood lymphocytes, chromosomal changes in Syrian hamster embryo fibroblasts, and in human B-cell lymphoblastoid cell line (WIL2-NS) using the micronucleus assay [106, 110, 111]. Rahman et al. demonstrated a significant increase in micronuclei induction in Syrian hamster embryo cells. The results showed that TiO_2 NPs (20 nm) induced chromosomal changes at 1.0 $\mu g/cm^2$ for 12 h, 24 h, 48 h, 66 h, and 72 h. Furthermore, transmission electron microscopy of the exposed cells revealed the typical chromatin compaction of apoptosis [106].

The study by Wang and colleagues also found that treatment with 130 $\mu g/ml$ TiO_2 NPs induced approximately 2.5-fold increase in the frequency of micronucleated binucleated cells and a significant reduction in the cytokinesis block proliferation index was observed by the cytokinesis block micronucleus (CBMN) assay [110].

The molecular mechanisms associated in the process of apoptosis induced by TiO_2 nanoparticles have been clearly elucidated. Zhao et al. showed that TiO_2 nanoparticles produced a concentration-dependent cytotoxicity for 24 h and apoptosis and necrosis induction in a mouse epidermal (JB6) cell line. Western blot analysis inhibited apoptosis-associated proteins (caspase-8, Bid, Bcl-2, Bax, caspase-3, PARP, and cytochrome c) activation by TiO_2 NPs treatment. Moreover, TiO_2 NPs also induced mitochondrial membrane permeability injury, which leads to cytochrome c release from the mitochondria to cytosol [113]. Their study was confirmed by Kang et al. that TiO_2 NPs activated caspase-9, caspase-3, and caspase-3-mediated PARP cleavage in human lymphocytes, as well as activated two MAPKs, p38 and JNK [111]. However, Shi et al. provide an evidence that apoptosis exposed to TiO_2 NPs was primarily through the intrinsic apoptosis pathway in BEAS-2B cells. They found that expression of caspase 9 was increased in a concentration-dependent manner, whereas the level of caspase 8 and t-Bid was not altered in the TiO_2 NPs treated cells [114]. The distinction between the two studies may be attributed to the difference in cell surface receptors between immune cells and airway epithelial cells [114].

The N-acetylcysteine (NAC) as an antioxidant and free radical scavenger is used to reduce cell damage [115]. Xue et al. assessed that the chemoprotective effect of NAC on TiO_2 NPs induced oxidative stress and apoptosis in human keratinocyte (HaCaT) cells. The results showed that TiO_2 NPs induced apoptosis and increased keratin 6 (K6) mRNA expression. NAC strongly suppressed intracellular ROS and NO production in TiO_2 NPs treated cells [116].

13.3 Conclusion and Outlook

It is difficult to directly compare potential hazards of nanomaterials in human and environmental exposure, due to a number of factors, including exposure time, nanomaterial preparation, cell type, and different physicochemical properties characterization of nanomaterials such as size distribution, crystalline structure, surface area, surface coating, and so on. It is a crucial question that needs to be solved, which includes the use of dispersants, impact of agglomeration/aggregation state, impurities and surface defects, and generally the identification of physico-chemical characteristics that are associated with oxidative stress, inflammation and cell signaling alterations. Furthermore, long-term animal studies evaluating the toxicity of nanomaterials should be performed for risk assessment. The molecular mechanisms of nanomaterials may be involved in disease, for example, cancer, cardiovascular diseases, and neurodegenerative diseases still remain a question. Finally, a continuing evaluation of the human health implications of exposure to nanoscale materials will be essential before the use of these materials can be fully realized.

References

1. Helland, A., Wick, P., Koehler, A., Schmid, K., Som, C., Reviewing the environmental and human health knowledge base of carbon nanotubes. *Environ. Health Perspect.* 2007, 115, 1125–1131.

2. Wiesner, M. R., Lowry, G. V., Alvarez, P., Dionysiou, D., Biswas, P., Assessing the risks of manufactured nanomaterials. *Environ. Sci. Technol.* 2006, 40, 4336–4345.

3. Oberdorster, G., Oberdorster, E., Oberdorster, J., Nanotoxicology: an emerging discipline evolving from studies of ultrafine particles. *Environ. Health Perspect.* 2005, 113, 823–839.
4. Maynard, A. D., Nanotechnology: assessing the risks. *Nano Today* 2006, 1, 22–33.
5. Lam, C. W., James, J. T., McCluskey, R., Hunter, R. L., Pulmonary toxicity of single-wall carbon nanotubes in mice 7 and 90 days after intratracheal instillation. *Toxicol. Sci.* 2004, 77, 126–134.
6. Arredouani, M., Yang, Z. P., Ning, Y. Y., Qin, G. Z., Soininen, R., Tryggvason, K., Kobzik, L., The scavenger receptor MARCO is required for lung defense against pneumococcal pneumonia and inhaled particles. *J. Exp. Med.* 2004, 200, 267–272.
7. Lanone, S., Boczkowski, J., Biomedical applications and potential health risks of nanomaterials: molecular mechanisms. *Curr. Mol. Med.* 2006, 6, 651–663.
8. Elder, A., Gelein, R., Silva, V., Feikert, T., Opanashuk, L., Carter, J., Potter, R., Maynard, A., Ito, Y., Finkelstein, J., Oberdorster, G., Translocation of inhaled ultrafine manganese oxide particles to the central nervous system. *Environ. Health Perspect.* 2006, 114, 1172–1178.
9. Oberdorster, G., Oberdorster, E., Oberdorster, J., Nanotoxicology: an emerging discipline evolving from studies of ultrafine particles. *Environ. Health Perspect.* 2005, 113, 823–839.
10. Imahori, H., Fukuzumi, S., Porphyrin- and fullerene-based molecular photovoltaic devices. *Adv. Funct. Mater.* 2004, 14, 525–536.
11. Oberdorster, E., Manufactured nanomaterials (fullerenes, C60) induce oxidative stress in the brain of juvenile largemouth bass. *Environ. Health Perspect.* 2004, 112, 1058–1062.
12. Takahashi, M., Kato, H., Doi, Y., Hagiwara, A., Hirata-Koizumi, M., Ono, A., Kubota, R., Nishimura, T., Hirose, A., Sub-acute oral toxicity study with fullerene C60 in rats. *J. Toxicol. Sci.* 2012, 37, 353–361.
13. Lehnert, B. E., Morrow, P. E., Association of Fe-59 oxide with alveolar macrophages during alveolar clearance. *Exp. Lung. Res.* 1985, 9, 1–16.
14. Shinohara, N., Nakazato, T., Tamura, M., Endoh, S., Fukui, H., Morimoto, Y., Myojo, T., Shimada, M., Yamamoto, K., Tao, H., Yoshida, Y., Nakanishi, J., Clearance kinetics of fullerene C(6)(0) nanoparticles from rat lungs after intratracheal C(6)(0) instillation and inhalation C(6)(0) exposure. *Toxicol. Sci.* 2010, 118, 564–573.
15. Markovic, Z., Trajkovic, V., Biomedical potential of the reactive oxygen species generation and quenching by fullerenes (C60). *Biomaterials* 2008, 29, 3561–3573.

16. Jacobsen, N. R., Pojana, G., White, P., Moller, P., Cohn, C. A., Korsholm, K. S., Vogel, U., Marcomini, A., Loft, S., Wallin, H., Genotoxicity, cytotoxicity, and reactive oxygen species induced by single-walled carbon nanotubes and C(60) fullerenes in the FE1-Mutatrade markMouse lung epithelial cells. *Environ. Mol. Mutag.* 2008, 49, 476–487.

17. Dhawan, A., Taurozzi, J. S., Pandey, A. K., Shan, W., Miller, S. M., Hashsham, S. A., Tarabara, V. V., Stable colloidal dispersions of C60 fullerenes in water: evidence for genotoxicity. *Environ. Sci. Technol.* 2006, 40, 7394–7401.

18. Sayes, C. M., Fortner, J. D., Guo, W., Lyon, D., Boyd, A. M., Ausman, K. D., Tao, Y. J., Sitharaman, B., Wilson, L. J., Hughes, J. B., West, J. L., Colvin, V. L., The differential cytotoxicity of water-soluble fullerenes. *Nano Lett.* 2004, 4, 1881–1887.

19. Isakovic, A., Markovic, Z., Todorovic-Markovic, B., Nikolic, N., Vranjes-Djuric, S., Mirkovic, M., Dramicanin, M., Harhaji, L., Raicevic, N., Nikolic, Z., Trajkovic, V., Distinct cytotoxic mechanisms of pristine versus hydroxylated fullerene. *Toxicol. Sci.* 2006, 91, 173–183.

20. Xu, A., Chai, Y., Nohmi, T., Hei, T. K., Genotoxic responses to titanium dioxide nanoparticles and fullerene in gpt delta transgenic MEF cells. *Part. Fibre Toxicol.* 2009, 6, 3.

21. Cui, H. F., Vashist, S. K., Al-Rubeaan, K., Luong, J. H., Sheu, F. S., Interfacing carbon nanotubes with living mammalian cells and cytotoxicity issues. *Chem. Res. Toxicol.* 2010, 23, 1131–1147.

22. Firme, C. P., 3rd, Bandaru, P. R., Toxicity issues in the application of carbon nanotubes to biological systems. *Nanomedicine* 2010, 6, 245–256.

23. Huczko, A., Lange, H., Carbon nanotubes: experimental evidence for a null risk of skin irritation and allergy. *Fullerene Sci. Technol.* 2001, 9, 247–250.

24. Huczko, A., Lange, H., Calko, E., Grubek-Jaworska, H., Droszcz, P., Physiological testing of carbon nanotubes: are they asbestos-like? *Fullerene Sci. Technol.* 2001, 9, 251–254.

25. Mercer, R. R., Scabilloni, J., Wang, L., Kisin, E., Murray, A. R., Schwegler-Berry, D., Shvedova, A. A., Castranova, V., Alteration of deposition pattern and pulmonary response as a result of improved dispersion of aspirated single-walled carbon nanotubes in a mouse model. *Am. J. Physiol.-Lung C* 2008, 294, L87–L97.

26. Mercer, R. R., Hubbs, A. F., Scabilloni, J. F., Wang, L. Y., Battelli, L. A., Schwegler-Berry, D., Castranova, V., Porter, D. W., Distribution and

persistence of pleural penetrations by multi-walled carbon nanotubes. *Part. Fibre Toxicol.* 2010, 7.

27. Teeguarden, J. G., Hinderliter, P. M., Orr, G., Thrall, B. D., Pounds, J. G., Particokinetics in vitro: dosimetry considerations for in vitro nanoparticle toxicity assessments. *Toxicol. Sci.* 2007, 95, 300–312.
28. Sato, Y., Yokoyama, A., Shibata, K., Akimoto, Y., Ogino, S., Nodasaka, Y., Kohgo, T., Tamura, K., Akasaka, T., Uo, M., Motomiya, K., Jeyadevan, B., Ishiguro, M., Hatakeyama, R., Watari, F., Tohji, K., Influence of length on cytotoxicity of multi-walled carbon nanotubes against human acute monocytic leukemia cell line THP-I in vitro and subcutaneous tissue of rats in vivo. *Mol. Biosyst.* 2005, 1, 176–182.
29. Ma-Hock, L., Treumann, S., Strauss, V., Brill, S., Luizi, F., Mertler, M., Wiench, K., Gamer, A. O., van Ravenzwaay, B., Landsiedel, R., Inhalation toxicity of multiwall carbon nanotubes in rats exposed for 3 months. *Toxicol. Sci.* 2009, 112, 468–481.
30. Cui, D., Tian, F., Ozkan, C. S., Wang, M., Gao, H., Effect of single wall carbon nanotubes on human HEK293 cells. *Toxicol. Lett.* 2005, 155, 73–85.
31. Han, Y. G., Xu, J., Li, Z. G., Ren, G. G., Yang, Z., In vitro toxicity of multi-walled carbon nanotubes in C6 rat glioma cells. *Neurotoxicology* 2012, 33, 1128–1134.
32. Meng, L., Jiang, A., Chen, R., Li, C. Z., Wang, L., Qu, Y., Wang, P., Zhao, Y., Chen, C., Inhibitory effects of multiwall carbon nanotubes with high iron impurity on viability and neuronal differentiation in cultured PC12 cells. *Toxicology* 2013, 313, 49–58.
33. Xu, H., Bai, J., Meng, J., Hao, W., Xu, H., Cao, J. M., Multi-walled carbon nanotubes suppress potassium channel activities in PC12 cells. *Nanotechnology* 2009, 20, 285102.
34. Pacurari, M., Yin, X. J., Zhao, J., Ding, M., Leonard, S. S., Schwegler-Berry, D., Ducatman, B. S., Sbarra, D., Hoover, M. D., Castranova, V., Vallyathan, V., Raw single-wall carbon nanotubes induce oxidative stress and activate MAPKs, AP-1, NF-kappaB, and Akt in normal and malignant human mesothelial cells. *Environ. Health Perspect.* 2008, 116, 1211–1217.
35. Bokara, K. K., Kim, J. Y., Lee, Y. I., Yun, K., Webster, T. J., Lee, J. E., Biocompatability of carbon nanotubes with stem cells to treat CNS injuries. *Anatomy Cell Biol.* 2013, 46, 85–92.
36. Chen, C. S., Soni, S., Le, C., Biasca, M., Farr, E., Chen, E. Y., Chin, W. C., Human stem cell neuronal differentiation on silk-carbon nanotube composite. *Nanoscale Res. Lett.* 2012, 7, 126.

37. Kolosnjaj, J., Szwarc, H., Moussa, F., Toxicity studies of carbon nanotubes. *Adv. Exp. Med. Biol.* 2007, 620, 181–204.

38. Porter, A. E., Gass, M., Muller, K., Skepper, J. N., Midgley, P. A., Welland, M., Direct imaging of single-walled carbon nanotubes in cells. *Nat. Nanotechnol.* 2007, 2, 713–717.

39. Zeni, O., Palumbo, R., Bernini, R., Zeni, L., Sarti, M., Scarfi, M. R., Cytotoxicity investigation on cultured human blood cells treated with single-wall carbon nanotubes. *Sensors-Basel* 2008, 8, 488–499.

40. Kisin, E. R., Murray, A. R., Keane, M. J., Shi, X. C., Schwegler-Berry, D., Gorelik, O., Arepalli, S., Castranova, V., Wallace, W. E., Kagan, V. E., Shvedova, A. A., Single-walled carbon nanotubes: geno- and cytotoxic effects in lung fibroblast V79 cells. *J. Toxicol. Environ. Health* 2007, 70, 2071–2079.

41. Shvedova, A. A., Castranova, V., Kisin, E. R., Schwegler-Berry, D., Murray, A. R., Gandelsman, V. Z., Maynard, A., Baron, P., Exposure to carbon nanotube material: assessment of nanotube cytotoxicity using human keratinocyte cells. *J. Toxicol. Environ. Health Part A* 2003, 66, 1909–1926.

42. Shvedova, A. A., Pietroiusti, A., Fadeel, B., Kagan, V. E., Mechanisms of carbon nanotube-induced toxicity: focus on oxidative stress. *Toxicol. Appl. Pharmacol.* 2012, 261, 121–133.

43. Fenoglio, I., Greco, G., Tomatis, M., Muller, J., Raymundo-Pinero, E., Beguin, F., Fonseca, A., Nagy, J. B., Lison, D., Fubini, B., Structural defects play a major role in the acute lung toxicity of multiwall carbon nanotubes: physicochemical aspects. *Chem. Res. Toxicol.* 2008, 21, 1690–1697.

44. Huang, C. C., Aronstam, R. S., Chen, D. R., Huang, Y. W., Oxidative stress, calcium homeostasis, and altered gene expression in human lung epithelial cells exposed to ZnO nanoparticles. *Toxicol. In Vitro* 2010, 24, 45–55.

45. Sharma, V., Shukla, R. K., Saxena, N., Parmar, D., Das, M., Dhawan, A., DNA damaging potential of zinc oxide nanoparticles in human epidermal cells. *Toxicol. Lett.* 2009, 185, 211–218.

46. Rasmussen, J. W., Martinez, E., Louka, P., Wingettt, D. G., Zinc oxide nanoparticles for selective destruction of tumor cells and potential for drug delivery applications. *Expert Opin. Drug Deliv.* 2010, 7, 1063–1077.

47. Sharma, V., Singh, P., Pandey, A. K., Dhawan, A., Induction of oxidative stress, DNA damage and apoptosis in mouse liver after sub-acute oral exposure to zinc oxide nanoparticles. *Mutat. Res.* 2012, 745, 84–91.

48. Jeng, H. A., Swanson, J., Toxicity of metal oxide nanoparticles in mammalian cells. *J. Environ. Sci. Health A Tox. Hazard Subst. Environ. Eng.* 2006, 41, 2699–2711.

49. Lai, J. C., Lai, M. B., Jandhyam, S., Dukhande, V. V., Bhushan, A., Daniels, C. K., Leung, S. W., Exposure to titanium dioxide and other metallic oxide nanoparticles induces cytotoxicity on human neural cells and fibroblasts. *Int. J. Nanomed.* 2008, 3, 533–545.

50. Horie, M., Nishio, K., Fujita, K., Endoh, S., Miyauchi, A., Saito, Y., Iwahashi, H., Yamamoto, K., Murayama, H., Nakano, H., Nanashima, N., Niki, E., Yoshida, Y., Protein adsorption of ultrafine metal oxide and its influence on cytotoxicity toward cultured cells. *Chem. Res. Toxicol.* 2009, 22, 543–553.

51. Deng, X., Luan, Q., Chen, W., Wang, Y., Wu, M., Zhang, H., Jiao, Z., Nanosized zinc oxide particles induce neural stem cell apoptosis. *Nanotechnology* 2009, 20, 115101.

52. Zhao, J., Xu, L., Zhang, T., Ren, G., Yang, Z., Influences of nanoparticle zinc oxide on acutely isolated rat hippocampal CA3 pyramidal neurons. *Neurotoxicology* 2009, 30, 220–230.

53. Wang, B., Feng, W. Y., Wang, M., Wang, T. C., Gu, Y. Q., Zhu, M. T., Ouyang, H., Shi, J. W., Zhang, F., Zhao, Y. L., Chai, Z. F., Wang, H. F., Wang, J., Acute toxicological impact of nano- and submicro-scaled zinc oxide powder on healthy adult mice. *J. Nanopart. Res.* 2008, 10, 263–276.

54. Wang, L., Ding, W., Zhang, F., Acute toxicity of ferric oxide and zinc oxide nanoparticles in rats. *J. Nanosci. Nanotechnol.* 2010, 10, 8617–8624.

55. Kao, Y. Y., Cheng, T. J., Yang, D. M., Wang, C. T., Chiung, Y. M., Liu, P. S., Demonstration of an olfactory bulb-brain translocation pathway for ZnO nanoparticles in rodent cells in vitro and in vivo. *J. Mol. Neurosci.* 2012, 48, 464–471.

56. Lee, C. M., Jeong, H. J., Kim, D. W., Sohn, M. H., Lim, S. T., The effect of fluorination of zinc oxide nanoparticles on evaluation of their biodistribution after oral administration. *Nanotechnology* 2012, 23.

57. Han, D. D., Tian, Y. T., Zhang, T., Ren, G. G., Yang, Z., Nano-zinc oxide damages spatial cognition capability via over-enhanced long-term potentiation in hippocampus of Wistar rats. *Int. J. Nanomed.* 2011, 6, 1453–1461.

58. Kao, Y. Y., Chen, Y. C., Cheng, T. J., Chiung, Y. M., Liu, P. S., Zinc oxide nanoparticles interfere with zinc ion homeostasis to cause cytotoxicity. *Toxicol. Sci.* 2012, 125, 462–472.

59. Sharma, V., Anderson, D., Dhawan, A., Zinc oxide nanoparticles induce oxidative DNA damage and ROS-triggered mitochondria mediated apoptosis in human liver cells (HepG2). Apoptosis: an international journal on programmed cell death 2012, 17, 852–870.

60. Heng, B. C., Zhao, X., Xiong, S., Ng, K. W., Boey, F. Y., Loo, J. S., Cytotoxicity of zinc oxide (ZnO) nanoparticles is influenced by cell density and culture format. *Arch. Toxicol.* 2011, 85, 695–704.

61. Bae, H. C., Ryu, H. J., Jeong, S. H., Lee, E. Y., Park, Y. H., Lee, K. G., Choi, B. H., Maeng, E. H., Kim, M. K., Son, S. W., Oxidative stress and apoptosis induced by ZnO nanoparticles in HaCaT cells. *Mol. Cell. Toxicol.* 2011, 7, 333–337.

62. Kumar, A., Pandey, A. K., Singh, S. S., Shanker, R., Dhawan, A., Engineered ZnO and TiO(2) nanoparticles induce oxidative stress and DNA damage leading to reduced viability of Escherichia coli. *Free Radic. Biol. Med.* 2011, 51, 1872–1881.

63. Meyer, K., Rajanahalli, P., Ahamed, M., Rowe, J. J., Hong, Y. L., ZnO nanoparticles induce apoptosis in human dermal fibroblasts via p53 and p38 pathways. *Toxicol. In Vitro* 2011, 25, 1721–1726.

64. Guo, D., Bi, H., Liu, B., Wu, Q., Wang, D., Cui, Y., Reactive oxygen species-induced cytotoxic effects of zinc oxide nanoparticles in rat retinal ganglion cells. *Toxicol. In Vitro* 2013, 27, 731–738.

65. Yin, Y., Lin, Q., Sun, H., Chen, D., Wu, Q., Chen, X., Li, S., Cytotoxic effects of ZnO hierarchical architectures on RSC96 Schwann cells. *Nanoscale Res. Lett.* 2012, 7, 439.

66. Moos, P. J., Chung, K., Woessner, D., Honeggar, M., Cutler, N. S., Veranth, J. M., ZnO particulate matter requires cell contact for toxicity in human colon cancer cells. *Chem. Res. Toxicol.* 2010, 23, 733–739.

67. Reddy, K. M., Feris, K., Bell, J., Wingett, D. G., Hanley, C., Punnoose, A., Selective toxicity of zinc oxide nanoparticles to prokaryotic and eukaryotic systems. *Appl. Phys. Lett.* 2007, 90, 2139021–2139023.

68. Zhang, X. Q., Yin, L. H., Tang, M., Pu, Y. P., ZnO, TiO_2, SiO_2, and Al_2O_3 Nanoparticles-induced toxic effects on human fetal lung fibroblasts. *Biomed. Environ. Sci.* 2011, 24, 661–669.

69. Yuan, J. H., Chen, Y., Zha, H. X., Song, L. J., Li, C. Y., Li, J. Q., Xia, X. H., Determination, characterization and cytotoxicity on HELF cells of ZnO nanoparticles. *Colloids Surf. B Biointerfaces* 2010, 76, 145–150.

70. Yang, S. T., Liu, J. H., Wang, J., Yuan, Y., Cao, A., Wang, H., Liu, Y., Zhao, Y., Cytotoxicity of zinc oxide nanoparticles: importance of microenvironment. *J. Nanosci. Nanotechnol.* 2010, 10, 8638–8645.

71. Osmond, M. J., McCall, M. J., Zinc oxide nanoparticles in modern sunscreens: an analysis of potential exposure and hazard. *Nanotoxicology* 2010, 4, 15–41.

72. Song, W., Zhang, J., Guo, J., Zhang, J., Ding, F., Li, L., Sun, Z., Role of the dissolved zinc ion and reactive oxygen species in cytotoxicity of ZnO nanoparticles. *Toxicol. Lett.* 2010, 199, 389–397.

73. Muller, K. H., Kulkarni, J., Motskin, M., Goode, A., Winship, P., Skepper, J. N., Ryan, M. P., Porter, A. E., pH-dependent toxicity of high aspect ratio zno nanowires in macrophages due to intracellular dissolution. *ACS Nano* 2010, 4, 6767–6779.

74. Valdiglesias, V., Costa, C., Kilic, G., Costa, S., Pasaro, E., Laffon, B., Teixeira, J. P., Neuronal cytotoxicity and genotoxicity induced by zinc oxide nanoparticles. *Environ. Int.* 2013, 55, 92–100.

75. Shi, J., Karlsson, H. L., Johansson, K., Gogvadze, V., Xiao, L., Li, J., Burks, T., Garcia-Bennett, A., Uheida, A., Muhammed, M., Mathur, S., Morgenstern, R., Kagan, V. E., Fadeel, B., Microsomal glutathione transferase 1 protects against toxicity induced by silica nanoparticles but not by zinc oxide nanoparticles. *ACS Nano* 2012, 6, 1925–1938.

76. Sayes, C. M., Reed, K. L., Warheit, D. B., Assessing toxicity of fine and nanoparticles: comparing in vitro measurements to in vivo pulmonary toxicity profiles. *Toxicol. Sci.* 2007, 97, 163–180.

77. Premanathan, M., Karthikeyan, K., Jeyasubramanian, K., Manivannan, G., Selective toxicity of ZnO nanoparticles toward Gram-positive bacteria and cancer cells by apoptosis through lipid peroxidation. *Nanomed.-Nanotechnol.* 2011, 7, 184–192.

78. Ostrovsky, S., Kazimirsky, G., Gedanken, A., Brodie, C., Selective cytotoxic effect of ZnO nanoparticles on glioma cells. *Nano Res.* 2009, 2, 882–890.

79. Yu, K. N., Yoon, T. J., Minai-Tehrani, A., Kim, J. E., Park, S. J., Jeong, M. S., Ha, S. W., Lee, J. K., Kim, J. S., Cho, M. H., Zinc oxide nanoparticle induced autophagic cell death and mitochondrial damage via reactive oxygen species generation. *Toxicol. In Vitro* 2013, 27, 1187–1195.

80. Wilhelmi, V., Fischer, U., van Berlo, D., Schulze-Osthoff, K., Schins, R. P. F., Albrecht, C., Evaluation of apoptosis induced by nanoparticles and fine particles in RAW 264.7 macrophages: facts and artefacts. *Toxicol. In Vitro* 2012, 26, 323–334.

81. Gojova, A., Guo, B., Kota, R. S., Rutledge, J. C., Kennedy, I. M., Barakat, A. I., Induction of inflammation in vascular endothelial cells by metal oxide nanoparticles: effect of particle composition. *Environ. Health Perspect.* 2007, 115, 403–409.

82. Brunner, T. J., Wick, P., Manser, P., Spohn, P., Grass, R. N., Limbach, L. K., Bruinink, A., Stark, W. J., In vitro cytotoxicity of oxide nanoparticles: comparison to asbestos, silica, and the effect of particle solubility. *Environ. Sci. Technol.* 2006, 40, 4374–4381.

83. De Berardis, B., Civitelli, G., Condello, M., Lista, P., Pozzi, R., Arancia, G., Meschini, S., Exposure to ZnO nanoparticles induces oxidative stress and cytotoxicity in human colon carcinoma cells. *Toxicol. Appl. Pharm.* 2010, 246, 116–127.

84. Akhtar, M. J., Ahamed, M., Kumar, S., Khan, M. A. M., Ahmad, J., Alrokayan, S. A., Zinc oxide nanoparticles selectively induce apoptosis in human cancer cells through reactive oxygen species. *Int. J. Nanomed.* 2012, 7, 845–857.

85. Ng, K. W., Khoo, S. P. K., Heng, B. C., Setyawati, M. I., Tan, E. C., Zhao, X. X., Xiong, S. J., Fang, W. R., Leong, D. T., Loo, J. S. C., The role of the tumor suppressor p53 pathway in the cellular DNA damage response to zinc oxide nanoparticles. *Biomaterials* 2011, 32, 8218–8225.

86. Ahamed, M., Akhtar, M. J., Raja, M., Ahmad, I., Siddiqui, M. K., AlSalhi, M. S., Alrokayan, S. A., ZnO nanorod-induced apoptosis in human alveolar adenocarcinoma cells via p53, survivin and bax/bcl-2 pathways: role of oxidative stress. *Nanomedicine* 2011, 7, 904–913.

87. Stern, S. T., Adiseshaiah, P. P., Crist, R. M., Autophagy and lysosomal dysfunction as emerging mechanisms of nanomaterial toxicity. *Part. Fibre Toxicol.* 2012, 9.

88. Choi, W., Pure and modified TiO_2 photocatalysts and their environmental applications. *Catal. Surv. Asia* 2006, 10, 16–28.

89. Wu, W., Xiao, X. H., Zhang, S. F., Ren, F., Jiang, C. Z., Facile method to synthesize magnetic iron oxides/TiO_2 hybrid nanoparticles and their photodegradation application of methylene blue. *Nanoscale Res. Lett.* 2011, 6.

90. Dunford, R., Salinaro, A., Cai, L., Serpone, N., Horikoshi, S., Hidaka, H., Knowland, J., Chemical oxidation and DNA damage catalysed by inorganic sunscreen ingredients. *FEBS Lett.* 1997, 418, 87–90.

91. Konaka, R., Kasahara, E., Dunlap, W. C., Yamamoto, Y., Chien, K. C., Inoue, M., Irradiation of titanium dioxide generates both singlet oxygen and superoxide anion. *Free Radic. Biol. Med.* 1999, 27, 294–300.

92. Hirakawa, K., Mori, M., Yoshida, M., Oikawa, S., Kawanishi, S., Photo-irradiated titanium dioxide catalyzes site specific DNA damage via generation of hydrogen peroxide. *Free Radic. Res.* 2004, 38, 439–447.

93. Konaka, R., Kasahara, E., Dunlap, W. C., Yamamoto, Y., Chien, K. C., Inoue, M., Ultraviolet irradiation of titanium dioxide in aqueous dispersion generates singlet oxygen. *Redox Rep.* 2001, 6, 319–325.

94. Shi, H., Magaye, R., Castranova, V., Zhao, J., Titanium dioxide nanoparticles: a review of current toxicological data. *Part. Fibre Toxicol.* 2013, 10, 15.

95. Bermudez, E., Mangum, J. B., Wong, B. A., Asgharian, B., Hext, P. M., Warheit, D. B., Everitt, J. I., Pulmonary responses of mice, rats, and hamsters to subchronic inhalation of ultrafine titanium dioxide particles. *Toxicol. Sci.* 2004, 77, 347–357.

96. Sayes, C. M., Reed, K. L., Warheit, D. B., Assessing toxicity of fine and nanoparticles: comparing in vitro measurements to in vivo pulmonary toxicity profiles. *Toxicol. Sci.* 2007, 97, 163–180.

97. Xia, T., Kovochich, M., Brant, J., Hotze, M., Sempf, J., Oberley, T., Sioutas, C., Yeh, J. I., Wiesner, M. R., Nel, A. E., Comparison of the abilities of ambient and manufactured nanoparticles to induce cellular toxicity according to an oxidative stress paradigm. *Nano Lett.* 2006, 6, 1794–1807.

98. Brown, C., Fisher, J., Ingham, E., Biological effects of clinically relevant wear particles from metal-on-metal hip prostheses. Proceedings of the Institution of Mechanical Engineers. Part H, *J. Eng. Med.* 2006, 220, 355–369.

99. Sager, T. M., Kommineni, C., Castranova, V., Pulmonary response to intratracheal instillation of ultrafine versus fine titanium dioxide: role of particle surface area. *Part. Fibre Toxicol.* 2008, 5, 17.

100. Geiser, M., Kreyling, W. G., Deposition and biokinetics of inhaled nanoparticles. *Part. Fibre Toxicol.* 2010, 7, 2.

101. Li, Y., Li, J., Yin, J., Li, W., Kang, C., Huang, Q., Li, Q., Systematic influence induced by 3 nm titanium dioxide following intratracheal instillation of mice. *J. Nanosci. Nanotechnol.* 2010, 10, 8544–8549.

102. Wang, J. X., Liu, Y., Jiao, F., Lao, F., Li, W., Gu, Y. Q., Li, Y. F., Ge, C. C., Zhou, G. Q., Li, B., Zhao, Y. L., Chai, Z. F., Chen, C. Y., Time-dependent translocation and potential impairment on central nervous system by intranasally instilled TiO_2 nanoparticles. *Toxicology* 2008, 254, 82–90.

103. Wang, J., Liu, Y., Jiao, F., Lao, F., Li, W., Gu, Y., Li, Y., Ge, C., Zhou, G., Li, B., Zhao, Y., Chai, Z., Chen, C., Time-dependent translocation and potential impairment on central nervous system by intranasally instilled TiO(2) nanoparticles. *Toxicology* 2008, 254, 82–90.

104. Hussain, S., Thomassen, L. C. J., Ferecatu, I., Borot, M. C., Andreau, K., Martens, J. A., Fleury, J., Baeza-Squiban, A., Marano, F., Boland, S., Carbon black and titanium dioxide nanoparticles elicit distinct apoptotic pathways in bronchial epithelial cells. *Part. Fibre Toxicol.* 2010, 7.

105. Val, S., Hussain, S., Boland, S., Hamel, R., Baeza-Squiban, A., Marano, F., Carbon black and titanium dioxide nanoparticles induce pro-inflammatory responses in bronchial epithelial cells: need for multiparametric evaluation due to adsorption artifacts. *Inhal. Toxicol.* 2009, 21 Suppl 1, 115–122.

106. Rahman, Q., Lohani, M., Dopp, E., Pemsel, H., Jonas, L., Weiss, D. G., Schiffmann, D., Evidence that ultrafine titanium dioxide induces micronuclei and apoptosis in Syrian hamster embryo fibroblasts. *Environ. Health Perspect.* 2002, 110, 797–800.

107. Liu, S., Xu, L., Zhang, T., Ren, G., Yang, Z., Oxidative stress and apoptosis induced by nanosized titanium dioxide in PC12 cells. *Toxicology* 2010, 267, 172–177.

108. Park, E. J., Yi, J., Chung, K. H., Ryu, D. Y., Choi, J., Park, K., Oxidative stress and apoptosis induced by titanium dioxide nanoparticles in cultured BEAS-2B cells. *Toxicol. Lett.* 2008, 180, 222–229.

109. Cai, R. X., Kubota, Y., Shuin, T., Sakai, H., Hashimoto, K., Fujishima, A., Induction of cytotoxicity by photoexcited TiO_2 particles. *Cancer Res.* 1992, 52, 2346–2348.

110. Wang, J. J., Sanderson, B. J., Wang, H., Cyto- and genotoxicity of ultrafine TiO_2 particles in cultured human lymphoblastoid cells. *Mutat. Res.* 2007, 628, 99–106.

111. Kang, S. J., Kim, B. M., Lee, Y. J., Hong, S. H., Chung, H. W., Titanium dioxide nanoparticles induce apoptosis through the JNK/p38-caspase-8-Bid pathway in phytohemagglutinin-stimulated human lymphocytes. *Biochem. Biophys. Res. Commun.* 2009, 386, 682–687.

112. Wu, J., Sun, J., Xue, Y., Involvement of JNK and P53 activation in G2/M cell cycle arrest and apoptosis induced by titanium dioxide nanoparticles in neuron cells. *Toxicol. Lett.* 2010.

113. Zhao, J., Bowman, L., Zhang, X., Vallyathan, V., Young, S. H., Castranova, V., Ding, M., Titanium dioxide (TiO_2) nanoparticles induce JB6 cell apoptosis through activation of the caspase-8/Bid and mitochondrial pathways. *J. Toxicol. Environ. Health Part A* 2009, 72, 1141–1149.

114. Shi, Y. L., Wang, F., He, J. B., Yadav, S., Wang, H., Titanium dioxide nanoparticles cause apoptosis in BEAS-2B cells through the caspase

8/t-Bid-independent mitochondrial pathway. *Toxicol. Lett.* 2010, 196, 21–27.

115. Moschou, M., Kosmidis, E. K., Kaloyianni, M., Geronikaki, A., Dabarakis, N., Theophilidis, G., In vitro assessment of the neurotoxic and neuroprotective effects of N-acetyl-L-cysteine (NAC) on the rat sciatic nerve fibers. *Toxicol. In Vitro* 2008, 22, 267–274.

116. Xue, C., Liu, W., Wu, J., Yang, X., Xu, H., Chemoprotective effect of N-acetylcysteine (NAC) on cellular oxidative damages and apoptosis induced by nano titanium dioxide under UVA irradiation. *Toxicol. In Vitro* 2011, 25, 110–116.

Index